RENEWALS. 691-4574

DATE DUE

NOV 2 2			
OCT 2 1			
DEC 0 7			
FEB 0 4			
FEB 1 9			
OCT 2 2			
MAR 5			

Demco, Inc. 38-293

ENERGY, COMBUSTION, AND ENVIRONMENT

McGraw-Hill Series in Energy, Combustion, and Environment

Consulting Editor

Norman Chigier

ENERGY, COMBUSTION, AND ENVIRONMENT

Norman Chigier

University of Sheffield
England

McGraw-Hill Book Company

New York St. Louis San Francisco Auckland Bogotá Hamburg
Johannesburg London Madrid Mexico Montreal New Delhi
Panama Paris São Paulo Singapore Sydney Tokyo Toronto

This book was set in Times Roman. The editor was Diane D. Heiberg.
The production supervisor was Diane Renda.
Kingsport Press, Inc. was printer and binder.

ENERGY, COMBUSTION, AND ENVIRONMENT

2 3 4 5 6 7 8 9 0 K P K P 8 9 8 7 6 5 4 3 2 1

Library of Congress Cataloging in Publication Data

Chigier, N   A
    Energy, combustion, and environment.

    (McGraw-Hill series in energy, combustion &
environment)
    Bibliography: p.
    Includes index.
    1. Power resources. 2. Power (Mechanics)
3. Combustion engineering. 4. Air—Pollution.
I. Title. II. Series.
TJ163.2.C484   621.402′3   80-25048
ISBN 0-07-010766-1

CONTENTS

PREFACE

Energy, Combustion, and Environment is a textbook in which combustion science and fuel technology are related to energy production and utilization. Taking into account that 90 percent of the total energy generated in the world is derived from the combustion of the fossil fuels: oil, coal, and gas, the book is mainly devoted to combustion. Since all engines and power systems have an impact on the environment, the emission of pollutants and their interaction with the atmosphere are examined and practical means of control are recommended.

This book has been written for undergraduate and postgraduate students in Departments of Aeronautical, Mechanical, Chemical, Energy, and Environmental Engineering. It is also directed to scientists and engineers working in government and industrial organizations on research, design, and operation of power plants. It will also be of interest to those involved in legislation and control of air pollution. Scientists and engineers who have had no special training but have recognized the need to turn their attention to the major international problems of energy and the environment will find a global picture of the subject in this book. The book caters for readers with a wide range of interests and scientific and technical background knowledge. The Introduction and the chapter on energy can be understood by readers with little scientific training. The data on coal, petroleum, and gaseous fuels will be found useful by designers, operators, and students. Scientists with backgrounds in chemistry and physics will find information on flame structure. Fluid dynamicists will read the chapters on the physics of turbulence, and turbulent combustion for an understanding of the intimate relationship between the fluid mechanics of turbulent flow and chemical reaction. The oil industry and users of liquid fuels will be interested in the processes of atomization, vaporization, and burning of liquid sprays. Environmentalists will be interested in the mechanism of formation of pollutants in flames and methods of control to reduce the pollution of the environment. The more practical-minded reader will look to the chapter on engine combustion chambers, in which the technology and design of gas turbine, combustors, and spark ignition and diesel engines is described. Physicists, chemists, and instrumentation engineers will recognize that laser technology provides a powerful diagnostic technique for detailed measurements of velocity, temperature, and species concentration in flames. The principles and optical arrangements of laser systems for anemometry and Raman spectroscopy, together with their related signal processing and digital analysis, are discussed in the final chapter.

The shortage of energy and its future supply are problems of major international concern. Energy conservation programs have been initiated by many governments and the impact of energy generation systems on the environment is recognized as posing a possible threat to the survival of mankind, flora, and fauna. In recognition of the importance of these problems, McGraw-Hill have started a new series of books covering the fields of Energy, Combustion, and Environment. Combustion has been selected as the area of prime interest, covering the fields of fuel science and technology. Power plants and alternative energy sources such as solar, wind, wave, tidal, geothermal, and heat pump systems will also be included in the subject areas. In each of these energy systems, the impact on the environment of emissions and pollution will be discussed. This book, which has the same title as the series of books, launches the program and serves as an introductory text.

Professors, lecturers, and students have been seeking textbooks designed and structured for course work. This book has evolved from lecture notes prepared for students taking major courses in combustion engineering and fuel technology. Courses have been taught to undergraduates for several years at Sheffield University by the author, who has also given special courses at Stanford University in California and the University of Lyon in France. As editor of the international review journal, *Progress in Energy and Combustion Science*, the author has developed experience in providing information that is pedagogical and tutorial in nature, taking into account the special requirements of the students. The several years that the author has spent working in industry, and the many years he has spent as a consultant to industry, have made him recognize the special problems facing the design, manufacture, and testing of engines and combustion chambers.

Many books have been written on combustion, fuel science, and technology. These are listed and discussed in the bibliography at the end of the book. Most of these books have been written as monographs or advanced books, based upon research studies. There are only a few books that have been written specifically as textbooks for students. The McGraw-Hill book *Fuels and Combustion* by Smith and Stinson, published in 1952, was an undergraduate textbook for engineering students. There has been a large increase in the knowledge of the subject and the book is now out of date. Gaydon and Wolfhard's fourth edition of *Flames—Their Structure, Radiation, and Temperature* (1979) is a good companion volume to this book. Students will find more advanced information on physical techniques and more detailed information on the structure of flames. Glassman's book *Combustion* (1977) is based on twenty years of teaching combustion at Princeton. The subjects of chemical kinetics, explosion, and detonation are covered more extensively than in this book. Readers with special interest in digital computation will find Spalding's book *Combustion and Mass Transfer* (1979) useful. *Combustion Aerodynamics* by Beér and Chigier (1972) was based primarily on the research experience of the authors in industrial furnaces. It marked a change in emphasis from chemistry to fluid mechanics in the science of combustion.

This new book *Energy, Combustion, and Environment* covers a wider field of topics than many of the other books in the field. More emphasis is placed on the physics and fluid dynamics compared with the more classical emphasis on the chemistry of combustion. It brings together science and technology, discusses practical design, and recommends solutions to practical problems. It discusses energy and the environment in very general terms. The fundamental bases of flame structure, turbulence, and turbulent combustion are treated scientifically. Combustion chambers, burners, and atomizers are treated from a technological and engineering standpoint. The reader is guided to the more specialized literature listed in a comprehensive set of references.

This book has been compiled and written over a number of years. It has been inspired by contact with research students, undergraduate students, and colleagues. The research team, under my supervision, in the Combustion Aerodynamics Research Laboratory at Sheffield University has been actively engaged in research under contract to government and industrial organizations. All the research fellows, Ph.D. students, and technicians in this team have contributed to the papers published in research and review journals, which have served as a basis for the writing of this book. My active participation in conferences, symposia, and workshops in the United States and Europe has provided me with many personal contacts, which allowed me to learn and discuss topics in depth. Undergraduate and postgraduate students who have attended my courses have provided feedback and comments which helped me shape and design the material for their comprehension and use. Friends and colleagues who have specialized expert knowledge have reviewed, corrected, and commented on individual chapters of this book. Specific acknowledgment to authors is made in the text and in the list of references. I wish to thank all the persons who have helped me, by their inspiration, comments, and criticism, to write this book. My special thanks go to Carole Sayliss, who has worked with me on this book from its conception. She has taken down in shorthand and typed the several drafts and final form of the manuscript. Without her help and encouragement this book would have taken many more years to be completed. My wife, José, and my three sons, Joseph, Benjamin, and Jonathan, have provided me with the home and family background conducive to the writing of this book. I wish to express my grateful thanks to every person who has helped me to bring this book to fruition.

Norman Chigier

ENERGY, COMBUSTION, AND ENVIRONMENT

ONE

INTRODUCTION

1.1 THE ROLE OF COMBUSTION IN ENERGY CONSERVATION AND POLLUTION CONTROL

Combustion of fossil fuels is the dominant source of energy supplied to industrial, transport, and domestic systems. Generation of electricity is mainly dependent upon the release of chemical energy, liberated in the form of heat as a consequence of combustion of fossil fuels. Worldwide recognition of the limited supply of fuels and the high degree of dependency of our society on energy has led to a large-scale effort in search of alternative energy sources. Nuclear power is already playing an important role in the generation of electricity, solar energy is used for domestic heating, and major development projects are taking place for the utilization of solar, geothermal, wind, and tidal energy. In parallel with these developments, fossil fuels have become more precious and expensive and, in the long term, industrial societies will require to conserve energy as a means of self-preservation. In this climate, wastage of fuel will need to be prevented and the overall energy efficiency will become the overriding criteria in the design of engineering systems.

The availability and relative importance of fossil fuels has changed with the passage of time. Primitive man was satisfied with the burning of wood for heating and cooking. Coal, originally from surface, and subsequently from underground mining, replaced wood as the dominant fuel in the early part of the twentieth century. The discovery and exploitation of petroleum changed the relative roles of fuels so that, today, transport by air, rail, and automobile are almost entirely dependent upon the supply of petroleum products. Electricity generation has followed the availability and price of fuels. Coal played a dominant role in electricity generation and, after several decades of decline, has now again emerged as a prime fuel. Gas derived from coal was initially used as a source of lighting and heating. Again, after several decades of decline, gas and liquid fuels derived from coal are expected to be the major base of future synthetic fuels. Natural gas was initially ignored as a nuisance and allowed to burn indiscriminately. Subsequently, it was piped and transported in liquefied form. Because of its relative cleanliness, ease of transport, and convenience for burning, it is now recognized as one of our most precious fuels. The history of petroleum has passed through the periods of discovery, exploitation, and dominance of the energy scene, and is now entering a stage of decline. Long-term economic plans are based upon increasing dependence upon coal; reduced availability of natural gas and high-grade petroleum; increased utilization of lower quality petroleum fuels; and gasification and liquefaction of coal. Fossil fuels are expected to have a limited and restricted life, with gradual replacement by nuclear, hydraulic, geothermal, wind, and wave energy sources. Countries dependent upon imports of fuel recognize their own high degree of vulnerability, which is leading them to seek for energy independence. The supply, price, and utilization of fuel has moved from relative insignificance to overriding importance in economic plans and stability.

The primary sources of air pollution are emissions from combustion systems. In large cities, the combination of exhaust from automobile engines and industrial power plants have led to major problems of photochemical smog. Legislation has been introduced in many countries for immediate restriction on emissions of pollutants from automobiles and trucks, as well as from industrial plants. Automobile manufacturers and designers of industrial plant are making radical reassessments of combustion systems in order to greatly reduce the emission of pollutants. Combustion scientists and engineers are faced with the problem of simultaneously achieving maximum energy efficiency and minimum emission of pollutants. Solutions to these problems require radical reexamination of traditional concepts and designs of combustion systems.

Fuel Efficiency

The efficient utilization of fuel requires maximizing the overall energy efficiency. This involves examination of combustion, thermodynamic, and mechanical efficiency. The thermal efficiency is the ratio of the heat gained by the load to the heat energy supplied in the fuel. A well insulated boiler can achieve combustion efficiency close to 100 percent and thermal efficiency of the order of 90 percent. Heat loss by flue gases, high-temperature exhaust gas, and in coolant systems reduce the thermal efficiency. When electrical heating is used, 70 percent of the primary energy is lost in the power generation and transmission stages, so that the most efficient electrical heating systems cannot operate above 30 percent, based on primary fuel usage.

Thermal efficiency is increased by improved insulation and recycling of gaseous effluents and liquid coolants. Thermal efficiency is influenced by rates of heat transfer from flames in combustion chambers to combustion chamber walls and, hence, on temperature distributions within the combustion chamber. The capability of controlling and changing temperature distributions within the gas flow of a combustion chamber can lead to improvements in overall thermal efficiency.

The initial concern of the designer of a combustion chamber is to achieve high combustion efficiency. This requires complete burning of all fuel, supplied to the system. Since unburned fuel is also considered to be a pollutant, there is a double incentive to complete the combustion of fuel, so as to increase combustion efficiency and reduce emission of pollutants. In many combustion systems, combustion efficiency close to 100 percent could be achieved, provided the fuel is maintained at a temperature sufficiently high and a residence time sufficiently long, in the presence of an oxidizer, for the combustion of the fuel to be completed. Systems with combustion efficiency lower than 100 percent are constrained by other design considerations. A refinery flare is a good example of a system which might have 100 percent combustion efficiency but zero thermal efficiency, since none of the heat released is utilized.

Heat transfer from flames to solid surfaces is composed of conductive, convective, and radiative heat transfer. Luminosity of flames and presence of solid and liquid particles can lead to significant changes in the ratio of radiative to convective heat transfer. Conduction of heat in the vicinity of solid surfaces can be enhanced by reduction in thickness of stagnant films and by increasing turbulence levels in gases close to the surfaces. Convective heat transfer can be greatly increased by direct impingement of flames on surfaces. The focus of attention has turned from a simple consideration of combustion efficiency to a recognition of the important role that flame structure can have on thermal and total energy efficiencies.

Interaction of Energy Conservation and Pollution Control

The problems associated with achieving maximum combustion efficiency are closely linked to those of pollution control. Completion of burning and oxidation of fuel increases combustion efficiency while, at the same time, reducing emission of these pollutants. Oxides of nitrogen, on the other hand, have presented major problems, due to the contradictory requirements of pollution formation and combustion efficiency. Most efficient combustion, in well mixed systems, is achieved with stoichiometric mixture ratios, resulting in maximum temperatures. These high temperatures are most favorable for the formation of oxides of nitrogen. In many practical combustion systems, compromises have to be made, so as to reduce oxides of nitrogen formation by working under conditions of poor combustion efficiency and, subsequently, attempting to complete the combustion of unburned fuel at lower temperatures, less favorable to the formation of oxides of nitrogen. Nitrogen oxide is particularly difficult to treat as a pollutant and, hence, it is important to restrict its initial formation. Sulfur is transient in the combustion process. Pollution control requires either removal of sulfur before burning or extraction from effluent. Temperature levels in the flame and on surfaces affect the chemical composition and physical structure of the sulfur compounds and, hence, the location of deposition.

There are three ways in which pollution can be controlled: (1) cleaning the fuel prior to combustion; (2) reducing the formation of pollutants during combustion; and (3) cleaning the exhaust gases. Each one of these methods must be carefully scrutinized in the search for maximum economy in total energy expenditure. Fuel preparation, such as fractionation, catalytic cracking, and desulfurization, all require significant proportions of energy input. Staged combustion, exhaust gas recirculation, and reduced temperature levels are methods used for reduction of formation and destruction of pollutants during combustion. Some of these methods can involve additional energy expenditure of up to 10 percent. For example, electrostatic precipitators are used for collection of particulates from exhaust gases, and these are used extensively in electricity generating stations and in industrial plants. The electricity consumed by these precipitators adds significantly to the total cost and running of plant. The combustion system can no longer be considered in isolation and must be considered as an integral part of a total system, for which an energy balance must be made. All energy expended during the process must be subtracted from the energy released by chemical reaction in the burning of fuel. This is leading to major constraints on designers of combustors. On the other hand, the increased cost of fuel allows the designer to make major design modifications, utilizing increased pressures for jet mixing and air and steam blast atomization.

Fluid Dynamics in Combustion

Fluid dynamics plays an important role in mixing and transport properties in flames, as well as in the supply and exhaust ducts of combustors. The most effective means of controlling flame shape and length is by changing the aerodynamic flow patterns through burners. Fuel and oxidant are initially separated and, subsequently, mixed by turbulent jet flows, swirlers, and generation of recirculation zones. Boundary layers formed along surfaces generate velocity profiles, which are closely linked with temperature and species concentration profiles. High-intensity turbulence increases mixing rates, flame propagation rates, and rates of vaporization, volatilization, and combustion of fuel.

The interaction between turbulence and chemical kinetics is one of the most challenging problems in combustion science. Turbulence enhances rates of chemical reaction by increasing transport properties for heat, mass, and momentum transfer. Profiles of velocity, temperature, and concentration, and their fluctuating components are closely interrelated. For chemical reactions with reaction times of the same magnitude as mixing times, the interrelationship between fluid mechanics and chemical kinetics greatly influences chemical reaction rates. On the other hand, chemical reaction and heat release result in local changes in pressure and volume. These, in turn, cause acceleration, expansion of jet flows, and increases in local kinetic energy of turbulence. Air temperature changes from room to flame temperature conditions result in increases of gas kinematic viscosity by a factor of 20. This can lead to laminarization of turbulent flows.

Swirl is used in many forms of combustion chamber to control mixing and flame structure. Adding bladed swirlers to burners improves flame stabilization and produces short, high-intensity flames. Swirling jets generate internal recirculation zones, which provide all the required conditions for flame stabilization in high-speed flows. Fuels with long reaction times, such as coal, can be completely burned in combustors with the same dimensions as used for gas burning by providing long residence times, high mixing rates, and high temperatures in the recirculation zone of jets with swirl.

Turbulent flow is highly structured and, by visualization, can be shown to be composed of large and small eddy structures. Large structures can remain coherent and be identified at distances far downstream of their origin. These large structures affect the degree of mixedness in combustion systems and the phenomenon of unmixedness can be explained by the presence and dynamics of large eddy structures. Visual observation, high-speed photography, and schlieren photography show that many flames are composed of flamelets. These flamelets are formed at the interfaces between mixing regions of fuel and oxidant, and reaction zones reflect and mirror eddy structure within the turbulent flow. Many new developments are taking place in this area, particularly in the utilization of diagnostic techniques for determination of time- and space-dependent turbulence characteristics.

Atomization and Burning of Liquid Fuels

The largest proportions of fuel burned are in liquid form. The bulk liquid is broken into fine droplets by atomizers and injected directly into combustion chambers or into airstreams. High-grade liquid fuels can be readily atomized into fine droplets and, due to their high evaporation rates, the fuel is rapidly vaporized soon after injection. As the quality and grade of fuel decreases, atomization, vaporization, and burning become increasingly difficult. Heavy fuel oils are so viscous that they require to be heated prior to atomization. In electricity generation boiler furnaces, tens of tons of oil are burned per hour and maximum drop sizes may be as large as 800 μm. Lower-grade fuels are composed of very heavy long-chain molecules and contain asphaltenes, metals, and components that are not burned. These lower-quality fuels are being introduced into a wider range of combustion chambers and major problems, associated with wetting of surfaces by fuel, incomplete combustion, poor oxidation rates, formation of soot, and emission of high levels of smoke and particulates, are being encountered. Deposition of carbon on cooled surfaces changes the emissivity and roughness of walls and can cause burning of surfaces.

Atomization of bulk liquid into fine droplets can be carried out by mechanical, aerodynamic shear, ultrasonic, and electrical methods. The hydrodynamics of breakup of liquid sheets, formation of droplets and interaction of droplets with airstreams, in the presence of flames, are all subjects of fundamental interest, which require detailed examination. Detailed modeling and calculations of ballistics of droplets require information on initial conditions of liquid sprays immediately after breakup of liquid films. The initial conditions of a spray are the size distribution of particles, as well as initial velocities and direction of flight of individual particles. Interaction of liquid particles with air and gas streams determines their trajectories. The temperature, fuel vapour pressure, and transport properties of the environment surrounding the droplets determines the rate of vaporization. Large droplets with high momentum can impinge directly on walls of combustion chambers and result in a buildup of carbon. Large droplets may pass directly through flames without completion of vaporization or burning. The interaction between particle and fluid mechanics of droplets in sprays in the presence of flame is another area where there is a lack of information and where detailed study is required.

Liquid droplet combustion has been studied extensively by carrying out experiments on single droplets suspended on wires or on porous spheres covered with liquid films to simulate the burning of single droplets. This is an area of combustion where close agreement has been reached between experiment and theoretical prediction, based upon heat transfer from envelope flames to the liquid fuel surface and mass transfer from the liquid surface to the flame. Most of these studies have been made with the assumption that the single droplet is surrounded by a uniform air environment with an envelope flame surrounding the droplet. It was initially assumed that results from these single-droplet studies could be used directly in spray combustion. Detailed investigation and probing of spray flames have shown that clouds of droplets have flame structures different from those of individual droplets. Vaporization of clouds of droplets leads to the formation of mixture, rich in fuel vapour, in which combustion may not be possible, due to oxygen "starvation." In addition, the presence of large numbers of droplets can produce quenching and result in low temperatures within the spray. Photographic evidence and detailed experimental and theoretical studies have demonstrated that, in many liquid sprays, the flame surrounds the entire spray and there is no evidence of separate flames surrounding individual droplets.

Combustion Chambers

The principal types of combustion chambers are: gas turbine combustors, internal combustion engines, and furnaces. Gas turbine combustors are almost universally used in aviation and are playing an increasingly important role in marine transport and as land-based generators of electrical power. Internal combustion engines have been classically subdivided into spark ignition and compression engines. Spark ignition engines are almost universally used in automobiles. Compression ignition engines, commonly known as diesel engines, are almost universally used for heavy transport by road and rail. A major fraction of fossil fuel is burned in furnaces for electricity generation and for both direct and indirect heating in industrial plant. Heating of office buildings, industrial plants, and homes is effected by radiator or ducted air heat exchange systems.

Gas turbine combustors, developed for aviation, need to satisfy the stringent requirements of high energy output, small volume and light weight. Aviation fuels must satisfy fire safety and energy efficiency restrictions. Contingency plans are being prepared for both military and transport aircraft to take into account the expected decline in grade and quality of aviation fuels. Marine and land-based gas turbines have less stringent weight and fire safety restrictions. Lower-grade liquid fuels are already being used in these chambers and solutions are being sought for the serious problem of emission of particulates which, by impingement on turbine blades, cause serious problems of erosion and corrosion. The possibility of burning coal directly in gas turbine combustion chambers is being investigated but the particulate emission problem may be so severe as to make the system impractical. The light-weight structures used in aviation gas turbine combustors cannot withstand direct impingement of liquid fuel on walls and, as gas turbine compressor exit temperatures and turbine inlet temperatures rise, overall mixture ratios are becoming less lean, with significant increases in average gas temperature within the combustion chamber. Heavier fuels are resulting in more luminous flames and the increased radiative heat transfer is requiring radical reexamination of wall cooling.

Internal combustion engines are the major source of pollution in a large city, such as Los Angeles. Modifications to engine design have been required by legislation to restrict emissions of particulates, oxides of nitrogen, and hydrocarbons. These modifications involved reduction of formation of oxides of nitrogen within the combustion chamber and cleanup of emissions by catalytic and thermal reactors. Many proposed solutions to the problem involve severe energy penalties. The future survival of the automobile industry is dependent upon satisfying the increasing and simultaneous constraints of improving fuel efficiency and reducing emissions. Increased compression ratios and direct injection of fuel into combustion chambers of spark ignition engines are narrowing the distinction between spark ignition and diesel engines. Swirl is used to control mixing and increase turbulence intensity. Staged combustion and exhaust gas recirculation have been introduced and more information is sought on fuel vaporization, turbulence intensity levels, and pollution formation in quench layers formed on cooled surfaces.

Industrial furnaces are generally run at close to atmospheric pressures. Combustion chamber volumes are large and residence times are long, compared with other forms of combustion chamber. Furnace walls are generally ceramic and reach high temperature levels. Coal and heavy fuel oils are commonly used and radiation from highly luminous flames plays a dominant role in heat transfer. Burner design and control systems, which traditionally have been crude, are becoming more sophisticated as the cost of fuel increases. The major revival of coal as a fuel for the future will require the replacement of oil and gas, since coal can be most conveniently and efficiently burned in furnaces.

Fluidized bed combustors have a very promising future for stationary applications. Fluidized beds are capable of burning coal and other low-grade fuels with low pollutant emissions. Reaction temperatures are generally below 1300 K; heat exchanger tubes are immersed directly into the bed and heat exchange rates are high because of the high thermal efficiency associated with turbulent movement of particles within the bed. Because of the substantially reduced temperatures, NO_x emissions are low unless the fuel contains substantial organic nitrogen. Addition of materials such as limestone leads to absorption of SO_x, so that fuels with high sulfur content can be burned with low SO_x emissions. The fluidized-bed combustor still involves major technological problems, particularly during startup and shutdown. Future developments are expected, using high-pressure fluidized-bed combustors. Generation of electricity in large power stations may ultimately be effected by coal burning in pressurized fluidized-bed combustors. Incineration of waste and other industrial combustion systems are particularly suited to fluidized-bed combustion.

Instrumentation and Measurement Techniques

Measurement of flame characteristics can be made by utilization of spectroscopy, probes, and laser diagnostic techniques. Overall flame temperatures and species concentration measurements have been obtained in uniform small flames by use of "line-of-sight" optical techniques. Probes for insertion into flames have been used for measurement of: temperature by thermocouple, suction pyrometer, and pneumatic pyrometers; species concentration by removal of gases through suction probes and subsequent gas analysis by chromatography; velocity by water-cooled pressure probes. Flames are inhomogeneous and nonuniform with significant spatial and temporal variations of physical and chemical components. The line-of-sight measurement techniques provide integral values and are unsuitable for determination of spatial variations. Probing is invasive and causes local interference. Cooled probes produce local quenching and chemical and physical changes of gases occur within suction probes. In recent years, laser optical probes have been developed to provide noninvasive, nonperturbing diagnostic techniques for making measurements with high spatial and temporal resolution. The laser anemometer, for velocity measurement, is the most advanced of these laser diagnostic techniques. Measurements of velocity, velocity fluctuations, and turbulence characteristics have been made. These are allowing the mapping and analysis of turbulence flow fields in flames, following the techniques developed for hot-wire anemometry in nonburning flow systems. Laser Raman spectroscopy has developed rapidly in recent years and the feasibility of making instantaneous point measurements of temperature and species concentration has been demonstrated. Solutions are being found to overcome the serious problems of radiation from particulate matter in luminous flames. Microthermocouples with digital compensation by online microprocessors have been specially developed. Time constants of the thermocouple are measured in the flame immediately prior to the temperature measurement. Use of very fine wire thermocouples allows measurement of temperature fluctuations and temporal variation of flame front locations.

Particles in flames cause special problems for measurement techniques. Particles impinge on probes, causing damage and deposition of solids and liquid films. Particles clog suction probes. Solid particles have high emissivities, causing luminosity and providing major interference to high-energy pulsed laser spectroscopic techniques. On the other hand, information on ballistics of fuel particles is required. Variation in size and particles indicates rates of vaporization, volatilization, and reaction. Trajectories of particles determine local fuel/air mixture ratios and, hence, combustion characteristics. High-speed photographic and holographic techniques have been developed for determination of spray characteristics. The laser anemometer has been developed for simultaneous measurement of individual particle size and velocity in sprays and particle-laden flames. Thermocouples have been used for determination of gas temperatures in the presence of particles. Many of these techniques are still under development, with the strong incentive of providing more detailed experimental information on the characteristics of flames in combustion chambers.

Interdisciplinary Nature of Combustion

Early studies of flame properties were made by chemists, who were interested in determining the chemical composition of flames. Physicists examined the light-emission characteristics and developed spectroscopic techniques to provide information on temperature and number density of species. Burning of fuels for heating and promoting chemical change has always been prominent in industry. Steel making, chemical processing, and the transport industries are all based upon combustion processes. The high degree of sophistication of these engineering processes, coupled with the more recent legislative restrictions on pollutant emissions and limited fuel supplies, have led to a realization in industry that a more scientific approach is required to solve the technological problems.

Fluid mechanics has grown in importance in the field of combustion. Supply of fuels and oxidant, mixing, and removal of effluents involve fluid dynamics. Turbulence plays a dominant role in the majority of combustion systems. Dialogs have been set up between experts in kinetics and aerodynamics during the past decade. The kineticists recognize that transport properties and chemical reaction rates can be greatly influenced by velocity gradients and turbulence intensity. The gas dynamicists recognize the importance of heat release and chemical reaction on fluid flow patterns and turbulence characteristics. Physicists have been able to make major contributions to combustion diagnostics by the introduction of advanced optical and spectroscopic techniques for making measurements in flames. Electronic processing of signals has become essential for the high-frequency response measurements required for studies in turbulent flames. Microprocessors and minicomputers are being used for data acquisition and data analysis.

Theoretical prediction and mathematical analysis plays a key role in science. It has been very rare, in the field of combustion, that hypotheses have been made and, subsequently, tested by experiment. Engineering and technological developments have often been made without a full understanding of the science, and knowledge of flame characteristics has almost entirely been determined by measurement. The advent of computers with large memories and the significant developments in computational fluid dynamics and chemical kinetics has led to the formulation of computational schemes for combustion systems. Modeling of physical and chemical phenomena still plays a major role in these computational schemes. These models require to be tested by experiment and the more complex computations of three-dimensional flows with chemical reaction are still in the stage of "postdiction", rather than prediction. Numerical computations provide an important framework for bringing together the various components of a combustion system. The combustion research community has increased by several orders of magnitude during the past few decades and important contributions are being made by chemists, physicists, engineers, and mathematicians.

1.2 CATEGORIES OF FLAMES IN COMBUSTION CHAMBERS

In the field of combustion, there is a wide range of categories of flames, many of which lack precise definitions. In considering the wide spectrum of flame types it is, however, possible to divide them into general categories. Before doing so, it is necessary to define combustion and flames and distinguish them from general chemical reactions. It is difficult to provide an absolute criterion for combustion which allows it to be clearly distinguished from all other forms of oxidation. We thus resort to phenomenological definitions based upon the particular physical situation and the consequent reaction.

The phenomenon of combustion may be most simply understood by examining the history of a stoichiometric mixture of methane and oxygen in an enclosed vessel. Under normal atmospheric temperature and pressure conditions, no measurable reaction can be detected and the mixture is considered to be in a metastable state. When the temperature of this mixture is raised to approximately 200°C, oxidation reactions begin to take place with the formation of intermediates such as methanol and formaldehyde, as well as products of combustion such as carbon monoxide and carbon dioxide. These reactions are slow, requiring times of the order of several minutes. Global reaction processes can be followed and reaction rate constants can be determined with high degrees of accuracy. Reaction rates (Fig. 1.1) show an initial acceleration and, after achieving a maximum, the

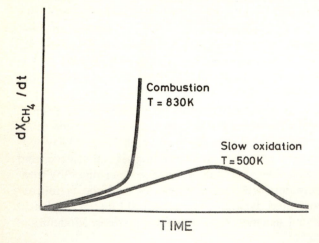

Figure 1.1 The distinction between combustion and slow oxidation. Reaction rate (rate of decrease of concentration with time) as a function of time for the oxidation of methane.

rate decreases as the reactants are consumed. This phenomenon is generally referred to as *slow oxidation*.

When the temperature of the mixture is raised to approximately 560°C, there is an initial short period during which the reaction phenomena are similar to those found under conditions of slow oxidation. After this initial delay time, however, the reaction rate suddenly increases and would become infinite if the reactants were not consumed. It is this "spurt" or "takeoff" in reaction rates which is the true characteristic of a combustion phenomenon. As a direct result of this rapid acceleration in reaction rates, the mixture loses thermodynamic equilibrium, resulting in the formation of atomic species and radicals.

Flames can generally be recognized by their emission of light due to the presence of excited radicals. This phenomenon is known as *chemiluminescence*. Combustion is also often accompanied by *chemi-ionization*, in which ionization results as a consequence of excitation under conditions of extreme thermodynamic disequilibrium.

From a theoretical point of view, it is convenient to consider *adiabatic flames*, in which all the reactions take place without heat loss. Hence, an adiabatic flame temperature can be calculated, which should, in principle, be the highest attainable temperature that the mixture could reach if no heat were lost to the surroundings. In practice, heat is lost from flames to the surroundings, which may be the walls of a combustion chamber or water-cooled surfaces in confined flames. Even in unconfined flames, heat is lost by radiation and convection to the surroundings.

Combustion is initiated by *ignition*. An ignition source can be a hot solid, a hot gas injected into the mixture, or an electrical spark. In each case, energy is supplied in order to initiate combustion reactions. After ignition, the reaction zone propagates through the mixture and the *leading edge* of this propagating reaction zone is called the *flame front*. If the outside of the vessel containing the mixture is heated so that the mixture temperature is gradually increased, it will ultimately reach a stage of *self-* or *autoignition*.

Flames may either be stationary or propagating. A laminar flame can be held *stationary* by adjusting the gas flow rate until the flame remains in a fixed position in space. In internal combustion engines and explosions, the flame *propagates* at a speed which depends on the velocity of sound under the conditions of high temperature and pressure which occur behind the shock wave and flame front. In gas turbines and industrial furnaces where the airflow is highly turbulent and where localized flow reversal occurs, the flame is considered to be *nominally stationary*; even though visual observation shows large-scale movement of the flame front, its average position is fixed. The subjects of *detonation* and stationary flames have, in the past, been treated separately, but more recent detailed investigations of flame structure show that the two phenomena are closely related. More attention is being devoted to the study of closed-vessel explosions as they occur in the internal combustion engine, with the aim of reducing emissions and improving the thermal efficiency of automobile engines. Explosions of dust and combustible gases are a special hazard in coal mines.

In industrial safety, separate consideration is given to the hazards due to burning and those due to explosion. Whereas explosions can lead to the total destruction of buildings, fires can, if detected soon enough, produce relatively little damage. The large concentrations of highly flammable and explosive materials in chemical plants and road, rail, and sea tankers have resulted in accidents causing huge explosions, which devastate entire plants and their neighborhoods, resulting in large-scale loss of life and property. The main discussion in this book will be on stationary flames, but some aspects of explosion and detonation will also be examined.

Flames are categorized according to the nature of premixing of the reactants. In a *premixed flame*, the reactants are completely mixed to the molecular level. In a *diffusion flame*, propagation is controlled by the velocity at which the reactants diffuse toward each other. Within these general divisions, there are imperfectly premixed flames, in which mixing may only be at the macro level but not completed at the molecular level. There are also either rich or lean premixtures, that depend on diffusion of one of the reactants before combustion is completed.

Completely premixed flames are seldom used in practice for reasons of safety. The bunsen burner, domestic cooker, and gas heater flames are examples of partial premixing. Premixed flames are kinetically controlled, and the rate of flame propagation, known as the *burning velocity*, is dependent upon the chemical composition and rates of chemical reaction.

In a *diffusion* flame, the fuel and oxidant are introduced separately into the combustion chamber and the rate of burning is mainly dependent upon the rates of diffusion of the fuel and oxidant. The diffusion flame is preferred in large-scale industrial practice and in gas tubines because safety is increased as the fuel and oxidant are kept separate, as well as providing greater flexibility in controlling flame length and shape and combustion intensity. Fuel and air may be partially premixed, with additional or seconday air supplied through a separate section of the burner. For burners with divergent sections, the fuel and oxidant are introduced separately at the diffuser throat but, at the exit of the diffuser, extensive mixing can already have taken place so that flames attached to these burners are not pure diffusion flames but partially premixed flames.

The term *staged combustion* is used when the combustion process can be separated into two or more distinct stages. Staged combustion may take place in separate interconnected combustion chambers, such as in the prechamber of an automobile engine. Within each stage, mixing rates, mean temperature, and concentrations of reactants and products are maintained at a particular level. In order to avoid stoichiometric mixture ratios where reaction results in high temperatures and high rates of formation of oxides of nitrogen, chambers have been designed to have a first stage where the mixture is rich, followed by a second stage where the mixture is lean. Staged or partially staged combustion can be achieved in a single combustion chamber by aerodynamically separating regions of the flame in which rich or lean mixture ratios are controlled.

Flames are also categorized according to their fluid mechanical properties. When the flow is laminar, the flame will generally be laminar, i.e., the flow of burned gases follows streamlines in the flow without turbulent diffusion. When the flow is highly turbulent, the flame is turbulent. In the transition zone between laminar and fully turbulent flows it is, however, possible for laminarization of a turbulent flow to be caused by combustion. This is mainly due to the large increase in kinematic viscosity with increase in temperature of gases, which leads to a local reduction in the Reynolds number. Again, many low-speed flames have regions which are mainly laminar but with a turbulent brush at the tail. As the Reynolds number is increased, this turbulent brush spreads farther upstream toward the burner.

Under laminar-flow conditions, diffusion of heat and mass are dependent on the molecular properties of the component gases. Under turbulent conditions, diffusion is dependent upon the scale and intensity of turbulence; molecular transport properties have a negligible effect on the overall rates of heat and mass transfer. Laminar flames are only found where burner dimensions are very small or flow rates are low, such as in domestic gas cookers or other low-energy gaseous fuel heating furnaces. In the vast majority of industrial applications, both the initial flow and subsequent flame are turbulent and the turbulence characteristics of the flow play a dominant role in the structure of the flame. Diffusion flames, even if highly turbulent, are mixing controlled, since rates of diffusion of both large- and small-scale eddies are slow compared with rates of chemical reaction. In calculations, it is often assumed that chemical reaction is infinitely fast and that, after diffusion between fuel and oxidant has been completed on the molecular level, chemical reaction takes place instantaneously. It is for this reason that the means of controlling flame shape and volume are mainly aerodynamic and the extent to which diffusion flame properties can be controlled by chemical kinetics is relatively small. For the special cases where the rate of mixing is very high and chemical reaction rates are slow, chemical kinetics can play an important role, even in diffusion flames. The formation and subsequent reaction of oxides of nitrogen is a good example where both the rates of mixing and chemical reaction play important roles in a diffusion flame.

In order to increase both combustion intensity and flame stabilization, *swirl* is frequently utilized. In a flame with swirl, the air, and sometimes the fuel, is given an overall rotating motion within the burner, either by introducing the air tangentially into the burner or by passing the air over vanes. Under these conditions, the flow emerging from the burner has circumferential velocity components which may be of the same or even greater magnitude than the axial and radial velocity components. The centrifugal forces and the associated variations in the pressure fields result in the formation of internal recirculation zones, together with considerable reductions in flame length and volume and, consequently, large increases in combustion intensity. These flames are generally associated with high rates of mixing and turbulence intensity. In nonswirling flames, the flow of fuel and air emerging from the burner is mainly axial and small radial components arise as the jet flame spreads radially outward. The circumferential components are negligibly small. In certain cases, however, where no attempt is made to impart a swirling motion to the flow, swirling motion may exist at the burner exit as a consequence of passage of the air through blowers and bends in pipes.

Even though an overall distinction can be made between, for example, stationary and propagating flames, intermediate cases arise in almost all of the broad divisions which have been discussed above. Many practical flames will have some partial premixing and a certain degree of swirl. For flames with Reynolds numbers near the critical value, both turbulent and molecular rates of diffusion need to be taken into account, and reaction rates and mixing rates are of comparable magnitude. In many flames, we can consider that the role of turbulent diffusion is dominant but it is seldom that the effects of molecular diffusion and chemical kinetics are completely negligible.

Flame Propagation, Explosions, and Detonation

When a combustible mixture is ignited by a spark, pilot flame, hot wire or pulse of radiation, a high temperature reaction zone, known as the *flame front*, is formed. This flame front propagates through the combustible mixture at a speed which is determined by the burning velocity, the flame area, and the expansion of hot gases behind the reaction zone. The presence of confining walls around the combustible mixture affects the pressure levels of the gas heated by the release of chemical energy. Flame propagation is arrested when the flame front reaches the walls, which act as a heat sink, resulting in *quenching* of the flame. In order to distinguish between the concepts of a stationary flame, explosion, and detonation, the various stages of propagation of a flame front in a tube filled with a combustible mixture of gases will be examined.

In slow-flame propagation, or deflagration, molecular or turbulent diffusion provides the energy feedback by which the hot zones ignite the cold reactants. Explosion usually denotes the consequences of the pressure rise resulting from confinement.

In general, an explosion is said to have occurred if energy is released over a sufficiently small time and in a sufficiently small volume so as to generate a pressure wave of finite amplitude traveling away from the source. This energy may have originally been stored in the system in a variety of forms; these include nuclear, chemical, electrical, or pressure energy. In the case of gaseous explosions, two mechanisms which often operate in conjunction can be distinguished: *thermal* and *chain branching*. Self-acceleration of a chemical reaction can occur as a result of heat release, if heat is produced faster than it is lost to the surroundings, since reaction rate coefficients vary exponentially with temperature; this is the thermal mechanism. Combustion also releases energy in the form of highly active free radicals and atomic species. *Chain reactions* are set up in which, for each intermediate consumed, another active intermediate is generated. When one reactive center reacts to give two or more reactive species, the chain is said to *branch* and the rate of reaction increases exponentially, leading to a *branching-chain explosion*. Termination of the reaction occurs, either by a recombination of two radicals to produce a stable molecule, or by reaction of a radical with a molecule to give either a molecular species or a radical of lower activity, which is unable to propagate the chain. This phenomenon is similar in principle to that which occurs in a nuclear explosion. In any explosive system, both the concentration of active species and the temperature will usually increase exponentially and, in most cases, both thermal and chain branching explosions occur simultaneously. *Chain initiation*, whereby free radicals are created from stable species, is a thermal process.

The most important characteristic of an explosion is the generation of pressure waves. Oppenheim (1972) has made detailed studies of the explosion and detonation processes, using streak-schlieren photography. The photographic records of the propagation of a stoichiometric H_2-O_2 flame show that, after ignition, in the early stages of propagation, the flame front has a structure which is *wrinkled-laminar* or *cellular*. The accelerating flame generates a pressure wave that precompresses the reactants, setting them in motion before they are engulfed by the flame. A positive feedback system is set up between the gas moving ahead of the flame and the flame front, resulting in increased acceleration and transition from laminar to turbulent flow. As burning velocities increase, so do pressure changes across the flame. For velocities less than the speed of sound, any pressure waves, caused, for instance, by changes in flame speed, will outstrip the deflagration front. For velocities in excess of about 1 km/s, however, it is possible for a shock wave and a reaction front to coalesce, giving rise to detonation, initiated by rapid compression of the reactants. The energy release by combustion prevents attenuation of the shock wave in the steady state. The consequence of pressure piling are highly directional. In *shock tubes*, the spheroidal wave front of a *detonative ignition* makes its first appearance at one of the walls. Upon collision, this front is transformed into (1) a *reactive detonation wave* propagating into the unburned mixture; (2) a *detonation wave* (an adiabatic shock) propagating into the burned gas; and (3) a decaying set of *blast waves* that bounce between the walls of the container.

Acceleration of the flame front can attain values of 10^6 m/s^2. The velocity of propagation of waves into the stationary reactants has a typical value of 2.5 km/s. With the local velocity of sound immediately behind the wave of 1.5 km/s, the process generates a flow field in the products with an effective particle velocity of the order of 1 km/s. For gases initially at atmospheric pressure, pressures up to 20×10^5 N/m^2 can be produced behind detonation waves. If the tube is closed at the end, the shock wave is reflected and, under these conditions, pressures of up to 100×10^5 N/m^2 can be generated. Whereas in stationary flames the density and, to a much lesser extent, the pressure *decreases* across the reaction zone, in a gaseous detonation very rapid increases in pressure and temperature arise. In a detonation, no compression of the unburned gas takes place once detonation has started. In an explosion, the unburned gas is compressed to high pressures before combustion occurs and no shock wave is set up. Explosion pressures can, thus, in some cases, be higher than the detonation pressures.

In order to establish a stable detonation, it is necessary for the flame front to coalesce with a shock wave. However, a situation may arise where a pocket of combustible mixture becomes trapped in a burning region and is compressed and heated there until self-ignition and an *internal explosion* occurs. This phenomenon, loosely but incorrectly termed "detonation," is responsible for "knocking" in an internal combustion engine.

The presence of flammable mixtures is a safety hazard and special precautions require to be taken in order to prevent the development of uncontrolled fires or explosions. In industrial safety, precautions have to be taken to prevent accidental ignition of potentially explosive mixtures. The presence of flame traps, to arrest flame propagation and safety release valves, can alleviate or reduce the impact of potential fires and explosions. For any given set of conditions, limits of flammability are defined, and mixtures will generally not ignite outside these limits. It can be shown, however, that if sufficient heat is applied to a mixture, and sufficient time is allowed for reaction to take place, there are strictly no true limits of flammability. In industrial practice, however, it is important to know the practical limits of flammability and detonation of particular reactant mixtures.

A final categorization is related to the phase or phases of the mixture. Most combustion takes place with reactants in the gaseous phase. Most liquid fuels initially vaporize before combustion takes place and, in the case of solids such as wood and coal, the volatile gases are initially liberated from the solid phase and burn after mixing with the surrounding oxygen. In coal combustion, and particularly char combustion, the oxidant is adsorbed on to the solid surface so that solid combustion takes place. In the detonation of solid or liquid explosives, combustion can take place in the solid and liquid phase.

ENERGY

2.1 ENERGY RESOURCES

Solar Energy

Solar energy originates from the thermonuclear reactions that are taking place continually within the sun. Hydrogen nuclei and the nuclei of other light elements, mainly helium, at temperatures of the order of 20×10^6 °C convert 4.4×10^6 tonnes of matter per second into 3.9×10^{26} joules of energy. The major portion of energy released from the surface of the sun is in the form of electromagnetic radiation. The wavelengths of radiation range across the full spectrum but 50 percent is in the infrared and 40 percent is in the visible light wavelength range. Out of the 3.9×10^{26} J/s radiated by the sun, the earth only intercepts 1.75×10^{17} J/s. Not all the intercepted solar radiation reaches the earth's surface; the atmosphere reflects back between 30 and 50 percent of the incident radiation.

The atmosphere, land surfaces, and the oceans absorb about two-thirds of the retained solar energy as heat energy. The hydrological cycle absorbs the remaining one-third of the radiation in evaporating water. A very small proportion, 0.2 percent, of the energy drives the atmospheric and oceanic convection currents and the ocean waves. The energy of the wind and waves is, therefore, solar in origin. Photosynthesis in plants only takes up 0.02 percent of the solar energy retained by the earth.

The intensity of radiation outside the earth's atmosphere is 1.4 kW/m^2. In passing through the atmosphere, this energy is absorbed, reflected, and scattered by atmospheric water, dust, and gases. The radiation transmitted through the atmosphere arrives at the earth's surface as direct radiation, or "sunshine." The scattered component reaches the earth's surface as diffused radiation. The ratio of direct-to-diffuse radiation depends upon atmospheric conditions, particularly clouds; on cloudy days all radiation is diffuse.

The amount of solar energy reaching any part of the earth's surface depends upon latitude. The two arid zones which encircle the earth between latitudes 15°N and 35°N, and 15°S and 35°S, receive the highest proportion of solar radiation (0.2 kW/m²). In the equatorial belt between 15°N and 15°S, high humidity and frequent cloud reduce the high rate of solar energy which would reach the earth's surface, so that the net amount of energy received per unit area in this zone is slightly lower than that received in the arid zone. The solar energy received at the earth's surface progressively decreases beyond latitudes 35°N and 35°S, with variations according to local atmospheric conditions. In latitudes beyond 45° the comparatively low elevation of the sun above the horizon and the short days result in low solar energy radiation during the winter months.

Fossil Fuels

The term *fossil fuels* refers collectively to all fuels derived through fossilization processes from previous living organisms. Fossil fuels include coal, peat, petroleum (crude oil), natural gas, and related materials such as tar sands and oil shale. Energy in fossil fuels is bound in the form of chemical energy due to photosynthesis as the result of solar energy. Biologically stored energy is released by oxidation at a rate approximately equal to the rate of storage. A minute fraction of vegetable and animal matter is buried under conditions of incomplete oxidation and decay. This material has been stored over millions of years and is the source of fossil fuels.

Coal is a combustible stratified rock formed from the remains of decaying vegetation. In the coal-forming ages, the climate was hot and humid and the carbon dioxide level in the atmosphere was probably higher than at present. These conditions encouraged rapid forest growth, providing large amounts of deposit vegetation in the swamps. This waterlogged environment prevented the normal conversion of dead vegetation into carbon dioxide and water. The damp climate assisted the formation of acid bogs, which are bacteriostatic. Slow bacterial decay took place until the bacteria could no longer survive under the acid conditions of the accumulated products of decomposition. This led to the formation of peat. Mature coal gradually develops as the products of bacterial decomposition are subjected to high temperatures and pressures, resulting from the continuous accumulation of silt, soil, and rock and from land movements. The succession of changes from the original vegetable matter to peat, then soft brown coal and lignite, to increasingly hard bituminous coals and, finally, anthracite, is known as *coalification.*

Petroleum is a general term meaning "rock-oil," indicating that it is found in the earth's crust. Petroleum occurs in sedimentary rocks deposited under marine conditions. Crude oil is formed by the gradual decay and compression of various marine deposits in an oxygen-deficient environment. Slow decomposition by anaerobic bacteria turns the marine remains into an amorphous material known as *sapropel*. As the fine-grained muds are buried and compressed by subsequent sediment accumulations, the sapropel is converted to petroleum compounds by physical and chemical processes at temperatures less than 200°C required for the preservation of the organometallic compounds found in most crude oils.

Large reservoirs of petroleum are found in areas where thick successions of marine strata were laid down. Migration from such a source bed to a suitable reservoir bed must be possible and a caprock is required to seal the escape of oil to the surface. The lateral migration of the petroleum must be prevented by traps within the reservoir beds. There are three forms of traps: (1) anticlinal, formed by folds in the earth's strata; (2) fault, produced when the earth's crust fractures under stress; and (3) salt dome, produced when deep beds, rich in salt, are forced upwards under high pressure.

Natural Gas

Natural gas is the term used for a mixture of predominantly hydrocarbon gases found in subsurface rock reservoirs. There are two forms of natural gas: (1) associated gas, the volatile portion of crude oil; and (2) nonassociated gas, unrelated to liquid oil accumulations and probably derived from vegetable matter. The liquids and gases, even in associated fields, are not necessarily derived from a common origin. Their different migration paths and processes may, however, lead to termination in the same rock reservoirs. These reservoirs are not hollow caverns, but formations of porous rock, in which gas and/or oil and water enter and usually separate out, accumulating in respective levels. The degree of separation depends on the relative proportions of liquids and gases, the viscosity of the petroleum, and the porosity of the rock. Associated gas is often found in the form of a gas cap overlying the oil-bearing strata.

Primary Energy Resources

The primary energy resources of the world are shown in Table 2.1. The estimates of proven recoverable amounts of primary energy resources are continuously being updated. Changes in these estimates are occurring, both as a result of more intensive exploration efforts and the worldwide recognition of the importance of making more accurate estimates. The intensive exploration of oil and gas offshore and on the outer continental shelves of many countries, as well as the development of resources in Alaska, have radically altered the estimates in recent years. The extensive oil deposits in rock shale and tar sand are more difficult to estimate and their availability for consumption is still dependent on the development of technology for their extraction. Increase in both the cost and the scarcity of the fuel accelerates the development of these technologies.

Nuclear energy has its origins in the nucleus of an atom. It is released as a result of the regrouping or rearrangements of nuclear particles in the processes of

Table 2.1 Primary energy resources of the world

| | Fossil fuels | | | | | | Hydroelectric | | Nuclear fuels | | |
| | Coal | | Oil | | Gas | | Capacity | | Conventional reactors 2% recovery | Breeder reactors total energy | |
	10^{15} MJ	%	10^{15} MJ	%	10^{15} MJ	%	MW	%	10^{15} MJ	10^{15} MJ	%
Asia	2.62	16.5	2.42	59.0	0.45	23.3	140,538	25.3	0.007	0.34	0.4
Africa	0.42	2.6	0.58	14.0	0.21	10.9	145,218	26.2	0.484	24.20	27.6
Western Europe	1.76	11.0	0.046	1.1	0.15	7.7	50,043	9.0	0.101	5.03	5.8
Eastern Europe	5.33	33.5	0.38	9.3	0.65	33.5	52,918	9.2			
North America	5.04	31.7	0.30	7.3	0.38	19.3	57,728	10.6	0.915	45.80	52.3
South America	0.075	0.5	0.37	9.1	0.08	4.0	95,628	17.4	0.028	1.39	1.6
Oceania	0.66	4.2	0.01	0.2	0.03	1.3	12,987	2.3	0.215	10.70	12.3
World	15.9	100.0	4.10	100.0	1.95	100.0	555,060	100.0	1.75	87.5	100.0

radioactive decay, fission, and fusion. In nuclear reactions, part of the nuclear mass is transformed into energy. The potential energy of the nucleus is sufficiently large to allow 1 kg of mass to be converted, theoretically, to 9×10^{16} J.

Nuclear fission refers to the process of breakup of an unstable, heavy nucleus into two lighter nuclei. It can occur spontaneously or through the absorption by the unstable nucleus of a stray neutron. A neutron interacts with a nucleus of uranium-235 to form uranium-236, which immediately splits into nuclei such as strontium-90 and xenon-143. Stray neutrons released in the fission reaction cause fission of further nuclei, setting up a fission chain reaction. Fission processes are sustained and controlled in nuclear reactors.

Estimates of available nuclear energy vary considerably. It is estimated that 840,000 tons of uranium oxide can be produced, but all present evidence indicates that an acute shortage of low-cost ores is likely to develop before the end of this century. Intensive efforts are being made to develop large-scale breeder reactors and, if these succeed, the energy potentially available from the fissioning of uranium and thorium is a few orders of magnitude greater than that from all the fossil fuels combined.

The term *renewable energy sources* refers to energy sources which do not rely on finite reserves of fossil or nuclear fuels. These sources are directly or indirectly due to the sun and include solar energy, wind energy, wave energy, and hydroelectric power. Tidal energy arises from the earth's rotation. Geothermal energy is usually included under this general heading. These sources are continuously renewable. The term *ambient energy sources* is used to describe sources of power that arise from the harnessing of natural forces and, in particular, those whose utilization leads to no net input of heat into the earth. The exploitation of hydroelectric power, wind energy, or wave energy entails the diversion of mechanical power that would otherwise be degraded to heat at the temperature of the surroundings. The term *alternative energy sources* is commonly used to describe the whole spectrum of energy sources that do not rely on fossil fuels or nuclear fission. A breakdown of power available from nonfossil-fuel sources is shown in Table 2.2.

The world's energy resources are shown in Table 2.3, in which the population, production, consumption, and gross domestic product are compared for major regions of the world.

Table 2.2 Power from nonfossil-fuel sources

Source	Power available on continuous basis (MW)
Photosynthesis	10^7
Ocean heat	10^7
Available wind power	10^6
Hydropower	10^5
Total wind power	10^{14}
Tidal	13×10^3
Geothermal	60×10^3

Table 2.3 World energy resources (1975)†

	Population (millions)	Production (10^{12} MJ)	Consumption (10^{12} MJ)	Consumption per capita (10^6 MJ)	Gross domestic product (billion $)
United States	214	61.2	77.4	0.380	1397
Western Europe	367	20.0	53.8	0.153	1200
Soviet Union	255	49.5	45.8	0.19	419
Eastern Europe	106	14.3	18.0	0.177	90
China	839	17.9	17.4	0.022	160
Japan	111	1.1	15.2	0.144	433
Latin America	324	12.6	13.4	0.043	135
South and South-East Asia	1306	10.0	11.6	0.009	31
Canada	23	8.06	8.74	0.393	129
Africa	292	13.7	5.38	0.020	
Middle East	109	44.9	4.3	0.043	
Australasia	21	3.5	3.3	0.162	54

† *The World Energy Book*, D. Crabbe and R. McBride (eds.), Kogan Page, London, 1978.

2.2 ENERGY CONSUMPTION

In 1850 wood was the dominant source of energy. By 1910 coal accounted for 75 percent of the total energy consumption. In the fifty years between 1910 and 1960 coal lost its leading position to oil and natural gas, which were relatively cheap and more convenient. Figures 2.1 to 2.3 show trends in energy consumption for the world, the United States, and the United Kingdom—qualitatively these trends are the same among other developed countries. It is interesting to note that, currently, 2.69×10^{13} MJ per year of wood are used as fuel. In some South-East Asian and African countries wood still represents 90 percent of energy consumption.

A rapid increase in energy consumption started in the middle of this century. During the period between 1925 and 1950, the average annual percentage rate of change in total energy consumption was 2.2 percent. From 1956 to 1966 the figure was 5.5 percent and, in the following decade, i.e., 1966–1976, it reached 7.4 percent. The total energy consumption of the world in MJ and the per capita consumption for the years 1966 to 1976 and the per capita consumption in 1976 are given in Table 2.4. This shows that both the world consumption and the per-capita consumption have increased by approximately 50 percent in less than 10 years; the latter has doubled during the past 30 years. In future, the natural tendency for consumption to grow at a faster rate will be modified by the increase in cost of the fuel and the extent to which conservation measures will be successful.

In the past, energy resources were used mainly for production of heat for physical comfort. Today more than half of all energy consumed in the developed

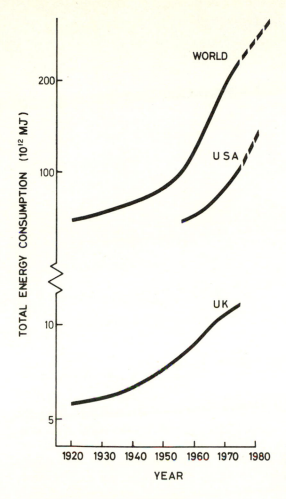

Figure 2.1 Total energy consumption for world, United States and United Kingdom between 1920 and 1975.

countries is used by industry (see Fig. 2.4). Hence, a close relationship exists between energy consumption and economic growth. Although there are anomalies, a relationship can be found between the per capita energy consumption for a given country and the gross national product (GNP) per capita, as shown in Fig. 2.5. The variations in the relationship between GNP and energy consumption for different countries are due to numerous factors. Climate plays an important role in the primary energy requirements of a country, with colder regions consuming more energy for heating requirements than more temperate regions. (In some regions of the United States this situation is reversed, more energy being consumed in the summer for air conditioning than in the winter for heating.) The type of industry dominant in a country and the mode of generating electricity have significant impacts on energy requirements. Some countries using lignite as a fuel with poor combustion properties for electricity generation have a high primary

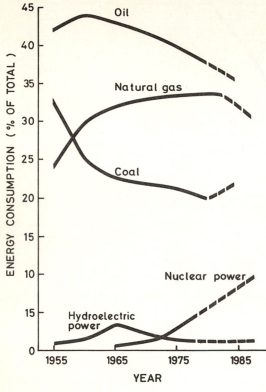

Figure 2.2 Energy consumption in United States.

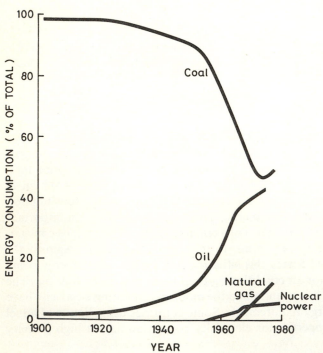

Figure 2.3 Energy consumption in United Kingdom.

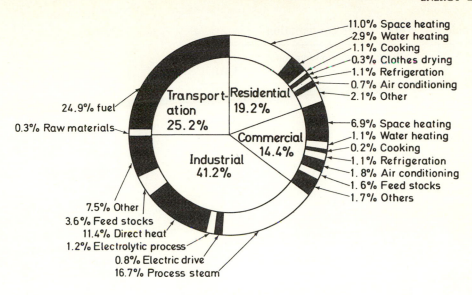

Figure 2.4 Patterns of energy consumption in the United States—consumer categories and major end uses.

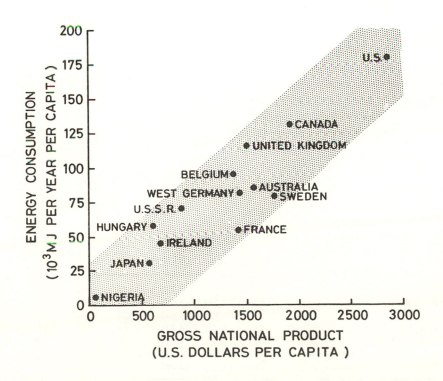

Figure 2.5 The relationship between energy consumption and gross national product, per capita.

Table 2.4 World energy consumption

World energy consumption 10^{12} MJ	
1966	195
1967	200
1968	214
1969	228
1970	240
1971	250
1972	259
1973	272
1974	274
1975	275
1976	289

Per capita consumption of petroleum products, 1976 10^8 MJ	
USA	1.57
Canada	1.52
Sweden	1.43
West Germany	0.98
Japan	0.90
France	0.90
Netherlands	0.90
Australia	0.85
UK	0.63
Argentina	0.36
Venezuela	0.36
India	0.04

energy consumption in relation to their GNP. It can be seen from Tables 2.4 and 2.5 that, despite the large increase in consumption of the less developed countries, the disparity between the developed countries and the developing countries in terms of energy consumption is still very large and is expected to remain so. In the world, 20 percent of the population consumes 90 percent of the energy. The

Table 2.5 World consumption of energy by regions ($\times 10^{10}$ MJ)

	1950	1960	1965	1970	1972	1975
North America	3284	4,377	5,297	6,781	7,209	8,635
Western Europe	1614	2,354	3,053	4,005	4,304	5,396
Japan	132	331	557	1,022	1,184	1,528
Soviet Union, Eastern Europe, and China	1506	3,675	4,204	5,213	5,649	8,135
Rest of world	643	1,028	1,383	1,888	2,099	3,809
World total	7180	11,763	14,494	18,911	20,444	27,503

Table 2.6 Production of primary energy resources ($\times 10^{12}$ MJ)

Year	Total energy	Coal and lignite	Oil	Gas	Hydro and nuclear
1965	152	67.1	55.5	25.9	3.17
1966	168	68.3	69.3	27.1	3.15
1967	173	65.3	74.5	29.2	3.52
1968	184	67.3	81.4	31.8	3.69
1969	195	68.7	87.6	34.9	3.96
1970	210	71.6	96.0	38.6	4.18
1971	218	71.0	101.5	41.3	4.45
1972	227	72.1	107.1	43.5	4.77
1973	242	73.4	117.3	45.8	5.02
1974	244	73.9	117.8	46.7	5.56
1975	257	79.5	121.4	49.9	6.67

United States share of world energy consumption has fallen from 38 percent in 1956 to 34 percent in 1966 and 28 percent in 1976. This is accounted for by the increase in consumption by China, Japan, the Middle East, and Latin America; their share has increased from 8.4 percent in 1956 to 18.5 percent in 1976.

The increase in world energy consumption has been accompanied by an increase in production of primary energy resources, made possible by advances in prospecting and production technology. The production of primary energy resources has increased at nearly the same rate as energy consumption, i.e., by about 50 percent in less than ten years (Table 2.6). Coal production has remained nearly constant, while production of oil and gas has increased by more than 50 percent.

2.3 STRUCTURE OF THE WORLD ENERGY MARKET

In the year 1950, world energy demand totalled about 7.0×10^{13} MJ. Coal contributed 4.3×10^{13} MJ to this total; oil 1.9×10^{13} MJ; natural gas 7.0×10^{12} MJ; and hydro power 1.1×10^{12} MJ. The main centers of energy consumption were self-sufficient in energy resources. Western Europe was dependent almost entirely on its large coal industries, while both the United States and the Soviet Union had large domestic supplies of coal, oil, and gas. The quantities of oil and coal in international trade were, at that time, relatively small as a proportion of total world demand. As economic growth accelerated in the post-war years, world energy demand moved ahead rapidly, with a near threefold increase in the period from 1950 to 1972. The greater part of this growth was met by rapid expansion in the use of oil and natural gas. During the period 1950 to 1970 almost two-thirds of world energy was consumed by North America, Western Europe, and Japan, as shown in Table 2.5.

The Soviet Union is, and the United States could be, almost entirely independent of imports of fuel. The discovery and development of massive additional oil reserves in the Middle East and Africa from the mid-1950s led to a ready availability of cheap oil supplies on world markets. Western Europe and Japan became the principal outlets for these supplies and, because of the competition from oil, a gradual erosion in the market for coal in Western Europe took place. The rapid growth in the use of oil has resulted in a steady increase in the overall dependence on imported energy in both Western Europe and Japan. In the period since 1961 oil imports into Western Europe have more than trebled and Japanese imports have increased almost sixfold. In 1972 the dependence on imported energy was as follows: Western Europe 61 percent, Japan 87 percent, the United States 12 percent, and the Soviet Union nil. Total oil imports into the major industrialized regions of the free world approached 5.8×10^{13} MJ a year in 1976.

Almost 66 percent of the world's present proven oil reserves are located in the Middle East and Africa. If the communist countries are excluded, the Middle East and Africa possess about 80 percent of the rest of the world's oil reserves. As the price of oil was arbitrarily increased fivefold in 1974, Western Europe and Japan came to realize their vulnerability and intensive efforts were made in order to become self-sufficient in energy supplies. The annual changes in production of primary energy resources between 1965 and 1975 are shown in Table 2.6.

2.4 ENERGY IN THE UNITED STATES

The American energy situation has changed significantly in recent years, as shown in Fig. 2.6. Throughout the 1960s there was a marked reduction in exploration for

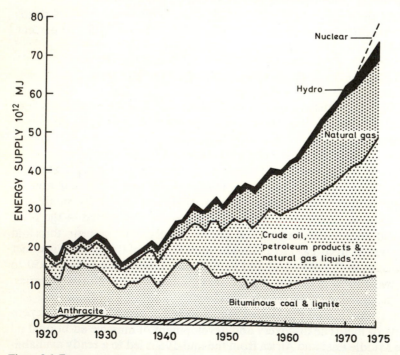

Figure 2.6 Energy supply in United States.

oil and gas compounded by a declining discovery rate. This had a serious effect on the availability of indigenous oil and gas. America's proven gas reserves reached a peak in 1967 and have since fallen steadily. In the case of oil, there has been a long-term decline in the real price of indigenous crude oil, partly resulting from the influence of imports.

In 1976 natural gas accounted for 28 percent of American energy demand, with consumption at 2.32×10^{13} MJ. American oil production reached a peak of 2.41×10^{13} MJ in 1970, falling back to 2.07×10^{13} MJ in 1976. By 1976 almost 44 percent of American oil requirements was being met by imports, a total of 1.62×10^{13} MJ. The United States has rapidly emerged as a major oil importer and, unless the trends in energy demand change, America's imports of oil in 1985 could be as much as 2.9×10^{13} MJ and account for as much as 53 percent of her oil requirements—30 percent of the overall energy consumption, compared with only 12 percent in 1972.

The shortage of gas in the United States, coupled with the restricted development in supplies of indigenous oil and in coal production, has led to an energy shortage. The cutback in coal production was due to the enforcement of improved health and safety standards in mines, environmental restrictions on burning high-sulfur coal, and restrictions on strip mining. In order to increase energy supplies, the development of indigenous energy resources is being accelerated. New incentives are being provided for oil and gas exploration; gas prices are being freed from federal regulations; rapid increases in the output of coal and nuclear power are being made, together with the development of shale-oil production. At the same time, substantial additional resources are being committed to energy research—particularly nuclear power, coal production and utilization, and the manufacture of synthetic natural gas (SNG) and oil from coal. The coal reserves in the United States are very large and estimates of undiscovered hydrocarbon reserves indicate that there is some scope for expanding oil and gas production in the future. Nevertheless, despite her considerable energy resources and an increasing use of coal and nuclear power, it appears likely that the United States will become increasingly dependent on energy imports.

2.5 WORLD ENERGY DEMAND

Predictions for energy demand made in the early 1970s, based on the rate of growth during the previous decade, showed that before the year 2000 the known reserves of fuel throughout the world could by no means satisfy the demands. The future demand for energy will depend upon political, economic, and social changes. An indication of predicted changes in world energy consumption is given in Table 2.7, where comparisons are given for the years 1970 and 1985.

The requirements for imported oil in the major industrialized countries are given in Table 2.8.

Table 2.7 World energy consumption 1970 and 1985 ($\times 10^{10}$ MJ)

	1970	1985
United States	6,582	11,380
Western Europe	3,319	6,537
Japan	1,022	2,905
All other areas	7,987	22,219
World total	18,910	43,041

Table 2.8 shows that the total import requirements for these countries will more than double between the years 1972 and 1985 and that the United States will remain a major importer of oil. The present total of world proved oil reserves would sustain current levels of production for some 35 years. In practice, as consumption will be increasing steadily, these present proved reserves are likely to be exhausted within 20 years. Even though more reserves are expected to be discovered, the ultimate recoverable oil reserves will only support the current rate of growth in oil demand for another few decades. The oil-producing countries of the world are most likely to follow the position of the United States, where the growth rate has reached a peak in extraction, and output levels have begun to decline. It is now clearly recognized that, if world energy requirements are to continue to grow and to be met in full, a gradual replacement of oil utilization by alternative fuels must take place. The lead times involved in the development of new energy technologies and in the establishment of additional energy production capacity are so considerable that oil will remain the dominant fuel in the world energy market, at least until the mid-1980s.

The main factors in the future development of the international oil market will be: (1) continuing increases in world oil demand, particularly in the import requirements of the major Western industrialized countries; (2) increasing dominance of the Organization of Petroleum Exporting Countries (OPEC) in the Middle East and Africa; (3) restriction of production and exportation of oil in order to conserve oil reserves; (4) increase in price levels and control of supply by

Table 2.8 Requirements for imported oil in major industrialized countries ($\times 10^{10}$ MJ)

	1972	1985
United States	1030	2,912
Western Europe	3046	4,928
Japan	1053	2,374
Total	5129	10,214

the OPEC countries, which will be dictated both by political and economic demands; and (5) the development of alternative fuels with increasing urgency as the price of oil rises and supplies are restricted.

2.6 NATURAL GAS

The production of natural gas has greatly increased during the last ten years. Exploration operations during the 1960s and 1970s have led to the discovery of huge fields of natural gas. In countries such as Holland, France, and the United Kingdom the initial discovery and subsequent exploitation of local reserves of natural gas have completely transformed the patterns of energy utilization in those countries, as well as making substantial impacts on their national economies.

The special advantages which natural gas has, compared with coal and oil, are: (1) facility of transport from a source at high pressure; (2) ease of burning; (3) low level of pollutant emission. In many industrial processes, and particularly in the field of heating, natural gas burning has a versatility and flexibility which is sufficiently great that extensive plans have been prepared for the gasification of coal and oil.

World proven reserves of natural gas are about one-half of the total for oil reserves. The proven recoverable amounts of natural gas in different regions of the world are shown in Table 2.1. This clean, premium fuel will be used in large quantities wherever it is found locally, and higher oil prices will stimulate an increase in international trade in gas. In the United States there is already a growing shortage of gas and this shortage is being met by substantial imports of gas and development of synthetic natural gas produced from coal or oil. Both the United States and Western Europe are importing liquified natural gas (LNG) from Africa, and the United States is expected to work with the Soviet Union in the development of the natural gas fields of Western Siberia.

Consumption of natural gas in the European Economic Community rose from 3.5×10^{11} MJ in 1960 to 7.13×10^{12} MJ in 1976. The share of natural gas in the total EEC energy demand increased from 2 percent in 1960 to 17 percent in 1976. The need for natural gas is expected to grow rapidly and the estimated requirements for the EEC in 1985 are 10^{13} MJ. There have been substantial reductions in the reserve assessments for France, Italy, and the Netherlands. Dutch gas reserves represent 60 percent of the Community total. In 1972 home sales of Dutch gas totaled 34 billion cubic meters (1.0×10^{12} MJ), and export sales 24 billion cubic meters (7.3×10^{11} MJ). This quantity of gas was sufficient to meet the home demand in Holland and also one-third of the gas requirements of neighboring EEC countries. Ninety percent of non-Dutch gas reserves lie in the Groningen Field.

There has been a fall in the availability of Dutch gas from the middle 1970s onwards, after having reached a gas production plateau in the year 1975. Imports play only a minor role in the gas supplies of EEC countries today. For the future, contracts for importing pipeline gas from the Soviet Union, and LNG from North

Africa, have been estimated to reach 8.1×10^{11} MJ in 1980. In view of the limitation of indigenous gas reserves in Western Europe, gas consumption will probably not rise to more than 15 percent of the total energy requirements in the EEC. The previous policy of encouraging bulk gas sales to the industrial and power station markets by low prices has already changed and governments are expected to restrict the use of gas in the future.

North Sea gas has brought about a complete change in supply to the British gas industry. In 1960 about 90 percent of the total gas manufactured in the United Kingdom was derived from coal, whereas by 1974 ninety-five percent of the total gas supplied in the United Kingdom was derived from natural gas—over 80 percent being supplied direct, the remainder being used instead of oil to manufacture town gas. By 1976 natural gas replaced all coal- and oil-based gas. During the seven years after the first natural gas was brought ashore from the North Sea, the total quantity of gas supplied to the United Kingdom was trebled. From supplying 7 percent of the country's energy needs in 1965, the gas industry is expected to supply up to 20 percent of the United Kingdom energy demand in 1980. Reserves of natural gas in the large fields in the southern North Sea have been estimated at about 3.7×10^{13} MJ.

Natural gas is generally found in close proximity to oil deposits and the processing and transportation of the liquid and gaseous fuels are carried out in parallel. Since natural gas is usually found under conditions of high pressure, the gas is piped through networks, across continents. Substantial amounts of natural gas are also liquefied and transported by sea or land and then regasified prior to distribution.

The hazards of handling liquefied natural gas (LNG) are common to handling of flammable cryogenic liquids. Large spills from ships or tankers cause hazards in addition to those normally associated with oil spills.

Liquefied natural gas has been considered as a fuel for engines, particularly because of the possibility of reduced emission of pollutants, as well as increased energy efficiency. In those countries, such as the United States, where there has been a shortage of natural gas, it has been recognized that gas should be considered as a premium fuel, reserved for applications in which it has special advantages. It does not seem likely that it will be ultimately used to any significant extent as a fuel in the transport industry.

2.7 OIL

Oil has proved to be one of the most versatile of the fossil fuels and, in the field of transport, it has so many advantages over alternative systems that there is every probability that it will retain its special role among energy sources. Many developments have taken place in the refining of crude oil in order to produce a very wide range of products, from the lightest gaseous hydrocarbons to very heavy residual fuel oils. The selection of a particular petroleum product for a practical combustor is dictated by the combustion volume, pressure, and temperature, as well as the total permissible weight of the system. The cost of the fuel and the feasibility of dealing with particulate matter are also taken into account.

For aircraft, the weight of the combustor and the weight of the fuel become the dominant factors. Aviation fuels are selected to provide very high heat release rates for burning in small pressurized chambers at almost 100 percent combustion efficiency. Since power failure can lead to disaster, ignition and flame stabilization must be guaranteed over the wide range of pressure, density, and inlet temperature that occur between ground level and altitudes of 20 km. Restricting the flammability of the fuel so as to avoid aircraft fires is of vital importance. For this reason, as well as for reasons of economy, special attention is being directed toward the utilization of heavier fractions of fuel oil for aviation but, currently, all aviation fuels are among the lightest and most flammable fractions of the hydrocarbon distillates.

The selection of fuel for automobiles is dictated by the burning characteristics for engines with a range of compression ratios. The convenience of transport by tanker, storage in large tanks, and pumping of fuels directly into automobile fuel tanks has very clearly led to the selection of a liquid rather than a solid or gaseous fuel. The relative importance of the cost of the fuel in the total cost of running automobiles has been very steadily increasing. This is leading to the investigation of the possibility of using heavier, and hence cheaper, fuels for all forms of transport. In heavy industry and agriculture the cost of the fuel has always been important, hence combustion systems have been developed for burning heavier residual fuels, where provision has been made for handling ash and particulate emissions. Consideration is currently being given to the burning of very heavy fuel oils containing asphaltenes which require to be heated to over 600 K for the viscosity to become sufficiently low to allow pumping and atomization. The impact of petroleum on the socioeconomic structure of the world has already become so important that, in many countries, fuel prices have a greater effect on the economy than food prices.

The estimated recoverable reserves of oil for various parts of the world are shown in Table 2.1. These estimates do not represent the potential reserves but those which have, so far, been proven to be recoverable. Major explorations are being undertaken in offshore and outer continental shelf areas, which are leading to the discovery of huge additional oil reserves. It can be seen from Table 2.1 that the reserves in Asia constitute 59 percent of world reserves, a large proportion of which comes from the oil-producing countries of the Middle East.

Exploration and Production of Oil

The major development in recent years has been exploration and production of petroleum in offshore regions. The proportion of offshore petroleum in worldwide production is estimated to reach 30 percent in 1980 and probably 40 percent by 1990. For countries such as the United Kingdom and Norway, the discovery of oil and gas in the North Sea has totally altered the national energy scene. These countries have changed from being major importers of oil to substantial exporters. Production of oil from offshore wells results in special environmental hazards associated with near-shore spills; the 1969 blowout in Santa Barbara, United States, in which approximately 10,000 tons of oil were released, caused severe damage to the marine ecosystem and shore life.

Transportation of Oil

Very large volumes of oil are transported by sea from oil-producing to oil-consuming countries. In 1972 over 10^9 tons of crude oil and 3×10^8 tons of refined products were transported by sea. The size of individual tankers has increased dramatically and the trend is towards utilization of ultralarge crude carriers (above 350,000 draft weight tons—DWT) and reduction in the number of small tankers. The number of accidents involving oil spills may decrease, but the environmental impact of any one particular accident was demonstrated by the Torrey Canyon, in which 50,000 tons of oil escaped, causing extensive damage in the United Kingdom and France. Most large tanker spills occur within 80 km of land and result from grounding or collision.

Oil spills due to marine transportation and offshore activities constitute about 35 percent of the total oil pollution in fresh and sea water. The remainder originates from disposal of dirty ballast and tank washings in the course of normal ship operations, as well as from refineries and industrial waste. It has been estimated that about 2×10^6 tons of oil find their way into the rivers and seas annually. The introduction of the "load on top" principle, in which new cargoes of crude oil are loaded on top of residues without tank washing, is leading to reduction in this form of pollution. New tankers are also being equipped with segregated water ballast tanks in order to minimize the mixing of oil and water.

Many countries have set up emergency task forces to deal with major oil spills. Some of the methods under consideration are burning, containment and removal, dispersal, and sinking. Certain strains of bacteria have been found that break down oil into harmless substances, and this biological method may ultimately be the most desirable solution to the problem. The acute toxicity of a self-dispersed crude oil has been found to decrease considerably in 24 to 72 hours but, if the oil is dispersed by a dispersant, the high toxicity remains almost unchanged over the same period.

Major waste-disposal problems arise from accidental spills from oil tankers, leakage from damaged pipelines, spills from overflowing tankers, discharges from tank washings, and direct waste from oil refineries. The waste is composed of the natural constituents of petroleum as well as substances added in the purification and separation processes. Sulfur and vanadium compounds from the crude petroleum, phenols formed during cracking and fractionation, and catalysts used in refining are some of the nonoil constituents. Most organic substances have an affinity for oil and may be present in oil films. Inorganic constituents are often water soluble and are usually found in the aqueous fraction of the refinery effluent. This oil fraction is the major toxic constituent discharged from a refinery. In order to reduce pollution, gravity methods are used for oil separation, as well as biological treatment with bacteria.

Most oil refineries have serious problems with odors. The principal malodorous compounds existing in crude oil formed during its processing are hydrogen sulfide and mercaptans. Ethyl mercaptan has a perceptible smell when present in a concentration of only one part per thousand million. Depending upon local wind conditions, a very small loss of hydrogen sulfide or mercaptans can create an unpleasant odor over areas of several square kilometers.

North Sea Oil

The discovery of oil and gas in the North Sea has temporarily changed the dependence of Europe on energy imports. The ultimate oil potential of the North Sea is not fully established, but it is estimated that by 1985, 9.0×10^{12} MJ per annum will be available. On present estimates, this will represent no more than one-sixth of West European oil requirements in 1985. Oil consumption increased in Western Europe by 40 million tons in the years 1972 and 1975, with a fall in consumption between 1973 and 1975, and after 1977. Unless a substantial number of new discoveries are made, the most probable outcome would appear to be a fairly rapid rundown of North Sea oil production in the late 1980s, with virtual exhaustion within 20 years of the fields that have been so far discovered.

2.8 SHALE OIL AND TAR SANDS

Shale is a relatively common sedimentary rock, laid down over a number of geological periods. It may contain a high proportion of carbonaceous material and, when shale contains upward of 33 percent organic matter, it is called *oil shale*. Oil is usually derived from oil shale by heating in special retorts to convert the organic matter to crude petroleum. Tar sands are sedimentary formations of loose-grained rock material bonded by heavy bituminous tar. Tar sands are generally extracted by strip mining. Oil is separated out by direct heating at temperatures of around 80°C, or by passing steam through the tar sands.

The shortage of crude oil supplies, combined with the rapid increase in cost of oil imports, has made the exploitation of shale oil and tar sands economical; several schemes for extraction have been put into practice. These hydrocarbons are similar to crude oil and can be refined. The majority of these hydrocarbon reserves lie in Colorado, Utah, Wyoming, and Canada, and large reserves have been found in Siberia. The reserves of shale oil and tar sands have been estimated to be twice the postulated figure for ultimate world recoverable reserves of conventional crude oil, but much of the shale oil is in very low-grade deposits. Estimates in 1973 were 10^{16} MJ of shale oil and 2×10^{15} MJ of tar-sands oil.

Shale can either be pulverized and burned directly in a specially designed furnace, or oil can be extracted from the shale by a chemical process. Plans have been put forward to improve extraction by underground nuclear detonations, followed by extraction of oil by conventional drilling methods. Shale is an inferior fuel to coal and contains a higher proportion of rock and metallic materials, which do not burn but become molten and solidify as ash or clinker during the combustion process. The emission of particulates and handling of ash and clinker are major problems. These problems can be overcome by using chemical processes to extract the oil from the crushed rock. This processing is usually undertaken as part of the mining operation. The shale has a larger volume after processing than before, and present technologies of producing oil from oil shale involve the use of large quantities of water. For each ton of oil produced, 90 tons of shale have to be handled. In tar sands, low-grade bituminous tar is mixed with sand deposits. When these deposits are heated, oil can be extracted. Tar sands are more easily worked than shale but recoverability is generally less than 15 percent.

The impact of shale oil and tar sands on the world energy market will not be significant before the end of the century. Total U.S. oil demands for 1985 have been predicted at 1.8×10^{11} MJ per day and the total production of nonconventional oils will probably not exceed 7.2×10^9 MJ per day by 1985.

2.9 COAL

Coal has had a long history as a primary energy source. It was the first among the fossil fuels to play a major role in energy consumption, and this reached a peak during the period 1920–1940. The special problems associated with mining of coal, coupled with the increasing availability of oil and gas, led to a decline in the importance of coal during the period 1940–1975. The recognition of the fossil fuel scarcity has led to a reappraisal of the importance of coal, which is currently considered as being a major source of energy supply. Electric power plants were converted from coal to oil firing during the period 1950–1970. Subsequently, many of these plants have been reconverted from oil to coal firing.

The coal reserves in the world, as compiled for the Survey of Energy Resources, World Energy Conference, 1974, are given in Table 2.1. This gives the proven recoverable coal. Intensive efforts are being made to improve the accuracy of these data. As a result of more exploration, there is improvement in the technology of detection and mining, as well as improvements in the techniques of making the estimates. It can be noted from Table 2.1 that the more highly developed countries with high energy consumption rates have over 70 percent of the world's available reserves of coal.

Coal is the world's most abundant energy resource. At the present rate of consumption, there are sufficient coal reserves to last for more than a century. As with other fuels, the proportion of proved reserves to estimated total resources is being constantly increased as further exploration takes place and more reliable geological data become available. World production and consumption of coal has continuously increased, and significant increases have taken place in the United States and the Soviet Union. In Western Europe, under the assumption that cheap oil supplies would be continuously available, coal production was reduced during the 1960s with the aim of phasing out coal mining. The sudden jump in the price of oil in 1974 has reversed this policy and production of coal in Europe is now considered to be viable.

Coal can be expected to make an increasingly important contribution to world energy supply. The world energy demand in the year 2000 has been estimated as 8.1×10^{14} MJ, and the U.S. Atomic Energy Commission has estimated a nuclear capacity in the year 2000 of 3,260,000 MW. Even with these optimistic estimates of the contribution of nuclear energy, fossil fuels will be required to provide 80 percent of the total fuel requirements at the turn of the century and the demand for fossil fuel will rise from 2.68×10^{14} MJ in 1976 to 6.19×10^{14} MJ. The world's reserves of conventional oil and gas cannot supply this demand, and the era of cheap oil supplies ended in 1974. With every country attempting to reduce its political and economic vulnerability to cessation of oil imports, the necessity to use indigenous coal supplies is no longer questioned.

In the United States rapid expansion of coal production has been initiated. Plans have been prepared in the United States for increase in coal production from 1975 levels of 1.63×10^{13} MJ to 2.7×10^{13} MJ in 1985. The United States is devoting \$10,000 million for energy research over the five years from 1975 and \$2000 million of this total will be devoted to coal research.

Rapid increase in coal production will be hampered by problems of invest-ment, work force, environment, and health and safety legislation. Energy policies of many West European countries during the 1960s were based on the assumption that abundant and cheap supplies of American coal would be available for export in the long term. This assumption proved to be false and it can be expected that the price will be high, and the availability limited, of all fossil fuels, including coal. The steady substitution of European coal by oil since the late 1950s led to the progressive closure of a large proportion of indigenous coal production capacity. In the former European Community of Six, coal production was reduced from 6.7×10^{12} MJ in 1957 to 3.69×10^{12} MJ in 1975. In the United Kingdom pro-duction fell from 6.13×10^{12} MJ to 3.85×10^{12} MJ over the same period. The reduction in coal production in Western Europe has been only broadly balanced by increases in other forms of indigenous energy production—natural gas, oil, nuclear, and hydropower—so that EEC energy production has remained at 1.35×10^{13} MJ. From the mid-1950s to the early 1970s the entire increase in Community energy demand has been met by imports. In 1976 coal consumption in the enlarged Community was 9.01×10^{12} MJ, out of which 1.75×10^{12} MJ was imported. The two principal coal consumers are power stations and coke ovens, which together use about 6.2×10^{12} MJ yearly. Coal is the major source of energy for the Community's power stations, accounting for 32 percent of total energy input in 1972. As the energy demand in the Community increases, the requirements for coal will be so great that the energy supply may well be depen-dent upon the production of coal.

Steel production in Western Europe is estimated to increase from 140 million tons in 1972 to 200 million tons in 1980. Coking coal is required for steel produc-tion, and the main supplier of coking coal on the international market is the United States, with an export of 37 million tons in 1971. The other main supplier of coking coal in the Western world is Australia, with some supplies coming through from Poland. Another potential source of supply is western Canada, but each of the countries supplying coal for export at the present time can be expected to revise both the price and quantity available for export so that these supplies cannot be considered reliable. World demand for coking coal has risen by 14 percent between 1969 and 1975—from 1.3×10^{13} MJ to 1.5×10^{13} MJ. Imports of coking coal into Europe face the problem of rapidly rising prices, coupled with decreasing availability, which will radically affect the price and availability of steel in Europe.

The fuel requirements for electricity power generation in the EEC totaled 8.9×10^{12} MJ in 1972—more than one-quarter of the total EEC energy demand. The contributions of various fuels and energy sources to electricity generation in the EEC in 1972 are shown in Table 2.9.

An increase of 5.4×10^{12} MJ in fossil fuels is required between 1970 and 1985, and this could be greater if the estimates for nuclear power are not fulfilled.

The European electricity supply industry has become heavily dependent on Middle East oil and, to a lesser extent, on East European coal. The Arab supply restrictions of October 1973 caused considerable dislocation to electricity generation in Western Europe. With the fivefold increase in oil prices in 1973–1974, indigenous coal has now become cheaper than oil both for the electricity power generation and industrial markets.

The coal-energy system consists of several steps involving extraction, processing, and combustion. The extraction of coal is carried out either by surface or underground mining. Underground coal mining is considered to be one of the most hazardous occupations, due to the fire and explosion hazards, land subsidence, and respiratory diseases, such as coal workers' pneumoconiosis. Surface (strip) coal mining is more common than underground mining. Strip mining has had a severe impact on the environment, both as a direct consequence of the removal of the coal and the cleaning and separation processes. Sulfur-bearing minerals associated with coal oxidize readily on exposure to air and water, leading to the formation of sulfuric acid, which contaminates underground waters and pollutes streams.

Table 2.9 Electricity generation fuel requirements—European Economic Community 1972 ($\times 10^{10}$ MJ)

Coal	288
Lignite	73
Oil	274
Natural gas	75
Other fuels	30
Hydro	108
Nuclear	48
Geothermal	3
Total	899

The extracted coal is usually processed to remove some of the impurities. A large coal-processing plant may clean 1 million tons of raw coal per year and produce about 1.5 tons of wastewater per ton of coal processed. During cleaning, about 24 percent of the raw mined product is discarded and about half of this amount is coal. Following the wet processing, the coal is dried by heating. Coal-dust has been a major problem during the dry processing, but the emissions of dust are being reduced by adding wet scrubbers and air recirculation to the cyclone separators.

Coal is mainly used for electricity power production: pulverized coal is burned in large boiler furnaces where steam is generated at high temperature and pressure. The high-pressure steam passes through steam turbines, where heat energy is converted into mechanical energy of rotation. The steam turbines drive the alternators for the generation of ac electricity. The thermodynamic efficiency of this system is increased by expanding the steam down to a low vacuum, where it is condensed for recycling through the boiler. Almost half of the total heat energy available in the steam is lost to the environment through the large quantities of cooling water that are used in the condensation process.

Environmental impacts of the use of coal in electricity generation arise during the precombustion, combustion, and steam turbine stages. The handling and storage of coal at the power plant site creates problems with coaldust. These are partially solved by enclosures around the coal stock and by careful layering and compaction of the coal.

The burning of coal results in the production of ash and slag. These waste materials can be used for producing building materials and ash can also be used as an aggregate in road construction.

Peat constitutes a valuable energy resource in some parts of the world, such as Scotland and Ireland, where peat is used as a domestic fuel and for the production of electricity. It can be found on dry land, where shallow open-cast mining is undertaken, and sometimes under water in marsh conditions. Peat has approximately one-quarter the calorific value of average coal.

2.10 NUCLEAR POWER

Intensive research and development has been carried out in the field of nuclear power for the past 30 years but, despite this, nuclear energy accounts for only 0.6 percent of total energy consumption in the world. In Western Europe it represents 2.0 percent of total energy consumption and in the United States it is 1 percent. The United Kingdom was the first country to adopt nuclear power and the first commercial station was commissioned in 1962. The nuclear program in the United Kingdom has not followed earlier predictions, due to both technical and economic difficulties. Most of the Magnox reactor stations of the first British program have been operating at reduced power since 1970, following the discovery of corrosion in some reactor components. The five advanced gas cooled reactor (AGR) stations of the second program have suffered cumulative delays through numerous design and construction difficulties. Because of these difficulties, no nuclear stations were ordered in the United Kingdom between 1970 and 1975. The total installed capacity of nuclear stations in the United Kingdom was 8870 MW in 1977. The nuclear contribution is expected to rise toward 80×10^{10} MJ by 1980.

In the EEC, installed capacity of nuclear stations totaled 10,900 MW in 1973. An increase in the nuclear power output in Western Europe to 130,000 MW by 1985 has been envisaged, but only 28,000 MW of capacity was installed by 1977. The EEC commission has assumed that electricity demand will treble between 1970 and 1985. Conventional power stations will be required to meet 60 percent of overall electricity requirements in 1985 and power station demand for fossil fuels will more than double, from 404×10^{10} MJ in 1970 to 834×10^{10} MJ in 1985. If, as seems likely, the nuclear program will not be fulfilled, the requirement for fossil fuels may be even greater. Fossil fuels will remain of paramount importance for power generation in Europe throughout the period to 1985 and the second European Nuclear Program shows continuing increases in the fossil fuel requirement between 1985 and the year 2000.

The development of nuclear power in the United States has been suffering from difficulties similar to those experienced in Europe. Public opposition to nuclear power on environmental and safety grounds is becoming very much more widespread. The dangers associated with nuclear power are: (1) high levels of thermal pollution through discharge of waste heat into surrounding waters (the level of heat discharged is 40 percent higher than for comparable fossil-fueled power stations); (2) the degree of low-level radiation emitted by a station when it is operating normally; (3) the difficulties of transporting nuclear fuels and radioactive waste; (4) the still unsolved problem of storing, in steadily increasing quantities, highly radioactive waste which, in some cases, will remain active for hundreds of years; and (5) the danger of a serious reactor mishap causing a major radiation emission.

The U.S. Department of the Interior's 1972 estimate of nuclear output in 1985 was 1160×10^{10} MJ. This contrasts with their previous estimate some 12 months earlier of 2020×10^{10} MJ. This very considerable reduction reflects the difficulties in constructing and operating nuclear stations and the growing concern over safety and siting. Estimates of nuclear contributions for major industrialized nations in 1985 are not expected to exceed 11 percent of their combined energy demands. For the world as a whole, the U.S. Atomic Energy Commission has estimated that installed nuclear capacity could rise to 636,000 MW in 1985 and to 3,260,000 MW in the year 2000. With world energy demand continuing to increase from an expected $43,000 \times 10^{10}$ MJ in 1985 to possibly $78,000 \times 10^{10}$ MJ at the turn of the century, such a massive growth in nuclear power—if achieved—would only raise its contribution to world requirements from 9 to 21 percent. The role of conventional fossil fuels will thus remain dominant.

All the various types of nuclear reactors currently in commercial operation or on order consume uranium. Estimates in 1973 indicated that the world's reserves of U_3O_8 available at a recovery cost of below \$30/kg will be exhausted by the end of the century if nuclear generation expands as anticipated. Attention is, therefore, being directed toward the breeder reactor, which recreates fissile materials and has a relatively limited net consumption of nuclear fuels. There are still considerable technological and economic and environmental problems which appear to prevent the development of commercially viable breeder reactors for some time in the future. More remote still is the prospect of nuclear fusion reactors, which would use a fuel (deuterium) that is readily available in almost unlimited quantities from the ocean, and which would pose no environmental threat apart from thermal pollution.

2.11 WORLD ENERGY PROSPECTS

The future energy requirements of the world will play a dominant role in the political and economic power of individual countries. During the first three-quarters of the twentieth century the world has moved from an agriculturally based to an industrially based society and the continuation of present-day technology and further development is entirely dependent upon the supply of energy. The balance of power, the price of fuel, and its availability changed so abruptly in the years 1973 to 1974 that predictions of future world energy requirements cannot be made with a great deal of certainty. On the other hand, it has become extremely important for countries to make predictions of their energy requirements and to prepare concrete plans for obtaining the amount of energy required.

Based on the assumption that, in the future, there will be growth, increase in world demand, increase in population, improvements in productivity and living standards in industrialized countries, and progressive industrialization of developing countries, world energy demand will continue to increase rapidly and, by 1985, 1.8×10^{15} MJ will be required.

In order to meet this growing demand, all available forms of energy supply must be increased. Nuclear power will probably emerge as the fastest growing fuel during the next 15–20 years. Nevertheless, by 1985 the overall demand for fossil fuels will be double the present level. Oil is the major fossil fuel in the world energy market today, with Western Europe and Japan almost totally dependent on imports for their requirements and the United States rapidly emerging as a major importer. Imported energy can, in the future, be expected to be neither cheap nor secure.

Western Europe and Japan have become excessively dependent on imported energy. Europe has to import 60 percent of its fuel, almost exclusively in the form of Middle East and African oil. North Sea oil and gas will help to alleviate the present critical situation, but production is expected to reach a peak at around 9.0×10^{12} MJ and will only cover one-sixth of West European oil requirements in the mid-1980s. Indigenous natural gas reserves are limited and, in many countries, there is insufficient supply to meet present-day demands.

Expansion of the nuclear power program will continue, but the occurrence of each accident, and the failure to solve the many technological problems associated with nuclear power generation and storing of nuclear wastes, have resulted in a major setback in the supply of nuclear power. Both the United States and the European Community have accepted the need for urgent and rapid increase in the mining of coal. The availability of coal, together with lignite, by-product gases, hydroelectricity, and nuclear power will be sufficient in the 1980s to allow at least two-thirds of total electricity generation to be based on indigenous energy resources.

Rising oil prices and restrictions in availability will stimulate an increased demand for coal in the industrial market. There will also be a great demand for synthetic natural gas and oil produced from coal. The increasing cost of energy has already begun and will continue to encourage the conservation and efficient use of fuels but, unless radical changes occur in the structure of society, the rate of expansion of world energy demand will continue to increase.

THREE

FOSSIL FUELS

3.1 INTRODUCTION

The physical and chemical properties of solid, liquid, and gaseous fossil fuels vary over a wide range and these variations have an important influence on flame characteristics. Fuel properties need to be taken into account at an early stage in the design of combustors. The particular phase of the fuel, degree of volatility, and chemical composition will affect the residence time, flow patterns, and, hence, overall dimensions of the combustor. Design of atomizers for liquid fuels and grinding plant for coal are influenced by the particular type of fuel, and this also affects the quality of emissions and their control. Because of the critical energy balances that are developing between conservation, fuel economy, and restriction of emissions, the role of fuel properties is becoming much more important.

The combustion of coal provides over 50 percent of the electrical power and over 25 percent of industrial process heat generated in the United States. Large-scale developments are taking place to expand the use of pulverized coal, both in direct-fired boilers and, eventually, in large gas turbine and magneto-hydrodynamic systems. Major developments have taken place in the liquefaction and gasification of coal. These have presented a new range of problems because of the difference in the molecular structure of these fuels as compared with natural gas and petroleum products.

The fluidized-bed combustor offers the possibility of burning coal with high sulfur content and at high pressures. If many of the major technological problems can be overcome, the fluidized-bed coal combustor could, during the next few decades, become the major source of electricity power generation.

The burning of coal in pulverized form will continue to play an important role in energy technology and emphasis is being concentrated on improving the energy efficiency and minimizing the emission of pollutants from pulverized-coal combustors. As the scarcity of coal increases, coals having lower calorific values and higher ash and moisture contents will require to be burned.

Nuclear power and coal are anticipated to become the major energy sources for base power generation during the next few decades. In refinery processing, the trend will be toward converting more of the residual fraction of crude oil to distillates. Less heavy fuel oil and more lighter fuels, such as motor gasoline, automotive gas oil, and aviation kerosine, will be produced. In the United States, due to the higher demand for motor gasoline, the conversion in refineries is at a higher level than in Europe and the rest of the world. The oil industry is already building processing facilities to convert residual fuels to distillates. Provision is also being made to allow adjustment of the proportions of heavy and light crude oil that are processed.

It is expected that motor gasoline will remain the dominant fuel for private automobiles and automotive gas oil the dominant fuel for diesel-powered commercial transport for at least the next 10 years. Users of automobiles with very high mileages may be attracted to the diesel engine passenger automobile, which has, at present, a higher initial cost but better fuel economy. The diesel engine vehicles are more expensive than similar gasoline engine vehicles, primarily because of the high cost of the fuel injection equipment, heavier duty battery, and starter motor. Provision also has to be made for the engine to withstand higher pressures in the compression-ignition diesel engine. The gasoline engine, however, provides a better performance, particularly for acceleration, than the equivalent diesel automobile. Barker (1977) estimates that, on the average, the fuel consumption of diesel engines is 25 percent lower than the gasoline engine on a fuel/volume basis but only 14 percent lower on a fuel/weight basis. Major efforts are being made to improve the fuel consumption of gasoline engines so that they may well approach the levels currently obtained in diesel engines.

There will be a significant increase in the availability of liquefied natural gas during the next decade. In the Middle East, the North Sea, and North Africa there is a surplus of natural gas and it has become commercially viable to liquefy, transport, and distribute this fuel. Liquefied petroleum gases (LPG) are excellent automotive fuels, having high octane quality without need for antiknock lead additives. They also burn with low levels of exhaust emissions. Conversion of automobiles requires the addition of LPG storage tanks. There is an improvement in thermal efficiency for LPG compared with gasoline engines but there is an increase in volumetric fuel consumption because of the lower calorific value of LPG.

Methanol has been advocated as a fuel which can be locally produced in place of imported oil. It has the advantages of high flexibility, in that it can be produced from both coal and oil and has very good emission characteristics. Comparative studies of performance of automobiles using methanol, both as a pure fuel and as a gasoline component, have indicated certain basic problems. These include high toxicity, low calorific value, high volatility, and incompatibility with many fuel system components. As long as crude oil derived fuels are available, there is little incentive to utilize methanol as an automotive fuel.

Hydrogen is an attractive fuel for the internal combustion engine because of the high thermal efficiency and very low level of exhaust emissions. As crude oil supplies diminish, it is possible, during the next century, that hydrogen may be the main alternative fuel. The safety problems associated with explosion characteristics of hydrogen are considerable, both in the supply and distribution system as well as within the vehicle. Provisions for safety against explosion will require use of cryogenic or hydride systems and these will add considerable weight to the vehicle.

As long as automobile engines specify the use of high-octane quality fuels, refineries will be required to meet this demand. The energy, and hence cost, required to refine fuels with high-octane quality is high. Also, with the expected increase in conversion of heavy fuel oil to distillates, there is a strong incentive for the replacement of current engines by engines that can run on middle-distillate fuels. The direct injection stratified charge engine and the spark-assisted diesel engine are able to provide satisfactory performance using low-octane, low-cetane-number, middle-distillate fuels with a boiling range of between 420 and 870 K. Examination of the total energy balance of the fuel system shows that the efficiency of the gasoline engine cannot be increased sufficiently through operation on high-octane gasoline to offset the increased energy required in the refinery for manufacturing this fuel.

Barker (1977) reports that a Texaco study in the United States showed a reduction in refinery fuel consumption from 8.6 percent of the crude oil throughput to 6.4 percent when production is changed from high-octane to middle-distillate fuels.

Gaseous fuels, and in particular natural gas, are considered as premium fuels for domestic heating and in industrial processes requiring a "clean" fuel. In the United Kingdom, gas derived from coal was initially used for street lighting and for domestic cooking. With the advent of North Sea gas, burners in all home and industrial installations had to be changed because of the considerable increase in the calorific value of natural gas. In the United States, as the shortage of natural gas develops and synthetic gas (from coal) production increases, users are faced with the alternatives of changing burners to accommodate lower calorific fuels or upgrading the fuel. Among the fuels, gas is generally the easiest to burn and produces the lowest emission level. Corrosion levels are usually lower and high levels of combustion efficiency can be achieved in small volumes.

In this chapter the principal properties of coal, petroleum, gas, and synthetic fuels are reviewed. The chemical composition of a selection of fuels is given in Table 3.1.

Table 3.1 The chemical composition of fuels

Percentage by weight

	C	H	O	N	S
Wood	50	6	43	0.5	0.5
Peat	58	5	35	1	1
Lignite	70	5	23	1	1
Low-rank coal	81	5	12	1	1
Steam coal	92	4	1	2	1
Anthracite	94	3	1	1	1
Low-temperature tar	83	8	8	0.5	0.5
High-temperature tar	92	5	1	1	1
Crude oil	86	13	..	...	1
Fuel oil	85	11	..	...	4
Kerosine	86	14			
Naphtha	85	15			
Natural gas	75	24.5	..	0.5	

3.2 COAL

Coal is a compact, stratified mass of organic material originating from plant debris which is usually found interspersed with smaller amounts of inorganic matter beneath sedimentary rocks. Chemical properties of coal depend upon: (1) the proportions of the different chemical constituents present in the parent plant debris; (2) the nature and extent of the changes which these constituents have undergone since their deposition; (3) the nature and quantity of the inorganic matter present.

The transformation of plant debris to coal is influenced by the following: (1) bacteria—the extent of decay which takes place before the deposit is covered by an impervious sedimentary layer; (2) temperature and time—operating mainly after bacterial action has ceased; (3) pressure—increasing with depth and accentuated by severe earth movements such as folding or buckling of strata.

Rank

Coals are classified by rank, which is a measure of the degree of change of chemical composition during the transition from cellulose to graphite. Coals are ranked according to their carbon content, so that a low rank indicates a small degree of change, as in peat, while a higher rank indicates a larger degree of change, as in anthracite.

Under the action of heat and pressure, water and oxides of carbon are expelled. The volatile matter content varies with rank—the lower the rank the higher the volatile matter content. Calorific value also increases with rank. Oxygen levels are generally low, with the higher ranks of coal. The approximate

Table 3.2 Age of coal related to geological periods

Geological		Approximate mean age (10^6 years)	Class
Era	Period		
Upper Paleozoic	Carboniferous	250	Anthracite
	Permian	210	Carbonaceous and anthracitic
Mesozoic	Triassic	180	Bituminous
	Jurassic	150	Bituminous
	Cretaceous	100	Subbituminous and bituminous
Tertiary	Eocene	60	Lignites and subbituminous
	Oligocene	40	Lignites
	Miocene	20	Lignites
Quaternary	Pleistocene	1	Peat

mean age of a range of coals and the related geological system is given in Table 3.2.

Coal rank is most simply related to the age of deposits with the highest ranking coals having the greatest age. Coals may be simply divided according to age in millions of years, with anthracite = 250, lignite = 40, and peat = 1.

Classification of Coal

Various classification systems have been developed for coal. The most general classification divides coals into lignites, bituminous, and anthracite coals, in which the basis of classification is age. Coals are also classified according to their usage such as coking, gas making, steam raising, and domestic coals. More precise classifications are based upon chemical composition but, in addition, coals are separated according to their size and ash content. We shall first divide coals into their general types.

Lignites The first stage of formation of coal from peat is referred to as lignite and this group of coals has the lowest rank. Lignites may be black, or brown and earthy, with a woodlike structure. When dried in air, they tend to disintegrate and the air-dried material has a moisture content between 15 and 25 percent. The dry, ash-free content of carbon varies from 60 to 75 percent and that of oxygen from 20 to 25 percent. They are commonly marketed in the form of briquettes.

Subbituminous coals A group of coals, having characteristics between lignites and bituminous, are referred to as subbituminous coals. They possess high moisture and volatile matter content but possess no coking properties. Carbon content ranges from 75 to 83 percent and oxygen content from 10 to 20 percent, on a dry, ash-free basis.

Bituminous coals Bituminous coals are black and banded in appearance. Because of the wide range of properties, bituminous coals are further subdivided into caking (see next section on physical properties) categories, ranging from non- to very strongly caking coals. Carbon contents range from 75 to 90 percent, with a change in volatile matter content from 20 to 45 percent. Bituminous coals are those used for general purposes in industry.

Semibituminous coals Semibituminous coals are a subdivision between bituminous and anthracite coals. Carbon contents range from 90 to 93 percent, volatile matter content from 10 to 20 percent, and oxygen from 2 to 4 percent.

Anthracites Anthracites are the oldest coals with the highest carbon content. Carbon contents are greater than 93 percent, volatile matter content is less than 10 percent. Anthracites have no caking tendencies.

Classification Systems

Three major classification systems are currently used. They are: (1) The American Society of Testing Materials (ASTM); (2) The British National Coal Board (NCB); and (3) the international classification of hard coals developed by the Economic Commission for Europe (ECE).

ASTM have adopted two methods for classifying coals. Coals containing less than 31 percent volatile matter (VM) on a dry-mineral-matter-free basis (dmmf) are classified only on the basis of *fixed carbon* (FC), where FC = 100 − VM percent. Coals with FC above 86 percent are referred to as anthracite, while coals with FC between 69 and 86 percent are referred to as bituminous, as shown in Table 3.3.

The remaining groups of coal are classified according to their calorific value, determined on a moist-mineral-matter-free basis, i.e., the calorific value of the coal containing its natural bed moisture, but not including any visible moisture on the coal surface. The classification includes three groups of bituminous coals with calorific value (CV) between 25.7 and 32.7 MJ/kg, three groups of subbituminous coals with CV between 19.4 and 30.3 MJ/kg, and two groups of lignites with CV less than 19.4 MJ/kg.

Some physical properties can be associated with each class of coal: anthracite is nonagglomerating, bituminous is agglomerating and/or weathering, while lignite may be consolidated, or unconsolidated in the case of brown coal.

Table 3.3 American Society of Testing Materials classification of coals of the United States and Canada

Class	Group	Fixed† carbon (%)	Calorific‡ value (MJ/kg)	Physical properties
Anthracite	Meta-anthracite	> 98		
	Anthracite	92–98		
	Semianthracite	86–92		Nonagglomerating
Bituminous	Low volatile	78–86		
	Medium volatile	69–78		
	High volatile A	< 69	> 32.7	
	High volatile B		30.3–32.7	
	High volatile C		25.7–30.3	Agglomerating or nonweathering
Subbituminous	Group A		25.7–30.3	Weathering and nonagglomerating
	Group B		22.2–25.7	
	Group C		19.4–22.2	
Lignite	Lignite		< 19.4	Consolidated
	Brown coal		< 19.4	Unconsolidated

† Dry, mineral free.
‡ Moist, mineral free.

The coal classification system devised by the British National Coal Board divides coals into groups and classes with numbers referred to as "rank" between 100 and 900. Classification is based throughout on the percentage of volatile matter content, on a dry-mineral-matter-free basis. General descriptions, such as steam and volatile coals, are used as well as reference to the caking properties. The classification is shown in Table 3.4.

The international system of classification is based partly on the British and partly on the American classification systems. Coals are first separated into 10 class numbers, 0–9. Coals with VM less than 33 percent on a dry-ash-free basis are divided into six classes, according to their volatile matter content. Coals containing more than 33 percent VM are divided into four classes on the basis of their moist-ash-free calorific value, determined at 30°C and 96 percent relative humidity. Coals are then divided into four groups on the basis of their caking properties. These groups are then further divided into subgroups, according to their coking properties.

Anthracite, bituminous, and higher rank subbituminous coals with gross calorific values more than 23.94 MJ/kg, on a moist-ash-free basis, are termed "hard" coals. The lower rank subbituminous and lignite coals are termed "soft" or "brown" coals. The international scheme of classification for brown coals is

Table 3.4 Coal Classification System used by the British National Coal Board

Group	Volatile matter† (%)	Description
100	< 9.1	Anthracite
200	9.1–19.5	Steam
300	19.6–32.0	Medium volatile
400	32.1–36.0 and above	Very strongly caking
500	32.1–36.0 and above	Strongly caking
600	32.1–36.0 and above	Medium caking
700	32.1–36.0 and above	Weakly caking
800	32.1–36.0 and above	Very weakly caking
900	32.1–36.0 and above	Noncaking

† Dry, mineral-matter-free basis.

based on two intrinsic principal characteristics, which indicate the value of brown coal as a fuel, and as a raw material for chemical purposes, namely (1) the total moisture on an ash-free basis, and (2) the tar yield on a dry-ash-free basis. Class numbers vary from 10 to 15, based on total moisture between 20 and 70 percent. Group numbers range from 00 to 40 for tar yields, on a dry-ash-free basis between 10 and 25 percent.

Physical Properties of Coal

The properties: moisture, ash, volatile matter, sulfur, and calorific value used in the assessment of coals for industrial purposes are given in Table 3.5. The average properties of solid fuels are listed in Table 3.6. The air requirements and the composition of dry and wet waste gases are also given in Table 3.6 for a range of coals and wood. The utilization of a coal is partly determined by physical properties, such as strength, caking power, and calorific value. Other properties, such as the optical reflectance, provide further means for ranking coals.

Table 3.5 Properties used in the assessment of coals for industrial purposes

Class	Moisture		Ash	Dry, ash-free basis		
	As mined (%)	Air-dried (%)	Air-dried (%)	Volatile matter (%)	Sulfur (%)	Calorific value (MJ/kg)
Lignite	20–40	15–25	7–12	40–50	1–3	21.0–29.4
Subbituminous	10–20	5–16	5–10	41–45	1–2	32.4–33.6
Bituminous	5–15	1–10	2–10	20–45	1–2	33.6–36.5
Semi-bituminous	1–13	0.6–1	2–5	10–20	1–2	36.4–36.9
Anthracite	2–3.5	1–3	1–2	4–9	1–2	35.6–36.4

Table 3.6 Average properties of coals and solid fuels

| | Anthracite | Coking coals | | General purpose coal | Charcoal | Coke | Wood |
		Medium volatile	High volatile				
Moisture, %	8	7	9	13	2	8	15
Ash, %	8	8	8	8	1	7	7
Carbon, %	78.2	75.8	71.6	65.7	90.2	82.0	42.5
Hydrogen, %	2.4	4.1	4.3	4.0	2.4	0.4	5.1
Nitrogen, %	0.9	1.3	1.6	1.4	0.7	0.9	0.5
Sulfur, %	1.0	1.2	1.7	1.7	0.8	0.8	0.4
Oxygen, %	1.5	2.6	3.8	6.2	2.9	0.9	36.5
Calorific value kJ/kg gross	29,680	30,850	29,560	26,770	33,730	28,660	15,830
net	28,960	29,770	28,400	25,560	33,150	28,380	14,360
Air requirements per kg, kg air	9.84	10.15	9.74	8.84	11.11	9.58	5.10
m^3 air at 273 K, 1.01×10^5 N/m^2	7.61	7.85	7.53	6.84	8.60	7.41	3.95
Waste gas at 273 K, 1.01×10^5 N/m^2 m^3/kg of fuel	7.84	8.17	7.89	7.23	8.77	7.55	4.64
CO_2 content of dry waste gas, %	19.5	18.5	18.3	18.4	19.9	20.6	20.3
Composition of wet waste gas CO_2, %	18.6	17.3	16.9	17.0	19.2	20.2	17.1
H_2O, %	4.5	6.5	7.3	8.0	3.3	2.0	15.6
Oxides of N and S, %	0.3	0.3	0.5	0.5	0.2	0.2	0.2
N_2, %	76.6	75.9	75.3	74.5	77.3	77.6	67.1

Density The variable density of coals is an important factor in coal preparation, enabling the separation of those coals with a high "dirt" content from other more valuable coals. Variations in rank, nature, and proportion of constituents, and moisture content, all affect the density of the coal. When density is determined using liquids such as water or alcohol, they provide different results than when helium is used as the displacement medium. Densities determined by helium displacement are less than those determined by water displacement for the same coals if their carbon contents are below 85 percent. For higher ranks, densities from water displacement are less than those for helium. These different results are reflections of the porosity of the coal, since helium is capable of penetrating deep into the ultramicroscopic pores. Thus, we distinguish between the *true density* of a coal, as established using helium as the displacement medium, and the *apparent density*, determined by using water. Corrections also need to be made for the presence of moisture and incombustible mineral matter. Normally only the apparent densities are determined, which are sufficient for most practical purposes. Figure 3.1 shows the variation in apparent density plotted as a function of carbon percent for various classes of coal. Densities for the soft coals are approximately 1.5×10^3 kg/m^3. As carbon content increases, the density passes through a minimum for bituminous coals with a carbon content of approximately 86 percent,

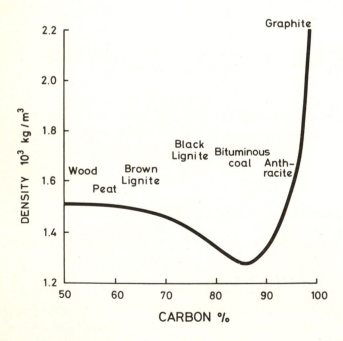

Figure 3.1 Density as function of carbon content for solid fossil fuels.

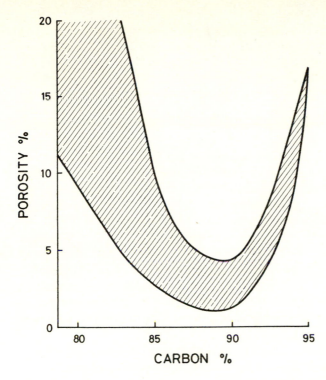

Figure 3.2 Porosity as function of carbon content for solid fossil fuels.

beyond which the density increases rapidly through the high rank bituminous and anthracite coals, reaching a value of 2.2×10^3 kg/m³ for graphite.

Porosity Coal is interspersed by minute pores of different size and capillaries, so that it possesses, in effect, an ultramicroscopic spongelike structure. The porosity, or volume percentage occupied by such pores, may be calculated from density measurements using helium and then mercury as the displacing media. Under normal pressures, helium alone will penetrate and fill the pore spaces so that the difference between the two density measurements is an indication of the porosity. Alternatively, water may be used in place of mercury if the coal specimen is first coated with a thin film impervious to water. The variation in porosity with carbon content for coals is shown in Fig. 3.2. In the lower rank coals, the porosity may exceed 20 percent but it rapidly diminishes to a minimum at the highest rank bituminous coals (carbon content 89 percent), above which point it increases toward the anthracite and hard coals.

If the mercury is subjected to pressure, it is possible to force some mercury into the pores, although even under high pressures not all are filled. Consequently, it is believed that two pore systems occur in coal, one formed by larger pores which are penetrated by mercury under pressure, and the other composed of ultrafine pores, which remain accessible only to helium.

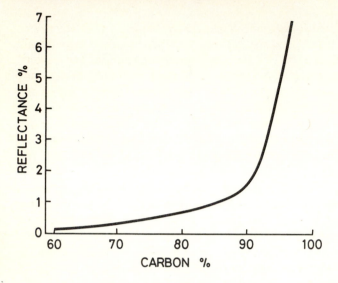

Figure 3.3 The maximum reflectance of vitrinites in cedar oil. (*After D. W. Van Krevelen 1961.*)

Strength The strength of a coal is related to its hardness and friability. Hardness is tested by the depth and area of penetration when a pyramidal or spherical inden-tor is pressed into the coal surface. Such tests reveal a maximum hardness for coals with a carbon content around 83 percent and a minimum at about 90 percent carbon. Anthracites with carbon contents over 93 percent behave as elastic materials.

Friability is the ease with which coal is fractured or crumbled. The friability of a coal has a considerable effect on the amount of degradation that occurs during its transportation and preparation. Friability is tested by determining the impact strength index. For such an estimation, a steel plunger is dropped for a constant number of times on to a coal sample at the base of a steel cylinder. The percentage of coal remaining in the initial size range after the test is the impact strength index. Coals can also be assessed by their grindability, as determined by the amount of work required to grind a specified coal sample to a uniform size. Coals with a carbon content of about 90 percent have a maximum grindability, which is in accordance with other strength determinations.

Reflectance The reflectance of a coal is the percentage of incident light reflected from a polished surface. It is determined by microscopic examination, under oil immersion, where a photocell measures reflected light. Reflectance increases con-tinuously with coal rank, as shown in Fig. 3.3, and, hence, provides a rapid method of quantitatively assessing the carbon content and other chemical characteristics of a coal. Moreover, it is apparently unaffected by weathering, unlike other properties.

Caking and coking properties A coal is said to have caking properties when, as a result of heating, there is a tendency for the coal to adhere so as to form a solid mass. When large cakes of coke are formed, they can damage grates and stokers. Coals intended for the manufacture of metallurgical coke require to have strong caking properties. For carbonization, coals with high gaseous thermal yields and moderate caking properties are required. These are usually found among the lower rank coals. In order to achieve a maximum yield of coke, coals with low volatile matter, i.e., high rank, are selected. Both the lowest and highest rank coals have poor caking qualities with good caking qualities found among the bituminous coals.

The caking property of a coal is determined by its capability, in powdered form, to swell and agglutinate on heating. Coke is the solid residue after liberation of volatile constituents from mineral coal. Usually there is a close relation between coking and caking properties of coal. The term *coking coal* is reserved for those coals which are used for the manufacture of metallurgical coke. Many coals are only weakly caking and, therefore, require blending with more strongly caking varieties before a good quality coke may be obtained from them. Several tests, each assessing slightly different characteristics, and therefore not exactly equivalent, have been designed to record the caking properties. The swelling of a coal is determined by the crucible swelling or dilatometer test and the agglutinating power by the Roga or Gray-King assays.

The swelling index of a coal is determined by rapidly heating a ground coal sample in a silica crucible for one-and-a-half minutes to 800°C. In the next minute, the temperature is raised to 820°C and maintained at this level until the volatile matter is burned off, or for two-and-a-half minutes, whichever is the greater period of time. The crucible is cooled and, if the residue is coherent, the coke button is removed and compared with a standard series of numbered profiles to determine the crucible swelling number. If the residue is noncoherent, the swelling number is zero. The contraction and dilation of a coal particle heated in a furnace for a range of temperatures is shown in Fig. 3.4.

Calorific value The calorific value of coal is determined by burning 1 gram of air-dried coal in a combustion bomb under pressure of 3 MPa of oxygen. The bomb is immersed in a water bath and the total heat evolved during combustion of the coal sample is determined from the rise in temperature of the surrounding water bath.

Calorific value is determined on two bases: (1) the gross calorific value, which includes all the condensed moisture, and (2) the net calorific value, which excludes the heat content of the moisture. Since, in the many practical combustion devices, the moisture in the combustion products remains in the vapor phase, the net calorific value is referred to as the practical heating value. The difference between the gross and net calorific value is about 1.2 MJ/kg; it is dependent on fuel hydrogen content and moisture content.

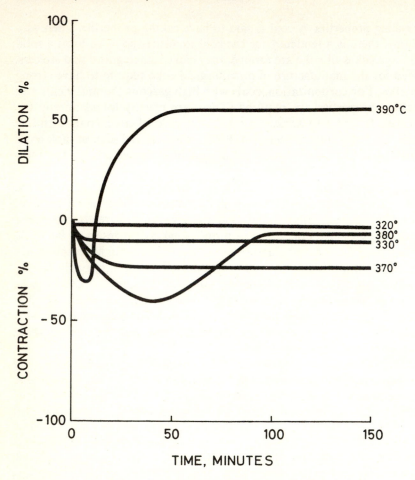

Figure 3.4 The contraction and dilation of a coal particle heated in a furnace for a range of temperatures.

The calorific values of coals with carbon contents below about 92 percent exhibit a marked relationship to their rank, as shown in Fig. 3.5. A slight decrease in calorific value occurs in the anthracites. In coals of similar rank, notably those with carbon contents less than 85 percent, the variations in calorific values may be attributed partly to differences in geological properties of the coal. Since calorific value provides a means of assessing the merits of coal as a fuel, it is used as a parameter in coal classification systems. The relation between hydrogen and carbon content, and also the relation between volatile matter and calorific values, is shown in Fig. 3.6.

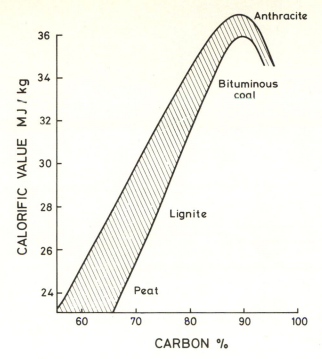

Figure 3.5 Calorific value as function of carbon content for solid fossil fuels.

Chemical Properties of Coal

The analysis of coal is carried out by two standard means: (1) a proximate analysis and (2) an ultimate analysis. The proximate analysis involves the determination of surface moisture, moisture in air-dried coal, ash, volatile matter, and fixed carbon. Sulfur content and calorific value are estimated.

The moisture content of a coal is divided into two parts: (1) inherent moisture, absorbed or adsorbed within the coal substance, and (2) surface moisture, which has been acquired during water spraying, wet washing, or exposure to rain and snow. If the coal is left exposed to the atmosphere, the surface moisture and some of the inherent moisture will evaporate until the residual moisture is in equilibrium with that of the surrounding air. The coal is then said to be in an "air-dried" condition.

Proximate analysis *Surface moisture* is determined by allowing crushed coal to dry for 24 hours at 297 K under a humidity of 75 percent. The percentage surface moisture is determined by the loss in weight due to evaporation. The coal is then considered to be in an air-dried condition and all subsequent analyses are carried out in this condition.

Inherent moisture is determined by heating several grams of air-dried coal for one hour at 380 K. In order to avoid oxidation of the coal, the heating is carried out by passing nitrogen over the coal or carrying out the heating under vacuum.

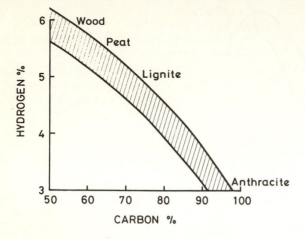

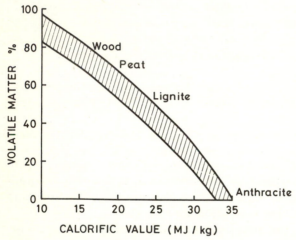

Figure 3.6 Relation between (a) hydrogen content, and (b) volatile matter and calorific value for solid fossil fuels.

The percentage of inherent moisture is determined by the loss in weight due to evaporation.

Moisture content in coal entering a combustion chamber requires heat for evaporation and sensible heat when it is raised to high temperatures. Since no heat release is associated with water vapor, high moisture contents are undesirable. When coal is pulverized, it is first dried artificially in order to remove the surface moisture.

The bulk density of coal is initially lowered as the quantity of surface moisture is increased. A minimum is reached, after which further wetting causes the particles to pack more closely and the bulk density increases. The optimum quantity of surface moisture depends upon the proportion of very fine particles present. Common values are 5 to 7 percent of surface moisture. Wetting is encouraged in the manufacture of metallurgical coke as a means of control of bulk density and also as a means of promoting certain gasification reactions.

Ash is determined by inserting 1 to 2 grams of dried coal in a furnace at a temperature of 700 K for 30 minutes, allowing a free flow of air. The sample is then transferred to a second furnace, at a temperature of 1050 K, where it is heated in an oxidizing atmosphere until the weight becomes constant. This procedure is to ensure that all combustibles are oxidized and that the residue contains only ash. The sample is then cooled and the percentage ash determined from the weight of the residue.

Ash is the powdered residue left after combustion. Since it makes no contribution to the industrial process, means are sought to minimize the quantities of ash and, hence, reduce the cost of transport to and away from the plant, as well as handling.

When coal is burned on an industrial grate, it should contain sufficient mineral matter to form a layer of ash on the fire bars, which acts as a protective blanket and prevents the fire bars from becoming overheated. Ash contents between 7 and 12 percent are suitable for industrial-scale combustion on grates. In pulverized-fuel firing, ash contents are of relatively minor significance in the combustion process. The resulting fly ash needs to be arrested but, with appropriate arrestors such as cyclones and electrostatic precipitators, ash contents up to 25 percent can be tolerated.

During carbonization, most of the inorganic mineral matter remains in the coke and, consequently, the ash content of coke is higher than that of the original coal.

Volatile matter is determined by inserting 1 gram of air-dried coal into a furnace at 1200 K for a period of 7 minutes. The sample is kept covered and cooled rapidly to prevent oxidation. The aim in this test is to liberate only the volatile matter without affecting the fixed carbon. The loss in weight is measured and, after deducting the percentage of inherent moisture, the percentage of volatile matter is determined relative to the initial weight of the sample.

Fixed carbon is determined by difference, assuming that the air-dried sample is composed of moisture, ash, volatile matter, and fixed carbon.

Ultimate analysis The ultimate analysis is a chemical means of determining the composition of coal by separating the principal constituents: carbon, hydrogen, nitrogen, sulfur, and oxygen.

Carbon is determined by combustion and the measurement of CO_2, according to the following reaction:

$$C + O_2 = CO_2$$

The concentration of CO_2 is determined by absorbing the exhaust gases in soda lime.

Hydrogen is also determined by combustion, according to the reaction

$$H_2 + \tfrac{1}{2}O_2 = H_2O$$

The H_2O concentration is determined by absorption of the exhaust gases in calcium chloride.

Nitrogen concentration is determined by heating the coal sample in concentrated sulfuric acid in the presence of a catalyst. Nitrogen is converted to ammonia. The ammonia is absorbed in an alkaline absorbent and subsequently recovered by steam distillation, followed by absorption in acid and determination by titration.

Sulfur is converted to sulfate by heating the coal in a mixture of magnesium oxide and anhydrous sodium carbonate. The concentration of sulfate is estimated gravimetrically by precipitation as barium sulfate.

Sulfur is present in coal as organic sulfur and iron pyrites (FeS_2). The organic sulfur is evolved as H_2S with smaller amounts of COS, CS_2, and thiophene. During combustion, these volatile sulfur compounds oxidize to form SO_2 and SO_3. Any sulfur retained by the coke is also eventually oxidized to SO_2 and SO_3. In carbonization and gasification processes, the H_2S remains in the product gas and requires to be subsequently removed.

During combustion or carbonization, the inorganic sulfur is converted to iron sulfide, much of which is in the form of Fe_3S_4, which has a low melting point. During combustion, the sulfide melts and oxidizes to ferrous oxide (FeO), which reacts with silicates that are present in the coal ash, to form clinker—the hard mass formed by the fusion of mineral impurities in the coal. Accumulations of clinker retard the free flow of combustion air and require to be periodically removed. During carbonization, the majority of the inorganic sulfur remains in the coke. Sulfur in metallurgical coke is undesirable, since it leads to metal embrittlement.

Sulfur oxides formed during combustion combine with water vapor to form H_2SO_4 and H_2SO_3. During cooling, when the temperature falls below the acid dew point, an acid mist forms and becomes deposited on cool surfaces, causing acid corrosion. SO_2 and SO_3 can also form sulfate deposits on surfaces and water tubes, which reduce the efficiency of heat transfer from the hot gases to the water.

Chlorine occurs in coal, mainly as NaCl, partly as KCl, and partly in an unidentified form. The total chlorine content is usually less than 0.3 percent. During combustion or carbonization, the NaCl is volatilized and subsequently dissociates, whilst nonalkali chloride is evolved as hydrogen chloride (HCl).

Sodium ions (Na^+), produced from the dissociation of NaCl, react with brickwork and mortar, causing damage to the interior walls of combustion chambers and coke ovens. They can also lead to the formation of rocklike deposits on surfaces within high-duty boilers. Any HCl evolved tends to dissolve in the water vapor present in the combustion products, leading to the formation of hydrochloric acid. Chlorine contents above 0.5 percent are very likely to cause serious damage to brickwork.

Oxygen concentration is determined by difference.

Physical and chemical data In the recording of physical and chemical data on coal, reference is generally made to the basis and conditions under which the data were determined. Analyses made on coal, including surface moisture, are referred to as *as received* or *as sampled*. When surface moisture has been removed, data are quoted on an *air-dried* or *dry* basis. Ash is mainly composed of mineral matter but it can include carbon in the form of sodium carbonates. On the other hand, some mineral matter is volatilized so that not all of the mineral matter is retained in the ash. Determination of volatile matter content and calorific value are reported on a *dry-ash-free* (daf) or *dry-mineral-matter-free* (dmmf) basis.

The volatile matter is calculated on a dry-mineral-matter-free basis, according to the following formula:

$$\text{Mineral matter} = 1.10A + 0.53S_T + 0.74CO_2 - 0.32$$

where A = ash, S_T = total sulfur, and CO_2 = carbon dioxide derived from carbonate. The percentages are determined from air-dried samples. For most British coals, mineral matter is approximately 1.15 of the ash percentage.

Coals from any particular source can have a wide variation in composition and many samples need to be analyzed. Data are sometimes quoted according to the degree of accuracy of the analysis. When coals are blended or mixed, the variation in properties of a particular batch of coal will be even greater.

Experimental Investigations

In addition to the standard tests that are generally carried out to establish the principal characteristics of coal, a number of experimental investigations have been carried out in order to provide more detailed analysis of coal composition and combustion characteristics.

Analysis of coal samples Coal samples have been chemically analyzed in heated reactor vessels. Chemical and physical analyses have been made of unreacted material and these have been compared with samples of partially reacted coal extracted from the reactor vessel for a range of residence times. Gases liberated from the coal during volatilization are analyzed separately from the chemical and physical analysis of the solid residue. Coal samples are tested for porosity and volatility. Physical characteristics are determined by examination by light microscopy and scanning electron microscopy. Experiments have been carried out on coal samples in which reaction rates have been measured at fixed gas temperatures or at fixed particle-heating rates. Some attempts have been made to measure particle temperatures, gas temperatures, and concentration distribution of gases surrounding the particles. There is some agreement on absolute magnitudes of overall chemical reaction rates, obtained in a number of different experiments, but there has been considerable controversy over interpretation of the data in order to determine the mechanisms of volatilization and reaction.

Extensive data on the pyrolysis of coals and chars in inert and oxidizing atmospheres is reported by van Krevelen (1961). These studies are generally carried out in small experimental furnaces or reactors, where temperatures are usually below 1000 K and heating rates are less than 1 K/s. In practical combustors, such as utility boilers, gas turbines, and magneto-hydrodynamics (MHD) combustors, peak temperatures can vary from 1500 to 2700 K and heating rates can be of the order of 10^5 K/s. Data from the experimental furnaces cannot be extrapolated to the higher temperature and heating rate conditions pertaining to large practical combustors. From the work which has been carried out, it is clear that the heating rate is an important factor in the changes of physical and chemical structure of coal during the combustion process.

There are a number of fundamental differences between the combustion of coal and other hydrocarbon fuels. Coal has a higher carbon/hydrogen ratio than other hydrocarbon fuels. This is because the chemical structure of coal is largely aromatic, while petroleum and natural gas consist mainly of aliphatic hydrocarbons.

Coal molecules have a high degree of polymerization. The molecules are arranged in an orderly fashion in layers, depending upon the coal rank. The polycyclic molecule layers are linked by intermediate —C—C and —C—O—C— bonds. Molecular weights of large-chain molecules can easily be of the order of 2000 and above.

The density of coal generally varies between 10^3 and 2×10^3 kg/m^3. Coal is highly porous, with internal surface areas in the range of 10 to 200 m^2/g. A 100-μm particle can have a total area of 100 mm^2. The high porosity results in swelling under conditions of high heating rates. Further significant changes in pore structure occur in the course of crushing, cleaning, drying, and pulverizing coal. Grinding of coal has been found, initially, to increase the porosity due to additional fracturing, while pulverizing to very fine sizes can lead to sealing of pores with the consequent decrease in reactivity.

Attempts have been made to understand the different mechanisms of reactions that occur under conditions of pyrolysis—endothermic decomposition in the absence of oxygen—and oxidation at temperatures and heating rates lower than those occurring in practical combustors. Under laboratory conditions, sufficient time may be available for relatively slow depolymerization reactions involving bond scission at the aliphatic and cross-link bridges to occur, resulting in the formation of high molecular weight liquid and gaseous tars and nonaromatic species. If sufficient time is available, these molecules may then undergo further slow gas-phase cracking reactions to form lower molecular weight species, or they may recondense within the coal-pore structure to form more energetic cross-link bonds. These slow processes occur in the absence of convection and rapid heating.

The pyrolysis and oxidation processes occurring in laboratory furnaces are radically different from those in practical combustors where small coal particles interact with high-temperature gases. The combustion of pulverized coal in furnaces occurs in two stages. Initially, during times in the range of 10–200 ms, there is a rapid devolatilization of the particle. The second stage involves the heterogeneous combustion of the devolatilized residual char. The time to complete the second stage is in the range of 0.5–5 seconds, depending upon the properties of the coal and the combustion environment.

In studies of combustion of coal particles, special efforts have been made to determine the relative influence of diffusion and chemical kinetics on the rates of reaction. Detailed studies have been made examining the influence of changes in coal and char particle size, porosity and composition, as functions of changes in the combustion environment. Changes in the oxygen partial pressure and the temperature have been found to have important influences on the observed chemical reaction rates.

Devolatilization Devolatilization is preceded by an induction or ignition delay time, which is dependent upon heating rate. The coal particle is rapidly heated to temperatures in excess of 1000 K. There is some evidence that particles can be superheated to temperatures of the order of 500 K higher than the gas temperature. Under these conditions, ignition occurs on or very close to the particle surface and is nearly coincident with volatile evolution.

During rapid devolatilization at higher temperatures and higher heating rates, larger fractions of lower molecular weight species, in addition to tars, are evolved. It has been suggested that these species derive from rupture of the $-H_2$, $-CH$, $-CH_2$, $-CH_3$, and $-OH$ bonds which appear on the periphery of the coal molecule, as well as a more complete disruption of the weaker cross-link bonds between layers. This is followed by rapid diffusion of these species from the particle surface and their subsequent gas-phase oxidation. During this stage, approximately 20 percent of the fuel-bound nitrogen is evolved, probably in the form of reactive CN, NH, and NH_2 radicals and NH_3, HCN, and NO species.

It has been suggested that the mechanisms of devolatilization are different for small particles, less than 100 μm, than for large particles. For small particles, heterogeneous surface reaction is presumed to be dominant, with rapid diffusion of the evolved species away from the particle. This devolatilization is kinetically limited. As particle size increases above 100 μm, diffusion becomes more important, until finally, for very large particles, devolatilization may become diffusion controlled.

Combustion of devolatilized char residue A complete description of the reaction mechanisms for char particles must take into account the diffusion and reaction processes taking place, both within the particle and on the particle surface. The reaction process is initiated by diffusion of the oxidizing species to the particle surface and, subsequently, throughout the internal pore structure. This is followed by chemisorption of the gaseous reactant on to the internal and external particle surface. After reaction has taken place on the particle surface, the gaseous reaction products require to be desorbed from the surface and diffused away from the particle surface. Reaction rates are initially dependent upon the amount of heat transferred by radiation, convection, and conduction from the surroundings to the particle surface. Subsequently, reaction rates become more dependent upon local temperature conditions, determined by reaction on the surface.

Coal is generally pulverized to sizes smaller than 100 μm. A typical size distribution is one in which 90 percent by weight of the particles are less than 75 μm. For such small particles, the principal component of the overall combustion time is the time required to burn the nonvolatile residue (the char) following devolatilization. This time is a strong function of the char properties, and hence of the original coal properties, as well as the preceding time and temperature history of the devolatilization process.

Chars are much less reactive than the parent coal, due to the loss of the more reactive species during devolatilization, as well as the disruption of the original organic structure. Chars are more porous—there are more large pores—than the original coal. This open-pore structure is generally believed to have a strong influence on the mechanisms and rates of char combustion. Chars derived from lower rank coals are observed to burn more rapidly than chars from high rank coals.

Char combustion involves both surface- and gas-phase oxidation processes with mass diffusion and chemical kinetics playing important roles. The relative importance of these processes varies with the dimensions of the char particle. For particles larger than 100 μm, at temperatures above 1200 K, mass transfer by diffusion of oxygen to the particle is considered to be the rate limiting process. At the surface, a fast heterogeneous reaction between carbon and oxygen takes place with the formation of CO. The CO oxidizes to CO_2 in the gas phase close to the particle. On the basis of this simple diffusion model, the rate of combustion is equal to the rate of diffusion of oxygen to the exterior surface of the char particle. The reaction is considered to occur on the exterior surface at a constant particle density but a decreasing particle size. The rate of diffusion is given by

$$R_{\text{diff}} \sim [(D_0/d)\rho_0(T/T_0)^{0.75} \ln F]$$

where D_0 is the average diffusion coefficient, d is the diameter of the spherical particle, T is the reaction temperature, ρ_0 and T_0 are the reference gas density and temperature, and F is a function of the oxygen mass fraction in the free stream. In this model, the oxygen concentration at the particle surface is taken to be negligible compared with the free stream concentration, and the combustion rate is independent of the total system pressure. Further, the combustion rate is weakly dependent on the temperature and inversely proportional to particle size. Because the system is diffusion controlled, the reaction rates become independent of the parent coal rank.

As the char particle size decreases, the rate of mass transfer increases. Thus, as the char particle size decreases, the influence of finite rate chemical kinetics becomes more important. Measured rates of char combustion for particles initially less than 100 μm have been found, in some cases, to be several orders of magnitude lower than the rates predicted by models based on simple diffusional control.

Scanning electron micrography of fly ash from coal-fired utility boiler exhausts indicate extensive formation of cenospheres. Samples of char, which have been examined after partial combustion, show only small changes in outside particle diameter, but increased porosity. Thus, the internal pore structure of the char can be actively involved in the burning process. For bituminous coal char particles, of the order of 40 μm, it has been suggested (Mulcahy and Smith, 1969) that chemical kinetic mechanisms must be based on partial or total penetration of the pore structure by the oxidizing species. The degree of penetration is a function of pore size which, in turn, is a function of the degree of reaction. The outside diameter of the char may remain nearly constant while reaction proceeds within the particle. As reaction proceeds internally, the ratio of internal to external surface area increases, the porosity increases, the particle density decreases, and, hence, the reaction rate accelerates.

Among the many experiments which have been carried out on the combustion of coal and char particles, there have been very few in which there was sufficient control or measurement to establish the rates of the various diffusional and reaction processes. Hence, there is a very wide scatter in reported values and conflicting evidence as to the relative importance of the diffusional and reaction processes. The suggestion that the rate control moves from diffusion to chemical kinetics as the particle size decreases has been based upon the classical arguments of increase in surface to volume ratio as particle size decreases. This argument may, however, be quite irrelevant for situations where the internal surface area in porous char is very much greater than that of the outer surface. There is, for example, some evidence to show that finer grinding of particles to reduce the exterior particle diameter and increase the total number of particles has little effect on the total rate of combustion.

Reaction mechanisms Previous studies of the pyrolysis and combustion of pulverized coal have focused attention on overall rates of burning and concentrations of products at the combustion chamber exhaust. By altering variables such as particle size and concentration, excess air, air preheat, and heat transfer, it has been concluded, mainly by deduction, that finite rate chemical kinetics can have an important influence on reaction rates of particles with diameters less than 100 μm.

A complete understanding of the combustion of pulverized coal requires a knowledge of the transport phenomena and chemical kinetics associated with individual coal particles during their passage through the combustion chamber. From the time when a particular particle enters a combustion chamber, it passes through a wide range of environments with temperatures and individual partial pressures varying from those of the cold air entering the combustion chamber through those of high-temperature stoichiometric-mixture regions of the flame, to those of the postflame region, where some remaining coal particles may be found in a sea of products. The heat transfer, by radiation from the combustion chamber walls and other coal particles, and by convection in the region of high-temperature gases, also varies over extremes.

The determination of the rate of chemical reaction of pulverized coal particles in combustion chambers requires, initially, a knowledge of the temperature and gas concentration environment through which the particle passes. This environment is, initially, influenced and may subsequently be determined by the rate of reaction of the particle itself. It is because of these difficulties that many studies of the reaction mechanisms have been made with individual particles reacting in environments where the temperature and gas concentrations were both monitored and controlled. The other alternative has been to study global characteristics of particles in combustion chambers and, from these, to attempt to make deductions about mechanisms for combustion of individual particles. Comprehensive information is not yet available on the relative rates and mechanisms of formation of the major products, CO, CO_2, and H_2O as a function of total pressure, local gas temperature and composition, particle temperature, and heating rate.

From kinetic studies, it is known that OH radicals and H_2O in the gas phase, adsorbed onto the particle, can have an important influence on the relative rates of conversion to stable products and pollutants. Most practical combustors utilize recirculation zones to improve flame stabilization. These recirculation zones have high concentrations of products so that oxidation rates of particles in the presence of high concentrations of H_2O and CO_2 need to be determined.

The temperature of individual particles is difficult to determine, yet it is clear that the temperatures achieved by the particles will have an important influence on the reaction rates. The level of turbulence intensity and scale of turbulence greatly affect the rates of heat and mass transfer within the boundary layers surrounding each individual particle. The extent to which particles follow local streamlines affects the drag and, hence, the trajectories. For particles larger than 50 μm, the drag to momentum ratio will not be sufficiently high for particles to follow gas streamlines and certainly not the higher-frequency velocity fluctuations. In dense sprays, coal particles are sufficiently close to each other to cause interaction, leading to local "starvation" of oxygen and buildup of combustion products. When, further, it is taken into account that each individual particle is influenced by the properties of the parent coal, including particle size, particle size distribution, porosity, ash composition, and rank, it is not surprising that very few definitive statements can be made about the rates and mechanisms of combustion of pulverized coal particles in practical combustion chambers.

3.3 PETROLEUM

Crude Oil

Crude petroleum is a naturally occurring, free-flowing liquid which has a specific gravity between 0.78 and 1.00. Petroleum is found in reservoirs and is recovered by drilling through rock surfaces. These naturally occurring oil reservoirs are usually under pressure so that the oil flows up the well pipe; otherwise, it is pumped mechanically. Crude oils are complex mixtures of a vast number of compounds and the composition of the crude oil varies very widely, according to its location.

The compounds in crude oils are essentially hydrocarbons or substituted hydrocarbons, in which the major elements are carbon (85–90 percent) and hydrogen (10–14 percent), with small amounts of sulfur (0.2–7 percent), nitrogen (0.1–2 percent), and oxygen (0–1.5 percent). Crude oils generally contain trace amounts of elements such as vanadium and nickel and may also be contaminated with chlorine, arsenic, and lead. The trace elements and contaminants are normally present in concentrations of the order of 10 ppm.

The hydrocarbon compounds in crude oil are made up of paraffins, naphthenes, and aromatics. Olefins are absent from crude petroleum but are a product of all cracking reactions. The straight-chain normal paraffins have much lower octane numbers than the highly branched paraffins and are, therefore, less desirable as gasoline fuel. In the higher boiling ranges of kerosine and diesel fuels, normal paraffins improve the burning qualities, smoke point, and cetane number but they have much higher melting points and, hence, their presence adversely affects the cloud point and pour point (see Glossary).

Two main classes of monocyclic naphthene compounds are present in petroleum:

cyclo-hexanes *cyclo*-pentanes

The important difference between these two classes of naphthenes is that the *cyclo*-hexanes (but not the *cyclo*-pentanes) are readily dehydrogenated using a platinum catalyst to yield aromatics, thereby resulting in a marked improvement in gasoline octane number.

Aromatics, the third main group found in the low-boiling distillates, improve the octane number of gasolines but adversely affect the burning properties of kerosines and gas oils. They can be removed relatively easily by solvent extraction.

The lower boiling olefins (C_2–C_4) are manufactured by the thermal cracking of light distillates. More complex hydrocarbons are formed by joining two or more naphthene rings together or by joining naphthene and aromatic rings. Many of the higher boiling portions consist of mixed molecules, such as decalin, naphthalene, and diphenyl.

Crude oil is broken up into fractions by distillation in a fractionating column. Vacuum is used at higher temperatures to avoid cracking and a true boiling point (TBP) is obtained by plotting the total yield of distillate against the still head temperature. A typical crude oil analysis is shown in Table 3.7.

The analytical tests applied to original crude oil and its fractions are largely those developed by the Institute of Petroleum (IP) and the American Society for Testing Materials (ASTM), in conjunction with the major oil companies.

Variations in Quality of Crude Oils

Specific gravity Crude oils in current production range from the very heavy asphaltic and naphthenic stock with specific gravities between 0.979 and 1.00, such as Boscan and Bachaquero from Venezuela or some of the heavy Californian crude oils up to the very light (specific gravity 0.7925 to 0.8155) such as Qatar (Middle East), Hassi Messaoud (Algeria), Moonie (Australia), and Cumarebo (Venezuela).

Table 3.7 Typical crude oil analysis (Kuwait export crude)

Column groups — Total Original crude oil; Atmospheric distillation: Distillates (Gas C_1–C_4, Gasolines (debutanized), Naphtha, Kerosine, Diesel oils) and Residues (Fuel oils).

	Total Original crude oil	Gas C_1–C_4	Gasolines (debutanized)	Gasolines (debutanized)	Naphtha	Kerosine	Diesel oils	Diesel oils	Fuel oils	Fuel oils
Boiling point range, °C		to 15	15–95	15–149	95–175	149–232	232–343	343–371	>343	>371
Yield on crude, % wt	100	1.77	6.05	13.55	11.6	12.25	17.0	4.15	55.45	51.3
Yield on crude, % vol.	100	2.52	7.85	16.65	13.4	13.55	17.5	4.1	49.8	45.75
Specific gravity, 15.5°C	0.869		0.663	0.703	0.749	0.785	0.843	0.885	0.967	0.975
Total sulfur, % wt	2.5		0.020	0.025	0.049	0.15	1.27	2.41	4.02	4.16
Mercaptan sulfur, % wt			0.015	0.018	0.018	0.006				
Paraffins, % wt			87.5	77	67.5	62				
Naphthenes, % wt			11	16	18.5	20				
Aromatics, % wt			1.5	7	14	18				
Octane number (research) clear			63.5	50	35					
Smoke point, mm						28				
Freezing point, °C						−54.5				
Aniline point, °C						15	21	23		
Diesel index						68	58	47		
Cloud point, °C							−12	11		
Pour point, °C	−32						−15	10	15	21
Wax content (BP), % wt	5.5					trace			9.1	8.7
Kinematic viscosity at 21°C, mm²/s	17.0					1.15	5	12.8		
38°C, mm²/s	9.6					1.00	3.53	9.85	1150	2185
50°C, mm²/s							2.79	6.8	480	850
60°C, mm²/s									260	436
99°C, mm²/s									44.6	64.4
Acidity, mg KOH/g	0.15					0.02	0.07	0.14	0.17	0.17
Total nitrogen, ppm	1200						40	290		
Total ash, % wt	0.006								0.011	0.012
Vanadium, ppm	27								49	53
Nickel, ppm	7								13	14
Carbon residue (CON), % wt	5.2								9.3	10.1
Asphaltenes, % wt	1.4								2.5	2.7
Molecular weight (av)		54	82	96	118	150	222	289	550	595

The average specific gravity of crude oils supplied to the United States and Canadian refineries is 0.845, while crude oil supplied to European refineries has an average specific gravity of 0.855. Specific gravity is easily determined with considerable precision and, when applied to fractions of fixed TBP range, can be used either alone or in conjunction with other tests as an indication of hydrocarbon composition. Low specific gravities imply paraffinic material, whereas high specific gravities indicate naphthenic stocks.

Sulfur Crude petroleum generally contains some organic sulfur, nitrogen, and oxygen compounds, with sulfur compounds forming the largest group of nonhydrocarbon compounds. Crude oils vary considerably in their sulfur content, ranging from some extremely low-sulfur crude oils, with less than 0.1 percent by weight of sulfur, up to highly sulfurous crude oils, with as much as 5 to 7 percent by weight. In the light distillates, sulfur may be present in one of the following forms:

(1) Hydrogen sulfide (H_2S) — a gas which dissolves in crude oil
(2) Mercaptans — C_2H_5—S—H ethyl mercaptan
(3) Monosulfides — R—S—R
(4) Disulfides — R—S—S—R
(5) Cyclic sulfides — tetrahydrothiophene

In the heavier fractions of crude oils, where most (80–90 percent) of the sulfur is usually found, the sulfur is present mainly in complex ring structures of the benzothiophene type. Since the sulfur atom is only a small part of a large molecule, the total content of sulfur-containing compounds in the higher boiling ranges of a sulfurous crude oil of, say, 3 to 5 percent sulfur-content may well amount to more than half of the total material. However, all that is required is removal of the sulfur atom. This can be achieved by breaking the C—S—C bonds by high-pressure destructive hydrogenation, the sulfur being removed as H_2S; sulfur removal does, however, entail a high cost.

Crude petroleum is classified according to sulfur content as follows:

(1) High sulfur—all Middle East and some United States and Canadian
(2) Moderate sulfur—Venezuela
(3) Low sulfur—North Africa and Nigeria, most United States and Canadian light crude oils, and many of those from Europe and East Indies

Total sulfur content distribution in typical light, medium, and heavy crude oils is given in Table 3.8 and sulfur content as a function of specific gravity is shown in Fig. 3.7.

Sulfur content is of such great importance because of corrosion and air pollution resulting from SO_x emissions. The sulfur content falls into the above broad geographic relationships and there are no means of predicting sulfur content. Sulfur content does not correlate with any other measured properties, nor, given the sulfur content of a total crude oil, is it possible to predict the sulfur distribu-

Table 3.8 Total sulfur content distribution in typical, light, medium, and heavy crude oils

	Middle East				Africa	
Crude oil gravity, °API	40°	31°	25°	44°	37°	27°
Total sulfur, % wt						
Crude	0.74	2.50	3.86	0.14	0.14	0.25
Gasoline	0.011	0.025	0.069	0.004	0.007	0.01
Kerosine	0.040	0.15	0.46	0.011	0.004	0.039
Diesel oil	0.59	1.27	2.11	0.12	0.07	0.17
Residual fuel	1.49	4.02	5.43	0.36	0.23	0.40
Vacuum distillate†	1.23	2.86	3.57	0.33	0.18	0.33
Vacuum residue‡	1.82	5.06	6.46	0.44	0.3	0.55
	Venezuela			Canada		
Crude oil gravity, °API	43°	30°	16°	40°	25°	7°
Total sulfur, % wt						
Crude	0.20	0.96	2.32	0.26	2.37	5.08
Gasoline	0.001	0.004	0.027	0.016	0.084	. . . .
Kerosine	0.001	0.042	0.174	0.030	0.34	0.78
Diesel oil	0.19	0.60	1.14	0.25	1.39	1.79
Residual fuel	0.51	1.60	2.78	0.52	3.39	5.36
Vacuum distillate†	0.47	1.12	2.16	0.44	2.39	3.68
Vacuum residue‡	0.73	2.12	3.12	0.66	4.26	6.13

† Potential stock for manufacture of lubricating oils.
‡ Potential stock for manufacture of road asphalt (bitumen).

tion in the various products. For instance, Zubair (South Iraq) crude oil, with a sulfur content of 1.95 percent by weight, yields "sweet" gasoline and kerosine fractions with very low sulfur contents, which need only the minimum of refining treatment. On the other hand, Gach Saran (Iran) crude oil contains less total sulfur, yet yields light distillates which need much more treatment. The distribution of sulfur for these two oils is shown in Table 3.9.

Although the greater part of the sulfur in most crude oils is firmly held in stable, nonreactive, ring structures, some crude oils contain higher proportions of reactive sulfur in the light distillates. These, such as crude oil from the Slaughter Field of West Texas, are generally termed "sour" and, although total sulfur content may be lower than that from the Middle East, such as Kuwait, more corrosion occurs and greater treatment costs are involved. Some crude oils also generate large amounts of hydrogen sulfide on distillation, due either to the presence of elemental sulfur or very unstable polysulfides.

In order to deal with sulfur-containing oils, special steels or equipment protected by ammonia or soda addition against the corrosive action of sulfur compounds are used. A range of processes is currently available which enables the

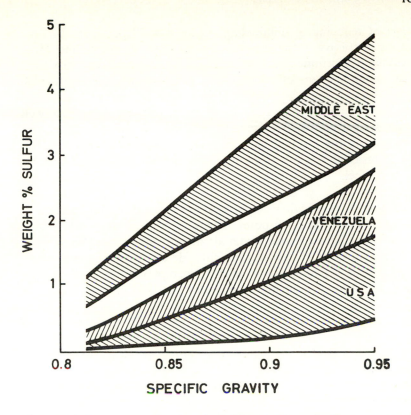

Figure 3.7 Sulfur content as function of specific gravity for crude oils.

removal or control to a satisfactory level of the sulfur compounds in any distillate product. These, however, add significantly to the cost of the fuel.

In the preparation of light straight-run distillates to be used as components of gasoline, it is necessary to remove all the hydrogen sulfide and mercaptans so as to avoid all odor and corrosion. Since sulfur compounds interact with tetraethyl lead (TEL), it is necessary to minimize the total sulfur content in order to achieve the maximum improvement in octane number from the addition of TEL. With the

Table 3.9 Distribution of sulfur (% weight)

	Zubair		Gach Saran	
	Total sulfur	Mercaptan sulfur	Total sulfur	Mercaptan sulfur
Crude	1.95	. . . .	1.58	
Gasoline	0.013	< 0.001	0.14	0.07
Kerosine	0.057	< 0.001	0.26	0.006
Diesel oil	1.11	< 0.001	1.01	< 0.001
Fuel oil	3.40	. . . .	2.58	

current tendency of reducing or eliminating addition of TEL, it may become possible to be less restrictive in the removal of total sulfur from gasoline. Most refineries use catalytic reforming, yielding hydrogen, which is subsequently used for reducing the sulfur content of middle distillates, such as aviation turbine kerosine.

Residual fuel oils contain most of the sulfur from the original crude oil, as well as high molecular weight nonreactive asphaltic materials. Vanadium, nickel, and nitrogen compounds in the residual fuel oils poison catalysts so that catalytic desulfurization of residual fuel oils is an expensive process for an end product which is a low-price fuel.

Wax content and pour point The flow characteristics of a crude oil are specified in terms of the wax content and pour point. The wax content is defined by an empirical test as the weight of solid material thrown out of solution when a mixture of oil with a solvent such as methylene chloride is cooled to a temperature of $-25°C$. This wax is generally found to be a complex mixture of paraffin hydrocarbons which, at least in the lower boiling ranges, consists almost entirely of the normal paraffins.

Crude oils having a low sulfur content tend to have high wax contents and yield residues of high pour point. Thus, the problem of control of the sulfur content of fuel oil is closely related to problems dealing with high wax contents and pour points. Technically, it has been found cheaper to solve the wax problem by wax cracking than to solve the sulfur problem by residue desulfurization.

Wax contents of crude oils range from less than 0.5 percent for the so-called "wax-free" oils up to wax contents of 40–50 percent for such extremely waxy crude oils as Bahia (Brazil) or Minas (Sumatra). The bulk of the world's crude oil supplied contains between 5 and 7 percent by weight of wax, which becomes equivalent to about 15 percent by weight in the higher boiling fractions. The high-sulfur Middle East crude oils are all in this class of moderate wax content, whereas the low-sulfur crude oils from North Africa generally have higher wax contents of 10–20 percent.

The wax in crude oil is generally looked upon as a nuisance and the pour point of an oil is generally fairly closely related to its wax content. In the transportation and handling of crude oils, the pour point must be below the minimum temperature at which the oil is likely to be handled in tankers, pipelines, and storage tanks. Crude oils containing much wax may require special heating facilities or dilution with less waxy stocks to ensure satisfactory pipeline handling under winter conditions. If, during a shutdown, solid wax crystals are allowed to grow, gellike structures are formed in the crude oil, which make it very difficult, if not impossible, to restart the flow with the pump pressures available.

Wax begins to appear in the upper half of the gas/diesel oil range (above $300°C$) and, hence, such products made from a waxy crude oil will tend to have high pour points. The bulk of the wax is present in the atmospheric residue and most of the wax distillates in the vacuum distillate range. The wax content is a property which, like sulfur, appears to be difficult to correlate with other properties or geochemical factors.

Classification of Crude Oils

Crude oils can be classified according to their hydrocarbon composition. Figure 3.8 shows the correlation between the midboiling point, in °C, and the specific gravity for distillate fractions. This graph provides a guideline for comparing crude oils and for assessing the characteristics of distillate products which are most influenced by hydrocarbon type. For a given boiling point, a high specific gravity is indicative of the presence of condensed molecules such as naphthenes, aromatics, and complex multiring compounds, whereas low specific gravity indicates open-chain molecules, i.e., paraffins. Figure 3.8 shows that, although some

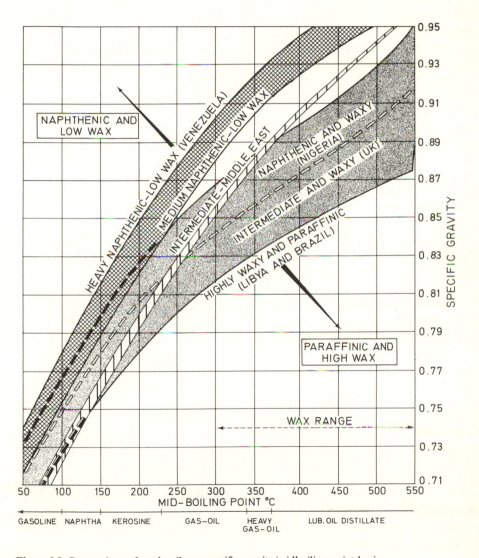

Figure 3.8 Comparison of crude oils on specific gravity/midboiling point basis.

crude oils are entirely naphthenic throughout the boiling range and others are entirely paraffinic, there are many which are both paraffinic and naphthenic in the gas-oil range. The very narrow bracket covering all Middle East crude oils indicates that these are all broadly of much the same hydrocarbon type. Crude oils can thus be fitted into a spectrum of decreasing paraffin content, coupled with increasing naphthene content, according to the specific gravities of standard kerosine and gas-oil fractions. Other correlated properties, such as octane number of gasoline, naphthene content of reformer feed stock, smoke point of kerosine, and diesel index of gas oil, all follow a similar sequence.

For the production of motor spirit, light distillates should be naphthenic/aromatic rather than paraffinic in order to achieve higher octane numbers and better yields from catalytic reforming. The middle distillates (160–350°C) should be paraffinic (i.e., low specific gravity) to ensure good burning qualities, such as smoke point, diesel index, and cetane number. However, high paraffinic content in the upper gas-oil range often implies high wax content and high pour points. Hence, a paraffin-intermediate oil, such as Middle East, is often the best compromise. For fuel oils (above 350°C) wax content and pour point become the prime considerations with a tendency to avoid very waxy paraffinic stocks.

There is generally much less variation in the aromatic contents of crude oils than in their paraffin and naphthene contents. Some crude oils from West Texas and from the East Indies are particularly rich in aromatics. Variations in hydrocarbon composition in crude oils are mainly considered in terms of changes in the paraffin and naphthene content.

The presence of oxygen compounds in crude oils is determined by a test for organic acidity and, if this is found to be high, may require treatment with caustic soda. Basic nitrogen compounds can sometimes occur in kerosine/gas-oil distillates, giving rise to changes in color. The ash content of most crude oils is generally less than 0.01 percent by weight. Presence of lead and arsenic in parts per billion can cause damage to platinum catalysts. Vanadium and nickel are present in all crude oils, with concentrations varying between 1200 ppm and 0.12 percent by weight. These metals also adversely affect the activity of catalysts. Vanadium, in combination with sulfur, is a major corrosion nuisance.

Petroleum Refinery Products

Crude oil is broken down by distillation in a refinery into a series of fractions. The simple fractions are referred to as "straight runs." Because of the variation in characteristics of crude oils from different locations, as well as the specific requirements for each type of fuel, refineries blend crude oils from different fields. Cracking processes permit heavier fractions to be broken down into lighter fractions. The final product marked by a refinery is a blend of several components, each possessing some, but not necessarily all, of the characteristics desirable in the final product.

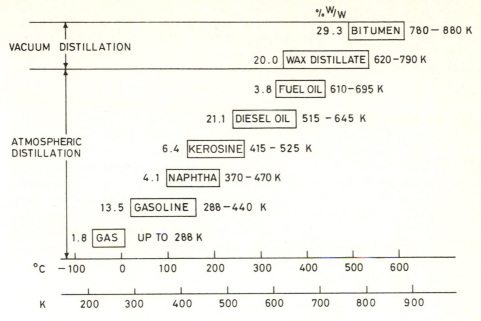

Figure 3.9 Fractions obtained from crude oil by fractional distillation.

The fractions obtained from crude oil by fractional distillation are shown in Fig. 3.9. This shows the boiling point ranges and percentage by weight of the atmospheric distillates (gas, gasoline, naphtha, kerosine, diesel oil, and fuel oil) and the vacuum distillates (wax distillate and bitumen).

During the refining of oils, the nitrogen concentrates in the heavy fractions. Residual fractions from a California crude oil may have as much as 1.1 percent by weight nitrogen, whereas light distillate oils generally have nitrogen concentrations under 0.1 percent.

The fuels produced by refineries reflect the demand in each particular country. The contrast between refinery yields in the United States and Western Europe is shown in Table 3.10.

Table 3.10 Comparison of refinery yields in the United States and Western Europe

	Yield % by weight	
	United States	Western Europe
Gasoline	42	18
Kerosine	7	4
Gas/diesel oil	22	26
Fuel oil	7	38
Others (including fuel for refinery operating and loss)	22	14

The main differences in the comparison shown in Table 3.10 are the higher percentage (42 percent) of gasoline required by the United States for automobiles, while in Western Europe there is a higher percentage (38 percent) of the refinery yield for fuel oil used for generation of electricity and as a major fuel in industry. Changes in this pattern are already occurring in Western Europe, where the fivefold increase in the cost of crude oil in the early 1970s resulted in dramatic economy consequences to many European countries. The development of North Sea oil and gas will also affect the ultimate usage of liquid fuels in European countries. As oil prices rise, it will become economically more attractive in Europe to increase the proportion of lighter fractions produced from crude oil. The refinery balance will also be affected by the lighter nature of North Sea crude oil. European refineries will install more hydrocracking plant to reduce the proportion of fuel oil produced and increase the proportion of the lighter fractions. Fuel oil which is presently consumed as a fuel for industry and, in particular, for electricity generation, will gradually be replaced by coal and liquids derived from coal, as well as by nuclear power.

In the future, closer cooperation will be required in the planning of refinery production and fuel usage. An overall view of the total availability of fuels and the total requirements for industrial and transport users will show the extent to which refinery production must be matched with fuel usage. Combustion engineers are currently examining the extent to which all combustion chambers can accept a wide range of fuels. In order to obtain maximum energy efficiency and fuel economy, a balance needs to be found between the economics of refinery production and modifications of combustion chamber design.

Motor Gasoline and Its Effects on Engine Performance

The principal requirements of a motor gasoline are: (1) sufficient volatility for the fuel to be completely vaporized prior to ignition under all operating conditions; (2) ability for combustion to take place without knock; (3) to provide good fuel economy; (4) to minimize deposition on surfaces of inlet system and combustion chamber; and (5) to provide complete burning and maintain emissions of pollutants at low levels.

Volatility Motor gasolines are mixtures of hydrocarbon liquids boiling in the range between 300 and 500 K. The volatility of a gasoline is usually expressed in terms of the volume percentage that distils over, at, or below certain fixed temperatures. The distillation curves for the majority of gasolines lie between the volatile and less volatile curves shown in Fig. 3.10.

Most gasolines contain small quantities of gaseous hydrocarbons, such as butane. The quantity of these gases is measured by the Reid vapor pressure (RVP) of the gasoline. The more volatile parts of the gasoline, as characterized by the percentage distilling below 343 K, influence the ease with which engines can be started under cold conditions. Most marketed gasolines are more volatile than would be considered absolutely necessary, in order to improve the ease of starting,

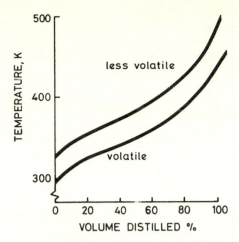

Figure 3.10 Distillation curves for motor gasoline.

preferably without the use of choke. There is an upper limit of volatility, governed by the tendency of more volatile gasolines to give rise to vapor-lock. In many countries the differences between summer and winter temperatures are so great that one grade of gasoline cannot meet the conflicting requirements and, hence, two or even more grades are marketed, depending on the time of year. The summer grade is usually less volatile, since cold starting is not a problem, and freedom from hot fuel handling problems plus good fuel economy are paramount. The winter grade is more volatile so as to provide good cold starting and warmup, and contains anti-icing additives.

Fuel economy The fuel consumption of a vehicle depends upon the compression ratio of the engine, engine size and performance, and efficiency of the carburetor. Under constant speed, fully warmed-up driving conditions, fuel consumption depends strongly on the specific gravity of the fuel. The higher the specific gravity, the lower the consumption. Under nonmotorway driving conditions, the influence of specific gravity is less important. In urban motoring, fuel consumption depends on the number of cold starts, average length of journey, and the volatility of the gasoline. Fuel economy is also influenced by the extent to which engines remain tuned. Detergent additives help to keep engines clean and running efficiently for longer periods.

Engine knock The compression ratio of engines has increased in order to improve the overall efficiency and the maximum obtainable power output. The extent to which the compression ratio can be raised is limited by the onset of detonation or knock and this depends on the antiknock quality of the fuel. The octane number of a fuel is an expression of the antiknock quality and is measured by comparing the performance of the fuel against the performance of mixtures of isooctane, with an octane number of 100, and n-heptane with an octane number of zero. Under standard laboratory test conditions, a 98-octane number gasoline is one which gives the same antiknock performance as a mixture, by volume, of 98 percent

Table 3.11 Octane numbers of gasoline components

Component	Motor octane no.	Research octane no.	Boiling range (K)
Clear gasoline	83	91	320–495
Light petroleum naphtha	65	65	340–420
Aromatics			
toluene	112	124	383
p-xylene	124	145	410
n-butyl benzene	116	114	455
1,4-diethyl benzene	138	151	457
1-methyl naphthalene	114	123	513
Methanol	87	110	337
Methanol + 10% water	93	114	

isooctane and 2 percent n-heptane. The actual octane number required for any particular engine depends both on its compression ratio and on certain mechanical features. Octane numbers of gasoline components are given in Table 3.11.

Octane numbers are measured under two conditions—relatively mild test conditions to give the research octane number (RON) and a more severe test condition giving the motor octane number (MON). The motor octane number is, thus, less than the research octane number. A direct measure of the knock characteristics of a fuel is made for cars on the road, from which the road octane number is determined. The road octane number is usually intermediate between the research and motor octane numbers.

Additives to gasoline Small quantities of corrosive substances and solid impurities are deposited on surfaces within the fuel system, carburetor, and combustion chamber. These greatly reduce the overall life, efficiency, and general reliability of the engine.

Most crude petroleum contains sulfur compounds and some of these, or their derivatives, remain in the various distillates during the course of refining. If these sulfur compounds are not removed, they burn in the combustion chamber of the engine to give the highly corrosive sulfuric and sulfurous acids. Sulfur impurities are, therefore, generally removed by chemical treating processes in the refinery. Other impurities, including water, are removed from the gasoline by settling and filtration. The formation of gummy materials is prevented by the use of antioxidants and deactivators.

The fitting of devices which return crankcase blow-by gases to the inlet system causes additional deposits, which can lead to the production of richer fuel-air mixtures. Deposits on inlet valves can also lead to power loss and valve sticking. In order to keep carburetors and inlet systems free from deposits, carburetor detergents are added to the gasoline. These additives are effective in preventing deposition, but have also been found to give significant reductions in emissions and improvements in fuel economy. Antiknock additives lead to the formation of deposits in the combustion chamber. Further additives, known as

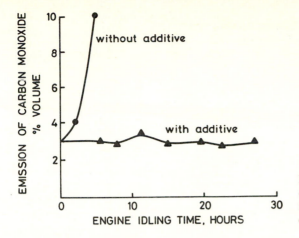

Figure 3.11 Effect of detergent additive on emission of carbon monoxide under idling conditions in a spark ignition engine.

scavengers, are added to the antiknock compounds to reduce the tendency for deposits to form and, hence, greatly extend the time before decarbonization is required. The effect of a detergent additive on emission of carbon monoxide under idling conditions is shown in Fig. 3.11.

Aircraft Fuels

The gas turbine in a jet engine has become the most common form of propulsion for aircraft, so that fuels for the aircraft industry are required to burn under the continuous combustion conditions of gas turbine combustion chambers. The safety requirements in aircraft are much more stringent than for land-based propulsion systems with the requirement to minimize the fire hazard, both under flight and crash conditions. From the safety point of view, the volatility requires to be low and the flash-point of the fuel should be high. Aircraft fly at altitudes above 10 km, where local ambient temperatures are low enough to freeze liquid fuel and also cause special ignition problems. In order to cater for these high-altitude conditions, the freezing point of the liquid should be low and there must be sufficient volatile components to allow ignition of the fuel with air ingested at the temperatures and pressures in the upper atmosphere.

The fuel most generally used in commercial aircraft is known as "Jet A," which is basically a kerosine fuel. Military aircraft in the United States and some commercial airlines outside the United States use the fuel known as "J.P.-4." Some of the characteristics of these two fuels are shown in Table 3.12. Jet A, with a flash-point above 313 K, is less volatile than J.P.-4 and has been found to provide more safety in crash and ground handling conditions. The larger fraction of natural petroleum is included in J.P.-4, which also includes low-boiling naphthas, available at low cost from the refineries. In order to meet the requirement of low freezing points, the high boiling point components must be reduced and replaced by lighter, more volatile components.

Table 3.12 Aircraft fuels

	Jet A	J.P.-4
Flash point (K)	313	
Distillation (K)		
10%	477	
50%	505	461
90%	550	516
Final boiling point	561	
Freezing point (K)	233	215
Volume % aromatics	20	25

3.4 GASEOUS FUELS

Natural Gas

The term "natural gas" is used for gaseous fuel which occurs naturally and is burned after a minimum of treatment or processing. Natural gases are mixtures of hydrocarbon and nonhydrocarbon gases, found in subsurface rock reservoirs where in many, but not all, cases they are associated with liquid petroleum. Crude oil reservoirs usually contain a wide variety of hydrocarbons, ranging from light gases to heavy liquids. Gas may be found in a "gas gap" above the oil in a reservoir or, in some cases, gas may be found in isolation, where it has formed after separation from oil during the course of migration or minor geological changes. Gas from a reservoir which has no oil is said to be "unassociated" or "nonassociated" in that its production is not significantly related to the production of oil. Oils are saturated with gases at the prevailing reservoir temperature and pressure. When the oil is released, during subsequent liquid petroleum production and processing, gases are liberated due to changes in temperature and pressure. In addition to gases from originally marine-based petroleum sources, natural gas from nonpetroleum origin is also found in the vicinity of freshwater coal deposits.

The following terms are used in natural gas industry:

dry or lean	high percentage of methane
wet	high concentrations of higher hydrocarbons (C_5–C_{10})
sour	high concentrations of H_2S
sweet	low concentrations of H_2S
residue gas	residue after extraction of higher hydrocarbons
casinghead gas	derived from oil well by extraction at the surface

A typical composition of natural gas is CH_4, 85 percent; C_2H_6, 10 percent; and C_3H_8, 3 percent. Higher hydrocarbons, C_5–C_{10}, may also be present. The major nonhydrocarbon constituents are CO_2, N_2, He, and H_2S. Methane is the major constituent of all natural gases, whether they be of associated or unassociated origin. Ethane is also found in natural gas, though in much smaller proportions. Depending on location, varying amounts of inert and noxious gases, such as nitrogen, carbon dioxide, helium, water vapor, hydrogen sulfide, and traces of many other gases are found. In associated gas, the ethane content is usually higher than in unassociated gas, and, additionally, there are also quite significant amounts of propane and the higher hydrocarbons, the proportions of which tend to decrease progressively as their boiling points increase.

The constituents of a selected number of natural gases are listed in Table 3.13. This table shows that there is a wide variation in composition with a resultant variation in calorific value. The proportion of methane is generally in excess of 70 percent and, for North Sea gas, is above 90 percent. The quality of natural gas is improved by processing. Higher hydrocarbons are stripped off to provide natural gas liquids and provide a source for liquefied petroleum gases or natural gasolines. Water vapor needs to be separated from the gas before it enters the distribution system, where, if it remained, it could form solid hydrates at low operating temperatures.

Hydrogen sulfide is corrosive, toxic, and has a foul odor. Its combustion products contain sulfur oxides, which are themselves corrosive to many materials and harmful to animal and vegetable life. Most countries have imposed limits on the amounts of total sulfur and hydrogen sulfide which may be permitted in fuel gases. Natural gases containing these impurities above the maximum permitted level must be treated before distribution. When the level of hydrogen sulfide

Table 3.13 Composition of natural gas from various sources

| Source | % by volume | | | | | | | | | Average C.V. MJ/m^3 |
	CH_4	C_2H_6	C_3H_8	C_4H_{10}	C_{5+}	N_2	CO_2	O_2	H_2S	
France, Lacq										
Crude	69.2	3.3	1.0	0.6	0.5	0.6	9.6	...	15.2	
Purified	97.0	←——— 3.0 ———→			....	....	.....	...	....	36.2
Canada,										
Alberta	90.0	←——— 8.0 ———→			....	....	0.5	0.2	1.0	39.0
Venezuela	88.5	←——— 2.9 ———→			....	4.6	3.8	0.2	0.3	39.1
Italy,										
Po Valley	98.0	.....	.....	......	....	1.0	.....	...	....	38.0
United States	70–95	3–18	←——5.0——→		....	1–14	0.1–7	...	....	39.1
North Sea										
(average)	90.5	3.9	0.9	0.3	0.2	3.9	0.3	...	....	37.4
Iran	73.0	21.5	.....	......	....	....	.....	...	5.5	39.0
Soviet Union	89–98	←—2.5—→		1.0	....	....	7.5	...	1.0	39.8
Groningen	81.3	2.9	0.4	0.2	....	14.3	0.9	...	....	32.6
Algeria	83.8	7.1	2.1	0.9	0.4	5.5	0.2	...	....	39.1
Lybia	64.5	21.0	8.4	4.2	1.9	....	.....	...	....	39.0
Australia,										
Bass Straits	93.0	2.0	1.0	......	....	3.0	1.0	...	....	36.8

present in the gas is sufficiently high, it becomes economic to recover the sulfur so that sour gas becomes a useful source of elemental sulfur.

Hydrogen sulfide is removed or recovered by a number of processes, most of which employ an absorbent solution (such as 25 percent diethanolamine in water), which reacts with hydrogen sulfide in an absorption tower. The resultant sour solution is then regenerated by heating, the solution is recycled to the absorber and the hydrogen sulfide passes forward to another plant for conversion into sulfur. The hydrogen sulfide is burned to yield sulfur dioxide, which, in the presence of heat and a catalyst, reacts with further hydrogen sulfide to produce elemental sulfur of high purity. Any remaining tail gases are flared off at the plant stack.

Inert constituents are, for the most part, ignored, but as their proportion increases, they lead to a reduction in calorific value of the gas and increases in transport and distribution costs. Inerts may materially affect the combustion characteristics of the gas and, where these become important, inerts may require to be removed or diluted by blending with other gases. When concentrations of helium exceed 0.2 percent by volume, the helium is extracted. Natural gas is currently the main commercial source for helium.

Liquefied Natural and Petroleum Gas

Natural gases are liquefied for distribution by tanker and for use as liquid fuels. The term "liquid natural gas" (LNG) is used for liquids consisting mainly of methane, while the term "liquefied petroleum gas" (LPG) is used for liquids containing mainly propane and butane. When LNG and LPG are compared with petroleum liquid fuel, they have the following advantages and disadvantages:

Advantages

extremely clean burning
good antiknock properties without requiring additives
specific energy/weight of fuel is 15 percent higher than for gasoline or kerosine
for gas turbines provides a smaller heat sink with a reduced flame volume
in rockets provides a higher specific impulse than standard liquid fuels

Disadvantages

density is approximately half that of conventional HC liquid fuels—requiring larger volumes in tankers
high cost of insulation in tankers to maintain low temperatures

Properties of Gaseous Fuels

Gaseous fuels are separated into the following major subdivisions: natural gas, liquefied petroleum gas, Lurgi crude gas, lean reformer gas, rich reformer gas, blast-furnace gas, coal gas, producer gas, and water gas.

The composition in percentage by volume is given in Table 3.14. Also listed in this table are the density, theoretical air requirements, the Wobbe number, the

Table 3.14 Properties of typical gaseous fuels[†]

	Lurgi crude gas	Leant[‡] reformer gas	Catalytic[§] rich gas	North Sea gas	Commercial propane
Composition (% by vol.)					
CO_2	25.6	16.7	21.0	0.2	
CO	24.4	2.2	1.0		
H_2	37.3	46.4	17.0		
N_2	1.8	....	....	1.5	
CH_4	10.3	34.7	61.0	94.4	
C_2H_6	....	....	....	3.0	1.5
C_3H_8	....	....	....	0.5	91.0
C_4H_{10}	....	....	....	0.2	2.5
C_5H_{12+}	....	....	....	0.2	
C_3H_6	....	....	....	....	5.0
$C_nH_m(\sim C_{2.5}H_5)$	0.6				
Density (kg/m³) at 15°C and 101.325 kPa	0.897	0.611	0.831	0.723	1.869
Theoretical air-requirement (vol./vol.; dry basis)	2.556	4.462	6.236	9.751	23.762
Wobbe no. (MJ/m³)	13.745	26.840	30.580	50.322	76.064
Weaver flame speed factor	41.35	39.50	20.318	14.13	16.788
Net calorific value at 15°C 101.325 kPa, dry; per vol. of fuel gas (MJ/m³)	10.64	16.79	22.57	34.82	86.43
Composition of waste gas (% by vol.) CO_2	19.0	10.3	11.6	9.6	11.7
H_2O	18.3	22.2	19.5	18.8	15.4
N_2	62.7	67.5	68.9	71.6	72.9

	Blast furnace gas	Coal[¶] gas	Producer gas (coke)	Carburetted water gas
Composition (% by vol.)				
O_2	....	0.4	....	0.4
CO_2	17.5	4	5	5.6
$C_nH_m(m/n \simeq 2)$	....	2	....	7
CO	24	18	29	30.5
H_2	2.5	49.4	11	37
CH_4	....	20	0.5	14
N_2	56	6.2	54.5	5.5
Density (kg/m³) at 15°C and 101.325 kPa	1.27	0.59	1.10	0.77
Theoretical air requirement (dry basis; vol./vol.)	0.631	4.060	1.000	4.270
Wobbe no. (MJ/m³)	3.118	25.937	5.289	23.875
Weaver flame speed factor	4.44	39.81	11.20	33.88
Net calorific value at 15°C 101.325 kPa, dry; per vol. of fuel gas (MJ/m³)	3.13	16.14	4.77	17.31
Composition of waste gas (% by vol.) CO_2	27.7	10.5	19.2	13.9
H_2O	1.7	20.5	6.7	16.9
N_2	70.6	69.0	74.1	69.2

† Ref.: *Technical Data on Fuel*, Rose, J. W., and Cooper, J. R., British National Committee, World Energy Conference, London, 1976.

‡ ICI lean gas, CO converted and enriched with CRG; naphtha feedstock at 20 bar.

§ Naphtha feedstock at 25 bar.　　　　　　　　　　¶ Continuous vertical retorts (steaming).

Weaver flame speed factor, the net calorific value, and the composition of waste gases.

Gaseous fuels in a particular burner can be interchanged by taking into account the following considerations: (1) at constant supply pressure the rate of flow of heat energy through a particular burner remains reasonably constant; (2) at constant supply pressure and flow rate of gas and oxidant, the size and shape of the resultant flame do not change significantly; (3) at constant air/fuel ratio and flow rate, the formation of partially oxidized intermediates does not exceed a certain maximum; and (4) at constant air/fuel ratio and flow rate, the formation of soot and carbon does not exceed a certain maximum.

The flow of gas through an orifice of fixed dimensions is a function of pressure difference, viscosity, and gravity. If the pressure difference and viscosity are maintained constant, the rate of flow becomes inversely proportional to the square root of the specific gravity. In order to maintain a constant supply of heat energy through a given burner, equipped with a pressure regulator, the Wobbe number, which is equal to the "calorific value/$\sqrt{\text{specific gravity}}$" should be maintained constant. The gross calorific value is normally used for calculation of the Wobbe number.

The Weaver flame speed factor is a relative burning velocity in which hydrogen is assigned the value of 100. The burning velocity can be measured in a number of different ways and is dependent on pressure, temperature, and humidity. The Weaver flame speed factor is defined as

$$S = \frac{\sum F_1}{\sum A_1 + 5I - 18.8 O_2 + 1}$$

where S is burning velocity, A_1 is the stoichiometric air requirement of each component multiplied by its mole fraction, F_1 is the Weaver flame speed factor of the component multiplied by its mole fraction, I is the mole fraction of inerts, and O_2 is the oxygen content in the fuel. The Wobbe number and the Weaver flame speed factor are the two main criteria used in the gas industry for interchangeability.

The average properties of separated gaseous fuels, including hydrogen, hydrocarbons, and carbon monoxide are given in Table 3.15. The properties include density, calorific value, combustion requirements, and combustion products.

Flammability Limits, Ignition Characteristics, and Explosion Characteristics of Gaseous Fuels

For practical purposes, the flammability limits can be fixed for a range of gaseous fuels burning in air and oxygen at atmospheric pressure and room temperature, as shown in Table 3.16. The upper and lower limits for combustion are expressed as a fuel percentage by volume. The lower limit is the same for mixtures with air and oxygen, while the upper limit is considerably higher for oxygen than that for air.

The ignition characteristics of gaseous fuels are usually given in terms of a minimum temperature at which autoignition takes place, as shown in Table 3.17.

Table 3.15 Average properties of separated gaseous fuels

	Hydrogen H_2	Methane CH_4	Ethane C_2H_6	Propane C_3H_8	Ethylene C_2H_4	Benzene C_6H_6	Toluene C_7H_8	Carbon monoxide CO
Density at 15°C and 1.01 × 10⁵ N/m² (dry), kg/m³	0.085	0.678	1.272	1.865	1.186	3.304	3.897	1.185
Net calorific value at 15°C and 1.01 × 10⁵ N/m², MJ/m³	10.22	33.95	60.43	86.42	55.96	134.05	159.54	11.97
Combustion requirements O_2 (% by volume)	0.5	2.0	3.5	5.0	3.0	7.5	9.0	0.5
Air	2.38	9.52	16.67	23.81	14.29	35.71	42.86	2.38
Combustion products in volumes/unit volume CO_2		1.00	2.00	3.00	2.00	6.00	7.00	1.00
H_2O	1.00	2.00	3.00	4.00	2.00	3.00	4.00	
N_2	1.88	7.52	13.17	18.81	11.29	28.21	33.86	1.88

Table 3.16 Flammability limits of gaseous fuels in air and O_2 at atmospheric pressure and room temperature†

		Fuel percentage by volume		
Fuel	Formula	Lower limit (air or O_2)	Upper limit (air)	Upper limit (O_2)
Methane	CH_4	5.0	15.0	61
Ethane	C_2H_6	3.0	12.4	66
Propane	C_3H_8	2.1	9.5	55
Ethylene	C_2H_4	2.7	36	80
Acetylene	C_2H_2	2.5	80	
Hydrogen	H_2	4.0	75	94
Carbon monoxide (wet)	CO	12.5	74	94

† *Technical Data on Fuel*, Spiers.

Table 3.17 Ignition characteristics of gaseous fuels†

Compound (formula)	Minimum autoignition temperature (K), $P = 1$ bar (in air)	Fuel/air ratio (as fraction of stoichiometric) at which lowest value of minimum ignition energy occurs	Minimum ignition energy for stoichiometric mixture (mJ)	Quenching distance (mm)
CH_4	813	0.88	0.34	2.1
C_2H_6	788	1.17	0.285	
C_3H_8	723	1.28	0.305	1.9
C_4H_{10}	....	1.47		
C_2H_4	763	....	0.096	1.1
C_2H_2	578	....	0.020	0.65
H_2	....	....	0.019	0.64

† *Technical Data on Fuel*, Spiers.

Table 3.18 Explosion characteristics of gaseous fuel[†]

Substance	Maximum safe % of O_2 (N_2 diluent)
CH_4	14.6
C_2H_6	13.4
C_3H_8	14.3
C_2H_4	11.7
H_2	5.9
CO (wet)	5.9

[†] *Technical Data on Fuel*, Spiers.

The fuel/air ratio, as a fraction of the stoichiometric value, at which the lowest value of minimum ignition energy occurs, and the minimum ignition energy for stoichiometric mixtures are also shown in Table 3.17, as is the quenching distance, within which ignition generally does not occur.

For purposes of safety, the explosion characteristics of gaseous fuels are given in Table 3.18, where the percentage of oxygen in a gaseous mixture is expressed as a maximum "safe" value. All of the data presented in Tables 3.16 to 3.18 have been determined empirically and mainly serve as guidelines for practical use in industry.

3.5 SYNTHETIC FUELS

The term *synthetic fuels* is used to describe fuels manufactured by chemical means from fossil fuels. The most important synthetic fuels are those derived from coal. Instead of burning the coal directly, it is made to undergo chemical change so as to provide a liquid or gaseous fuel. By heating, boiling, oxidation, hydrogenation, reforming, and cracking, a wide range of fuels can be produced from the original coal.

Production of synthetic fuels requires energy, mainly in the form of heat, and it has become increasingly important to consider the total amount of energy consumed in producing a fuel which, in itself, is meant to be an energy source. There are other considerations, such as the preference for using liquid fuels in transportation, rather than solid coal. Also, the convenience of using gas for domestic and commercial heating is much greater than that of using coal. Because of the large amounts of energy required to produce these fuels, the whole system requires to be reexamined from an energy conservation point of view. In the automobile industry, high-octane fuels have been demanded in order to obtain high performance from automobile engines. The cost of producing synthetic high-octane liquid fuels from coal is cost-prohibitive. The tendency now, therefore, is to reconsider the basic combustion requirements of both automobile and aircraft

Table 3.19 Hydrogen/carbon ratio of fossil fuels

	Hydrogen/carbon ratio	Resources (United States) 10^{15} MJ
Petroleum	2.00	0.7
Tar sands	1.75	0.1
Oil shale	1.90	20.0
Coal	0.75	90.0

engines and find ways of achieving efficient combustion with heavier fuels. It is also becoming important to reserve certain fuels for transportation and to burn coal and the heavier fuels in industrial plant, which are not so dependent upon fuel composition for efficiency in performance of the combustion device. One of the more important differences in chemical composition between coal and petroleum is the net hydrogen/carbon ratio. The variation of this ratio with type of fuel is shown in Table 3.19.

The amount of petroleum remaining worldwide is estimated to be of the order of 4×10^{15} MJ. The current liquid fuel consumption in the United States is of the order of 0.04×10^{15} MJ per year. These and other statistics have led to the universal conclusion that the shortage of petroleum will increase and will reach major proportions toward the end of the century. Examination of Table 3.19 shows that oil shale is available in large quantities and also produces a liquid with chemical properties close to those of petroleum. Thus, in those regions where oil shale is available, major efforts are being made to overcome the technological problems of separation of the shale from the oil. It is, however, predicted that a major source of liquid fuels in the future will be from liquefaction of coal.

Conversion of Coal to Liquid Fuel

Coal has, on the average, a hydrogen/carbon ratio of 0.75 and liquids derived directly from the coal by liquefaction and separation will also have the same low H/C ratio. The conversion of a low H/C ratio liquid fuel to the higher value of 2.0, as in petroleum, requires the addition of hydrogen in special hydrogenation processes. The quantities of hydrogen required are of the order of 1780 m^3 to produce 1 m^3 of liquid products. This hydrogen may be produced from coal or, alternatively, hydrogen may be produced electrochemically from nuclear or ocean thermal power. An energy balance of the overall process of producing liquid from coal with similar chemical properties to petroleum shows that the combination of hydrogenation, conversion to a lower molecular weight, and hydrocracking results in an overall efficiency of 43 percent for the hydrogen used. In general, it has been found that the heating value of the hydrogen produced is in the range of 55–60 percent of the heating value of the starting coal, so that hydrogen production and hydrogenation is extremely wasteful of energy. Utilization of heat exchange between entering and leaving streams and other energy conservation schemes can lead to some reduction in these losses.

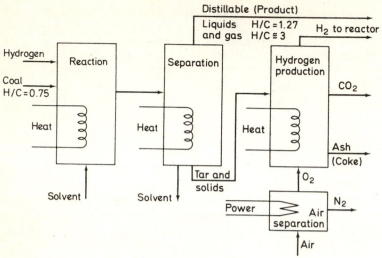

Figure 3.12 Generalized coal liquefaction process.

The overall process involved in liquefaction of coal is shown in Fig. 3.12. A mixture of coal, hydrogen, and solvent is heated in a reactor. This mixture is then separated by filtering out the solids and the high boiling tar. The liquid products are distilled to give liquids with H/C ratio = 1.27 and gas with H/C ≃ 3. The yield of liquid from such a process can be varied by varying the quantity of hydrogen added, the pressure, temperature, and time employed, as well as by changing the catalysts and solvents. The gas produced consists mainly of methane. The tars and solids are heated in the presence of oxygen as a means of producing hydrogen, which is recycled in the process. Also, some of these tars and solids can be used for power generation in the furnace of a boiler.

Methanol and paraffinic fuels are manufactured by first converting char or coal to a mixture of carbon monoxide and hydrogen. The efficiency of this conversion can be as high as 62 percent. The catalytic reaction of the CO-H_2 mixture to form methanol or liquid hydrocarbons is extremely exothermic but, if this heat is usefully employed in the process itself, overall efficiencies between 40 and 50 percent can be achieved.

Boiling Range of Synthetic Fuels

The boiling range of a fuel is one of its important specifications. In the spark ignition engine which uses gasoline with a boiling range varying from slightly above room temperature to 500 K, the fuel satisfies the requirement of providing adequate evaporation in the carburetor and intake manifold. For cold weather conditions ignition becomes dependent on boiling of the lighter components, so that fuels are restricted to a maximum boiling point in order to ensure complete evaporation before ignition. Provision of separate heating for cold weather conditions and improvement of atomization, as well as increase in residence time, can reduce the restriction on the maximum boiling point.

In aircraft engines, the maximum boiling point of the fuel is primarily set by the requirement that the fuel must not freeze in cold weather. On the other hand, in order to reduce the hazard of explosion, fuels are required to have a sufficiently low volatility that the vapors formed above the liquid are outside the limits of flammability under normal storage and operating conditions. The range of boiling temperatures for various fuels is as follows: gasoline, 321–494 K; Jet A, 450–560 K; and diesel oil, 450–616 K.

Refineries are currently providing the large quantities of heavy fuel oils, with high-boiling components, used in power generation and industrial furnaces, while reserving the lighter fractions for transportation. As coal and nuclear energy become the dominant fuels for power generation, refineries will be required to decrease the proportions of heavy fuel oils in refinery production. The boiling range of liquid hydrocarbons can be adjusted by hydrocracking and catalytic-cracking and the hydrogen-carbon content can be increased by hydrogenation.

Refining

Refining of crude petroleum is carried out under conditions of "low conversion," in which a moderate-octane fuel is produced and which only involves an energy consumption of 3 percent of the energy content of the fuel. For "high conversion," a substantial degree of boiling-range conversion and octane-number improvement is carried out, requiring an energy consumption of the order of 10 percent. The high-conversion process caters for requirements of automobiles with high com-pression ratios needing high-octane fuels. Energy balances currently being made in refineries show that there is substantial room for energy conservation within refineries and that it is possible to extract more fuel from crude oil than was previously considered to be economic.

Energy requirements are a function of the degree of conversion to gasoline. The total refining energy cost for production of gasoline from coal synthetic crude oil is of the order of 30 percent. This is approximately three times as high as that required for refining a light petroleum crude oil. This energy loss for production of gasoline from coal is so high that reductions in the energy loss are being sought by blending products from shale and petroleum. Also, attempts are being made to relax the boiling range and aromatic requirements for fuels in combustors. Thus, the feasibility of using liquid fuel derived from coal, in transportation, is depen-dent upon both improving the efficiency of the conversion process and also im-proving the ability of combustors to consume lower-grade fuels.

Changing Patterns in Fuel Consumption

The patterns of fuel consumption are governed both by the supply of fuels and also by the industrial and transportation requirements. Until 1975, there was steady growth in the development of power generation, transportation, and industry to provide the fuel consumption pattern shown in Table 3.20. A pattern of consumption was developed under conditions of plentiful supply of petroleum

**Table 3.20 Liquid hydrocarbon consumption (petroleum)
United States**

		$10^6 m^3$/day	%
Power generation		0.238	9
Industrial		0.286	11
Commercial and residential		0.382	15
Transportation			
Automotive	1.033		
Aircraft	0.159		
Other	0.223		
Total		1.415	54
Petrochemical and other		0.302	11
Total		2.623	100

and natural gas with only small energy contributions from nuclear power and a coal industry which was in a state of decline.

In power generation, direct burning of oil and gas is being replaced by burning of pulverized coal. Nuclear power generation went through a period of rapid growth in the 1960s and, provided waste disposal and environmental problems can be solved, is expected to grow substantially during the next few decades. The proportion of petroleum used for power generation will, therefore, decrease. In industry, coal will replace oil and gas in furnaces wherever this does not interfere with the chemical or metallurgical processes. If cheap sources of nuclear or coal generated electricity are available, electrical processes can displace many oil-burning processes. In commercial and home heating, solar heat, heat pumps, electricity, and supply of synthetic gas will provide options which should lead to large-scale reductions in the consumption of liquid fuel. In the field of transportation, liquid fuel burning will remain dominant but gradual changes will be made from petroleum refined products to those derived from coal. As the demand for heavy fuel oils for electricity generation declines, additional cracking and processing will be required from refineries to cater for requirements in transportation. The petrochemical industries will continue to rely on petroleum as the prime feedstock for these processes.

As can be seen from Table 3.20, in the United States, within the transportation sectors, automotive use currently accounts for about 73 percent of the total. More stringent legislation, penalizing the use of large cars with high rates of fuel consumption, will lead to a reduction in the total amount of liquid fuel consumed. Aircraft fuels currently account for 12 percent of the total. For the past 10 years fuel consumption for air transportation has grown at 4 percent per year, versus 3.9 percent per year for automotive consumption. It has been predicted that air transportation growth will exceed the rate of ground transportation fuel consumption growth. It has been predicted that aircraft might consume 15 percent of transportation fuels in the 1990s and as much as 30 percent by the year 2000.

Synthetic fuels pose special problems for the aircraft industry. Future fuel supplies will tend to be rich in higher boiling components and the supply of volatile components is expected to diminish, since liquid fuels derived from coal, shale oil, and tar contain relatively small proportions of these components. Gas turbine combustion systems will require to burn fuels of higher molecular weight than those for which they are now designed. Currently, the percentage of aromatics is restricted to 20 percent in Jet A fuel. This restriction was imposed in order to avoid problems associated with high aromatic content fuels, which are: (1) smoke formation under high power conditions: (2) overheating of the combustion chamber due to flame radiation; and (3) formation of solid carbonaceous deposits which cause fuel spray distortion and result in damage to turbine blades when deposits become detached. These effects can be directly correlated with the hydrogen-to-carbon ratio. Figure 3.13 shows smoke number as a function of H/C ratio and Fig. 3.14 shows maximum liner temperature as a function of H/C ratio in a gas turbine combustor in a series of tests carried out by NASA with a variety of fuels in a single JT8D combustor. These results show that both the smoke number and the maximum liner temperature under cruise conditions increase almost linearly as the hydrogen-to-carbon ratio is decreased from a value of 2.2 to 1.4. The increase in smoke number indicates the greater formation of soot, which is not completely oxidized, while, at the same time, the rise in liner temperature

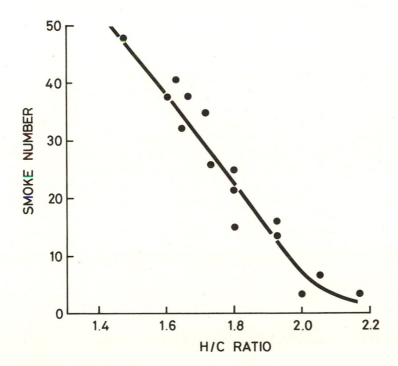

Figure 3.13 Smoke number as function of H/C ratio in JT8D gas turbine combustor.

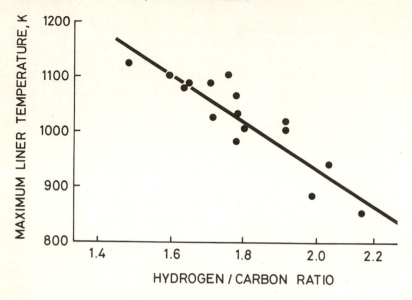

Figure 3.14 Maximum liner temperature as a function of H/C ratio in a JT8D gas turbine combustor under cruise conditions.

results from increased flame radiation, due to formation of larger numbers of particles. This relatively simple correlation allows fuel composition to be related to combustion performance.

The use of liquid fuel derived from coal in aircraft will either require a change of the fuel composition to satisfy present specifications or, alternatively, the aircraft combustion systems will have to be modified to cope with the new fuels. For liquids derived from coal, where the hydrogen-to-carbon ratio is between 1.27 and 1.53, this corresponds to between 50 and 80 percent aromatics. If combustion chambers are redesigned, so as to allow longer residence time in high-temperature oxidizing regions of the flame, oxidation of carbon particles can be completed. Radiation from flames may be reduced by finer atomization of the fuel and more rapid burning, so as to reduce the volume of the luminous region of the flame. The problem associated with the fuel being rich in higher boiling components, with the resultant higher freezing points, will require provision being made for heating of the fuel and insulation of storage tanks and distribution lines, so as to prevent freezing under any ambient conditions the fuel may encounter.

Automotive Fuel Consumption

Many alternative fuels are being sought for automobiles but the efficiency and performance of the spark ignition engine, using high-octane fuels, has reached such high levels that it appears that liquid fuel derived from petroleum will continue to be the main source of supply for automobiles for several decades. Automobiles are expected to remain the major consumers of liquid fuel and the

Table 3.21 Comparison of fuel consumption in automobile engines

Engine	Compression ratio	Mileage/gallon
Otto	8	18.3
Otto	12	21.0
Diesel	15	20.7
Gas turbine	. . . .	20.8
Stirling	. . . .	26.3

fraction of total liquid fuel used for automobiles is expected to increase as coal and other fuel supplies increase. Emphasis within the automobile industry has been turned toward improving vehicle fuel economy. A study carried out at the Jet Propulsion Laboratory in California on compact cars (1500 kg) has been made for "mature" technology. This makes use of projected advances in technology but does not include the ceramic high-temperature components which belong to "advanced" technology. Each engine is fueled with gasoline and mileage is compared for engines giving the same power output. The results of this study are shown in Table 3.21.

The standard Otto-cycle engine with a compression ratio of 8 has a low efficiency. The efficiency has been improved by raising the compression ratio to 12 but this requires the use of high-octane fuels. The high-compression-ratio Otto engine can provide a higher performance than the diesel engine and stratified charge engine. The performance of the diesel engine in this case is below that quoted by other sources, since the mature diesel engine has been required to provide the same acceleration performance as that of the spark ignition engines. The gas turbine and Stirling engines provide high efficiency but they have not yet achieved the degree of perfection and performance of the spark ignition engine.

The choice of an engine for automobiles for the future will be dictated by the ability to utilize available fuels at high efficiency and with a low emission of pullutants. In refineries, production of gasoline requires boiling range conversion plus reforming of some of the components to increase the octane number by converting naphthenes and paraffins to aromatics. The production of distillate (diesel) fuels requires mainly distillation and relatively little boiling range conversion. The refining efficiency is improved by increasing the proportion of distillate fuels produced. In order to minimize the energy expenditure within the refinery and reduce the cost of the final product, a strong case has been made for the adoption of distillate fuels by the automobile industry. These distillate fuels can be used by the diesel, gas turbine, and Stirling engines but not, at present, by the spark ignition engine.

In the refining of liquid fuel derived from coal, the first consideration is the extent to which hydrogen should be added to increase the hydrogen-to-carbon ratio by aromatics hydrogenation and to reduce the boiling range by hydrocracking. A strong case is made for automobiles to accept fuels with a lower hydrogen-to-carbon ratio in order to avoid the cost and energy expenditure of hydrogenation. The diesel engine requires fuels to be highly paraffinic because of the difficulty of igniting aromatics and the tendency to increase smoke emission. The use of fuels with lower hydrogen-to-carbon ratio points toward engines with longer residence times, so that these fuels can be best burned in the continuous flow engines such as the gas turbine and the Stirling engines. With the spark ignition engine, where combustion is intermittent, it is more difficult to cope with fuels with low hydrogen-to-carbon ratio.

Methanol and methanol-water blends have been shown to provide the combination of high-compression-ratio operation and low emissions. Methanol is a high hydrogen-to-carbon ratio fuel and its manufacture results in energy losses. For fuels derived from coal, there appears to be little or no net gain in overall thermal efficiency in the utilization of methanol as compared with gasoline.

Production of Liquid Fuels from Coal

Several processes have been developed for converting solid fossil fuels to liquids with properties similar to those of liquid fuels derived from petroleum, and these are shown in Fig. 3.15.

Carbonization Carbonization is the destructive distillation at low temperatures of solid fuels such as coal, lignite, oil shale, and tar sands. This yields crude tars and oils that may be converted to liquid fuels. Carbonization processes recover only about 15 percent of the total initial energy of the coal. Low-temperature carbonization of coal is more efficient for producing combustible gases and by-products other than liquid fuels.

In the commercial distillation of oil shale, crushed shale is fed into retorts that crack the organic material, kerogen, with gas or steam at 600–800 K in order to produce crude oil similar in character to petroleum but containing higher percentages of sulfur, nitrogen, and oxygen. The retorting process can also be carried out underground without requiring mining. The crude oil is then refined by processes similar to those developed for treating petroleum.

Coal hydrogenation When hydrogen is added to coal at elevated temperatures and pressures, the hydrogen reacts with the coal molecules to produce a fuel with a higher hydrogen/carbon ratio than the original coal. Coal hydrogenation consists of two steps: (1) liquid-phase hydrogenation at about 200 atm and 730 K of a coal and oil paste, to which a catalyst, usually a compound of tin or iron, is added; and (2) vapor-phase hydrogenation of middle oil—the major product from the first step—under similar conditions of pressure and temperature, on a bed of catalyst such as molybdenum sulfide. Hydrogenation can also be carried out in a single step at about 800 K and also in fluidized beds of coal under pressure.

COAL UTILIZATION / CONVERSION PROCESSES

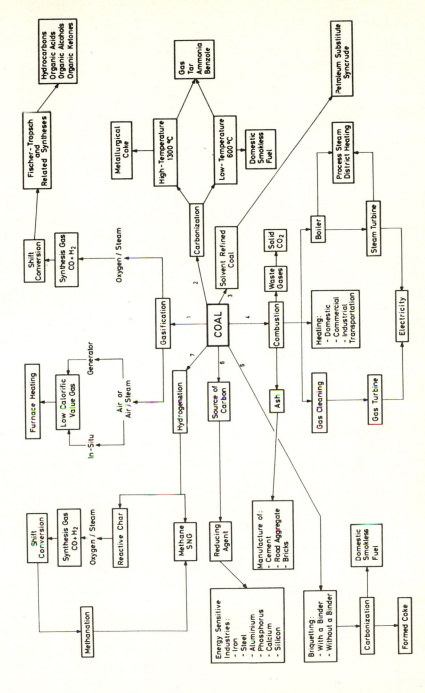

Figure 3.15 Coal utilization/conversion processes. (*By courtesy K. Littlewood, University of Sheffield.*)

Production of Gaseous Fuels

Fuels that exist as gases at room temperature and atmospheric pressure are known as gaseous fuels and are composed of one or more simple gases in varying proportions. They may also include some inert gases and oxygen. The most common simple combustible gases are hydrogen (H_2), carbon monoxide (CO), methane (CH_4), ethane (C_2H_6), ethylene (C_2H_4), propane (C_3H_8), propylene (C_3H_6), butane (C_4H_{10}), butylene (C_4H_8), benzene (C_6H_6), and acetylene (C_2H_2). The inert gases are generally carbon dioxide (CO_2) and nitrogen (N_2). The properties of the most common commercial gases, given in Table 3.22, are described below, approximately in the order of increasing calorific value.

Blast-furnace gas In blast-furnace operations, gas is derived from various reactions occurring in the furnace. The principal combustible constituent is carbon monoxide, resulting from the partial combustion of coke according to the reaction

$$2C + O_2 = 2CO$$

The blast-furnace gas contains a relatively high percentage of inert gases. Since blast-furnace gas has the lowest heating value of all commercial gases, it is usually not transported outside the steel plants. It is mainly used for preheating air for the blast furnace, as fuel for driving the blowers, underfiring the coke ovens, generating steam, heating the soaking pits, and other miscellaneous purposes.

Producer gas Producer gas (synthesis gas) is formed by blowing air or a mixture of air and steam through an incandescent fuel bed, composed of coke, coal, lignite, peat, or wood. When air alone is blown through incandescent carbon the principal reactions are $C + \frac{1}{2}O_2 = CO$, $CO + \frac{1}{2}O_2 = CO_2$, and $CO_2 + C = 2CO$. If all the carbon dioxide were converted to carbon monoxide, the resulting gas would be composed of CO, 34.5 percent and N_2, 65.5 percent. In practice, complete conver-

Table 3.22 Properties of processed gaseous fuels

	Composition (%)									Gross calorific value (MJ/m³)	Theoretical flame temperature (K)
	CO_2	O_2	CO	H_2	CH_4	N_2	C_2H_4	C_6H_6	C_3H_6		
Blast-furnace gas	11.5	...	27.5	1.0	....	60.0	.....	.....	.....	3.43	1727
Producer gas											
Anthracite	8.0	0.1	23.2	17.7	1.0	50.0	.....	.....	.....	5.33	1944
Bituminous	4.5	0.6	27.0	14.0	3.0	50.9	.....	.....	.....	6.07	2019
Coke	6.4	0.0	27.1	13.3	0.4	52.8	.....	.....	.....	5.03	1928
Blue/water gas	5.4	0.7	37.0	47.3	1.3	8.3	.....	.....	.....	10.69	2294
Carbureted water gas	4.3	0.7	32.0	34.0	15.5	6.5	4.7	2.3	.....	19.90	2311
Coal gas											
Horizontal retort	2.4	0.8	7.3	48.0	27.1	11.4	1.3	1.7	.....	20.19	2255
Steamed vertical retort	3.0	0.2	10.9	54.5	24.2	4.4	1.5	1.3	.....	19.82	2280
Coke-oven gas	2.2	0.8	6.3	46.5	32.1	8.1	3.5	0.5	.....	21.39	2261
Reformed natural gas	1.4	0.2	9.7	46.6	37.1	2.9	1.3	.....	0.8	38.00	

sion of CO to CO_2 does not take place. The reaction of incandescent carbon with steam is as follows: $C + H_2O = CO + H_2$ and $C + 2H_2O = CO_2 + 2H_2$. Since the steam reactions are endothermic, the proportion of steam added to the air is such that the heat liberated by the carbon–oxygen reactions is balanced by the carbon–steam reactions. A secondary reaction of considerable importance in producer gas operation is $CO + H_2O = CO_2 + H_2$ (the water gas equilibrium equation).

Producer gas has a low calorific value and is generally used for industrial operations such as under-firing coke ovens; heating open-hearth furnaces; glass melting furnaces; lime, brick, and ceramic kilns. The producer plant is generally integrated with the industrial plant and the gas is not transported over any appreciable distances.

Blue gas or water gas Blue gas or water gas is produced by alternately blasting an incandescent fuel bed with air and with steam. The respective reactions between the carbon and air and the carbon and steam are similar to those taking place in producer gas operation but follow each other cyclically instead of occurring simultaneously. The cycle is thermodynamically balanced so that heat liberated during the airblast and stored in the fuel bed is sufficient to provide the heat absorbed during the steam blast. The gases produced during the airblast, containing nitrogen admitted with the air, are discharged to the atmosphere, while the gases generated during the steam blast provide the useful fuel gas. Blue gas may also be produced in continuous operation by substituting oxygen for air in gas producers. This results in a producer gas that is practically nitrogen-free. Blue gas is used for heating in forge-welding operations, but the major uses of blue gas, without enrichment, are as synthesis gas for the production of ammonia and methanol and other chemical operations.

Oil gas Gas produced entirely from the cracking of crude petroleum oil is referred to as oil gas. Oil gas is produced in a cyclic operation, in which oil atomized by steam is sprayed on previously heated checker bricks, where the oil is cracked and the steam reacts with some of the deposited carbon to form water gas. The checker bricks are heated by the combustion of oil with an excess of air, which also consumes the carbon that failed to react with the steam. Oil gases with a very wide range of calorific values can be produced so that these may be interchanged with carbureted water gas, coal, coke oven gas, or natural gas. Oil gas may also be produced by catalytic cracking methods, either in a continuous or in cyclic operation. In continuous operation, oil vapors and steam flow through a bed of catalyst inside externally heated tubes. In cyclic operation, the catalyst is alternately heated by the combustion of some of the oil and cooled by the cracking of the oil and the reaction of the carbon with the steam. Oil gas may also be generated continuously by partial combustion with air or oxygen.

The composition of oil gas varies widely depending on the fraction reformed and on process techniques. The main components are CH_4, H_2, and CO_2. Maximum calorific value is around 25 MJ/m^3, but is often less than half this and so it requires to be blended or further processed. Uses include: domestic heating, production of substitute natural gas, and production of carbureted water gas.

Carbureted water gas When blue gas or water gas is mixed with hydrocarbon gases produced by thermal cracking of oil, this is known as carbureted water gas. Prior to the advent of natural gas this was used for domestic and commercial cooking and heating. Essentially the manufacture of carbureted blue gas is the combination of the manufacture of blue gas and oil gas accomplished simultaneously in connected pieces of apparatus. Carbureted water gas is often blended with natural gas and liquefied petroleum gases. In countries where there is a plentiful supply of natural gas, the generation of carbureted water gas has been discontinued. As natural gas supplies begin to decline, carbureted water gas may well return as the main supply for domestic and commercial heating and cooking.

Coal gas Coal gas is produced by destructive distillation of bituminous coal, during which it is heated in retorts in the absence of air. The quality and yield of gas vary with the type of coal, the temperature and time of carbonization, and the type of retort. Vertical retorts are often operated with the introduction of steam for the purpose of producing blue gas by the reaction with hot coke. This, in effect, cools the coke before discharging and provides a larger yield of gas, although the calorific value of the gas is lower than that obtained from horizontal retorts.

Coke-oven gas This gas is produced by the distillation of bituminous coal. The distillation products leaving the coke ovens are drawn through hydraulic mains and coolers by exhausters and passed through tar extractors, saturators, light-oil scrubbers, purifiers, and other equipment for removal of tar, ammonia, light-oil and sulfur compounds. The uses of coke-oven gas include bright annealing, controlled furnace atomosphere, heat treatment of steel, and glass melting.

Reformed natural gas Natural gas that is decomposed by thermal or catalytic cracking, or partial combustion processes to a gas with a lower heating value, is referred to as reformed natural gas. Reformed natural gas can be mixed with carbureted water gas, oil gas, coal-, or coke-oven gas. Thermal cracking may be carried out in the fuel bed of a blue-gas generator or in the checker brick of an oil-gas generator. Catalytic cracking or partial combustion is accomplished in a manner similar to that used for cracking or partial combustion of oil.

Acetylene Acetylene is manufactured by the pyrolysis of methane, LNG, or naphtha by processes such as flame cracking, electric arc cracking, and regenerative cracking on a hot refractory solid. By-products include hydrogen and ethylene. Acetylene is mainly used with oxygen for cutting and welding operations. Acetylene gives the highest flame temperature of commonly used fuels.

The Future Role of Synthetic Fuels

A number of different scenarios have been prepared attempting to predict the utilization of fuel for the remainder of this century. Political, economic, and environmental factors, as well as the willingness to change existing systems, will all influence the manner in which fuel will be burned. It seems clear that a major trend will be to reduce the consumption of fuel derived from liquid petroleum, due to its increased scarcity and higher cost, and concentrate on the utilization of liquid gaseous fuels derived from coal. A fundamental problem is associated with the lower hydrogen-to-carbon ratio of liquid derived directly from coal. The questions being faced today are the extent to which these liquids derived from coal should be refined and the extent to which combustion chambers should be modified to accept fuels with lower hydrogen-to-carbon ratios. Refining costs are approximately proportional to refining energy consumption. Longwell (1977) has shown that the elimination of restrictions on aromatics content, hydrogen-to-carbon ratio, and boiling range could reduce refining energy cost by as much as 20 percent of the heat of combustion of the liquid fuel being processed. The total cost of refined products can be reduced by one-third, and a strong argument is being put forward to change power plants and engines to make them capable of burning fuels with low hydrogen-to-carbon ratio. The major problem which needs to be faced in the combustion chamber is to increase the oxidation of soot formed in the early stages of combustion and, hence, reduce emissions of soot.

In automotive systems, the use of a continuous flow system, such as the gas turbine and the Stirling engine, provides the residence time under high-temperature oxidation conditions required to achieve soot burnout. Currently, direct injection of liquid into diesel and stratified charge engines does not allow for complete oxidation of the soot. Improvements can be expected in performance as a result of finer atomization and attempts to increase the residence time of soot particles in high-temperature flame regions. In the spark ignition, Otto cycle engine, high octane number fuels can be supplied, provided that the problems associated with high aromatic content can be solved. Also, in aircraft engines, the ability to burn highly aromatic, wide boiling range fuels has been demonstrated under certain conditions, so that there is hope that these fuels will be acceptable to the aircraft industry.

FOUR

COMBUSTION AND FLAMES

4.1 FLAMMABILITY LIMITS

Flames can propagate through gaseous mixtures only within certain limits of composition. Limits of flammability can be determined experimentally by simply enclosing a mixture in a reactor and igniting the mixture with a spark. The flame will propagate throughout the reactor, only within the upper and lower limits of mixture ratio. The energy of the spark is an important parameter and it has been demonstrated by Weinberg (1975) that, if sufficient energy is supplied to a mixture, the limits of flammability can be very greatly extended. For industrial safety, limits of flammability are prescribed on the basis of prevention of flame propagation and explosion of mixtures ignited by flames or sparks with energies commonly encountered in industrial practice. Figure 4.1 shows the flammability limits for mixtures of methane and oxygen diluted with nitrogen or carbon dioxide. These curves show the following: adding diluent narrows the limits of flammability; increase in temperature widens the limits; changing the diluent from N_2 to CO_2 has only a small effect. Flammability limits for a range of fuel gases mixed with air or oxygen are given in Table 3.16.

Sustained combustion can be supported in a fuel-air mixture at a particular temperature and pressure condition within certain limits of equivalence ratio. Flammability limits are usually determined for quiescent conditions, and these require to be modified under flow conditions. Flammability limits vary with the temperature of the gas mixture. The size of the test container can also influence flammability limits, particularly when the test container is sufficiently small for walls to quench reactions. In flowing systems, heat is convected away from the

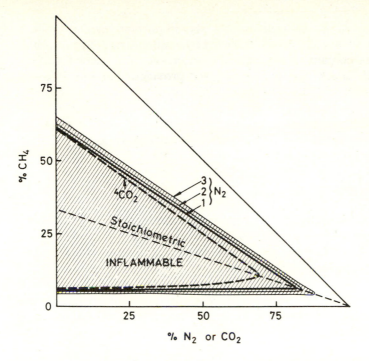

Figure 4.1 Flammability limits for mixtures of methane and oxygen diluted with nitrogen or carbon dioxide. Numbers on the graph refer to initial temperature: 1—293 K; 2—433 K; 3—648 K; 4—293 K.

ignition source and this can reduce the flammability limits. Flammability limits can be widened in recirculation zones where conditions approach those of a well mixed reactor.

An accurate definition of flammability limits cannot be made without considering the energy and mechanism of ignition. Many of the variations in data quoted in the literature are due to differences in methods and energies of ignition. Distinction needs to be made between the possibility of igniting a mixture at a particular location and the subsequent propagation of a flame throughout the mixture from the ignition source. A more precise definition of flammability limits is "Composition limits outside which the mixture will not permit the flame to propagate indefinitely, however powerful the source of ignition that is applied." Such a definition draws a definite distinction between flammability and *ignitability*, so that ignition energy considerations become of experimental, but not theoretical, importance.

4.2 IGNITION

A flammable mixture can be ignited by a heat source and result in flame propagation, provided that the heat balance conditions for propagation are satisfied. The energy supplied by the ignition source, gas temperatures, flame volume, and presence of quench wall surfaces play important roles in ignition. Ignition of a flammable mixture can be achieved by a number of different means: injection of hot gas containing free radicals; insertion of a heated surface; or use of an electric spark. Ignition takes place provided that the following conditions are satisfied: (1) the quantity of energy of the ignition source is sufficiently high to overcome the activation barrier; (2) the energy released in the gas volume exceeds the minimum critical value for ignition; and (3) the duration of the spark or other ignition source is sufficiently long to initiate flame propagation but not too long to influence the rate of propagation.

Spontaneous Ignition

When a gaseous mixture of fuel and oxidant is maintained at ambient temperature, reaction rates are extremely slow and the mixture is in a metastable state. As the temperature is raised, slow oxidation commences and, as a result of the exothermic reactions, temperatures will rise as long as the rate of heat release is greater than the heat loss through the walls of the container. As the mixture temperature is raised further, the reaction rate suddenly increases, giving rise to rapid combustion reactions. This condition is referred to as *spontaneous ignition* and the mini-

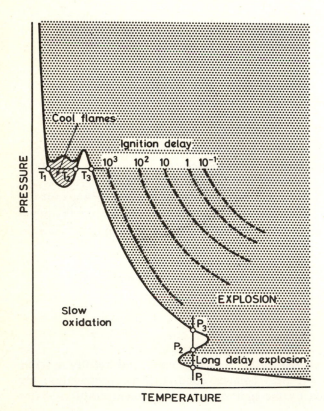

Figure 4.2 Slow oxidation, spontaneous ignition, and explosion as a function of pressure and temperature variation in hydrocarbon-air mixtures. (*De Soete and Feugier, 1976.*)

mum temperature at which rapid combustion reactions are initiated is called the *spontaneous ignition temperature*. The factors influencing the spontaneous ignition temperature of a given mixture are the balance between heat release and heat loss, as well as the supply of reactants.

Figure 4.2 shows the regions of slow oxidation, cool flames, spontaneous ignition, and explosion as a function of pressure and temperature for hydrocarbon-air mixtures. Spontaneous ignition occurs at lower temperatures when the pressure is increased and, for any given pressure, the ignition delay decreases as the temperature is raised. These phenomena are of particular importance in diesel compression ignition engines and also in consideration of the explosion hazard of a mixture. In Fig. 4.2, three spontaneous ignition temperatures, T_1, T_2, and T_3, are given for a given pressure, representing a region of cool flames. Also in this figure, three pressures, P_1, P_2, and P_3, are shown for a given temperature in the region of spontaneous ignition with long delay. The delay in spontaneous ignition is due to the time required to build up a concentration of radicals sufficient to allow the transition from slow oxidation to rapid combustion.

In practice, spontaneous ignition temperatures require to be known for gaseous mixtures which may escape from systems and form explosive mixtures with air. In many chemical industrial operations, special care is required to ensure that sufficient heat is continuously removed from the system in order to prevent the temperature within the reactor exceeding the critical value for spontaneous ignition.

Cool Flames

Cool flames occur under particular pressure and temperature conditions, as shown in Fig. 4.2. Under these conditions of pressure and temperature, combustion is not completed and intermediate products such as CO and CH_2O occur. Cool flames also require an induction period before ignition but, because combustion is incomplete, the flames are less exothermic than normal flames and, hence, they are called *cool flames*. An increase in pressure or temperature of the mixture outside the cool flame region results in normal spontaneous ignition.

The S-shaped curve in the low-pressure region of Fig. 4.2 shows the presence of three pressure limits for the transition between slow oxidation and combustion for a stoichiometric hydrogen-oxygen mixture. At low pressures, hydrocarbon-oxygen mixtures are converted almost entirely to water-oxygen and carbon monoxide–oxygen mixtures. This transformation requires a long delay time. Spontaneous ignition, therefore, becomes dependent on the mixtures of oxygen with the intermediates, H_2 and CO, rather than with initial hydrocarbon. Measurements of pressure and species concentration as a function of time of an *n*-octane–air mixture under adiabatic compression show that the ignition delay is divided into two distinct phases. At the end of the first stage, there is a slight increase in pressure corresponding to the formation of partial products of combustion (CO and CH_2O), characterizing a cool flame. At the end of the second stage, there is a sudden rapid increase in pressure causing spontaneous ignition, accompanied by the formation of complete products of combustion (CO_2 and H_2O). The addition of tetraethyl lead and other organometallic compounds inhibits the onset of detonation by raising the spontaneous ignition temperature and increasing delay times.

Ignition by Electric Spark

The release of electrical energy between electrodes results in the formation of a plasma in which the ionized gas acts as a conductor of electricity. Spark durations vary from the order of one to several hundred microseconds. The electrical energy liberated by the spark is given by

$$E = \int_0^\theta VI \, dt \qquad (4.1)$$

where V is the potential, I is the current, θ is the spark duration, and t is time.

Electrical energy is very rapidly transformed into thermal energy and, because the temperature of the ionized gas is generally above 3000 K, the local autoignition delay time is short compared with the spark duration, θ. Ignition can only take place if the electrical energy exceeds the critical value, E_c, and if this energy is liberated within a critical volume, v_c. It will also be a function of the spark duration.

Ignition by Heated Surfaces

A flammable mixture can be ignited by a hot surface provided that sufficient heat is transferred from the surface to the gas mixture. During the process of heat transfer from the surface to the gas, reactions are initiated as the temperature rises and the combination of additional heat transfer from the surface and heat release by chemical reaction can lead to ignition of the mixture.

For heat transfer by conduction only, the energy equation becomes

$$\rho c_p(\partial T/\partial t) = k\nabla^2 T + qk_0 \exp(-E/RT) \qquad (4.2)$$

where k is the conductivity of the gas; q is the global volumetric thermicivity; and k_0 is a kinetic parameter.

For one-dimensional heat conduction, we have

$$\rho c_p(\partial T/\partial t) = k\partial^2 T/\partial x^2 + qk_0 \exp(-E/RT) \qquad (4.3)$$

In experiments where a heated cylinder is introduced into a propane-air mixture, it is found that approximately 11 ms is required for ignition. During this time, heat released by chemical reaction is negligible. This time interval before ignition is referred to as the *ignition delay*, which is primarily dependent upon the physical delay due to heat conduction.

De Soete (1973) calculated the ignition delay and compared this with experiments for stoichiometric mixtures of propane and air at atmospheric pressure. The instantaneous temperature of each elementary volume of the mixture is calculated by solving Eq. (4.3) and, hence, determining T as a function of x and t. The relation between the chemical delay time, θ, and the diffusion time, τ, is given by the expression

$$1 = \int_0^\tau dt/\theta(T_x, t) \qquad (4.4)$$

Both experiment and calculation show that τ has a minimum value at a distance x_a from the surface. Close to the surface both θ and τ increase in value because T decreases with decrease in x. Ignition takes place close to but not at the surface.

Spontaneous Ignition Temperature

The entire region within the flammability limits of a fuel-air mixture can be divided into two subregions, separated by the spontaneous ignition temperature (SIT). For liquid fuels, this parameter is determined using standardized tests where liquid fuel is dropped into an open-air container heated to a known temperature. The spontaneous ignition temperature is defined as the lowest temperature at which visible or audible evidence of combustion is observed. Typical values for a range of liquid fuels are shown in Table 4.1. This data shows that, as the length of an n-paraffin chain is extended, the SIT is reduced. These data are specific to a particular experiment and cannot always be used as hazard criteria for other physical conditions. SIT decreases rapidly with pressure until approximately 2 atm, with very small changes above this pressure.

Table 4.1 Spontaneous ignition temperatures

Fuel	SIT (K)
Propane	767
Butane	678
Pentane	558
Hexane	534
Heptane	496
Octane	491
Nonane	479
Decane	481
Hexadecane	478
Isooctane	691
Kerosine (JP-8 or Jet A)	501
JP-3	511
JP-4	515
JP-5	506

Ignition delay time

The ignition delay time is defined as the time, after ignition, required for a given fuel-air mixture to achieve "significant" reaction. The onset of "significant" reaction can be defined in terms of the rate of change of temperature dT/dt and it can be shown that the ignition delay is exponentially related to the initial temperature T_i

$$t_{ign} \sim \exp\left(E_a/RT_i\right)$$

Ignition delay time is not strongly dependent on variation in mixture ratio but there is a strong dependence on pressure. For liquid fuels, ignition delay times are of the order of 50 ms at 700 K and of the order of 10 ms at 800 K.

Below the spontaneous ignition temperature, mixtures can be ignited by a heat source which allows temperature to locally exceed the SIT. The most common method of achieving this is by spark discharge.

Critical ignition volume

Combustion waves are generally spherical or cylindrical; the flame volume and combustion surface increase from a point or line source during propagation. When the rate of increase of the combustion volume exceeds a critical value, the combustion wave cannot continue to propagate without the supply of additional energy. During flame propagation, successive layers of gas are raised to a reactive state by heat and mass transfer from the flame front. A positive exchange of heat between the flame and the fresh mixture is achieved when the rate of increase of flame volume is less than the rate of increase of volume of burned gases. This is the condition that governs the critical volume for ignition. For a "point" spark source, the flame initially has a spherical form. The critical ignition volume is determined from the calculation of the rate of change of flame volume with respect to radius compared to the rate of change of volume of burned gases. The critical volume is determined from

$$\frac{d}{dr}\left(\tfrac{4}{3}\pi[(r+e)^3 - r^3]\right) \leq \frac{d}{dr}\left(\tfrac{4}{3}\pi r^3\right) \tag{4.5}$$

that is, $2er + e^2 \leq r^2$, or

$$r_c = e + \sqrt{2} \tag{4.6}$$

where e is the thickness of the flame front and r_c is the radius of the critical spherical volume for ignition.

When the electrodes are sufficiently far apart, the flame initially has the form of a cylinder with length e equal to the separation distance between the electrodes. Thus, for a cylindrical flame, $r_c = e$, the critical ignition volumes are:

(1) For a spherical flame

$$v_c = 4\pi e^3(1 + \sqrt{2})^3/3 \tag{4.7}$$

(2) For a cylindrical flame

$$v_c = \pi e^2 d \tag{4.8}$$

Minimum ignition energy

Minimum ignition energies are calculated on the assumption that sufficient energy must be supplied from the exterior to the critical volume to raise the mixture to the flame temperature T_f. The critical energy is thus

$$E = v_c \rho c_p (T_f - T_i) \tag{4.9}$$

ρ and c_p are the density and specific heat of the mixture and $T_f - T_i$ is the difference between flame and initial temperatures. Only a fraction ζ of the total spark energy E is utilized for ignition energy because of heat being transferred to the exterior of the critical volume.

The theoretical minimum ignition energy is calculated on the assumption that the energy is released from the ignition source instantaneously and homogeneously throughout the critical volume. If the ignition duration is too long, energy will be transferred by thermal conductivity outside the critical volume and ignition cannot be achieved because the energy within the critical volume is insufficient to raise the mixture to the flame temperature. For minimum ignition energies of the order of several millijoules, the energy must be released in several microseconds in order to ensure that there is no significant transfer of energy away from the ignition source.

Minimum ignition energies are determined experimentally by introducing two electrodes into a combustion bomb, in which both the ignition energy and the separation distance between the electrodes can be varied. Successful ignition is determined by measurement of the sudden increase in pressure as the mixture explodes. The energy released is calculated from the expression

$$E = CV^2/2 \tag{4.10}$$

where C is the capacity of the condenser and V is the potential of the discharge. Minimum ignition energy varies as a function of distance between the electrodes; the quenching distance corresponds to the theoretically infinite energy required to ignite the mixture. Beyond the quench distance, minimum ignition energy remains constant with separation distance and then begins to gradually increase. In the region where the minimum ignition energy remains constant, the ignition source behaves as a "point"; as the distance is increased, the source becomes linear and, hence, the minimum ignition energy increases. For a point source, E_m is independent of separation distance, whereas for a line source, it is proportional to separation distance.

The quench distance increases as pressure is decreased. For mixtures of methane, ethane, and butane with air, minimum ignition energies are required at, or close to, stoichiometric mixture ratios; E_m increases rapidly as the mixture ratio becomes rich or poor.

The Influence of Gas Flow on Ignition

In gas turbine combustion chambers, industrial furnaces, and some types of engines, the fuel-air mixture is not constant. As gas flow velocity increases, heat is convected away from the ignition source and turbulence intensity increases diffusion rates. Ballal and Lefebvre (1975a) carried out an experimental study to investigate the effects of pressure, velocity, mixture strength, turbulence intensity, and turbulence scale on minimum ignition energy and quenching distance, d_q. They carried out their tests in a closed-circuit tunnel in which a fan was used to drive propane-air mixtures at subatmospheric pressures at velocities up to 50 m/s. Turbulence intensity and scale was varied by inserting perforated plates upstream of the electrodes. Figure 4.3 shows the influence of velocity on quenching distance and Fig. 4.4 shows the influence of velocity variation on minimum ignition energy. Both d_q and E_m have minimum values near stoichiometric mixture ratios where flame temperatures are maximum. This tends to suggest that thermal/diffusion processes control the flame propagation. As mixture ratios are increased, the ratio of convective transfer to diffusional transfer changes. As velocity increases, the spark kernel, during the initial period of its development, becomes more

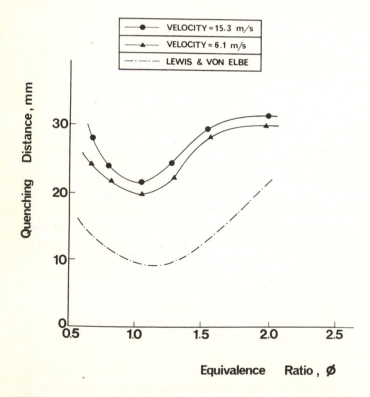

Figure 4.3 Influence of velocity and mixture strength on quenching distance. (*Ballal and Lefebvre, 1975a.*)

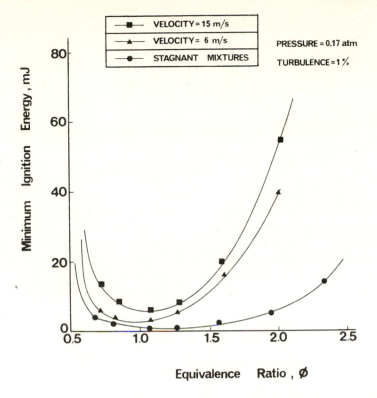

	VELOCITY = 15 m/s
	VELOCITY = 6 m/s
	STAGNANT MIXTURES

PRESSURE = 0.17 atm

TURBULENCE = 1 %

Figure 4.4 Minimum ignition energy variation with equivalence ratio for different mainstream velocities. (*Ballal and Lefebvre, 1975a.*)

diluted with cold mixture, thus requiring more energy to compensate for the loss
of heat. However, increased flow velocity leads to displacement of the spark in the
downstream direction, thereby reducing the loss of heat and active radicals to
the electrodes. In an arc discharge, energy is distributed almost symmetrically
between the electrodes. With a glow discharge, as much as 60 percent of the energy
is concentrated in the vicinity of the cathode. Therefore, the displacement of
the spark away from the electrodes is much more beneficial for an arc discharge
than for a glow discharge.

Figure 4.5 shows that quenching distance increases with reduction in pressure.
Ballal and Lefebvre show that, at a velocity of 6 m/s, this increase is less than that
observed with stagnant mixtures. The effect of pressure change can be largely
explained as due to changes in mean free path and velocity of molecules which
change the physical properties of the gas as a function of pressure.

Turbulence can affect ignition processes in several different ways. Flame propagation is augmented by wrinkling of the flame front or breakup into individual
flamelets with resultant increase in flame surface area. Within the flame zone,
transport of radicals and other active species is accelerated. However, at the same
time, turbulence increases the loss of heat by diffusion to the surrounding un-

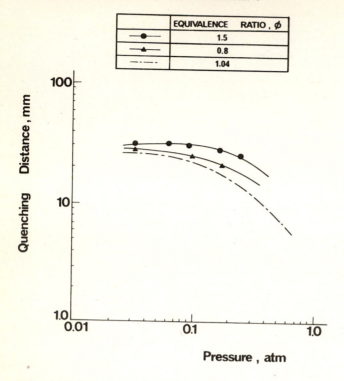

	EQUIVALENCE RATIO, ϕ
●——	1.5
▲——	0.8
-----	1.04

Figure 4.5 Influence of pressure and mixture strength on quenching distance. (*Ballal and Lefebvre, 1975a.*)

burned gas. Figures 4.6 and 4.7 show that the heat loss effect is overriding, so that both quenching distance and ignition energy increase with turbulence intensity. Ballal and Lefebvre also showed that turbulence scale has an important influence on d_q and E_m. The effects of turbulence intensity and turbulence scale are inter-related. At low turbulence intensities, increase in turbulent scale causes wrinkling of the flame front. At high turbulence intensities, the flame becomes sufficiently stretched so that individual flamelets are formed. The effect of turbulence scale on d_q and E_m varies, therefore, as a function of the turbulence intensity.

Ignition of Liquid Fuel Sprays

Ignition in liquid fuel fired combustors is affected in several ways by the presence of liquid droplets. Local mixture ratios are a function of the degree of vaporization so that, if the ignition source is located in a dense spray region, the mixture ratio based upon the vaporized fuel and air concentrations can easily be outside the limits of flammability. The presence of droplets can have a local quenching action, and energy derived from the ignition source can be absorbed as latent heat for vaporization of the liquid fuel without there being sufficient heat to raise the mixture to the flame temperature. Thus, the location of the ignition source, degree

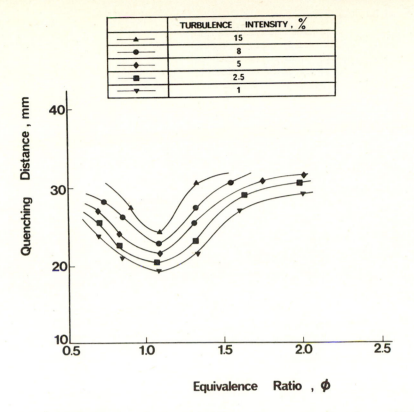

	TURBULENCE INTENSITY, %
—▲—	15
—●—	8
—◆—	5
—■—	2.5
—▼—	1

Figure 4.6 The effect of variation of turbulence intensity on quenching distance. (*Ballal and Lefebvre, 1975a.*)

of atomization, fuel spray characteristics, and fuel volatility become additional factors which require to be considered in the ignition of flowing mixtures containing droplets. Subba Rao and Lefebvre (1973) and, later, Rao and Lefebvre (1976) measured minimum ignition energies in kerosine sprays injected into flowing airstreams under conditions of atmospheric pressure and temperature. Measurements were performed using optimum values of spark duration and with the spark gap adjusted to slightly exceed the quenching distance. The main aim of the experiments was to determine the minimum ignition energy which was defined as the total electrical energy dissipated from the weakest spark that would ignite a flammable mixture when the spark gap is set at the quenching distance and the spark duration is the optimum value. Minimum ignition energies were measured under controlled conditions of fuel/air ratio, fuel drop size, and flow velocity.

Spark discharge characteristics are affected by flow velocity, quenching distance, and spark duration. Increase in air velocity requires an increase in gap voltage due to the stretching of the spark, which increases its resistance and, hence, the energy release. The arc becomes displaced by the airflow in a downstream direction so that heat loss to the electrodes is reduced. Quenching distances were determined by gradually reducing the gap length until no further

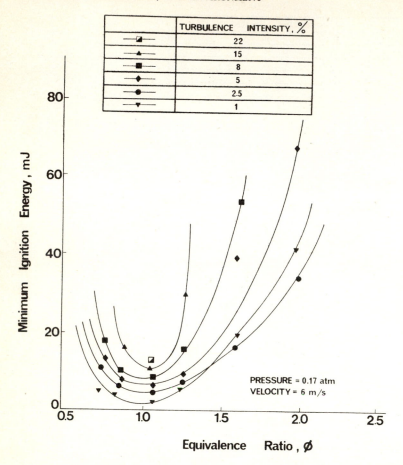

TURBULENCE	INTENSITY, %
◪	22
▲	15
■	8
◆	5
●	2.5
▼	1

PRESSURE = 0.17 atm
VELOCITY = 6 m/s

Equivalence   Ratio , ∅

Figure 4.7 The effect of variation of turbulence intensity on minimum ignition energy. (*Ballal and Lefebvre, 1975a.*)

sparking was possible. The quenching distance was obtained from the measurement of the gap width corresponding to the minimum value of ignition energy. For atmospheric pressure conditions, quenching distances were found to be always less than 3 mm. If the spark duration is too short, a large proportion of the available energy is dissipated by the formation of a strong shock wave. If the spark duration is too long, energy is wasted by being added to the mixture after the ignition process has been established. Optimum spark duration increases with reduction in flow velocity and with any change in mixture strength toward the stoichiometric value. For propane-air mixtures and velocities up to 50 m/s, optimal spark durations are in the range from 5–60 μs. In sprays, optimum spark duration was found to vary between 30 and 80 μs.

In liquid sprays, the most important parameters affecting minimum ignition energy are droplet diameter, gas velocity, and fuel/air ratio. All these parameters

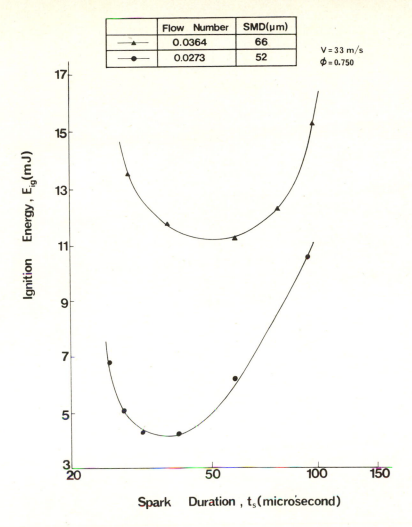

	Flow Number	SMD(μm)
▲	0.0364	66
●	0.0273	52

V = 33 m/s
φ = 0.750

Figure 4.8 Ignition energy variation with spark duration for two values of mean droplet diameter (SMD). (*Rao and Lefebvre, 1976a.*)

are interrelated. Rao and Lefebvre (1976) examined the separate effects of each of these parameters. The pronounced effect of mean droplet diameter (SMD) on minimum ignition energy, E_{min}, is shown in Fig. 4.8. This shows that, by reducing the SMD from 66 to 52 μm by improvement of atomization quality, E_{min} is reduced threefold. Figure 4.9 shows that, for any given value of SMD, increase in air velocity requires an increased E_{min}. This is due partly to increased heat losses from the spark kernel as turbulent diffusion increases with velocity, but also because at higher velocities the spark discharge is required to heat a larger amount of fresh mixture throughout its duration. Both these factors tend to

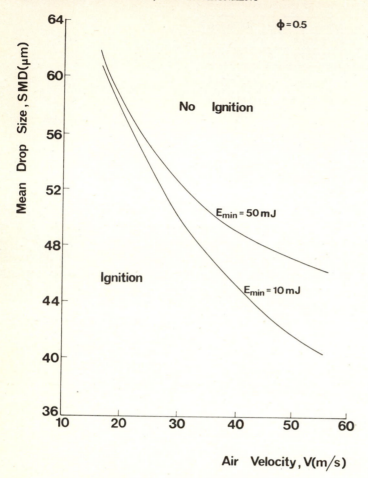

Figure 4.9 Influence of flow velocity on ignition limits for two levels of spark energy in liquid fuel sprays. (*Rao and Lefebvre, 1976.*)

reduce the flame temperature and, hence, the rates of fuel evaporation and chemical reaction also. However, if additional energy is provided in the spark to compensate for these losses, then sufficient inflammable mixture is again produced to effect ignition. Increase in spark energy is effective in overcoming the adverse effects in air velocity, only for the case of small drop size, i.e., below 50 μm. Figure 4.10 shows that, for constant values of SMD and air velocity, E_{min} is appreciably reduced by an increase in fuel/air ratio. The benefit that can be derived from an increase in spark energy is less when the initial value of the spark energy is high. This is partly due to the reduced efficiency of conversion of electrical energy into useful thermal energy at high spark energies, as well as due to the fact that extension of the weak ignition limits reduces the amount of fuel evaporation. The principal conclusions of Rao and Lefebvre are summarized as follows:

1. For each experimental condition there is an optimum value of spark duration for minimum ignition energy.
2. Optimum spark duration increases with increase in droplet diameter (SMD) and reduction in air velocity. It is fairly insensitive to changes in fuel/air ratio.
3. Increase in flow velocity requires increased energy for ignition.
4. Even slight improvements in atomization quality can greatly reduce the amount of energy required for ignition and appreciably extend the weak ignition limits.
5. When increased air velocity causes a reduction in SMD, weak ignition limits can be extended by an increase in air velocity. However, for SMD less than 60 μm, if SMD is maintained constant, then increase in air velocity has a detrimental effect on weak ignition limits.
6. For low values of SMD, the weak ignition limit is extended by increasing spark energy. For high values of SMD, however, very large amounts of additional energy are required to effect even a small extension of the weak ignition limit.
7. The use of high energy sparks is an inefficient method of achieving ignition. A more efficient way of improving ignition performance is by improving fuel atomization. Provided the fuel is well atomized, very low energy sparks are capable of igniting high velocity mixtures.

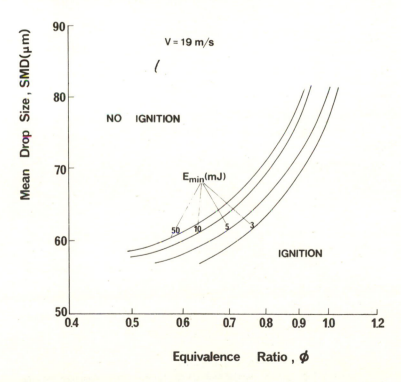

Figure 4.10 Ignition limits obtained at constant flow velocity and various levels of spark energy. (*Rao and Lefebvre, 1976.*)

4.3 BURNING VELOCITY

Burning velocity is determined by measuring the local inclination of the cold flow to the flame front and the magnitude of the velocity U prior to preheating (Fig. 4.11). When a nozzle with an area contraction ratio of more than 4 is used as a burner, the velocity profile of the gases emerging from the nozzle is almost completely uniform and the flame front forms a normal, perfectly geometric, cone. The gas velocity U can be determined from the ratio of the measured volumetric gas flow rate and nozzle exit area or, alternatively, it may be measured directly by a pitot tube, hot-wire, or laser anemometer. The angle of the flame front α can be determined from either direct or schlieren photography and the burning velocity is calculated from

$$S_u = U \sin \alpha \qquad (4.11)$$

As the diameter of the nozzle is reduced, boundary-layer effects begin to play a role and, thus, it is preferable to use a large-diameter burner. This is the simplest and most reliable method of measuring burning velocity. Because of the danger of flashback, a flame trap must be inserted in the nozzle as a safety precaution.

The direct measurement of local burning velocity can be made by the particle-track method. Lewis and von Elbe (1961) measured flow lines through a natural gas-air flame, using a rectangular burner (Fig. 4.12). Small magnesium oxide particles were added to the gas stream and the flame was illuminated intermittently from the side. The velocity and direction of the particles were determined from the photographs and the results showed that, in the vicinity of the burner rim, the burning velocity decreases toward zero and that burning velocities in the tip region are higher than in the remainder of the flame, where the burning velocity is uniform. These measurements showed the inaccuracies that can arise from using the flame-area method.

The *soap-bubble* method, in which a combustible mixture is blown into a soap bubble and ignited centrally, is also used for determining burning velocities. The surface of the bubble expands freely as the expansion proceeds at a constant

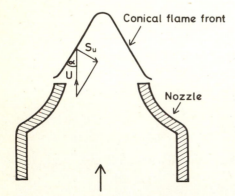

Figure 4.11 Conical flame front formed above a nozzle with a uniform velocity profile. S_u—burning velocity, normal to flame front; U—velocity of gas prior to preheating; α—angle between flame surface and gas flow.

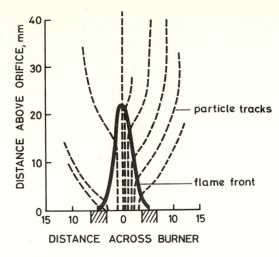

particle tracks

flame front

DISTANCE ACROSS BURNER

Figure 4.12 Flow lines through the inner cone of a natural gas–air flame from particle-track measurements. (*Lewis and von Elbe, 1961.*)

pressure. The expansion of the burned gases causes a bodily movement of the gases, through which the flame front is progressing and the velocity that is measured is the *spatial* velocity. The flame front is spherical and its change in diameter is photographed as a function of time. The burning velocity is the ratio of the spatial velocity to the expansion ratio of the bubble, that is, $(D/d)^3$ where d and D are the initial and final values of the diameter of the bubble. This method provides a direct measurement of the velocity of propagation of the flame. An alternative to the soap-bubble method is the *spherical-bomb* method, in which a spherical vessel is filled with a flammable mixture. Simultaneous records are made of the size of the spherical shell of burned gas and of the pressure in the vessel. The burning velocity is determined from

$$S_u = \frac{dr_b}{dt} - \frac{a^3 - r_b^3}{3P\gamma_u r_b^2}\frac{dP}{dt} \qquad (4.12)$$

where a = radius of bomb, r_b = radius of shell of burned gas, P = pressure, t = time, and γ_u = ratio of specific heats of the unburned mixture.

Gaydon and Wolfhard (1970) reviewed methods of measuring burning velocity, and concluded that the simple nozzle method, with correction for the boundary layer, is the best and most practical method, provided that the base of the flame is excluded from the determination. Maximum burning velocities for a range of hydrocarbon fuels are given in Table 4.2, and the variation of burning velocities with temperature for hydrocarbon/air mixtures is shown in Fig. 4.13.

When the burning velocity is determined, on the basis of the velocity relative to the flow of cold gas before preheating, only a small increase in S_u is found with increase in initial gas temperature. When the burning velocity is, however, referred to the actual gas, at the preheated temperature, Dugger et al. (1955) showed that the effect of preheating CH_4, C_3H_8, and C_2H_4 air mixtures from 200 to 600 K led to increases of S_u from 0.2 m/s to 1.5 m/s, as shown in Fig. 4.13. Kuehl (1961)

Table 4.2 Maximum burning velocities for hydrocarbon-air flames

Fuel	$S_{u_{max}}$ † (m/s)	% fuel at $S_{u_{max}}$
Methane	0.338	9.96
Ethane	0.401	6.28
Propane	0.390	4.54
Butane	0.379	3.52
Pentane	0.385	2.92
Hexane	0.385	2.51
Heptane	0.386	2.26
Ethylene	0.683	7.40
Propylene	0.438	5.04
Benzene	0.407	3.34
Acetylene	1.5	

† $S_{u_{max}}$ is burning velocity measured at atmospheric pressure and temperature (without preheat) for the fuel/air mixture ratio that generates the maximum burning velocity.

showed that, for propane-air mixtures, the burning velocity rises to 2.5 m/s at 800 K. If gas mixtures are preheated to over 800 K, preflame reactions occur which tend to lower the burning velocity. The influence of initial temperature and changes in oxygen/nitrogen ratio on burning velocity of methane flames is shown

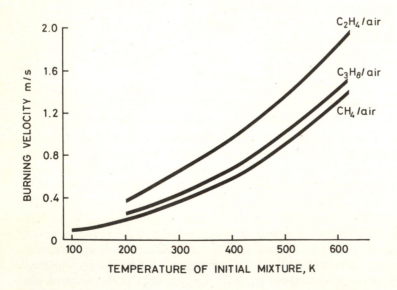

Figure 4.13 Effect of initial temperature on maximum burning velocity for hydrocarbon-air mixtures. (*Dugger et al., 1955.*)

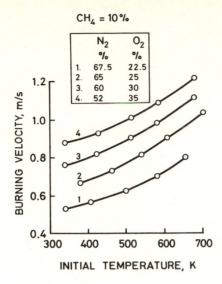

Figure 4.14 Variation of burning velocity with initial temperature for mixtures of methane-oxygen-nitrogen.

in Fig. 4.14. Heimel and Weast (1957) studied flames of benzene, heptane, and isooctane and found

$$S_u = B + CT_i^n \tag{4.13}$$

where T_i = initial temperature; B, C, and n = empirical constants with n lying between 2 and 3. They derived a theoretical expression relating the temperature variation of burning velocity with the activation energy of the chemical reaction.

For hydrocarbon-air flames, the values of S_u rise as pressure is reduced. For high-temperature flames, with oxygen, S_u varies little with pressure. For cooler flames with air, the pressure dependence can be expressed as a simple power law

$$S_{u_a} = S_{u_b}(P_a/P_b)^n \tag{4.14}$$

where S_{u_a} and S_{u_b} are the burning velocities at pressures P_a and P_b. Values of n are negative for cooler flames, rising to a small positive value at high temperatures.

The influence of hydrocarbon structure on burning velocity is shown in Table 4.3. Empirical formulas indicate that burning velocities are related to the number and types of chemical bonds in the hydrocarbon. Acetylene-air has a particularly high value of S_u of 1.5 m/s.

Burning velocities usually have maximum values well on the rich side of stoichiometric mixture ratios. The presence of inerts lowers burning velocity; for H_2-O_2-N_2 mixtures, S_u increases from 3 m/s for 21 percent O_2 to 11.8 m/s for 97 percent O_2. Studies of variation in burning velocity with the mixture strength show that the point of maximum burning velocity moves to less rich mixtures when the gas is preheated. Many organic halides, which are well known as inhibitors and fire extinguishers, reduce the speed of flames. These inhibitors generally interfere with chain propagation by absorbing superequilibrium radicals. The heavier organic halides are the most effective, with bromides more active than iodides or chlorides. For stoichiometric ethylene-air mixtures, addition of 2 percent methyl bromide reduces S_u from 0.66 to 0.25 m/s. For rich mixtures, the effect is even greater. Most hydrocarbons have a strong inhibiting action on the burning velocity of hydrogen-air flames—$3\frac{1}{2}$ percent butane reduces the flame speed from 2.75 to 0.15 m/s. Powdered metallic salts also cause a marked reduction in S_u. For methane-air flames, addition of 10 g/m³ of Na_2CO_3 reduces S_u from 0.65 to 0.15 m/s. The main effect is due to the inhibition by metal atoms after the evaporation of the salt (Rosser, Inami, and Wise, 1963).

The Influence of Free Radicals on Burning Velocity

The reaction zone of a flame has a structure which is far from being homogeneous. Reactants are transformed into products of combustion within a distance of several tenths of a millimeter, with temperature gradients of the order of 10^5 K/mm. Under these conditions the flame front is rich in free radicals which diffuse toward the unburned mixture. Because of the steep temperature and free radical concentrations within the flame zone, free radicals and heat diffuse rapidly toward the unburned gases. Flame propagation is governed by the combined diffusion of free radicals and heat. Unless the activation energy of branching is very low, sufficient heat and free radicals are required to initiate flame propagation. The velocity of flame propagation is controlled by the diffusion velocity of free radicals from the flame front to the unburned mixture. Following the analysis presented by de Soete and Feugier (1976), the time for free radicals to diffuse across a reaction zone of thickness δ_r is shown to be equal to the reaction time which is defined as

$$t_r = \delta_r / V_n \qquad (4.15)$$

where V_n is the diffusion velocity. The diffusion coefficient for free radicals D_m is related to the distance δ_r and reaction time t_r by

$$D_m = \delta_r^2 / 2t_r \qquad (4.16)$$

Hence

$$V_n \sim D_m / \delta_r \qquad (4.17)$$

For a flame front with a constant burning velocity, the reaction rate and the concentration of free radicals remain constant during propagation. Radicals which are lost from the reaction zone by diffusion and recombination require to be replaced through branching reactions. On average, each radical that diffuses across the flame front has at least one collision resulting in one branching reaction. Since radicals may also be destroyed by a chain-breaking reaction, the probability of branching (reduced by the probability of breaking) must be equal to unity over the reaction thickness δ_r. This is the condition required for a stationary flame. The total probability of branching is equal to the probability of branching in a given collision multiplied by the number of collisions Z of the radical during its passage across δ_r. The total probability of chain breaking is also dependent on the number of collisions.

The concentration of free radicals is controlled by the rate of chain branching, which increases the concentration, and by the rate of destruction in the gas phase and on walls, which reduces the concentration. The rates of formation and destruction depend upon the Arrhenius exponential factor $\exp(-E/RT)$. The rate of formation of radicals is given by the algebraic sum of the three elementary reaction rates

$$\frac{dX_R}{dt} = Z_2\, \delta X_R - Z_2 F_t X_R^2 - Z_2 \beta_s X_R \tag{4.18}$$

where X_R = concentration of free radicals; Z_2 = frequency of bimolecular collisions; and

$$\delta = X_A(X_Y/X_R) \exp(-E_r/RT)$$

$$F_t = Z_3 X_Y X_X / Z_2 X_R^2$$

$$\beta_s = (Z_s/Z_2) \exp(-E_t'/RT)$$

Z_3 = frequency of trimolecular collisions; Z_s = frequency of wall collisions. The activation energy of chain breaking at the wall E_t is usually small (4–8 kJ/mol) so that the increase in mole fraction of free radicals depends essentially on the chain-branching step. The autoignition delay θ is inversely proportional to the concentration of free radicals and is given by

$$\theta \sim \exp(E_r/RT) \tag{4.19}$$

The increase in concentration of free radicals X_R cannot continue indefinitely. X_R must reach a maximum value, which it may retain for a short period, and then decrease as reactants are consumed. This maximum concentration $(X_R)_{max}$ is attained when $dX_R/dt = 0$, i.e.

$$Z_2\, \delta(X_R)_{max} - Z_2 F_t (X_R)_{max}^2 - Z_2 \beta_s (X_R)_{max} = 0 \tag{4.20}$$

or

$$(X_R)_{max} = (\delta - \beta_s)/F_t \tag{4.21}$$

Combustion of a mixture, for a given temperature and pressure, requires that the critical concentration of free radicals $(X_R)_c$ must be less than $(X_R)_{max}$. The condition for autoignition can be written as

$$(X_R)_c \le (\delta - \beta_s)/F_t \tag{4.22}$$

For nonadiabatic conditions, the theoretical critical concentration of free radicals is much lower than for adiabatic conditions.

The condition for a flame to be stationary becomes

$$Z(\delta - F_t X_R) = 1 \tag{4.23}$$

The number of collisions Z for a radical crossing the reaction zone with thickness δ_r is equal to the number of collisions per second $(Z_2 = c/\lambda)$ multiplied by the diffusion time $t_r = \delta_r^2/2D_m$. The condition for stationary flames then becomes

$$\frac{c}{\lambda}(\delta - F_t X_R)\frac{\delta_r^2}{2D_m} = 1 \qquad (4.24)$$

Combining the above expressions with that for burning velocity leads to

$$V_n = \left[\frac{2D_m(V_r - V_t)}{X_R}\right]^{1/2} \qquad (4.25)$$

where V_r is the chain-branching velocity $= (c/\lambda)\,\delta X_R$, and V_t is the chain-breaking velocity $= (c/\lambda)F_t X_R^2$.

These expressions show the relation between the reaction kinetics and the physical phenomena of flame propagation. By this relation, information on reaction mechanisms such as activation energies and the type of branching can be obtained from measurements of burning velocity. These theoretical considerations show that reductions in flame temperature must lead to reduction in burning velocity and that the temperature of the products of combustion and the burning velocity reach maximum values near stoichiometric mixture ratios. Dilution with inert gases reduces flame temperature and burning velocity because of changes in heat capacity and also because of reductions in mole fractions of fuel and oxidant and because of reductions in chain-branching reactions. Conversely, preheating of gases leads to increases in burning velocity and flame temperature (Figs. 4.13 and 4.14).

Quenching

The presence of walls in a combustion system generally results in heat losses from the flame and hot gases to walls which are usually at lower temperatures than the gases. Solid surfaces provide a means for breaking the chain propagation as free radicals diffuse toward the surfaces where they recombine more easily than in the gas phase. Free radicals can be retained on a solid surface for longer periods than are required for collision in the gas phase, and thus the probability of recombination with other radicals is greater on or near solid surfaces than in the surrounding gas. The surface also acts as a "third body," draining the energy liberated by the recombination. These effects of quenching lead to nonadiabatic conditions in the system and the effect increases with surface-to-volume ratio of combustion chambers.

Experiments carried out on spark ignition of mixtures between parallel plates show that, with sufficiently powerful sparks, a flame develops in the immediate neighborhood of the spark but does not propagate through the mixture unless the plates are separated by more than the *quenching distance*. Quenching distance, as a function of equivalence ratio for hydrocarbon-air mixtures, is shown in Fig. 4.15. The effect of quenching can also be studied by burning a flame on a burner and measuring the diameter below which the flame will not flash back. Alternatively, one can adjust the ambient pressure at which a given burner diameter no longer allows flashback and this determines the *quenching diameter* for a certain ambient

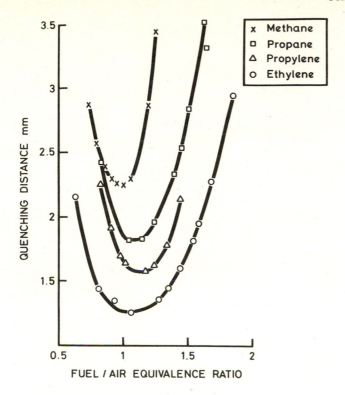

Figure 4.15 Quenching distance as function of equivalence ratio for hydrocarbon mixtures with air. (*De Soete and Feugier, 1976.*)

pressure. The quenching diameter increases as the gas pressure is reduced and is roughly inversely proportional to the pressure. The nature of the burner surface does not appear to influence the quenching properties. Heating of the walls reduces the quenching distance, which is roughly inversely proportional to the square root of the absolute temperature. Quenching effects may, however, also be due to the removal of active centers by diffusion to the walls. Hydrogen-atom diffusion to the walls, coupled with the surface recombination $H + O_2 = HO_2$, has been postulated as a likely mechanism for H-atom removal. The stabilization of the flame depends critically on the quenching effect. In order to prevent flashback, flame traps or holders are introduced into burners. A fine gauze is often used as a flame trap and the critical flame speed, above which a particular gauze will not arrest the flame, is inversely proportional to the width of the mesh of the gauze.

4.4 FLAME STRUCTURE

One of the main aims of the combustion scientist is the determination of flame structure. This requires determination of the dimensions of various zones in the flame and the temperature, velocity, and species concentrations throughout the system. The probing methods developed by Fristrom and Westenberg (1965), the spectroscopic studies of Gaydon (1974), and the optical techniques developed by Weinberg (1963) have all led to an improved knowledge of the structure of flames. Developments in analytical models have taken place in parallel with experiments and, for the simplest types of flame, such as a laminar gaseous flat flame, calculations of flame dimensions and profiles are found to be in good agreement with experimental data. We shall first examine the structure of the "one-dimensional" flame, in which all significant variations occur in the direction of the flow of the gases, perpendicular to the flame front.

A one-dimensional flame can be divided into the four zones: (1) unburned gas, (2) preheating, (3) reaction, and (4) burned gas, as shown in Fig. 4.16. The boundaries are designated, using the temperature profile as a reference basis. The premixed gases approaching the flame have a uniform velocity, temperature, and concentration, with constant physical properties, in the unburned gas zone. In the preheating zone, temperature increases are due solely to heat conduction from the flame, while chemical reaction rates and heat release are negligible. All chemical reaction and heat release occur within the reaction zone. The gases emerging from the flame enter the burned gas zone, in which velocity, temperature, and species concentration are uniform.

In principle, the preheating zone commences at the point where the temperature begins to rise, the reaction zone commences at the point at which exothermic reactions just begin to be significant, and the reaction zone ends when combustion has been completed. Each one of these criteria is imprecise. Theoretically, temperatures rise exponentially across the boundary between the unburned and preheat-

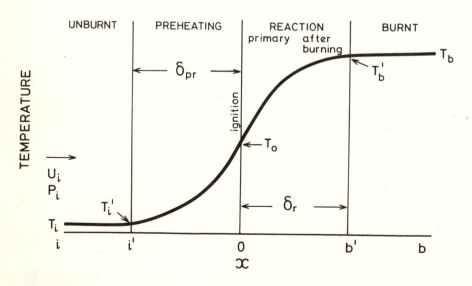

Figure 4.16 Structure of a one-dimensional premixed flame.

ing zones and approach the temperature of the burned gases asymptotically. Experimentally, the boundaries are also difficult to define, since they are dependent upon the accuracy of the temperature-measuring instrument in regions where there are slow exponential changes in temperature. The *initial preheating zone boundary temperature*, T_i', is related to the temperature of the unburned gas, T_i, far upstream of the flame, by

$$\frac{T_i' - T_i}{T_0 - T_i} = \frac{1}{100} \tag{4.26}$$

where T_0 is the temperature at the ignition point (Fig. 4.16). In the preheating zone, we consider that heat release by chemical reaction is not significant and that heating of the gases takes place only by conduction from the main reaction zone, upstream into the unburned gases. Temperature measurements made with thermocouples in low-pressure flames confirm the existence of a zone in which the gases are solely heated by heat conduction and reaction rates are negligible (Gaydon and Wolfhard, 1970).

The boundary between the preheating and reaction zones is the *ignition plane*—the plane where the temperature has risen to a value at which exothermic reactions "just begin to be significant." If the presence of a radical such as OH is taken as an indication of the presence of flame, the boundary between the preheating and reaction zones can be defined by the position where the concentration of the OH radical reaches 1 percent of the maximum value. The *ignition plane, moment of ignition*, and *ignition temperature* T_0 occur at this boundary. Ignition temperatures may vary between 700 and 1500 K with values between 1100 and 1300 K for methane-air flames. There appears to be no connection between the ignition temperature and the spontaneous-ignition temperature, as measured in experiments with long induction periods. Additives have been found to alter spontaneous-ignition temperatures without appreciably affecting the limits of flammability, the burning velocity, and the ignition temperature.

The *reaction zone* is subdivided into (1) primary reaction zone, in which the major portion of the hydrocarbons are consumed, where both the rates of reaction and temperature gradients are high, and (2) after-burning region, where the conversion of intermediates, such as CO and H_2 to CO_2 and H_2O, occurs relatively slowly and the temperature rise is small. The emission of light from the reaction zone of hydrocarbon flames is mainly from OH, CH, and C_2 radicals. The thickness of the *effective reaction zone* δ_r is defined by specifying a *final reaction zone boundary temperature* T_b' which is, in turn, defined by reference to the ignition temperature T_0 and the final temperature of the burned gases T_b in the equation

$$\frac{T_b - T_b'}{T_b - T_0} = \frac{1}{100} \tag{4.27}$$

T_b' is thus the temperature at position δ_r where the temperature reaches 99 percent of the value of the temperature increase in the reaction zone. An *equivalent* *thickness of the reaction zone* δ_r^* dependent upon the assumption of a linear temperature gradient, is defined by the equation

$$\left(\frac{\partial T}{\partial x}\right)_0 = \frac{T_b - T_0}{\delta_r^*} \tag{4.28}$$

The thickness of both the preheating and reaction zones can be calculated for the one-dimensional flame from the equations of conservation of mass and energy. The small changes in pressure across the flame are neglected. The subscripts are used as follows: *i* refers to the initial conditions in the unburned gas; *b* refers to the final values in the burned gases, and 0 refers to conditions at the ignition plane. The origin of the *x* coordinate is taken at the ignition plane.

From the equation of continuity, the total mass flux per unit area remains constant across the flame, i.e.,

$$\rho v = \rho_i v_i = \rho_u S_u \tag{4.29}$$

The term S_u has traditionally been used for the burning velocity based on conditions of the gases in the unburned state.

The energy equation for one-dimensional flow is

$$\frac{\partial}{\partial x}\left(k\frac{\partial T}{\partial x}\right) - \frac{\partial}{\partial x}(c_p T \rho v) + QR = 0 \tag{4.30}$$

where T is absolute temperature, k is the thermal conductivity, Q is the heat of reaction, R is the rate of reaction, and c_p is the specific heat. The first term in Eq. (4.30) is the heat transfer by conduction, the second term is the heat transfer by convection, and the third term is the heat release by chemical reaction. In the preheating zone, we have assumed heat release by chemical reaction to be negligible and, thus, Eq. (4.30) reduces to

$$\frac{\partial}{\partial x}\left(k\frac{\partial T}{\partial x} - c_p T \rho v\right) = 0 \tag{4.31}$$

Integrating with respect to *x* yields

$$k\frac{\partial T}{\partial x} - c_p T \rho v = k_1 \tag{4.32}$$

where k_1 is a constant of integration.

Introducing the boundary conditions

At $x = -\infty$: $\qquad\qquad \dfrac{\partial T}{\partial x} = 0$

At $x = 0$: $\qquad\qquad \dfrac{\partial T}{\partial x} = \left(\dfrac{\partial T}{\partial x}\right)_0$

we have

$$k_1 = -c_{pi} T_i \rho_i v_i \qquad (4.33)$$

and

$$k_0 \left(\frac{\partial T}{\partial x} \right)_0 = c_{p0} T_0 \rho_0 v_0 - c_{pi} T_i \rho_i v_i \qquad (4.34)$$

Introducing the continuity equation (4.29) and the integration constant (4.33) into Eq. (4.32) yields

$$\frac{\partial T}{\partial x} = \rho_i v_i \left(\frac{c_p T - c_{pi} T_i}{k} \right) \qquad (4.35)$$

Introducing the boundary condition at $x = 0$, we have

$$\left(\frac{\partial T}{\partial x} \right)_0 = \frac{\rho_i v_i}{k_0} (c_{p0} T_0 - c_{pi} T_i) \qquad (4.36)$$

Both the thermal conductivity k and the specific heat c_p vary with temperature. This variation is given by the following empirical equations

$$c_p = A + BT \qquad (4.37)$$

and

$$k = a + bT \qquad (4.38)$$

Introduction of Eqs. (4.37) and (4.38) into Eq. (4.35), followed by integration and use of the boundary conditions, yields the thickness of the reaction zone. A rough approximation of δ_{pr} can be obtained by taking average values of the specific heat $\bar{c}_p$ and thermal conductivity k, and assuming that these values are constant in the preheating zone. Equation (4.35) then yields, after integration,

$$T - T_i = (T_0 - T_i) \exp \left(\frac{\bar{c}_p \rho_i v_i x}{k} \right) \qquad (4.39)$$

Introducing the boundary condition, $x = -\delta_{pr}$, $T = T_i'$, and Eq. (4.26), results in

$$\frac{1}{100} = \exp \left(\frac{\bar{c}_p \rho_i v_i}{k} [-\delta_{pr}] \right)$$

or

$$\delta_{pr} = \frac{4.6 \, \bar{k}}{\bar{c}_p \rho_i S_u} \qquad (4.40)$$

since $v_i = S_u$. Equation (4.40) shows that the thickness of the preheating zone is inversely proportional to the burning velocity.

We calculate the equivalent thickness of the reaction zone by equating (4.28) for the reaction zone with Eq. (4.36) for the preheating zone, assuming a continuous change across the ignition plane, yielding

$$\left(\frac{\partial T}{\partial x}\right)_0 = \frac{\rho_i v_i}{k_0}(c_{po}T_0 - c_{pi}T_i) = \frac{T_b - T_0}{\delta_r^*}$$

or

$$\delta_r^* = \frac{T_b - T_0}{c_{po}T_0 - c_{pi}T_i}\frac{k_0}{\rho_i S_u} \tag{4.41}$$

The thickness of the equivalent reaction zone is also inversely proportional to the burning velocity.

In order to calculate the thickness of the effective reaction zone δ_r, account needs to be taken of the heat release due to chemical reaction, and the calculation becomes more complex. The thickness of the luminous zone can be taken as an approximate value of the thickness of the reaction zone. For hydrocarbon flames at atmospheric pressure, $\delta_{pr} = 0.7$ mm and $\delta_r = 0.2$ mm.

Fristrom and Westenberg (1965) have made detailed measurements in a flat flame and shown that, for this case, the flame structure approximates the ideal one-dimensional flame. By introducing a stabilizer grid into the incoming stream,

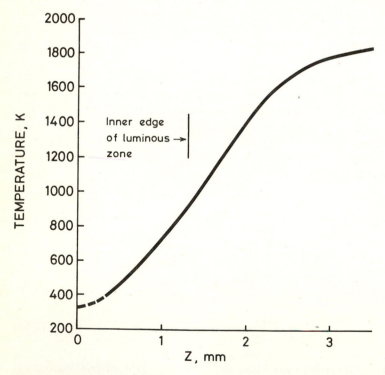

Figure 4.17 Temperature profile in one-dimensional premixed laminar flat flame. (*Fristrom and Westenberg, 1965.*)

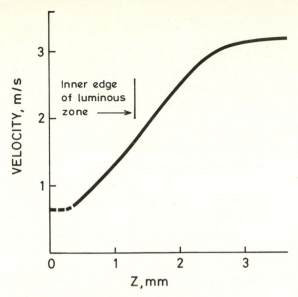

Figure 4.18 Velocity profile in one-dimensional premixed laminar flat flame. (*Fristrom and Westenberg, 1965.*)

the gas stream velocity is made lower than the free-space flame velocity so that the flame tends to burn back toward the screen. As it approaches the screen, heat is lost by conduction and the flame speed is lowered until a balance is obtained between the gas and flame speeds. The measurements made by Fristrom and Westenberg (1965) in a flat methane-oxygen flame on a 32-mm-diameter screened burner are shown in Figs. 4.17 to 4.19. The flame conditions were: CH_4, 7.8 percent;

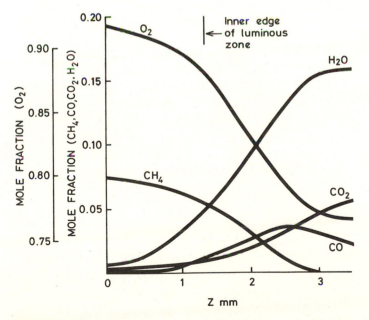

Figure 4.19 Concentration profile in one-dimensional premixed laminar flat flame. (*Fristrom and Westenberg, 1965.*)

O_2, 91.5 percent; $T = 350$ K and $P = 10^4$ N/m². Temperatures were measured both by thermocouples, with corrections for radiation, and by a pneumatic probe. Velocity was measured by the particle-track method and was found to be in agreement with values calculated from the continuity equation using the measured temperature and species concentration profiles. The concentration of stable species was measured by sampling and mass spectrometer analysis. CO, H_2, and CH_2O exist in the flame because they are intermediates in methane combustion. Carbon monoxide and H_2 are stable intermediates and, depending upon equivalence ratio, may become final products of combustion.

The principal effect of ambient pressure on flame structure is to change the thickness, at the various zones. For many flames, the distance scale changes approximately proportionally to the molecular mean free path, i.e., flame thickness is proportional to inverse pressure. Flame velocity is approximately proportional to the square root of the overall reaction rate. The lower limit depends on burner size, the onset of instability, heat losses, or some fundamental property of the flame equations. In practice, convective instability is most important at the lower limit and the lowest observed flame velocities are in the range of 10 mm/s. The upper limit is set by the velocity of sound which marks the transition to supersonic flow and the onset of a detonation wave. In detonations, an appreciable fraction of the total energy of the flame is carried by the kinetic energy of the flow. True flames have been observed with flame velocities as high as 0.6 that of sound. For low-velocity flames, 1 m/s, the pressure drop across the flame is less than 1 percent of the initial pressure and they are usually treated as constant-pressure systems. The pressure drop across the flame front is proportional to the square of the velocity change.

Energy flux due to thermal conduction is the product of thermal conductivity and temperature gradient, and is always directed upstream with respect to the convection flow. Since thermal conductivity is independent of pressure and the gradient of temperature is approximately an inverse function of pressure, the energy flux due to thermal conductivity (at constant flame velocity) varies inversely with pressure. The effects of change of flame velocity on temperature gradient, conductive energy flux, and pressure drop have been shown to be of the form

$$S_{u_{(const\ press)}} \propto \frac{1}{\text{flame thickness}} \frac{1}{T} \frac{dT}{dz} \begin{pmatrix} \text{conductive} \\ \text{energy flux} \end{pmatrix} \begin{pmatrix} \text{pressure} \\ \text{drop} \end{pmatrix}^2$$

while the effects of changing pressure, for flame velocity equal to 1 m/s, on the same flame parameters are given by

$$P \propto \frac{1}{\text{flame thickness}} \frac{1}{T} \frac{dT}{dz} \begin{pmatrix} \text{conductive} \\ \text{energy flux} \end{pmatrix} \begin{pmatrix} \text{pressure} \\ \text{drop} \end{pmatrix}$$

Rates of diffusion are different for each species in the flame. The contribution of diffusion to the total mass flux of each species is the product of its concentration gradient and diffusion coefficient at any point in the flame. The gradient varies from point to point and the diffusion coefficient of a given species depends upon temperature, pressure, and the concentration of various species in the gas mixture. Reactant species diffuse with the stream, product species diffuse against the stream, and intermediate species diffuse in either direction depending on the position relative to their concentration maximum.

4.5 LAMINAR DIFFUSION FLAMES

In a diffusion flame (Fig. 4.20) no mixing of the fuel and oxidant takes place prior to emission from the burner. In the simplest form of diffusion flame, a gaseous fuel is passed through an orifice into stagnant air surroundings. Diffusion takes place across the interface between the fuel jet and the surrounding air. A mixing layer starts at the burner rim and spreads both outward into the surrounding air and inward into the fuel jet. The concentration of oxidant is uniform outside the mixing layer and the concentration of fuel is uniform at the fuel jet nozzle exit. Concentration profiles of fuel and oxidant across the mixing layer develop as a consequence of diffusion in the radial direction. From these concentration profiles, lines can be drawn of the rich and lean flammability limits, and also the stoichiometric mixture line. The stoichiometric line will converge toward the axis and the flammability limit lines will diverge outward from the stoichiometric line as

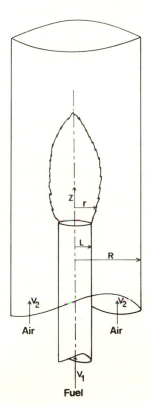

Figure 4.20 Physical model of a gaseous laminar jet diffusion flame in concentric coflowing annular airstream. (*Mitchell, 1975.*)

we proceed downstream from the burner exit. The mixing layer will tend to spread linearly with distance downstream. Preheating zones occur on both the fuel-rich and fuel-lean sides of the reaction zone. When the fuel jet is surrounded by a stream of air, the flame front will converge onto the axis if there is an excess of air, and the flame will diverge if the air quantity in the surrounding stream is below the stoichiometric requirements. The lines of maximum temperature and maximum luminosity will (in general) coincide with the stoichiometric mixture line.

Laminar diffusion flames are obtained by using small-diameter nozzles and low velocities, so that the Reynolds number at the nozzle exit can be maintained well below the transition level from laminar to turbulent flow. In order to prevent disturbance due to draughts, laminar diffusion flames are most effectively studied by surrounding the gas jet with a low-velocity airstream. The system is most stable when the mean velocity of the fuel and airstream are the same, so that velocity gradients only arise as a result of buoyancy forces. Diffusion thus occurs only under the driving force of the concentration gradients. This was the system initially investigated in the classical work of Burke and Schumann (1928).

Diffusion flames are flames in which the mixing rate is sufficiently slow compared to the reaction rate that the mixing time controls the burning rate. In a perfectly premixed system, sufficient energy and time have been expended to create a mixture so that there are no inhomogeneities at any level down to the molecular scale. An imperfectly premixed system is one in which the reactants have been brought into contact with each other but insufficient time or energy has been expended for perfect mixing. In a partially premixed system, insufficient fuel or oxidant is supplied for stoichiometric reaction—the mixture is introduced through the burner with an excess of fuel or oxidant. The complete burning of a fuel-rich mixture requires mixing by diffusion. The term "diffusion flame" is reserved for those flames in which there is a total separation between fuel and oxidant so that all burning must be preceded by mixing. A continuous spectrum of flames exists between the perfectly premixed flame and the diffusion flame.

Consider a stream of air, separated by a wall, flowing in parallel alongside a stream of fuel. When the fuel and oxidizer come into contact at the end of the thin separating wall, a mixing region of thickness δ is formed. The length l is defined as the distance which molecules of fuel and oxidant travel before the reaction rate exceeds an ignition threshold level. If the reactants have a constant velocity v, and the diffusion coefficient between the fuel and oxidant is D, the ratio between diffusion time and residence time is given by

$$\frac{t_D}{t_R} \propto \frac{\delta^2 v}{Dl} \tag{4.42}$$

When the velocity of the fuel and oxidant molecule varies along the diffusion trajectory, the residence time of the particle is determined by integration of the velocity from the origin of the mixing region to the ignition plane. In a perfectly premixed flame, t_D/t_R is zero; in a diffusion flame, the rate of diffusion is so slow compared with the reaction (residence) time that the ratio is much greater than one. The candle flame is an example of a laminar flame in which the mixing process occurs slowly by molecular diffusion and the flame structure is governed by molecular heat and mass transfer. In low-velocity gas streams, as found in domestic cookers and some industrial gas heating appliances, velocities are sufficiently low for the flow to be laminar and diffusion to be molecular.

Flame Height

In laminar diffusion flames, flame height is governed by fuel and oxidizer flow rates and burner dimensions. If the flow is laminar throughout the flame, a thin reaction zone is formed around the stoichiometric mixture line and the flame thickness is governed by diffusion, convection, and reaction processes. Under laminar flow conditions, buoyancy, thermal convection, and density gradients generate flow with velocities which may be of the same order of magnitude as that of the main flow. At very low fuel rates, meniscus-shaped flames can be formed and, as the fuel-flow rate is carefully increased, these flames lift off the burner but remain attached at one point. At higher fuel-flow rates, the flame is completely lifted from the burner and vortex-shaped flames are observed. When diffusion flames are lifted above the burner, this phenomenon is due to the time required for the fuel emerging from the burner to entrain, mix, and become heated sufficiently to ignite.

Flame height increases linearly with nozzle velocity when the flame is completely laminar. For a fixed air velocity, increase in the fuel/oxygen ratio causes an increase in flame height. For very small flames, flame height is independent of pressure. For larger flames, the flame thickness is decreased when the pressure is increased. This leads to an increase in the slope of the concentration gradients, which leads to a stage when chemical reaction rates become important. Under such conditions, flame height has been found to increase with pressure. Changes in inlet temperature have been found to cause only small changes in flame appearance but reactions such as those involving oxides of nitrogen are well known to increase formation rates and, hence, emission as a direct result of increase in inlet temperature.

Hydrocarbon Diffusion Flames

A hydrocarbon diffusion flame can be subdivided into two readily identifiable distinct zones—the main reaction zone, which is generally blue, and the luminous zone, which is yellow. The major portion of the gaseous reaction takes place in the main reaction zone, while the luminous zone indicates the presence of carbon particles. Carbon particles are formed whenever the carbon-atom-to-oxygen-atom ratio is greater than unity, but are also frequently formed at lower C/O ratios. In a diffusion flame, the mixture ratio is very rich on the fuel side of the flame, leading to the formation of carbon as soon as the temperature becomes sufficiently high to decompose the fuel. These carbon particles penetrate into the region containing the combustion products, H_2O and CO_2, and the intermediates, CO and H_2. Provided the temperature is above 1000 K, these carbon particles will react with water vapor and carbon dioxide to form CO. The carbon particles will thus react with combustion products or oxygen if the temperature is sufficiently high and the residence time of the carbon particles in the high-temperature zone is sufficient. These carbon particles can easily be quenched by contact with low-temperature oxidant and, thus, leave the flame as solid soot particles. Increasing the fuel-flow rate in a partially premixed or diffusion flame leads to the formation of soot and smoke.

Spectroscopic analysis of laminar diffusion flames shows that absorption increases as the flame surface is approached. Mass spectrometric studies of methane diffusion flames show that methane is decomposed by pyrolysis. The following reaction mechanisms were suggested by Mitchell (1975) to be representative of the major reaction paths which could lead to formation of some of the observed species:

$$2CH_4 \rightarrow C_2H_6 + H_2 \rightarrow C_2H_4 + H_2 \rightarrow C_2H_2 + H_2 \rightarrow C + H_2$$

The mechanisms are as follows:

Initiation

$$CH_4 \rightarrow CH_3 + H$$
$$CH_4 \rightarrow CH_2 + H_2$$

Second step-bimolecular reactions

$$CH_2 + CH_4 \rightarrow C_2H_6$$
$$CH_3 + CH_3 \rightarrow C_2H_6$$

Third step—dissociation of ethane to ethylene

$$C_2H_6 \rightarrow C_2H_4 + H_2$$

Fourth step—ethylene reaction with methylene

$$CH_2 + C_2H_4 \rightarrow C_3H_6$$

Pyrolysis of ethylene

$$C_2H_4 \rightarrow C_2H_2 + H_2$$

Final decomposition step—acetylene decomposes to carbon and hydrogen

$$C_2H_2 \rightarrow 2C + H_2$$

All the above pyrolysis products have been observed in the luminous zone of methane-air diffusion flames. The pyrolysis products account for about 1 mole percent of the carbonaceous materials found in the inner mantle of methane-air diffusion flames. Oxygenated hydrocarbons have concentrations of the order of 1 ppm, and these are completely consumed as they diffuse to the flame surface.

The Primary Reaction Zone

The reactions occurring in the primary reaction zone of methane-air diffusion flames are essentially the same as those occurring in premixed methane-air flames. Experimental studies of Dixon-Lewis and Williams (1967), Fristrom (1963), and Mitchell (1975) on premixed methane-air and methane-oxygen indicate that a sufficient methane oxidation mechanism consists of the following steps:

1. Reaction of methane with radicals H, O, and OH to form the methyl radical

$$CH_4 + H \rightarrow CH_3 + H_2$$

$$CH_4 + O \rightarrow CH_3 + OH$$

$$CH_4 + OH \rightarrow CH_3 + H_2O$$

2. Oxidation of the methyl radical to formaldehyde

$$CH_3 + O_2 \rightarrow CH_2O + OH$$

$$CH_3 + O \rightarrow CH_2O + H$$

$$CH_3 + OH \rightarrow CH_2O + H_2$$

3. Reaction of formaldehyde to give the formyl radical

$$CH_2O + M \rightarrow HCO + H + M$$

$$CH_2O + OH \rightarrow HCO + H_2O$$

$$CH_2O + O \rightarrow HCO + OH$$

$$CH_2O + H \rightarrow HCO + H_2$$

4. Reaction of the formyl radical to form CO, which finally oxidizes to CO_2

$$HCO + OH \rightarrow CO + H_2O$$

$$HCO + H \rightarrow CO + H_2$$

$$HCO + M \rightarrow CO + H + M$$

$$HCO + O \rightarrow CO + OH$$

$$CO + OH \rightarrow CO_2 + H$$

where M is any flame species.

The following chain branching reactions, and linear combinations of these, are necessary to propagate the reaction sequence:

$$H + O_2 \rightarrow OH + O$$

$$O + H_2 \rightarrow OH + H$$

$$H + H_2O \rightarrow H_2 + OH$$

The following recombination reactions serve to terminate the sequence in the postflame regions:

$$O + O + M \rightarrow O_2 + M$$

$$H + H + M \rightarrow H_2 + M$$

$$H + OH + M \rightarrow H_2O + M$$

$$H + O + M \rightarrow OH + M$$

The above reaction sequence establishes the thickness of the primary reaction zone. This reaction zone completely encloses the luminous region of the over-ventilated flames. The thickness of the reaction zone is about 2 mm at the maximum flame diameter and decreases as the flame tip is approached. The blue radiation from this region is partially O_2, Schuman-Rung, and partially $CO + O$ radiation.

Models of Laminar Diffusion Flames

In the examination of premixed flames, we have been able to use a one-dimensional model. All diffusion flames require to be considered as two or three dimensional. For systems which are axisymmetric, without swirl, a cylindrical coordinate system can be adopted, confined to the axial z and radial r coordinates. In the formulation of models simplifying assumptions have been made concerning the "significant" chemical reaction equations, the variation of physical properties such as conductivity and specific heat, and the fluid dynamics of the system. These simplifying assumptions were made in order to allow the equations to be solved. The development of experimental diagnostic techniques has allowed accurate determination of flame structure; the significant chemical reactions and the expressions for the rate constants have been established and it is now possible to solve complex simultaneous differential equations by numerical methods and large computers. A brief résumé will be given of some of the earlier models, which fall into one of the following two categories: (1) infinitely thin flame surfaces; (2) broad reaction zones, the thickness of which is determined by the extent to which chemical reactions allow fuel and oxygen to coexist.

The physical model of a gaseous laminar jet diffusion flame in a concentric coflowing annular airstream is shown in Fig. 4.20. The gaseous fuel emerging from the central circular nozzle forms a jet, spreading radially outward as a result of molecular diffusion with the surrounding airstream. The rate of spreading of the jet is also influenced by the difference in the initial velocity of the fuel stream V_1 and the annular airstream V_2. After ignition, the mixture burns as a steady laminar flame. The shape of the flame is similar to that of a candle flame, initially divergent and, subsequently, converging to the tip of the axis, where all the fuel has been consumed.

The theoretical model is formulated on the basis of the conservation equations for mass, momentum, energy, and species. In the various models that have been formulated, there has been general agreement in the selection of the basic equations. The models differ in the assumptions and in the method of solution of the equations. It has been necessary to introduce physical models and empirical information on physical constants in order to predict the flame shape and length.

The conservation-of-species equation for the steady-state axisymmetric conditions is given by

$$v_r \frac{\partial C_i}{\partial r} + v_z \frac{\partial C_i}{\partial z} = \frac{1}{r} \frac{\partial}{\partial r}\left(D_i r \frac{\partial C_i}{\partial r}\right) + \frac{\partial}{\partial z}\left(D_i \frac{\partial C_i}{\partial z}\right) + \dot{R}_i \tag{4.43}$$

where C_i is the concentration of chemical species i, v_r and v_z are the radial and axial velocity components respectively, r and z are the radial and axial directions respectively, D_i is the molecular diffusion coefficient, and $\dot{R}_i$ is the rate of formation or consumption of chemical species i.

The total number of species taking part in the chemical reactions of hydrocarbon-oxygen flames has not yet been fully determined. Information on rates of formation and consumption, as well as the values of diffusion coefficients of individual species over the range of temperatures found in flames, is only known with a limited degree of accuracy. Further, the extent to which the heat release, convection, radiation, and diffusion of chemical species influences velocity flow fields affects the values of velocity in Eq. (4.43).

Infinitesimally Thin Flame Surface Model

In the infinitesimally-thin-flame-surface model, all chemical reaction is confined to a surface which separates the fuel and the oxidant. Fuel diffuses to the flame from one side, oxidant diffuses to the flame from the other side, and concentrations of both fuel and oxidant decrease to zero as a result of being consumed. Products are formed instantaneously as the result of chemical reaction at stoichiometric ratios. These products diffuse through the fuel and oxidant with concentrations decreasing asymptotically to zero. The flame sheet acts as a source of reaction products and a sink for reactants. These models are sometimes referred to as global, one-step reaction models.

Burke and Schumann (1928) assumed that the fuel and oxygen have the same diffusivity and derived the following differential equation for predicting flame shape and height

$$\mathbf{v} \cdot \nabla C_{fu} = \nabla \cdot (D \nabla C_{fu}) \tag{4.44}$$

where C_{fu} is fuel concentration. In the one-step global reaction model, oxygen is considered to combine with fuel in a fixed ratio to form a neutral product so that, for purposes of mathematical analysis, oxygen is regarded as a "negative fuel." A single differential equation is sufficient to describe the distributions of both fuel and oxygen. The flame front is the surface where the concentration of the fuel is zero and the flame height is the distance from the nozzle at which the fuel concentration is zero on the axis.

In order to simplify the solution of the equations, Burke and Schumann (1928) made the following assumptions: (1) radial convection and axial diffusion are negligible; (2) the velocity of the fuel and air are the same and remain constant in the region of the flame; and (3) the coefficient of interdiffusion of the two gas streams is constant.

For a flame whose length-to-diameter ratio is large, the first assumption is valid but, for short flames, the rate of axial diffusion becomes comparable to that of radial diffusion. The second assumption cannot be valid, since heat release and temperature change cause significant changes in velocity. The large temperature variations also lead to large variations in the molecular diffusivity, so that the second and third assumptions are known not to be valid. It has been suggested, however, that the effects of temperature variation may counterbalance so that the variation in the ratio of diffusion coefficient to velocity may be small. On the basis of the above assumptions, Eq. (4.44) can be written as

$$\frac{\partial C_{fu}}{\partial z} = \frac{D}{V_1}\left(\frac{\partial^2 C_{fu}}{\partial r^2} + \frac{1}{r}\frac{\partial C_{fu}}{\partial r}\right) \tag{4.45}$$

For the case of methane reacting with oxygen, we have

$$CH_4 + 2O_2 \rightarrow 2H_2O + CO_2$$

The rates of consumption of fuel and oxygen are then related by

$$\dot{R}_{CH_4} = \tfrac{1}{2}\dot{R}_{O_2} \tag{4.46}$$

also

$$C_{fu} = C_{CH_4} - C_{O_2}/2 \tag{4.47}$$

The initial conditions are

$$C_{fu} = C^o_{CH_4} \qquad \text{from } r = 0 \text{ to } r = L \text{ at } z = 0$$

$$C_{fu} = -C^o_{O_2}/2 \qquad \text{from } r = L \text{ to } r = R \text{ and } z = 0$$

and the boundary conditions are

$$\frac{\partial C_{fu}}{\partial r} = 0 \qquad \text{when } r = 0 \text{ and } r = R$$

A solution to Eq. (4.45) which satisfies the initial and boundary conditions is

$$C_{fu} = \frac{L^2}{R^2}C_O - \frac{C^o_{O_2}}{2} + \frac{2LC_O}{R^2}\sum_\lambda\left[\frac{1}{\lambda}\frac{J_1(\lambda L)J_0(\lambda r)\exp\dfrac{-D\lambda^2 z}{V_1}}{[J_0(\lambda R)]^2}\right] \tag{4.48}$$

where $C_O = C^o_{CH_4} + C^o_{O_2}/2$ and the superscript o denotes conditions at the inlet and λ assumes the value of all the positive roots of the equation $J_1(\lambda R) = 0$ where J_0 and J_1 are zero- and first-order Bessel functions. The shape of the flame is determined by finding those values of r and z which satisfy Eq. (4.48) when $C_{fu} = 0$. The height of the flame is given by the value of z when $r = 0$.

When Burke and Schumann (1928) carried out their analysis and experiments, they were restricted by the lack of knowledge of mathematical solutions to the basic differential equations, the small amount of information available on diffusivities of gases at high temperatures, and the unsophisticated instrumentation available for making measurements in flames. In their determination to obtain good agreement between theory and experiment, they made a number of assumptions which are now known not to be valid. More sophisticated analytical and numerical methods are available today for the solution of the basic differential equations so that it is not necessary to make the initial assumptions of Burke and Schumann (1928). Experimental data on variation of diffusion coefficients as a function of concentration and temperature allow a set of empirical equations for diffusion coefficients to be introduced into the basic equations. Burke and Schumann (1928), in their initial comparisons between theory and experiment, found that incorporating the values of the diffusivity of gases at ambient conditions gave inaccurate results. By assuming a constant value for the diffusion coefficient of 49.2 mm²/s for methane-air and 67.1 mm²/s for CO-air flames, they obtained good agreement between theory and experiment. It must be recognized that there was no scientific justification for the selection of these constant values of diffusion coefficients, since the numbers were chosen in order to force agreement between theory and experiment.

Savage (1962) obtained a solution of the Burke-Schumann equation (4.48) in dimensionless form, from which concentration profiles for a variety of methane-air flames can be conveniently determined. Barr (1949) extended the work of Burke and Schumann by allowing the fuel and air to have constant but different velocities. By postulating that the air and fuel maintain their respective velocities, he was assuming no momentum transfer between fuel and airstreams. Fundamentally, it is untenable that mass transfer occurs by diffusion, while no momentum transfer takes place between the fuel and air streams. Barr also selected a diffusion coefficient that allowed agreement between prediction and experiment. For a variation of inlet air velocities between 44.5 and 130.5 mm/s and fuel velocities between 8.5 and 26.1 mm/s, Barr found that the empirically adjusted diffusion coefficient only varied between 18 and 26 mm²/s.

Hottel and Hawthorne (1949) and Wohl, Gazley, and Kapp (1949) extended the work of Burke and Schumann by considering free, unconfined diffusion flames. In adjusting their theoretical predictions with their measurements, they used two empirically determined constants. They concluded that flame height varied approximately as the square root of fuel-flow rate, and attributed this nonlinear relation to variations in the diffusion coefficient. They concluded that diffusivity governed the time required to complete combustion of the fuel and considered that the linear relationship determined by Burke and Schumann was a special case, valid for short, unconfined flames.

Powell and Browne (1957) questioned the interpretation that the nonlinear relation for length of open (free) flames was due to variation of the diffusion coefficient along the flame. They used mixing similarity in their analysis and concluded that the assumption of a mean diffusivity was justified for both closed and open diffusion flames. For the cases where flame shape and height did not agree with the Burke and Schumann analytical predictions, Powell and Browne ascribed these differences as being primarily due to changes in flow pattern, due to density differences in the fuel and oxidant streams.

Fay (1954), in his analysis of unconfined laminar jet diffusion flames, made three basic assumptions: (1) natural convection is negligible; (2) diffusion coefficients of all species are the same; (3) Prandtl and Schmidt numbers are both unity. He assumed the following expressions for concentration and temperature as linear functions of velocity

$$c_i = a_i + b_i u \tag{4.49}$$

and

$$c_p T = e + fu \tag{4.50}$$

where c_i is the concentration of species i; a, b, e, and f are arbitrary constants determined by the conditions; c_p is the heat capacity, assumed constant; T the temperature; and u the velocity in the axial direction. These simple expressions relating concentration and temperature to velocity are convenient to use in analyses, but there is no scientific foundation for assuming such simple linear relationships. Fay arrived at the following equation for flame length

$$L = \frac{3}{8\pi} \frac{1 + \varepsilon\alpha(1 + \beta)}{\varepsilon\alpha(1 + \beta)} \frac{M}{\mu_r} \frac{\rho_\infty}{\rho_r} \left[I\left(\frac{\rho_\infty}{\rho_f}\right) \right]^{-1} \tag{4.51}$$

where ε = mass fraction of oxygen in the atmosphere; α = mass ratio of fuel to oxygen in a stoichiometric mixture; β = mass ratio of diluent to fuel in the jet; M = total jet mass flow rate; μ = molecular viscosity; ρ = density. The subscripts r, f, and ∞ represent conditions at the reference point, the flame surface, and ambient conditions, respectively. The function $I(\rho_\infty / \rho_f)$ is shown graphically in the paper by Fay. According to the Fay equation, (4.51), flame length increases linearly with exit velocity and with the square of the nozzle diameter.

Many of the assumptions which have been made in the past have been shown to be not only invalid, but the changes of factors such as diffusion coefficients, which were assumed to be constant, have dominant effects in changing flame shape. Temperature changes from 300 to 2000 K result in one order of magnitude change in diffusion coefficient and viscosity. Density differences are sufficiently large that buoyancy forces are comparable with, or even larger than, rates of change of momentum found in low-velocity laminar diffusion flames. More sophisticated and meticulous experimental results show the variations of flow temperature and species concentrations throughout the flame system in much greater detail than earlier experimental studies. The expressions derived by the authors cited above can still be used for engineering purposes.

Broad Reaction Zone Models

In the broad reaction zone models, flames are no longer considered to be infinitely thin. The distributions of individual species concentrations, temperature, and velocity are being determined experimentally with an increasing degree of accuracy and in an increasingly large variety of diffusion flames. The broad reaction zone models account for gradients within the zone and momentum across the flame. Sternling and Wendt (1974) developed a model in which the thickness of the reaction zone was determined by the diffusion flux into the flame front. The zone was treated as a well stirred region. By this means, they could utilize kinetic reaction mechanisms which had previously been determined for premixed systems. The flame was divided into five separate zones. On one side of the central reaction zone is a region of bulk fuel with uniform gas composition and, on the other side, there is bulk oxygen with a uniform composition. The concentration of fuel decreases across zone B. Within the reaction zone, R, both oxygen and fuel are present, and their concentrations only reach zero at the opposite outer edges of the reaction zone. In this model, therefore, we have regions of concentration change, due to diffusion in zones A and B and reaction taking place in zone R, where the profiles of fuel and oxygen overlap. The earlier models considered reactions to be very fast, but not infinitely fast, and Sternling and Wendt determined a reaction volume for the combustion of H_2 and HCN mixtures. Their predicted NO concentrations were not in exact quantitative agreement with experimental data but they were able to determine the correct functional dependence of combustion intensity and bound nitrogen conversion on NO emissions. There is, in fact, no need to assume that reactions are fast and it is possible to take into account more detailed variations of concentration distributions according to the true rates of reaction for both fast and slow reactions.

Clark (1967, 1968, 1969, and 1974), in a series of studies, developed models of flat diffusion flames based on singular perturbation methods. Perturbation velocities, enthalpy, concentrations, and temperature were superimposed on the free stream values of these quantities. The convective terms in the conservation equations were linearized according to Oseen and the derived equations were split on the basis of velocity components, which had irrotational and solenoidal parts. Irrotational parts, or waves, were defined to account for disturbances which were dominantly due to pressure and temperature changes. Solenoidal parts were defined to account for the viscous effects of shear. He showed the development of a series of waves for pressure, temperature, viscosity, and composition. By using similarity in these four wave equations, solution could be obtained of a general form. Clark applied this procedure to irreversible and frozen-flow problems, where the time needed for a reaction-inducing collision to occur was so much longer than the time needed for an element of reactant to flow through the flame, that virtually no creation or destruction of a species occurred, and the chemical terms in the composition wave vanished.

Clark considers reaction-broadening in which a finite reaction time is allowed for an irreversible reaction. Under conditions of equilibrium-broadening, a finite equilibrium constant is allowed for an infinitely fast reaction. The method of matched asymptotic expansions is used to solve the system of equations. The method involves the construction of asymptotic series for the dependent variables which are valid for certain overlapping regions in the independent variable field. The outer solution is analogous to the Burke-Schumann flame-sheet model of a fast, irreversible reaction, and the inner solutions provide information on flame-zone structure. Predictions of the structure of methane-air diffusion flames were found to be consistent with experiments for fuel-lean systems. The ambiguities which were found to exist in the fuel-rich structure could be traced to the choice of reaction rate constants.

The Mitchell-Sarofim Model

Mitchell (1975), together with Sarofim, developed a theoretical model for laminar diffusion flames. This model eliminates the restrictions of the classical Burke-Schumann model in that allowances are made for natural convection effects and variable thermodynamic and transport properties. The predicted fields of temperature, velocity, and species concentration were subsequently compared with a carefully controlled experimental study of a laminar methane-air diffusion flame. The very good agreement found between prediction and experiment demonstrates that some of the original assumptions of Burke and Schumann are not valid. The Mitchell-Sarofim model demonstrates the high degree of correlation that can be obtained between prediction and measurement in laminar diffusion flames.

The theoretical model relates the mixing of fuel, air, and combustion products with flame characteristics. The fundamental conservation equations for steady-state, axisymmetric conditions are as follows:

Continuity

$$\nabla \cdot \bar{G} = 0 \tag{4.52}$$

Momentum

Radial direction $$\qquad \nabla \cdot (\bar{G}v_r + \bar{\tau}_r) + \frac{\partial P}{\partial r} = 0 \tag{4.53}$$

Axial direction $$\qquad \nabla \cdot (\bar{G}v_z + \bar{\tau}_z) + \frac{\partial P}{\partial z} - \rho g = 0 \tag{4.54}$$

Energy

$$C_p(\nabla \cdot \bar{G}T) - \nabla \cdot (k\nabla T) - \sum_j \rho D_j C_p^j(\nabla T \cdot \nabla W_j) - Q_{\text{gen}} = 0 \tag{4.55}$$

Species

$$\nabla \cdot (\bar{G}w_j - \rho D_j \nabla w_j) - \dot{R}_j = 0 \qquad j = 1, N \tag{4.56}$$

where $\bar{G}$ = mass flux vector; v_r and v_z = velocity components in radial and axial directions respectively; P = total pressure; ρ = mass density; T = temperature; k = gas thermal conductivity; Q_{gen} = rate of energy generation by reaction; w_j = mass fraction of species j; D_j = diffusion coefficient of species j; $\dot{R}_j$ = rate of creation of species j; and τ = shear stress.

The solution to the preceding set of nonlinear partial differential equations leads to the determination of composition, velocity, temperature, and pressure throughout the system. Allowance is made for variable thermodynamic and transport properties, for the generation of energy due to chemical reaction, and for the rate of formation and destruction of species due to chemical reaction. The equations are solved subject to the following boundary conditions: (1) specified inlet conditions; (2) symmetry about the centerline; (3) wall impervious to matter (and heat if adiabatic); (4) constant wall temperature (if not adiabatic); (5) no slip at the wall; and (6) zero axial gradients at the exit.

Vorticity is defined as

$$\omega = \frac{\partial v_r}{\partial z} - \frac{\partial v_z}{\partial r} \tag{4.57}$$

The stream function is defined as

$$rG_r = -\frac{\partial \psi}{\partial z} \qquad rG_z = \frac{\partial \psi}{\partial r} \tag{4.58}$$

Introduction of the vorticity and stream function into the momentum equations allows elimination of pressure as a dependent variable and replacement of the radial and axial components of the mass flux vector by a single stream function equation. Integration over finite areas is used to obtain the finite difference representation of the above system of equations. The resulting nonlinear algebraic equations were solved by computer using an alternating-direction implicit iterative scheme.

For chemical reaction, the Burke-Schumann flame-sheet concept was adopted, in which chemical reaction is assumed to be infinitely fast within a flame sheet and reaction is considered to be frozen outside this reaction zone. Heat is only released within the flame sheet, and the magnitude of the heat generation term depends upon the fluxes of fuel and oxygen toward the sheet. With this assumption, a single species conservation equation can be used to locate the surfaces of the sheet. A single global reaction is used in place of the array of possible reactions that can occur at either surface. The global reaction is of the form

$$[1 \text{ g}] \text{ fuel} + [s \text{ g}] \, O_2 \rightarrow [(1+s)\text{g}] \text{ products} \tag{4.59}$$

Reaction rates are related by

$$\dot{R}_{\text{fuel}} = \frac{1}{s} \dot{R}_{O_2} = -\frac{1}{1+s} \dot{R}_{\text{products}} \tag{4.60}$$

If it is assumed that the diffusion coefficients of fuel and oxygen are the same, the oxygen conservation equation can be divided by s and subtracted from the fuel conservation equation to give

$$\nabla \cdot [\bar{G}(w_{\text{fuel}} - w_{O_2}/s) - \rho D_{\text{fuel, mix}} \nabla(w_{\text{fuel}} - w_{O_2})] - [\dot{R}_{\text{fuel}} - \dot{R}_{O_2}/s] = 0 \tag{4.61}$$

Defining a new variable

$$w^* = w_{\text{fuel}} - w_{O_2}/s \tag{4.62}$$

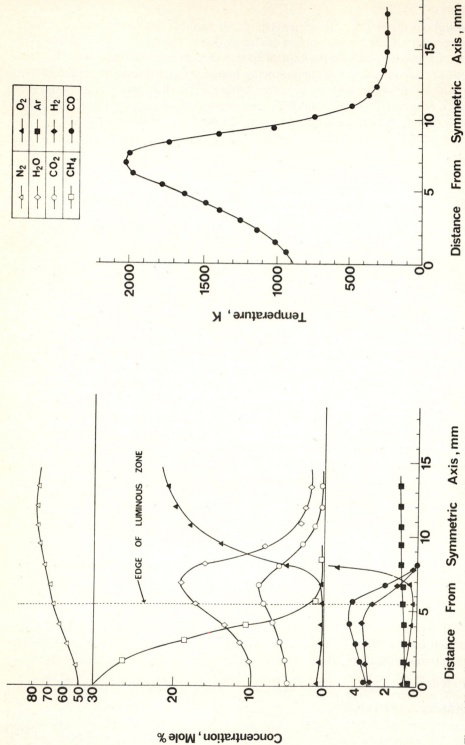

Figure 4.21 Radial concentration and temperature profiles in laminar methane-air diffusion flame at $z/d = 1.89$. (*Mitchell, 1975.*)

and noting that the last term in Eq. (4.61) is zero, Eq. (4.61) can be rewritten as

$$\nabla(\bar{G}w^* - \rho D_{fuel, \, mix} \nabla w^*) = 0 \qquad (4.63)$$

The flame-sheet assumption does not allow fuel and oxygen to coexist.

Positive values of w^* represent fuel concentration and negative values represent oxygen concentration divided by $-s$. Depending upon the value of s, the locus of $w^* = 0$ represents either the surface where the value is consumed, or the stoichiometric oxygen surface. When s is related to the stoichiometric oxygen coefficient i of the global reaction

$$s = \frac{M_{O_2}}{M_{fuel}} i \qquad (4.64)$$

where M_{O_2} and M_{fuel} are the molecular weights of oxygen and fuel respectively. When solving Eq. (4.63) for the distribution of w^*, the diffusion coefficient of fuel is used for $w^* > 0$ and that of oxygen for $w^* < 0$.

Mitchell (1975) made an experimental study of a vertical, cylindrical, methane-air laminar diffusion flame in a steady atmospheric air environment. Fuel flowed through an inner tube and air flowed through an outer concentric tube. Temperature measurements were made using 75-μm Pt/Pt 13 percent Rh thermocouples with radiation and conduction corrections. An 80-μm quartz probe was used for removal of gas samples and subsequent analysis. Radial profiles of concentration and temperature at $z/d = 1.89$ are shown in Fig. 4.21. Methane is seen to diffuse to the primary reaction zone, just outside the luminous flame edge, where it is completely consumed. A stoichiometric amount of oxygen diffuses to the reaction zone, where the oxygen concentration decreases to virtually zero—traces of oxygen penetrated the flame surface near the flame base. Water and CO_2 maxima coincide with the temperature maximum at the thin blue reaction zone. The intermediate species, CO and H_2, have flat profiles inside the luminous flame surface and fall to zero in the reaction zone. Profiles of concentration and temperature along the symmetric axis are shown in Fig. 4.22. The temperature in the reaction zone increases with axial distance until the adiabatic flame temperature is approached and then slightly decreases until the flame tip is reached. This is due to the higher radiative and conductive heat losses to the burner surface in the lower regions of the flame and to the decrease in burning velocity near the flame tip. The concentration profiles show that complete combustion of methane takes place at two distinct surfaces which bound the reaction zone.

The fuel-rich side of the reaction zone is bounded by the luminous flame surface, where methane is rapidly converted to CO, CO_2, H_2, and H_2O. The oxygen-rich side of the reaction zone is bounded by the surface, through which oxygen, in stoichiometric proportions, diffuses. The distance between these two surfaces is determined primarily by the burnout of CO to CO_2. Changes in gas composition outside these surfaces are principally due to the interdiffusion of reactants and products. These experiments confirmed that diffusion is the predominant mechanism of material transport to the reaction zone and that chemical

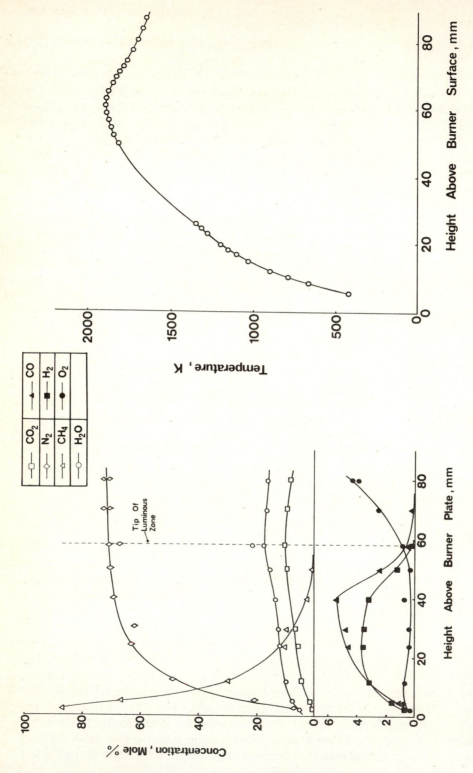

Figure 4.22 (*a*) Concentration and (*b*) temperature profiles along the symmetric axis in laminar methane-air diffusion flame. (*Mitchell, 1975.*)

146

reaction within the reaction zone is rapid. The Mitchell-Sarofim model utilizes this flame-sheet concept to locate either the luminous flame surface or the surface through which a stoichiometric amount of oxygen diffuses. The model, however, eliminates the restrictions of the classical Burke-Schumann model by making allowances for natural convection effects and variable thermodynamic and transport properties.

The Mitchell-Sarofim model was used to predict the shape and height of methane-air flames, as shown in Fig. 4.23. The global reaction of methane was assumed to be

$$CH_4 + 2O_2 \rightarrow CO_2 + 2H_2O \tag{4.65}$$

Since the oxygen stoichiometric coefficient $i = 2$, the fuel-oxygen mass-fraction which determines the stoichiometric oxygen surface is

$$w^* = w_{CH_4} - w_{O_2}/4 \tag{4.66}$$

Direct comparison between predictions and experiments of flame height are shown in Fig. 4.23. Experimentally, the flame height was measured at the distance above the burner exit at which the luminous zone of the diffusion flame closed at the burner axis. Flame height was also determined from the location at which the concentration of CO on the symmetric axis falls to zero. Mitchell found that the experimentally determined flame heights could be matched with the predictions by using an oxygen stoichiometric coefficient of 1.76. Justification for using an oxygen stoichiometric coefficient less than 2 was based on the assumption of incomplete combustion of the fuel at the flame surface and the indication that the reactions at the flame surface result in an increase in the total number of moles. Good agreement could also be found between predicted and experimentally determined flame shape (Fig. 4.23). The flame shape measurements were based upon the location of the visible edge of the blue reaction zone where the burnout of CO and the combustion of CH_4 to CO_2 and H_2O is complete. The small differences between predicted and experimentally determined concentration and temperature profiles at several heights above the burner surface were attributed to the neglect of energy-absorbing endothermic reactions inside the flame core and to the finite rates of chemical reaction in the reaction zone.

The model of Mitchell and Sarofim was sufficiently accurate for a prediction to be made of the aerodynamic flow field. The velocity distributions and stream function profiles for the case of methane and air streams entering the system at 4.5 and 9.88 cm/s, respectively, are shown in Fig. 4.24. The predicted velocity profiles show a peak in velocity at the flame surface. If increases in velocity were due solely to expansion, the peak in the velocity profile would be expected to coincide with the peak in the temperature profile. It was found, however, that as the distance above the burner port was increased, the peak in the velocity profile approached the symmetric axis faster than the flame surface. The velocities computed for the symmetric axis were greater than the velocities arising from natural convection. The streamline distributions in Fig. 4.24 show the presence of a large recirculation zone set up between the hot flame surface and the cooler shield wall. Air is

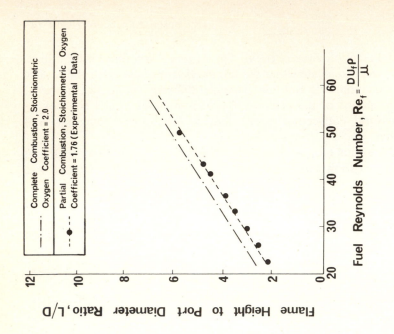

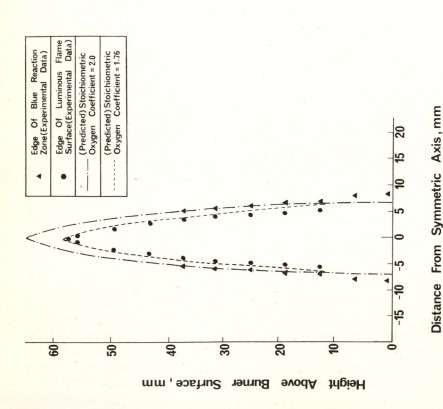

Figure 4.23 Comparison of predicted and experimentally determined flame shape and height for laminar methane-air diffusion flames. (*Mitchell, 1975.*)

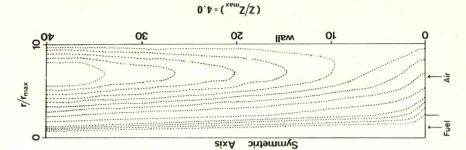

$(Z/Z_{max}) = 4.0$

$\psi_{axis} = 0.0$
$\psi_{wall} = 1.0$
$\psi_{max} = 2.18$

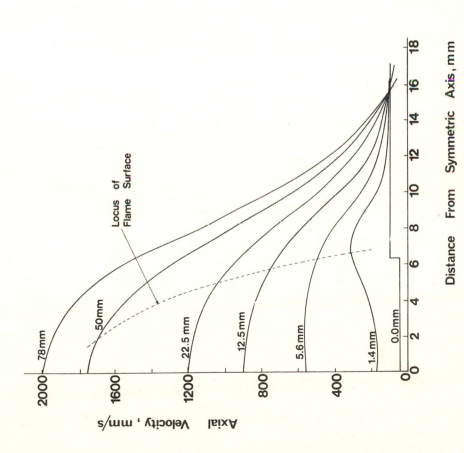

Figure 4.24 Predicted velocity and stream function profiles for laminar methane-air diffusion flame. (*Mitchell, 1975.*)

149

entrained into the system at the shield outlet to balance the momentum of the inlet fuel and air streams and the frictional losses at the shield wall. The presence of this recirculation zone reduces the total area available for flow of the feed and combustion gases. This results in the very high degrees of acceleration of the flow along the symmetric axis. The presence of the recirculation cell was initially verified by particle-track studies and the magnitude of the velocities has subsequently been determined by laser anemometer measurements. This close agreement between prediction and measurement represents the present state of the art for laminar diffusion flames.

THE PHYSICS OF TURBULENCE

5.1 DESCRIPTION OF TURBULENT FLOW

Turbulence is a subject that is treated in many books on fluid mechanics. A simple introduction to turbulence is given by Bradshaw (1971), in which he discusses the physics of turbulence and relates this to measurable quantities. Tennekes and Lumley (1972), in their book *A First Course in Turbulence*, discuss the nature, origin, dynamics, and statistical descriptions of turbulence. Scorer (1978), in *Environmental Aerodynamics*, provides definitions, properties, and applications of turbulence to flows on a very large, atmospheric scale. Hinze (1959), Batchelor (1953), and Bradshaw (1976) have written books which deal with turbulence in a more complex mathematical manner. In the first section of this chapter the basic concepts of turbulence are described, with particular emphasis on physical structure. This is followed by a presentation of the equations for momentum and energy and discussion of homogeneity and symmetry. The concepts of self-preservation and similarity are examined for free turbulent shear flows and this is followed by detailed descriptions of coherent structures.

Concepts of Turbulence

Turbulence can be characterized by the following phenomena:

Irregularity. The contours of turbulent flow surfaces and velocity variation with time at a point are irregular. Flow structure is often difficult to separate from random irregular background noise but most turbulent flows have a structure dominated by large-scale eddies that retain elements of coherence. This structure is detected by continual sampling, in which continuous signals are decomposed into separate time intervals, according to some specified discriminating criteria.

Diffusivity. In stationary fluids, the transfer of momentum, heat, and mass is governed by diffusion of molecules. In laminar flows, transport phenomena, other than by convection, are governed by molecular diffusion. In turbulent systems, the overall rates of transfer of momentum, heat, and mass are increased by several orders of magnitude, as the result of rapid mixing and stirring. Prandtl suggested that turbulent diffusivity could be considered analogous to molecular diffusivity; fluid particles, consisting of agglomerates of molecules, transfer heat, mass, and momentum across mixing lengths with velocities equal to the turbulent fluctuating velocity. In structured turbulent flow, large eddies convect molecules of fluid but diffusion is effected by molecular diffusivity across the interfaces of these large eddies.

Reynolds number. Transition from laminar flow to turbulent flow is characterized by the Reynolds number. As Reynolds number is increased, laminar flows become unstable and break down into turbulent flow. This shows that the inertial terms in the equation of motion increase in magnitude, compared with the viscous terms. In principle, turbulent flows are described by the equations of motion but, since these contain nonlinear inertia terms resulting in nonlinear partial differential equations, these generally cannot be solved. The characteristics of turbulence depend upon Reynolds number, unless that number is suitably high everywhere in the flow.

Three-dimensionality. Turbulence is a three-dimensional phenomenon, so that, even when there is a dominant main flow direction, turbulent fluctuations are significant in all three dimensions. Flows in pipes show motions of individual eddies in all directions, with eddies crossing the axis of symmetry.

Dissipation. Energy is required in order to generate turbulence and, if energy is not continuously supplied to a flow, the turbulence decays, due to the action of viscosity. A distinction is made between dissipation of energy in turbulent flows and dispersion of energy caused by wave motions which are not turbulent.

Liquid and gaseous phase. In fluid dynamics, liquid and gas flows are treated together, despite the large difference in density and viscosity. The major characteristics of turbulent flows are not controlled by the molecular properties of the fluid; liquid and gaseous flows having the same Reynolds number are considered to be dynamically similar, provided that there are no significant density inhomogeneities. Turbulent flows are generally subdivided into boundary layer, jet, wake, and pipe flows. The differences between these various types of flows are mainly in the large eddy structure. The small eddy structure appears to be common to all turbulent flows.

Bradshaw (1971) defines turbulence as follows: "Turbulence is a three-dimensional time-dependent motion in which vortex stretching causes velocity fluctuations to spread to all wavelengths between a minimum determined by viscous forces and a maximum determined by the boundary conditions of the flow. It is the usual state of fluid motion, except at low Reynolds numbers."

Turbulent flow can be distinguished from laminar flow by observation, as was done in the classical experiments of Reynolds in pipe flows. Laminar flow is undisturbed, with smooth streamlines in which visualization techniques indicate that the flow is made up of a series of sliding laminae. Laminar flows are either independent of time, or very simply dependent on time, and variations in space are comparatively simple. In accelerating or decelerating laminar flows, changes in velocity and direction are slow and small disturbances are damped by the action of viscosity. When small disturbances do arise, the flow still remains tranquil. Gentle oscillations of a simple sinusoidal or other periodic form can be tolerated in some laminar flows.

The transition from laminar to turbulent flow can be triggered as oscillations of velocity about the mean begin to grow in space and time. As the oscillations grow they gradually change from a simple periodic form to an apparently random, eddying motion, covering a continuous range of wavelengths and frequencies. A fully turbulent flow is often thought of as being disorganized, disordered, formless, unsystematic, and chaotic. However, the present indications are that there can be a continuous transition from laminar to fully turbulent flow, in which there is a gradual change from ordered to unordered and only finally to disordered flow.

Randomness is an essential characteristic of turbulence. As the transition from laminar to turbulent flow proceeds, the orderliness, periodicity, and regularity progressively break down. The velocity still remains a continuous function of space and time but the simple periodic dependence of the flow on independent variables disappears. Statistical correlations between motions at different points in the flow can be distinguished but, as transition proceeds, the probability distribution of velocity and other flow characteristics at a given point tend toward a gaussian, or normal, probability distribution characteristic of many random processes. The amount of information on the variation in space and time of quantities required for a complete description of a turbulent flow is so great that it is essential to use statistics for the organization of data, except for very small periods of time and variations in space. The understanding of turbulence has suffered greatly from an indiscriminate use of statistics. As more time-resolved information is becoming available, it is clear that, in some situations, an indiscriminately long time average in a flow, which is random but structured, fails to detect important physical processes.

The simplest statistical property is the average, or mean, with respect to time of a quantity at a point in space. It is generally useful to know the mean of a quantity but it must be recognized that the mean can be a very crude description of the phenomena so that, in a flow with large-scale variation of dimensions and velocities of eddies, the mean velocity may have little relation to the physical movements in the region considered. It has become necessary to provide more detailed and complex statistical descriptions of flows, such as rms, probability density functions, and conditional sampling, in addition to the mean values. On the basis of mean velocity, no quantitative difference is shown between laminar and turbulent flows, either experimentally or analytically, yet clearly there are important fundamental differences between turbulent and laminar flows. In turbulence, the term "statistical" relates to probabilities or average properties. A stochastic process is governed by the laws of probability and, hence, relates to random processes.

Statistical averaging may either be in time or space. An average with respect to time is defined mathematically as

$$\bar{f} = \underset{T \to \infty}{\text{Limit}} \frac{1}{2T} \int_{-T+t}^{T+t} f(t') \, dt' \tag{5.1}$$

where t is the time at the midpoint of the averaging period and $2T$ is the total time selected for averaging. Depending upon the type of flow, T may be selected so that the average $\bar{f}$ becomes independent of time and, hence, the flow becomes "statistically stationary." This can be checked by verifying if there is any change in $\bar{f}$ when T is increased or the average is taken at another period T. During the transition from laminar to turbulent flow, the time-averaged quantity may still be time dependent when a specific averaging time T is selected. For turbulent flows with a mean periodicity or flows with acceleration or deceleration at a point, such as starting jets, $\bar{f}$ is dependent on time t, i.e., the flow is nonstationary. For these conditions, time averages are not generally useful and ensemble averaging is used.

Before an attempt is made to take an average of a particular flow, a cursory examination needs to be made in order to establish whether the flow is statistically stationary or nonstationary. This requires an estimation as to whether an average will be constant over a given time period or whether it will vary with time. This estimation is, of course, dependent upon the overall time period that is ultimately selected for taking the average. For example, the flow in a pipe across which the differential pressure is maintained constant could be expected to be stationary, whereas flow from a reservoir through a pipe in which the quantity of fluid is gradually depleted could be expected to be nonstationary with a steady decrease in mean velocity. Flows in systems with rotating or alternating machinery, such as rotating engines or piston engines, have cyclic variations governed by the moving parts of the system. In all turbulent flows, information is also required concerning the deviations of the fluctuations from the mean. This is described in terms of a range of statistical parameters, such as rms, skewness and flatness factors, higher-order correlations, and probability functions. The term "ensemble averaging" is used for nonstationary flows, in which a mean is taken of a series of experiments.

Reynolds Stresses

Fluctuations in velocity and other properties of a turbulent flow produce changes in pressure forces and components of momentum. Internal resistance in pipes increases substantially as flow changes from laminar to turbulent. The increase in flow resistance is directly related to fluctuations in velocity. This can be examined by considering the rate at which the x-component of momentum passes through one of the faces $dy\,dz$ of an elementary control volume. The x-component momentum flux is given by

$$\rho(U + u)^2\,dy\,dz = \rho(U^2 + 2Uu + u^2)\,dy\,dz \qquad (5.2)$$

When this expression is averaged with time, it becomes equal to $\rho(U^2 + \overline{u^2})\,dy\,dz$. By definition, fluctuations have zero means but, since the mean square of the fluctuation velocity is nonzero, a specific mean momentum flux is associated with the velocity fluctuations. Momentum flux is a product of mass flow and velocity and, hence, a flow with fluctuating velocity will have an increased momentum flux compared to a nonfluctuating flow. The relationship between velocity and momentum flux is nonlinear and this leads to special problems in mathematical analysis.

In laminar flows, mean velocity is related to pressure change through viscosity. In turbulent flow, an additional apparent stress, $-\rho\overline{u^2}$, normal to the face $dy\,dz$ arises associated with the velocity fluctuations. This stress is directly related to momentum flux. Additional normal stresses $-\rho\overline{v^2}$ and $-\rho\overline{w^2}$ arise in the y and z directions, associated with velocity fluctuations v and w. The components $\overline{u^2}$, $\overline{v^2}$, and $\overline{w^2}$ are of the same order of magnitude in fully turbulent flows.

The rate at which x-component momentum passes through the face $dx\,dz$ of the control volume is the product of the mass flow in the y direction, $\rho(V + v)\,dx\,dz$, and the velocity in the x direction $U + u$. This product, after averaging, yields the mean momentum flux $\rho(UV + \overline{uv})\,dx\,dz$. The mean shear stress $\sigma_{xy} = -\rho\overline{uv}$ acts on the face $dx\,dz$. The additional turbulent stresses, both normal and shear, are called Reynolds stresses.

Turbulence produces additional fluxes of quantities other than momentum. Temperature fluctuations, θ, cause enthalpy fluctuations $\rho c_p \theta$ per unit volume, giving rise to enthalpy flux $\rho c_p \overline{\theta u}$ in the x direction and corresponding enthalpy flux in the y and z directions. This turbulent enthalpy transfer is additional to enthalpy transfer by molecular conduction. Turbulent mass transfer is treated mathematically, similarly to turbulent momentum and enthalpy transfer. Concentration fluctuations c_i result in mean mass flux of individual species for each component. In turbulent flows, Reynolds stresses and turbulent fluxes of enthalpy and mass transport of individual species are much larger than viscous stresses and other molecular transport rates.

Transport processes in turbulent flow have been related to transport processes described by simple kinetic theory by considering turbulent eddies to be analogous to molecules and "mixing lengths" to mean free paths. This analogy is not valid for at least two reasons: (1) turbulent eddies are continuous and contiguous, whereas gas molecules are discrete and collide only at intervals; (2) molecular mean free paths are generally small compared to the dimensions of the system. The dimensions of the largest turbulent eddies can be of the same magnitude as the transverse dimensions of the system, and distance traveled by the largest eddies can be greater than such dimensions.

The fundamental basis for relating turbulent transport to the gradient of mean quantities seldom applies in turbulent flows. Nevertheless, many modelers of turbulent flow have used turbulent eddy diffusivities, analogous to molecular diffusivities for transport of mass momentum and enthalpy. It has been shown that these turbulent diffusivities are generally not constant within any one system, nor are they simply related between one system and another. It is interesting to note that Bradshaw (1971) went so far as to state that these turbulent diffusivities are not even *discoverable* functions of the local variables, but dependent on the previous history of the flow which carries the turbulent eddies. This philosophy was ignored by the vast majority of modelers in turbulent flow.

Because of the absence of a fundamental basis for the description of the transport processes in turbulent flows, more understanding is required of the behavior of fluctuating quantities. Advances in the understanding and analysis of turbulent flows of engineering interest will be made by close coordination between experimental observation and measurement and theoretical analysis.

Vortex Stretching

Turbulent eddies have both translational and rotational motion. The net rate of rotation (or average angular velocity) about the z axis of fluid element is

$$\frac{1}{2}\left(\frac{\partial v}{\partial x} - \frac{\partial u}{\partial y}\right)$$

The z-component of vorticity is defined as twice this angular velocity. The velocity is distinguished from the rate of shear strain $(\partial u/\partial y) + (\partial v/\partial x)$. The vorticity is a measure of rotation, while the rate of shear strain is a measure of deformation. If, in addition to a rotation about the z axis, the fluid element is under the influence of a rate of linear strain in the z direction $\partial w/\partial z$, the element will be stretched in the z direction and its cross section in the xy plane will be smaller. If we take the case of an element of circular cross section in the xy plane and neglect viscous forces, the conservation of angular momentum requires the product of the vorticity and the square of the radius to remain constant. The integral of the tangential component of velocity round the perimeter, called the circulation, remains constant in the presence of viscous forces. During the stretching process, the kinetic energy of rotation increases at the expense of the kinetic energy of the w-component motion that does the stretching and the scale of the motion in the xy plane decreases. An extension in one direction can increase the length scales and increase the velocity components in the other two directions which, in turn, stretch other elements of fluid with vorticity components in these directions. The length scale of the motion that is augmented gets smaller at each stage. Stretching in the z direction intensifies the motion in the x and y directions, producing smaller scale stretching in x and y and intensifying the motion in the y, z, and z, x directions respectively. Qualitatively, an initial stretching in one direction produces nearly equal amounts of smaller scale stretching in each of the x, y, and z directions after a few stages of the process. The small-scale eddies in turbulence do not share the preferred orientation of the mean rate of strain. They tend to have a more universal structure. The cascade of energy moves from large to smaller scales; discontinuities of velocity would develop if it were not for the smoothing action of viscosity. Viscosity finally dissipates into thermal internal energy. The energy that is transferred to the smallest eddies does not play any essential part in the stretching process as such.

An element of fluid may be considered as part of a line vortex with its axis in the z direction. Any flow with vorticity can be considered to be made up of large numbers of infinitesimal vortex lines. A vortex sheet is a layer of locally parallel vortex lines. Laminar shear layers may be considered as a stack of elementary vortex sheets with vortex lines. Turbulence is a tangle of vortex lines or partly rolled vortex sheets, stretched in a preferred direction by the mean flow and in random directions by each other. Turbulence always has all three directions of motion, even if the mean velocity has only one or two components. If the fluctuating velocity component in one direction were everywhere zero, the vortex lines would necessarily all lie in this direction and there would be no vortex stretching, no transfer of fluctuation energy to smaller scales, and the motion would not be considered to be turbulent. If it were not for the diffuse effect of viscosity, vortex lines or sheets would move with the fluid; the effect of viscous diffusion is seen in the slow growth of molecular shear layers. In turbulent flow, viscous diffusion of vorticity is negligible, except for the smallest eddies. Fluid that is initially without vorticity can acquire it only by viscous diffusion. Once acquired, vorticity can be increased many orders of magnitude by vortex stretching. Pressure fluctuations do not directly affect vorticity in incompressible flow.

The rate of supply of kinetic energy to the turbulence is, in the absence of body forces, the rate at which work is done by the mean rate of strain against the Reynolds stresses in the flow, as it stretches the turbulent vortex lines. In laminar flow, viscous stresses caused by molecular motion dissipate mean flow kinetic energy directly into thermal internal energy. In turbulent flow, the eddies extract energy from the mean flow and retain it for a while before it reaches the small dissipating eddies. Turbulent kinetic energy per unit volume is introduced into the eddies that contribute to the Reynolds stresses in direct proportion to their contributions. The stress-producing eddies are the larger ones, which are best able to interact with the mean flow. Vortex stretching tends to make smaller eddies lose all sense of direction, so that the flow becomes statistically isotropic, and their contribution to the Reynolds shear stress is zero. The smaller eddies are much weaker than those that produce most of the Reynolds stress because most of the energy that reaches them is immediately passed on to the smallest eddies and there dissipated by viscosity. The size of the dissipating eddies depends on the viscosity and the speed of the flow. Typically, their wavelength is less than 1 percent of the radius or transverse dimension of the flow. Viscous stresses are usually so small, compared with turbulent stresses, and since the parts of the eddy structure that depend on viscosity are so small and so weak compared with the stress-producing part of the turbulence, we can, for many purposes, neglect viscosity, regarding it only as a property of the fluid that produces energy dissipation in very small eddies. For flows in process of transition from laminar to turbulent and the flow very close to a solid surface, viscous forces may not be neglected. Viscous and turbulent stress are of the same order and viscosity directly affects the eddies that produce the Reynolds stress. The main characteristics of turbulence are due to three-dimensional vortex stretching and are not caused by viscous forces.

Turbulence is generally unaffected by compressibility if the pressure fluctuations within the turbulence are small compared with the absolute pressure, i.e., the fluctuating Mach number must remain small. Since velocity fluctuations are usually a small percentage of the mean velocity, the foregoing condition is satisfied for all mean-stream Mach numbers less than about 5.

Bradshaw (1971) ascribes the following properties to turbulence:

1. Apparent mean stresses in turbulent flow are determined by the velocity fluctuations which depend on the whole history of the flow and not on the mean flow at the point considered.
2. Viscous stresses are usually small compared with turbulent stresses. Viscosity affects only the smallest eddies and the turbulent stresses are usually independent of viscosity.
3. If the fluctuating Mach number is much less than unity, turbulence is not directly affected by compressibility.

5.2 METHODS OF ANALYSIS

The Navier-Stokes equations of motion are generally considered valid for all flows. These equations are nonlinear, so that each individual flow pattern has certain unique characteristics that are associated with its initial and boundary conditions. No general solutions to the turbulent flow equations are available. It is not possible to make accurate quantitative predictions from the analysis of the equations of motion without relying on empirical data. The time-averaged turbulent flow equations are intractable unless assumptions are made in order to make the number of equations equal to the number of unknowns. This is known as the closure problem of turbulence theory.

In newtonian fluids, the Stokes law shows that stress is related to the rate of strain by viscosity. Some turbulence theories postulate that stress is related to rate of strain by turbulence-generated viscosity, which is supposed to play a role similar to that of molecular viscosity in laminar flows. This approach is based upon a superficial resemblance between the mechanism of transfer of heat and momentum by molecular motion and transport as the result of turbulent fluctuations. These phenomenological models replace molecular viscosity by eddy viscosity and mean free path by mixing length.

Continuum Flow

The essence of the continuum approximation in turbulent flows is that flow velocities and other continuum properties can be defined as averages over regions of space and intervals of time that are large compared with the scales of the molecular motion and small compared with the scales of the continuum flow. Mixing on micro scales has a special significance in combustion, where chemical reaction takes place between individual molecules. Thus, flows which may be considered to be well mixed from the point of view of continuous fluids may still require additional energy and time to complete mixing on the molecular level. If the largest eddies of a turbulent flow have a characteristic size L and a characteristic velocity V, the smallest scales of motion are of size $l_0 = (v^3 L/V^3)^{1/4}$, of duration $t_0 = (vL/V^3)^{1/2}$ and with characteristic velocity $v_0 = (vV^3/L)^{1/4}$, where v is the kinematic viscosity. Deviations from the continuum values due to molecular fluctuations are small if both

$$nl_0^3 \approx nL^3 \left(\frac{VL}{v}\right)^{-9/4} \gg 1 \tag{5.3}$$

and

$$\frac{c_1^2 t_0}{v} \approx \frac{c_1^2}{V^2}\left(\frac{VL}{v}\right)^{1/2} \gg 1 \tag{5.4}$$

where n is the number density of the molecules and c_1 is the root mean square of one component of the molecular velocity.

For standard conditions, $nl_0^3 \simeq 3 \times 10^{12}$ and $c_1^2 t_0/v \simeq 2 \times 10^5$. The velocity and size of eddies have orders of magnitude near those of the overall system, and it is generally found that only when velocities are greatly in excess of the molecular velocities does the continuum approximation fail to describe the turbulent motion.

Homogeneity and Symmetry of Turbulent Flows

The analysis of any turbulent flow can be greatly simplified if it can be established, or assumed, that all or part of the flow can be considered to be either homogeneous or symmetric about some central axis. In theoretical analyses, it is simply a question of making the appropriate assumptions. For real flows, an initial superficial examination may suggest that parts of the flow could be considered homogeneous or symmetrical, such as flows in cylindrical pipes, around spherical objects, or between rotating cylinders. A test for homogeneity or symmetry, made by measurement, is dependent upon the frequency response of the measuring instrument relative to the frequency of velocity fluctuations, and also the period and method of averaging that is used. Flows in cylindrical pipes and in round jets are in the mean, symmetrical about the central axis. Turbulent flows are composed of large eddy structures, with dimensions comparable to the width of the system and, hence, the flows are not symmetrical. Eddies generated at wall boundaries or interfaces will cross from one boundary to another without recognition of a line of symmetry. Streamlines describing these eddy movements and velocity measurements based on conditional sampling, triggered by eddy movements, will show no semblance of symmetry or homogeneity.

Free turbulent flows have been considered to have a tendency to "forget" their origin and details of their initiation. Fully developed jet flows are characterized by their momentum flux, so that jets with the same momentum issuing from round nozzles are hardly distinguishable, in time-averaged measurements far downstream, from rectangular or irregularly shaped nozzles. This "forgetfulness' does not necessarily apply to individual eddies which originate from the boundaries of the nozzle and can still be distinguished well into the fully developed region of a flow.

For practical purposes, we continue to use a system of cartesian coordinates to describe flows with symmetry along a plane and cylindrical polar coordinates for axisymmetric systems. Most turbulent flows are strongly inhomogeneous with respect to variation of one coordinate and many flows are nearly unidirectional over a significant part of the flow. From a consideration of the relative magnitudes of velocity components and velocity gradients, flows can be shown to be unidirectional, isotropic, or homogeneous with respect to one or more coordinates. Fully developed pipe flows have significant variations only in the radial direction. Flows between concentric rotating cylinders and in highly swirling flows are dominated by tangential velocity components. In jet and wake flows, variations in radial directions are of greater magnitude than those in axial directions, while variations in the tangential direction are small.

The separation of a parameter into a mean and fluctuating value is artificial, yet there is a physical difference between the parts of the kinetic energy and entropy densities that are associated, respectively, with the mean fields and the fluctuation fields. Transfer from mean value to fluctuation quantities is normally irreversible and is the first stage in a cascade process, ending in transformation or destruction by molecular transport processes. The details of the energy and entropy transfer processes are of vital importance for the full understanding of turbulent flow. A homogeneous turbulent flow is one in which the mean values of functions of the fluctuations and of the mean velocity gradients are independent of position in the flow. Most turbulent flows are inhomogeneous but experimental and theoretical studies of homogeneous flows have provided some insight to the understanding of turbulence.

In homogeneous turbulence with uniform gradients of mean velocity, the equation for the balance of turbulent kinetic energy shows that energy is generated by working of the mean flow against the Reynolds stresses and that energy is dissipated as heat by working of the turbulent velocity gradients against the viscous stresses. The eddies containing most of the energy contribute most to the Reynolds stresses and receive most of the energy that is transferred from the mean flow. The rate of energy dissipation is proportional to the mean square of the velocity gradient, which is determined by eddies much smaller than those containing most of the energy. Analysis of the spectrum equation shows that turbulent energy flows from the region of comparatively small wave numbers, where most of the energy is produced and resides, toward much larger wave numbers, where the rate of viscous dissipation is sufficient to convert the flow to heat.

Isotropic turbulence is defined by the condition that all mean values of functions of the flow variables are independent of translation, rotation, and reflection of the axes of reference. Such flows are difficult to obtain experimentally, though experimental approximations have been achieved in the wake of uniform grids. Batchelor (1953) has given a detailed theoretical description of isotropic and homogeneous turbulence. Some of the principal conclusions which have a bearing on the physical structure of turbulent flows are given below.

Turbulent eddies of different sizes affect one another only if their sizes are comparable. The interaction between the large, energy-containing eddies and the much smaller viscous eddies takes place over a number of intermediary stages. These stages are completely independent, except that the viscous eddies dissipate the energy lost by the large eddies. Motion on the large scale is essentially inviscid and is independent of the Reynolds number. Decaying turbulence reaches some sort of moving equilibrium when one set of eddy structures gives way to another set, similar in all respects except for changes in their common scales in velocity and length. In flows that are self-preserving, all aspects of the motion, except those directly influenced by viscosity, have similar forms at all stages. The differences in the similar forms can be described wholly by changes of velocity and length scales, which are functions of time or of position in the flow direction.

Inhomogeneous Shear Flow

Most flows of practical importance are neither isotropic nor homogeneous so that, in general, shear flows need to be treated as inhomogeneous. Strong similarities are found between jets and wakes, which together are considered to be free turbulent flows. There are distinct differences between free turbulent flows and flows bounded by or influenced by walls. In free turbulence, inhomogeneity arises from spreading of the flow into the ambient nonturbulent fluid while, in the case of wall turbulence, the physical restriction of the solid boundary causes inhomogeneity. An inhomogeneous shear flow can be divided into the following parts:

1. The mean velocity field $U(x)$
2. The large-eddy motion, $u'(x)$
3. The main turbulent motion, $u''(x)$

In most turbulent flows, variations of mean velocity are considerably greater than the turbulent fluctuations, and the large-eddy contribution to the velocity gradient is a small perturbation of the mean flow gradient. Turbulent shear flows are considered to be made up of a main turbulent motion of scale slightly smaller than the scale of the lateral inhomogeneity and a group of large eddies of lateral scale comparable with the width of the flow. By this means, the motion is divided into separate components—those existing and formed by nearly uniform shear and those whose existence and form are determined by the inhomogeneity of the flow. Large eddies, as a distinct group, are connected with the folding of the interface between turbulent and nonturbulent fluid.

Local and Asymptotic Invariance

If the characteristics of the turbulent motion at a point in a flow field appear to be controlled, mainly by the immediate environment, the flow may be considered to be locally invariant. When time and length scales of the flow vary slowly downstream, the turbulence time scales can be small enough to permit adjustment to the gradually changing environment, so that the turbulence is dynamically similar everywhere if nondimensionalized with local length and time scales. For flows which are in a state of dynamic equilibrium, the local energy input approximately balances the local losses. If the energy transfer mechanisms in turbulence are sufficiently rapid, the effects of past events do not dominate the dynamics, so that the equilibrium is governed mainly by local parameters, such as length scales and times. In these types of self-preserving flows, the origin of the flow and upstream conditions are not relevant. The concept of a structured turbulent flow consisting of eddies originating in mixing layers or in wall regions and, subsequently, remaining coherent as they move downstream, is in conflict with the concept of self-preservation and local invariance. Since experimental evidence for flows with high Reynolds number indicates that local similarity is established, it may be that the coherent structures change sufficiently as they move downstream to maintain this similarity. Although many attempts have been made to demonstrate that some flows are self-preserving and locally invariant, it is not generally possible to separate a local flow from its previous history.

The concept of asymptotic invariance is generally used for very high Reynolds number flows. Turbulent flows approach the limiting condition when Reynolds number tends to infinity. This limit process is applied to conditions where molecular viscosity has effects which are tending to disappear asymptotically. Hence, turbulent flows at high Reynolds number are considered to be independent of viscosity. The flows are supposed to have "Reynolds number similarity." A proof of Reynolds-number similarity is given by the coincidence of superimposed profiles of velocity, nondimensionalized by characteristic velocity and time scales. These flows are termed "fully developed turbulent flows," which are generally found at large distances from the origin of the flow.

In many free turbulent flows there is a tendency toward self-preservation and a region of developing flow can be defined where the transverse distributions of mean quantities change with distance downstream, followed by a self-preserving region in which the transverse distributions retain the same functional form. Self-preservation may only be possible as an asymptotic condition which is not achieved within the range of observation or examination. For a flow to be self-preserving in form, the variation of a mean value quantity M must be of the form

$$M = M_1 + m_0 \, f(y/l_0, z/l_0) \tag{5.5}$$

where l_0 is a scale of length and m_0 is a scale of quantity. Both l_0 and m_0 are functions of the downstream distance, x, alone. M_1 is usually a fixed initial condition but, for some cases, this is considered as a reference level of the quantity in the free stream. All free turbulent flows are considered to be in a state of development toward self-preservation and, in some flows, self-preservation is achieved. The term "similarity" is also used for a flow in which there is similarity of profiles of transverse distributions, normalized by appropriate length and initial reference scales. Total self-preservation is achieved when profiles of all mean quantities, including velocity, stress, temperature, and concentration, have similarity profiles. The axial distance at which self-preservation is achieved is not necessarily the same for all measured quantities. For example, in jet and wake flows, self-preservation of the mean velocity profile is achieved well before self-preservation of shear stress distributions.

The principle of moving equilibrium is related to the concept of self-preservation. In any developing flow, turbulent fluid is carried along by the mean flow and the instantaneous conditions at any section of the flow are determined to a considerable extent by the conditions at some earlier time, when the volume of fluid now at the point of observation was some distance upstream. In order to explain the change in position and size of eddies, we require to consider the history of these eddies from the time of their initiation. The principle of self-preservation asserts that a moving equilibrium is set up in which the conditions of the initiation of the flow are largely irrelevant and so the flow depends on one or two simple parameters and the flow is geometrically similar at all sections. At the time that the concepts of self-preservation and the principle of moving equilibrium were first postulated, it was hypothesized that there should be a tendency for all flows to move toward a state of equilibrium and self-preservation. This hypothesis is open to question; in free jet and wake flows, where eddies are seen continuously to grow, coalesce, and merge, the moving equilibrium and self-preservation conditions may never be achieved.

An important corollary of the principles of self-preservation and moving equilibrium is that the flows are independent of the conditions at the origin. This assertion has been referred to loosely as the "forgetfulness" of the flow or, more explicitly, that the conditions at the initiation of the flow are largely irrelevant in the self-preserving region. The term "fully developed turbulent flow" is used for pipe flows to describe the condition in which there are no changes in mean velocity profile with distance downstream the pipe. This terminology has also been used for free turbulent flows where, despite the continuous increase in width and change in mean velocity, the flow in said to be fully developed when profiles have become similar. In all turbulent flows, there is an initial region which is dominated by the influence of the conditions at the origin of the flow. In a jet flow, the origin is nominally considered to be in the plane of the exit of the orifice, through which the flow emerges; in a wake flow, it is the plane at the downstream extremity of the obstacle generating the wake; in free shear flows, it is the point of initial contact of the separate streams. The initial conditions of free turbulent flows are determined by flow conditions upstream of the origin, so that the boundary layer thickness, intensity of turbulence, wall roughness, degree of swirl, and magnitude of transverse velocity components in the approach flow will have an influence which starts from being dominant at the origin and declines as the free turbulent flow develops. In flows such as swirling jets or recirculation zones in the wake of bluff bodies, the region of major interest can be the initial region, with little attention devoted to the far downstream, fully developed, self-preserving region.

Specification of Flow Field

Specification of a flow field can either be made according to Lagrange or Euler. In the lagrangian specification, the trajectories and velocities of individual particles are examined and recorded in the form, $x(x_0, t_0; t)$, i.e., the position at time t of a particle which was initially at position x_0 at time t_0. This type of information can be obtained by particle track high-speed flash photography in which the changes in velocity are recorded along the trajectory of an individual particle. In the eulerian specification, there is a fixed coordinate system and the velocity of particles is given at various points in space. This is recorded as $u(x, t)$, i.e., the velocity of the particle which is at position x at time t. The eulerian specification has dominated fluid mechanics for many years because of the theoretical simplifications and the experimental convenience of using probes at fixed points in space while measuring the velocity as a function of time. One of the limitations of the eulerian specification is the difficulty of distinguishing structure within flows which are averaged with respect to time. This disadvantage is being partially overcome by the use of conditional sampling, in which discrimination is used in the averaging procedure.

The distribution function for the eulerian specification is $F[u(x, t)]$ and the statistical or ensemble average of $M[u(x, t)]$, a function of the flow field in space and time, is given by

$$\langle M[u(x, t)]\rangle = \int F[u(x, t)]M[u(x, t)]\, dV \qquad (5.6)$$

i.e., the ensemble average for a given volume and time is obtained by integrating the product of the distribution function and the function of the flow field over the function space.

Many turbulent flows have, for convenience, been considered to be statistically stationary with respect to time. The coordinate system is fixed in space and the velocity components at a fixed point are taken to be stationary random functions of time. The concept of statistically stationary flows requires to be used with considerable care, since many flows with periodic or other significant variations with time can be made to be statistically stationary by averaging over a long time. For statistically stationary flows, the ergodic hypothesis leads to the simplified mean value with respect to time [Eq. (5.1)].

Recognition of symmetry in a flow can greatly facilitate the description because of the elimination of changes in one of the coordinate directions. The statistical quantities in flows in pipes, jets, and wakes have a tendency to be symmetrical about the axis, with the most significant variations in velocity occurring in the radial direction, a smaller variation occurring in the axial direction, and no variation in the azimuthal direction. If the flow variables are stationary random functions of any space or time coordinate, mean values are independent of that coordinate and the flow becomes statistically homogeneous for that coordinate. Even though all flows are physically three dimensional, we use the concept of a one-dimensional flow when mean values only change along one coordinate; two-dimensional flows are those in which mean values change along two coordinates, while there is no change along the third coordinate.

Equations for Momentum and Energy Based on Mean Values

Velocity variation in turbulent flow is generally expressed as the sum: $U_i + u_i$, where U_i is the mean velocity and u_i is the fluctuation from the mean value, which has a mean value equal to zero. The mean value of the continuity equation is given by

$$\frac{\overline{\partial(U_l + u_l)}}{\partial x_l} = \frac{\partial U_l}{\partial x_l} + \frac{\partial \overline{u_l}}{\partial x_l} = \frac{\partial U_l}{\partial x_l} = 0 \tag{5.7}$$

The velocity fluctuations satisfy the continuity equation

$$\frac{\partial u_l}{\partial x_l} = 0 \tag{5.8}$$

The equation for the mean velocity can be written

$$\frac{\partial U_i}{\partial t} + U_l \frac{\partial U_i}{\partial x_l} + \frac{\overline{\partial u_i u_l}}{\partial x_l} = -\frac{\partial P}{\partial x_i} + \frac{g_i(T - T_a)}{T_a} + v \frac{\partial^2 U_i}{\partial x_l^2} \tag{5.9}$$

where $P + p$ is the pressure difference from the ambient pressure, $T + \theta$ is the potential temperature, and T_a is the ambient potential temperature. The terms p and θ are the fluctuations of pressure and temperature respectively with mean values equal to zero. The equation for the mean velocity can be rewritten in the following form

$$\frac{\partial U_i}{\partial t} + U_l \frac{\partial U_i}{\partial x_l} = \frac{g_i(T - T_a)}{T_a} + \frac{\partial}{\partial x_l}\left[-\delta_{il} P + v\left(\frac{\partial u_i}{\partial x_l} + \frac{\partial u_l}{\partial x_i}\right) - u_i u_l\right] \tag{5.10}$$

Writing the equation of momentum in the above form shows that the mean flow is accelerated by forces arising from the mean buoyancy, the gradient of the mean pressure, or the viscous stresses developed by the mean flow alone, and by the virtual force which is the gradient of the Reynolds stress $-u_i u_j$. In combustion systems where there are large temperature variations, the buoyancy term can become significant, particularly when mean velocities are low enough for the buoyancy and acceleration terms to be of comparable magnitude in the above equation. The term involving gradients of mean pressure is generally neglected on the assumption that it is of lower order of magnitude than other terms in the equation, despite the lack of accurate information on mean or fluctuating pressure variations across turbulent flow fields. In fully developed turbulent flows, the high magnitude of the Reynolds numbers indicates that viscous forces are not more than second-order importance, and these are often neglected. In combustion systems, the value of the kinematic viscosity can increase by a factor of 20 as a result of temperature rise and the influence of viscous forces is likely to be more important than in flows without chemical reaction and temperature gradients.

The kinetic energy of turbulence has particular importance in combustion systems as the prime source of energy for the mixing process. The very high rates of heat release obtained in high intensity combustion flames, as compared with those obtained under laminar flow conditions, are related to the distribution of turbulent kinetic energy. In engineering fluid mechanical systems, reductions in kinetic energy are referred to as pressure or energy losses, with the implication that useful energy is being dissipated and lost. The "hydraulic" efficiency of fluid flow pipes, for example, is increased by reducing pressure losses and dissipation of energy by turbulence. In flows with chemical reaction, energy is required to promote turbulent interaction between eddies in order to break up large eddies into smaller eddies as a means of accelerating the macro and micro mixing processes. Turbulent kinetic energy, instead of being considered as a loss, requires to be considered as the main driving force for the mixing process. The equation for the kinetic energy of turbulence can be obtained by multiplying the equation for the total velocity

$$\frac{\partial(U_i + u_i)}{\partial t} + (U_l + u_l)\,\frac{\partial(U_i + u_i)}{\partial x_l}$$

$$= -\frac{\partial(P + p)}{\partial x_i} + \frac{g_i(T - T_a + \theta)}{T_a} + v\,\frac{\partial^2(U_i + u_i)}{\partial x_l^2} \qquad (5.11)$$

by the velocity fluctuation and taking the mean value so as to give

$$\frac{\partial}{\partial t}\left(\tfrac{1}{2}q^2\right) + U_l\,\frac{\partial(\tfrac{1}{2}q^2)}{\partial x_i} + \frac{\partial}{\partial x_l}\left(\overline{pu_l} + \overline{\tfrac{1}{2}q^2 u_l}\right) + \overline{u_i u_l}\,\frac{\partial U_i}{\partial x_l} = \frac{g_i}{T_a}\overline{\theta u_i} + \overline{v u_i\,\frac{\partial^2 u_i}{\partial x_l^2}} \qquad (5.12)$$

The term $q^2 = u_i u_i$ is the kinetic energy of the velocity fluctuations per unit specific volume. Each of the terms in Eq. (5.12) can be interpreted as follows:

1. Rate of increase of turbulent energy
2. Gain of energy by the mean flow through advection
3. Transport of turbulent energy by turbulent pressure gradients and by turbulent convection
4. Production of turbulent energy by working of the mean flow on the turbulent Reynolds stresses
5. Gain of energy through working of the buoyancy forces
6. Transformation of fluctuation energy to heat plus a smaller amount of energy diffusion by the working of viscous stress fluctuations

The equation for the total kinetic energy of the flow is given by

$$\frac{\partial}{\partial t}[\tfrac{1}{2}(\overline{q^2} + U_i^2)] + U_l \frac{\partial}{\partial x_l}[\tfrac{1}{2}(\overline{q^2} + U_i^2)] + \frac{\partial}{\partial x_l}[\tfrac{1}{2}\overline{q^2 u_l} + \overline{u_i u_l}\, U_i + PU_l + \overline{p u_l}]$$

$$= v\left[U_i \frac{\partial^2 U_i}{\partial x_l^2} + \overline{u_i \frac{\partial^2 u_i}{\partial x_l^2}}\right] + \frac{g_i}{T_a}[(T - T_a)U_i + \overline{\theta u_i}] \qquad (5.13)$$

Free Turbulent Shear Flows

Free turbulent flows are bounded by ambient fluid, which is generally irrotational and nonturbulent. The turbulent fluid is separated from the ambient fluid by a fairly well defined intermittency surface. This intermittency surface has indentations which may become as large as the width of the flow. A fixed probe, measuring velocity continuously as a function of time, shows the flow to be made up of alternating periods of turbulent and nonturbulent fluctuations, and this is called an intermittently turbulent signal. Within the intermittency surface, the turbulence is roughly homogeneous in scale and turbulent intensity. Free turbulent flows nearly always spread into the surrounding fluid so that the overall flow is necessarily inhomogeneous in the mainstream direction, as well as in the transverse direction.

In the initial and developing regions of a free turbulent flow, a bounding surface can be clearly distinguished as an interface separating the turbulent flow from the surrounding fluid. At any one instant of time, this bounding surface is convoluted with indentations having widths of the same order of magnitude as the width of the turbulent flow. Superposition of many of these instantaneous bounding surfaces shows that, in the mean, the boundary increases linearly with distance downstream. Continuous time records of velocity at a point in the flow show the phenomenon of intermittency in which periods of relative quiescence are interspersed with bursts of activity. Near the boundary of the turbulent flow, the intermittency can be directly associated with the presence of fluid from the turbulent region, giving rise to sharply defined intervals of intense rapid turbulent fluctuation and intervals of relatively smooth change, associated with the presence of surrounding fluid. The distinction between flows on either side of the bounded surface is enhanced when the anemometer is used to measure velocity gradients or vorticity, rather than velocity. When entrainment of nonturbulent ambient fluid occurs across a well defined bounding surface whose shape and position changes continuously within the surface, the flow is fully turbulent with intense, fine-scale fluctuations of velocity gradient and vorticity and the position of the surface can be defined to within a diameter of the smallest eddies of the turbulence. Outside the surface, velocity fluctuations arise from the pressure fluctuations induced by the turbulent eddies.

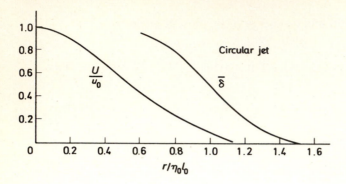

Figure 5.1 Variations of intermittency factor across a circular jet. (*Townsend, 1976.*)

Intermittency is defined by the proportion of time that a detector situated at a point (x, y, z) is within the turbulent fluid, as given by the equation

$$\int_{z}^{\infty} P(\zeta, x) \, d\zeta = \gamma(x, z) \tag{5.14}$$

where $P(\zeta, x)$ is the probability density function for a displacement ζ at a position (x, y) and γ is the intermittency factor which is measured by constructing an intermittency signal; $\delta(x, t)$ is defined to be unity if the anemometer is within the fully turbulent flow and zero if it is in the ambient fluid. Variations of the intermittency factor across a circular jet are shown in Fig. 5.1.

The boundary-layer approximation, which has been used for flows over solid surfaces, has also been applied to free turbulent flows. This boundary-layer approximation allows considerable simplification in the equations of motion. The approximation can be used in flows where the gradients of mean values in the transverse plane are considerably less than the gradients of mean values in the main flow direction. A length scale for variations of mean values in the transverse direction can be fixed, as a proportion of the width of the flow, and this is designated l. The length scale for variation of mean quantities in the main flow direction L is an order of magnitude greater than l.

The ratio of the longitudinal to the lateral scale, L/l, is least for jets issuing into still fluid, and its value increases with the velocity of the surrounding fluid. For jets, L/l is about 8, so that the boundary-layer approximation implies neglect of terms that can be as high as 12 percent of those retained. For flows which are axisymmetric, the equation for U becomes

$$U \frac{\partial U}{\partial x} + V \frac{\partial U}{\partial r} + \frac{\partial}{\partial x} \left[\overline{u^2} - \overline{v^2} + \int_{0}^{\infty} \frac{\overline{v^2} - \overline{w^2}}{r} \, dr \right] + \frac{1}{r} \frac{\partial \overline{uvr}}{\partial r} = U_1 \frac{dU_1}{dx} \tag{5.15}$$

where U is the mean velocity in the main flow direction x; V is the mean velocity in the transverse direction r; u and v are fluctuating velocities in the main flow and transverse directions respectively; U_1 is the mean velocity of the ambient flow in the x direction.

The equation for the integrated flux of momentum is given by

$$\frac{d}{dx}\int_0^\infty U(U-U_1)r\,dr + \frac{dU_1}{dx}\int_0^\infty (U-U_1)r\,dr + \frac{d}{dx}\int_0^\infty \{\overline{u^2} - \tfrac{1}{2}(\overline{v^2}+\overline{w^2})\}r\,dr = 0$$

$$(5.16)$$

The conservation of turbulent kinetic energy is given by the equation

$$U\frac{\partial(\tfrac{1}{2}\overline{q^2})}{\partial x} + V\frac{\partial(\tfrac{1}{2}\overline{q^2})}{\partial y} + \overline{uv}\frac{\partial U}{\partial y} + \frac{\partial}{\partial y}(\overline{pv}+\tfrac{1}{2}\overline{q^2 v}) + \varepsilon = 0 \qquad (5.17)$$

$$\text{Advection} - \text{generation} + \text{diffusion} + \text{dissipation} = 0$$

Each one of the terms in the above equation has been measured using hot-wire anemometry with the exception of the pressure-velocity product, which is obtained by difference. Measurements have been made in simple wakes, two-dimensional jets, axisymmetric jets, mixing layers, and for several kinds of boundary layer. The lateral distributions of the four terms are shown in Fig. 5.2 for a round jet from the measurements of Wygnanski and Fiedler (1969). Turbulent kinetic energy balances in wakes, jets, and mixing layers show the following common features:

1. Generation of turbulent energy is greatest near the position of maximum shear and is zero along an axis of symmetry.
2. Viscous dissipation is distributed over the whole flow and is not especially large in regions of strong generation.
3. There is an overall gain in advection of turbulent energy in jets and wakes, while there is an overall loss in mixing layers or constant pressure boundary layers.

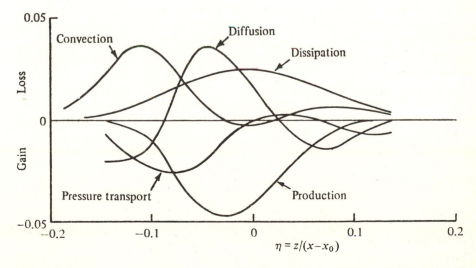

Figure 5.2 Balance of turbulent kinetic energy in self-preserving jet. (*Wygnanski and Fiedler, 1969.*)

4. The distribution of dissipation is fairly uniform. The nearly uniform distribution of dissipation is matched by the distribution of turbulent energy which is diffused from its place of production at least as efficiently as the momentum of the mean flow.

5.3 CORRELATIONS

Spatial Correlations

The correlation between the same fluctuating quantity measured at two different points in space is referred to as a *spatial correlation*. Spatial correlations provide information on the length scales of the fluctuating motion. For two points with coordinates $\mathbf{x}$ and $\mathbf{x} + \mathbf{r}$ and velocity component u, the *covariance* is defined as $\overline{u(\mathbf{x})u(\mathbf{x} + \mathbf{r})}$. This has the dimensions of (velocity)^2. The covariance is nondimensionalized to give the *correlation coefficient*, defined by

$$\frac{\overline{u(\mathbf{x})u(\mathbf{x} + \mathbf{r})}}{(\overline{u^2(\mathbf{x})} \cdot \overline{u^2(\mathbf{x} + \mathbf{r})})^{1/2}}$$

The dimensionless quantity $\overline{u(\mathbf{x})u(\mathbf{x} + \mathbf{r})}/\overline{u^2(\mathbf{x})}$ is more convenient to measure and is often referred to as the correlation coefficient. Correlation coefficients are written in the form $R_{11}(r_1, r_2, r_3)$ for the u-component correlation with variation in coordinates (x, y, z). The symbols R_{22} and R_{33} are used for the v and w correlations. In most of the experimental studies in which spatial correlations have been measured, variations in only one direction were studied of one velocity component, such as $R_{22}(r_1, 0, 0)$, which is the v correlation with separation r_1 in the x direction.

The correlation with separation r is a measure of the strength of eddies whose length in the direction of the vector $\mathbf{r}$ is greater than the magnitude of r. Eddies smaller than the separation distance do not contribute to the correlation. Separate correlations can be measured for each of the three velocity components which will, in general, be different for the same $\mathbf{r}$. Use of multiple probes to measure spatial correlations in several directions simultaneously would provide measurements of length scales from which three-dimensional geometry of eddies could be identified. The length scale of the energy-containing eddies is a length of order $\int_0^\infty R \, dr$. This is referred to as the *integral scale*.

Time Correlations

Correlation between the same fluctuating quantity measured at different times, at the same point in space, is referred to as an *autocorrelation*. Measurement of autocorrelations requires an instrument with a time-delay mechanism, such as a tape recorder with movable heads or a digital sample-and-delay system. Autocorrelations are simpler and more convenient to measure than spatial correlations, which require measurements to be made of correlations as the separation distance between probes is progressively increased. For conditions of low turbulence intensity, the autocorrelation of one velocity component with time delay τ, written as $R_{22}(\tau) \equiv \overline{v(t)v(t + \tau)}/\overline{v^2}$, will be the same as the spatial correlation with separation $- U\tau$ in the x direction (the direction of the mean velocity). This is based on the assumption that eddies or vortex lines do not change appreciably in shape as they pass a given point. This hypothesis of slow rate of change of shape of turbulent eddies is referred to as the *Taylor hypothesis*. When measurements of correlations with separation in the direction of the mean velocity are made using probes, interference effects can arise due to the downstream probe being in the wake of the upstream probe. The autocorrelation is used when the Taylor hypothesis is valid, i.e., low turbulence intensity.

Space-Time Correlations

Spatial correlations provide information about the instantaneous flow patterns in a frozen eulerian flow field. Information on growth and development of eddies can be obtained from examination of records of velocity as a function of time from two separate probes in the flow field. These provide space-time correlation functions which are sensitive to changes of flow pattern that arise either by displacement of individual eddies or by changes in the structure of the eddies. Townsend (1976) proposed that, in turbulent flow, individual eddies move with respect to the surrounding fluid with a self-induced velocity that depends on their structure. The centre of an eddy moves with a velocity compounded of the local mean velocity, the instantaneous local velocity of the larger eddies in which it is imbedded, and the self-induced velocity. The time-delay correlation function for a single velocity component and spatial separation in the direction of mean flow is shown in Fig. 5.3. The function $R_{11}(\mathbf{x}; r, 0, 0, \tau)$ is shown in Fig. 5.3, for several fixed time

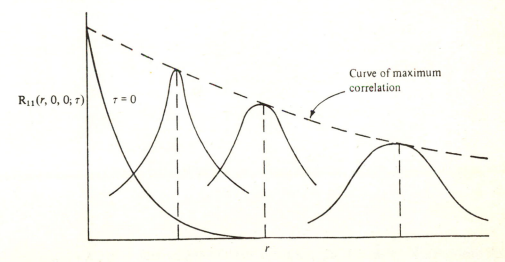

Figure 5.3 Time-delay correlations, as functions of streamwise separation for various time intervals. (*Townsend, 1976.*)

delays at several positions r. This figure shows that, with increasing time delay: (1) the maxima of the correlation function decreases, as the separation distance r increases; (2) the position of the maximum r_m increases nearly proportional to time delay—the ratio r_m/τ is defined as the convection velocity; (3) the radius of curvature at the maximum becomes greater; and (4) the variation about the maximum may become asymmetric. An eddy, moving with the convection velocity, but undergoing changes and loss of identity, is described by the variation of the maximum correlation with time delay and is, thus, shown by the first two factors. The last two effects indicate differences in behavior of smaller-scale components.

The smaller eddies of turbulent flow hold less energy than the larger ones and the major contribution to velocity differences comes from eddies of size comparable with the separation. Velocity differences can be interpreted as a mixture of the influence of eddies of size r, at $\mathbf{x}$, on the flow velocity at the point $\mathbf{x} + 0.5\mathbf{r}$. The covariance between velocity differences of eddies, separated by distance s in space and τ in time, measures the changes in eddies of size r and is termed the *structure function*. When the velocity pattern for eddies of size r is known, a single eddy with center at x_0 has the velocity differences

$$u_1(\mathbf{x}) - u_1(\mathbf{x} + \mathbf{r}) = af(\mathbf{x} - \mathbf{x}_0, \mathbf{r}) \tag{5.18}$$

where f is assumed to be the same for all eddies of size r and a is the velocity amplitude of the eddy. For a particular eddy, the amplitude and center position are functions of time and the contribution of one eddy to the structure function is

$$\{a(t)a(t + \tau)\}\{f[\mathbf{x} - \mathbf{x}_0(t)]f[\mathbf{x} + \mathbf{s} - \mathbf{x}_0(t + \tau)]\}$$

The first factor depends on a change in eddy amplitude and the second depends on translation of the eddy. The whole structure function is given by the probability function

$$P[a(t), a(t + \tau), x_0(t), x_0(t + \tau)]$$

This defines the probability of observing particular values of the amplitudes and center positions in one realization of the flow. Intrinsic changes in the eddies are described by the autocorrelation coefficient for individual eddy amplitudes with time delay τ. Movements of the eddy centers in the time interval are described as the combination of a translation in s space of $\overline{x_0(\tau) - x_0(0)}$ and a diffusive spread in s space by random movements of the eddy centers. A convection velocity is defined as $\mathbf{U}_c = \mathbf{S}_m/\tau$; the magnitude of σ, the standard deviation of the center displacement about its mean value, can be found by comparing the shapes of the structure functions for $\tau = 0$ and for the time delay τ. With a knowledge of σ, the autocorrelation coefficient for the amplitude of individual eddies can be calculated.

The smaller eddies are carried around by the larger ones and the convection velocity of the eddy centers of these smaller eddies is of the same magnitude as the local velocity. If u_0 and L_0 are the scales of velocity and length for the main turbulent motion, velocity changes will be small for times short compared with L_0/u_0.

The correlation between different components of velocity are referred to as *cross-correlations*. Time-delayed cross-correlations can be of the form $\overline{u(t)v(t + \tau)}$. Correlations of more than two velocity components are referred to as *higher-order correlations*. Combinations of three components of velocity at different points in space can provide a wide range of higher-order correlations. These higher-order correlations have rarely been measured but are relevant to sections of turbulence theory. When the Navier-Stokes equations are multiplied by products of velocities at different points before time averaging, a range of conservation equations is derived. The turbulent energy equations can be obtained by multiplying the x-component Navier-Stokes equation by the u-component velocity at the same point, adding the corresponding y and z equations to it and then taking the time average. Improvements in signal processing, data acquisition, and analysis have reduced the difficulty of making higher-order correlation measurements.

5.4 EDDY STRUCTURE

Townsend (1976) proposed that turbulent flow could be broken up into eddies, which are defined as flow patterns with spatially limited distributions of vorticity, having comparatively simple forms. The complete turbulent flow is made up of a superposition of many such eddies of different kinds, sizes, and orientations. Such eddies can be identified by flow visualization of tracer fluid injected either initially or at discrete points in the flow. Developments in high-speed photographic and flow visualization techniques have produced some remarkably clear photographs of discrete eddies in a variety of turbulent flow fields. These provide very convincing evidence of the existence of such eddies and it is possible to follow the formation, growth, rollup, coalescence, and subsequent disintegration of eddies at the various stages of their motion through a flow field. Identification of eddies can also be made by the examination of velocity variations as a function of space or time. A series of mean-value functions have been used to establish the relationship between the form of the functions and the presence of particular forms of eddies. The mean-value function most commonly used to establish the spatial structure of turbulence and its evolution in time is the covariance between velocity components measured at two separated points in the flow. This is referred to as the double-velocity correlation function, defined as follows

$$R_{ij}(\mathbf{x}; \mathbf{r}, \tau) = \overline{u_i(\mathbf{x}, t)u_j(\mathbf{x} + \mathbf{r}, t + \tau)}/(\overline{u_i^2})^{1/2}(\overline{u_j^2})^{1/2} \qquad (5.19)$$

where $u_i(\mathbf{x}, t)$ is the instantaneous value of the ith component of the velocity fluctuation at the position $\mathbf{x}$ and time t, $\mathbf{r}$ is the separation distance from the point $\mathbf{x}$ and τ is the difference in time from t. As the separation between the two points increases from zero, so the correlation function decreases from the value 1.0 until a distance is reached where the correlation function is equal to zero. On the assumption that the motion in one part of the fluid is statistically independent of the motion in a sufficiently distant part, a value of $\mathbf{r}$ can be found where R_{ij} is negligibly small and this is taken to be a typical dimension of the largest eddy. The complete correlation function as a function of position in all directions and of time has not yet been measured because of the experimental complexity but some measurements have been made of the simultaneous correlation function with $\tau = 0$ and of the time-space correlation with spatial separation in the direction of the mean flow.

Townsend (1976) proposed the following forms for the velocity distribution within an eddy

$$u_1 = -\frac{\partial}{\partial x_2}[\exp{(-\tfrac{1}{2}\alpha^2 x^2)}][\cos l_1 x_1 \cos l_2 x_2 \cos l_3 x_3]$$

$$u_2 = \frac{\partial}{\partial x_1}[\exp{(-\tfrac{1}{2}\alpha^2 x^2)}][\cos l_1 x_1 \cos l_2 x_2 \cos l_3 x_3] \qquad (5.20)$$

$$u_3 = 0$$

where

$$\alpha^2 x^2 = \alpha_1^2 x_1^2 + \alpha_2^2 x_2^2 + \alpha_3^2 x_3^2 \qquad (5.21)$$

These velocity distributions represent a finite, three-dimensional array of eddies. Townsend (1976) showed that, if the turbulent flow contains these eddies with their centers distributed randomly but statistically uniformly in space, the correlation functions take on the following forms

$$R_{11}(\mathbf{r}) = -\frac{\partial^2}{\partial r_2^2} f(\mathbf{r})$$

$$R_{22}(\mathbf{r}) = -\frac{\partial^2}{\partial r_1^2} f(\mathbf{r}) \qquad (5.22)$$

$$R_{12}(\mathbf{r}) = R_{21}(\mathbf{r}) = \frac{\partial^2}{\partial r_1 \, \partial r_2} f(\mathbf{r})$$

where

$$f(\mathbf{r}) = A \exp{(-\tfrac{1}{4}\alpha^2 r^2)}[\cos l_1 r_1 + \exp{(-l_1^2/\alpha_1^2)}]$$
$$\times [\cos l_2 r_2 + \exp{(-l_2^2/\alpha_2^2)}][\cos l_3 r_3 + \exp{(-l_3^2/\alpha_3^2)}] \qquad (5.23)$$

A is a constant specifying the intensity of the eddy system and

$$\alpha^2 r^2 = \alpha_1^2 r_1^2 + \alpha_2^2 r_2^2 + \alpha_3^2 r_3^2 \qquad (5.24)$$

Using the above formulations, Townsend constructed three simple artificial eddy structures, as shown in Fig. 5.4. For the simple eddy, type A: $l_1 = l_2 = l_3 = 0$. For the periodic array of eddies, type B: $\alpha_1 = \alpha_2 = \alpha_3 = 0$, the motion is periodic in space and infinite in extent. The finite array of simple eddies, type C, is for the condition $l_2 = l_3 = 0$. Townsend then derived the correlations for these eddies by using Eq. (5.22).

For simple eddies, consideration of the correlation function for large separations provides information on the size and form of these eddies. Turbulent flows

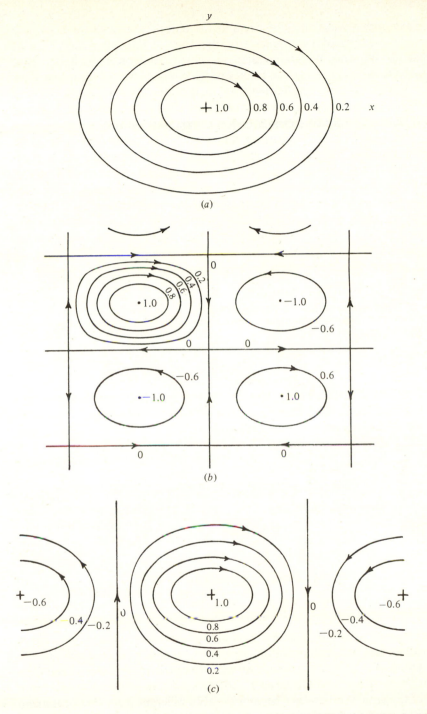

Figure 5.4 Simple eddy structures constructed by *Townsend* (1976). (a) isolated eddy (type A); (b) periodic array of eddies (type B); and (c) finite row of simple eddies (type C).

contain eddies with a wide range of size and the correlation function is used to assign energy to eddies of a particular size or range of sizes. Townsend (1976) argued that the influence of large eddies on velocity differences within smaller eddies may be considered negligible. Eddies of scale much smaller than a particular eddy contribute only to a small extent to velocity differences within the particular eddy. Thus, in considering the influence of individual eddies on adjacent eddies, it is only necessary to consider eddies of comparable size.

From the analysis of the spectrum functions of the three types of eddies, initially proposed by Townsend, he concluded that, due to the initial assumption that a physically acceptable eddy must be of finite extent, eddies are more likely to be of types A and C, rather than type B. The limitations of inferring velocity patterns from observed spectra and correlations have been stressed by a number of investigators, since they usually involve preconceptions of doubtful validity. After a period of many years, during which a wide range of concepts of turbulence were formulated, the foresight of Townsend in proposing some of the original concepts of eddy structure has come to be recognized.

The Taylor Approximation of Frozen Flow

The time-delay correlation function can be converted to a spectrum function to give information about eddies of various sizes. The decrease of maximum autocorrelation with time is caused mainly by the random movement of eddy centers and, if random displacements are small compared with the eddy diameters, the changes in the structure function can be shown to be those produced by simple translation by a convection velocity. In homogeneous turbulence, the convection velocity is the same for all sizes of eddies and equal to the mean-flow velocity. The Taylor approximation asserts that

$$u_i(\mathbf{x}, t) = u_i(\mathbf{x} - \mathbf{U}\tau, t + \tau) \tag{5.25}$$

for not too large values of τ. This approximation is used for calculating one-dimensional spectrum functions from measured frequency spectra. If a velocity pattern in the neighborhood of a probe is moving with velocity $U + u$, Fourier components of frequency ω are derived from components whose wave numbers satisfy

$$\mathbf{k} \cdot (\mathbf{U} + \mathbf{u}) = k_1 U_1 + \mathbf{k} \cdot \mathbf{u} = \omega \tag{5.26}$$

The approximation also allows the determination of mean values of functions of turbulent velocity gradients from single anemometers, using electrical circuits to perform time differentiation. The use of the frozen flow approximation can be applied to all flows whose variations of mean velocity and fluctuating velocity are both small, compared with the average velocity over the whole flow. This condition is not satisfied in turbulent jets and boundary layers where the convection velocities of the larger eddies may be considerably different from the local mean velocity. The small eddies that determine the velocity gradients are, however, generally convected with the local velocity of the fluid.

In ordinary shear flows, neither the turbulent motion nor the gradients of mean velocity are spatially uniform but the turbulence is found to be approximately homogeneous under conditions of nearly uniform strain. In shear-free turbulence, the greater part of the turbulent energy resides in a group of larger eddies, which are stable persisting structures, which transfer energy to smaller eddies without causing any significant change to the structure. When the larger eddies are subjected to either irrotational distortion or plane shear, the eddies remain persistent and retain their coherent structure.

The rate of energy dissipation in turbulent flow is determined by the structure and intensity of the larger eddies. The conversion of mechanical energy to heat is carried out by much smaller eddies of almost negligible total energy. The larger eddies break up to form eddies of slightly smaller size, which become unstable and, in turn, break up into even smaller eddies. The energy transfer process has been referred to as a cascade of instabilities. If the rate of energy flow down the cascade of eddy sizes is limited by the capacity of the first instability, the smaller eddies must adjust their motion to pass on the imposed energy flow to the smallest eddies that dissipate as heat. This cascade hypothesis for the structure of the smaller eddies, that contain only a small part of the whole turbulent energy, has been used by Kolmogorov. He argued that, if the Reynolds number of the flow was sufficiently large, the smaller eddies must be in a state of absolute equilibrium, in which the rate of receiving energy from larger eddies is very nearly equal to the sum of the rates of loss to smaller eddies by breakup and by working against viscous stresses. Kolmogorov separated the motion of the smaller eddies, which he considered to be nearly isotropic, from the larger eddies which may be inhomogeneous and anisotropic. The theory of local isotropy asserts that the motion of the smaller eddies depends only on the energy flow from the energy-containing eddies and on the fluid properties. The differences between various types of turbulent flows are essentially differences in the larger eddies so that, once the energy loss from the larger eddies is known, the structure of the smaller eddies can be automatically determined. It is this hypothesis of local isotropy and similarity which leads to the conclusion that the essential differences in the wide range of turbulent flows can be found from the examination of the differences in the large-eddy structure.

5.5 COHERENT STRUCTURES

Several studies have clearly demonstrated that turbulent flows contain structures or eddies possessing identifiable characteristics, existing for significant lifetimes, and producing recognizable and important events. The work of Roshko (1976) has led to the identification of large coherent structures in several turbulent shear flows. Further, studies in jets, boundary layers, mixing layers, and other shear flows also show the presence of these structures, and that the development of the flow is controlled by the interactions of these structures. The understanding of the properties of these structures provides insight into the physical processes occurring in turbulent flows, such as entrainment, transport, mixing, noise production, gustiness, and intermittency.

The Turbulent Mixing Layer

The experiments of Roshko (1976) and coworkers have provided the clearest identification of the structure of turbulent flow in a mixing layer. Brown and Roshko (1974) examined the plane mixing layer between a stream of nitrogen and a stream of helium, in which the low-density gas had a high velocity and the high-density gas had a low velocity. Their initial intention was to examine the effects of variation of velocity ratio and density ratio in a two-dimensional-plane mixing layer. Because of the large difference in optical indices of refraction of helium and nitrogen, Brown and Roshko (1974) were able to detect the interface between the two streams and its subsequent rolling up, as shown in Fig. 5.5. The conditions for this experiment were $U_2/U_1 = 0.38$, where U_2 and U_1 are the initial mainstream velocities of the nitrogen and helium respectively. The Reynolds numbers, based on the full length of the layer, are 1.2, 0.6, and 0.3×10^5 respectively from top to bottom. The bottom picture, with the lowest Reynolds

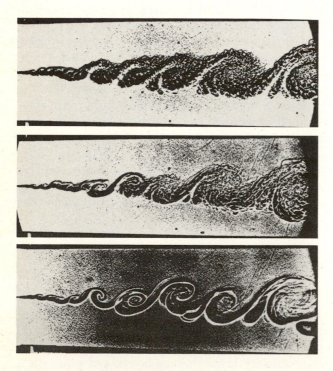

Figure 5.5 Two-dimensional mixing layer between helium (upper) and nitrogen streams. $U_2/U_1 = 0.38$. Reynolds numbers based on full length of layer are 1.2, 0.6, and 0.3×10^5 respectively, from top to bottom. (*Roshko, 1976.*)

number, shows a remarkably clear picture of the interface between the two streams. The interface rolls up so as to form eddies, or vortices, which have the appearance of breaking waves or rollers. The interface remains distinct as it rolls up and is finally diffused within the eddy. The eddies clearly grow in size and, as the Reynolds number is increased, fine-scale turbulence develops, but the basic structure has the same pattern as for the lower Reynolds number case.

From the first experiments of Brown and Roshko (1974), it was thought that the phenomenon they had observed was attributable to the large density difference between helium and nitrogen. The experiments were repeated using different gases of the same density on the two sides of the mixing layer: nitrogen on the high-speed side and a mixture of helium and argon on the low-speed side. With this mixture, optical visualization was still possible and the shadowgraphs of the flow showed a very similar picture of the interface as in the nonuniform density study.

Coherence and Lifetimes

Brown and Roshko (1974) examined the high framing-rate motion pictures of the mixing layer and, from an identification of the vortex-like structures on each frame, they determined the trajectories of individual vortices and their spacing. They concluded that all the vortices moved at nearly constant speed, which was approximately the average of U_1 and U_2. The birth of a new vortex was found to coincide with the demise of two or more old ones.

It has long been established, in turbulent mixing layers, that the scale of any feature of the flow increases with increasing distance downstream. This has followed from the general similarity property of the flow, which requires that all mean length scales be proportional to the distance from the origin. Brown and Roshko found that the sizes of the vortices and their spacings both increased with distance downstream. Roshko (1976) argued that the mean spacing $l(x)$ must increase with x, smoothly because there is nothing special about any particular value of x, and linearly (like the thickness) because of the particular similarity property for the flow. Analysis of the individual frames of the motion pictures indicated that spacings between individual vortices were fairly constant, changing only during the interaction effects. Roshko reconciled the two apparently contradictory features by pointing out that, in the vortex pairs passing any particular value of x, there is a distribution of the spacings about a mean value, $\bar{l}$. The mean value, $\bar{l} = 0.31x$, was found to be close to the most probable value.

The possible presence of large-eddy structures in wakes was studied by Townsend (1976) and Grant (1958). They made simultaneous measurements of the velocity fluctuation $u(t)$ at two points separated by a distance ξ. They measured the correlation $R(\xi) = u(x)u(x + \xi)$ for different directions and different velocity components. From these measurements, they inferred the presence of organized, large-eddy structures in the wake. Townsend further developed theories concerning the formation, size, and subsequent lifetime of these eddy structures.

Favre et al. (1957) used the method of space-time correlation in which the time interval, as well as the space interval, were varied

$$R(\xi, \tau) = \overline{u(x, t)u(x + \xi, t + \tau)} \tag{5.27}$$

From the streamwise, time-delayed correlations, the first evidence was obtained for the presence of large eddies moving at a convection velocity, U_c. Whereas, classically, space-time correlations were used to provide evidence of the decay of large eddies, Roshko (1976) has reinterpreted the results, showing that individual eddies do not decay, but their lifetimes are varied. The correlation envelope is considered to be the probability $P(\tau)$ for an eddy to survive to an age, $\tau = \xi/U_c$. The probability function for the lifespan of an eddy to exceed a value $L = U_c\tau$ is denoted by $P(L)$, or $P(L/\bar{L})$, where L is the average lifespan. The measured space-time correlation envelopes for mixing layers are fitted by the exponential function

$$P(L) = \exp\left(-L/\bar{L}\right) = \exp\left(-\lambda/\bar{\lambda}\right) \tag{5.28}$$

where $\lambda = L/x = \tau U_c/x$ and $\bar{\lambda}$ is interpreted as the average normalized lifetime and also the most probable value. Typical values for $\bar{\lambda}$ are between 0.35 and 0.5, leading to the conclusion that the average lifespan of an eddy is about $0.4x$ from its point of origin at x. Analyses of the motion pictures of Brown and Roshko (1974) led them to infer an average lifespan, $L = 0.43x$. On the basis of this comparison, Roshko has concluded that the space-time correlation envelope may be considered to be a life-expectancy curve of an individual vortex.

Roshko further discusses the possible relationship between the existence of coherent structures and the energy spectrum of velocity fluctuations. He explains that the presence of coherent vortex structures was not earlier revealed by measured energy spectra, due, in part, to the broad dispersal of vortex spacings, which produce a correspondingly broad peak in the spectrum. In addition, the velocity energy spectrum receives contributions to small wave numbers from the pairing events which may tend to overlap or submerge the broad peak. In previous attempts to explain energy spectra, there have been difficulties in explaining the sources of the energy contributions to the low wave numbers, which contain most of the energy. In addition to the important scales furnished by eddy spacings and eddy lifespans, even larger scales are introduced by vortex coalescence events and the resulting disruptions of order along the shear layer. Even though most of the evidence which has been presented so far has been based upon two-dimensional studies, it is recognized that three-dimensional effects may be very significant.

Shear Layer Growth by Vortex Interaction

The growth of shear layers has previously been explained as being due to turbulent diffusion. Turbulent diffusion was considered to be an extension of laminar diffusion on a molecular scale to turbulent diffusion on the scale of fluid particles. Concepts of turbulent diffusion and mixing require to be reexamined in the light of the evidence that has been brought forward by Roshko and coworkers. The spacings between individual pairs of vortices were found to remain approximately constant until coalescence occurred, when the spacing between the newly formed vortices increased. It can be deduced from this that the interaction which accomplishes this coalescence must be an important contributor to the growth of the mixing layer. Winant and Browand (1974) made a careful study of the interaction process and described "pairing" as the dominant mode of interaction and the principal mechanism for growth. In pairing, neighboring pairs of vortices rotate around each other and amalgamate into a larger one. Examples of pairing were shown for low Reynolds numbers by Winant and Browand (1974) and, subsequently, for much higher Reynolds numbers by Roshko (1976). In flows with large density differences, individual vortices can be followed over their lifespans, but pairing by orbiting was not evident. Suggestions have been made of other modes of amalgamation, such as elongation and accretion onto another vortex.

Entrainment

When fluid emerges from a nozzle in the form of a jet, the fluid particles passing through the nozzle become dispersed in the surrounding flow. This dispersion is confined within the limits of the jet boundary. The width of the jet boundary increases linearly with distance away from the nozzle exit. Fluid emerging from the nozzle can be distinguished from surrounding fluid by temperature, molecular species, concentration, or velocity difference. By the processes of exchange of momentum and mixing, nozzle fluid is dispersed and diluted. Nozzle fluid remains concentrated around the axis of the jet with dilution and dispersion occurring to a greater degree at the outer edges. The mass flow rate of fluid within the jet boundaries increases with distance away from the nozzle exit. Surrounding fluid is drawn into, and accelerated across, the jet boundaries. Entrainment is defined by the quantity of surrounding fluid which crosses the jet boundaries and then, subsequently, is accelerated in the main flow direction of the jet. When the surrounding flow is nonturbulent and irrotational, the phenomenon of entrainment can be described as the incorporation of nonturbulent fluid into the turbulent region or, conversely, the diffusion of the turbulent region into the ambient flow. When the ambient flow is turbulent, the definition of entrainment must be associated with the designation of a boundary, separating the main shear flow from its surroundings.

The studies of Roshko and coworkers are remarkably clear for the specific case of a two-dimensional mixing layer from which the shadowgraphs show a distinct interface between turbulent and nonturbulent fluid which rolls up to form separate eddies. In this rollup process, individual eddies are seen to engulf fluid from the surroundings. Roshko describes the process of entrainment by free stream fluid being drawn in between vortices and ingested into the shear layer, where it is digested and made turbulent by the action of the smaller eddies.

In the past, superficial boundaries of the mixing region have been designated by conical surfaces along the edges of the mixing layer with the apex of the cone at the origin of the mixing layer. These boundaries were fixed on the basis of measured mean profiles of velocity, concentration, or temperature and based on some arbitrary selection of the value of the mixing layer concentration at the edge of the layer. This definition is simple, but ignores the physical process by which the surrounding fluid is entrained across the boundary. Earlier attempts at explaining the process of entrainment were based upon extrapolations from the kinetic theory of gases by developing physical models, in which molecules were replaced by fluid particles or eddies and mean free paths were replaced by mixing lengths. Entrainment was described simply as a diffusion process governed by coefficients of turbulent eddy diffusivity, viscosity, and conductivity. The physical process of entrainment is more like a "gulping" or "engulfing" process, as shown by shadowgraphs of the interface between mixing layer and surrounding fluid, rather than the previous concepts of diffusion as a "nibbling" process.

The entrainment process, as accomplished by the large-scale engulfing action of the large eddies, determines the growth, mean characteristics of the flow, mean velocity profile, shear stress distribution, mean transport, and dissipation. The further "turbulization" by smaller eddies is merely a stage in the dissipation of the energy that has been extracted from the mean flow. Photographs of interfaces show that the ambient fluid makes deep incursions into the mixing region. Part of this ambient fluid, found deep within the mixing layer, appears to be undisturbed and indistinguishable from the bulk ambient flow. The boundary between ambient fluid and mixing layer fluid can be well defined at any one instant of time by a photograph of the interface. The quantity of entrained fluid, determined from integration of mean velocity profiles, can be expected to be very different from that determined on the basis of boundaries of the instantaneous visualized flow. There should, however, be some correlation between these values, which must converge at larger distances from the origin of the layer.

Intermittency provides further insight into the physical process of entrainment. When a probe, measuring velocity, temperature, or concentration, with a high-frequency response is inserted near the edge of a mixing region, the trace recording the velocity, temperature, or concentration as a function of time is found to be made up of bursts of activity, separated by periods of relative quiescence. Intermittency is defined as the proportion of time that the flow is quiescent. The intermittency factor has a value of one in the quiescent ambient fluid and decreases to a value approaching zero as the activity increases toward the center of the mixing region. The extent to which portions of ambient fluid penetrate into the mixing layer is indicated by the intermittency factor.

In turbulent flows, the standard deviation from the mean is a measure of the intensity of turbulence. This standard deviation can be determined from variations of temperature, concentration, and velocity as a function of time. The standard deviation of temperature, velocity, and concentration from the mean provides a measure of the intensity of turbulence, which is zero for stationary and laminar flows. In Roshko's experiment, where helium and nitrogen were used, the concentration measurements with a high-frequency-response aspirating probe showed that the excursions of helium concentration across the mixing layer were very large, and that unmixed fluid from one side penetrated deep into the other side of the mixing region. The intermittency in the flow was strikingly evident. Since concentration changes can take place only by short-range intermolecular diffusion, measurements of concentration and temperature are more sensitive to the mixing process than measurements of velocity fluctuations, which can be induced in the nonturbulent or irrotational parts of the fluid by those in the turbulent parts.

Mixing

The mixing process across the boundaries of mixing regions appears to be separated into two distinct processes, firstly, the large-scale entrainment process previously described, followed by the more intimate mixing process. Mixing is defined in terms of the respective concentrations of fluids. Depending upon the frequency response and accuracy of the measuring instrument, we can distinguish between time-average and instantaneous mixing and also separate mixing on the molecular, micro, or macro scales. Entrainment is considered as mixing on a macro scale; subsequently, mixing takes place on the micro and, finally, on the molecular scale. Turbulence intensity does not provide a clear indication of mixing, although there is some correlation between variations in turbulence intensity and changes in rates of mixing. Since combustion and chemical reaction take place at the molecular level, rates of chemical reaction can be taken as a measure of the extent of complete mixing between initially separated fluids.

Figure 5.6 shows a trace of the measurement of concentration of helium as a function of time where the sampling point is on the nitrogen side of the layer, as measured by Roshko (1976). The trace shows bursts of helium concentration, separated by periods of zero concentration. This is a clear demonstration of the intermittent nature of the flow. When the concentration has a value of one, the helium is unmixed with the nitrogen, which has a concentration with the value zero. In attempting to relate the concentration to the degree of mixing, we require to know the ultimate concentration that could be achieved at a particular point if the two fluids were perfectly mixed. This requires designating an arbitrary volume and, after allowing entry of fluids across the boundaries of this volume, provision of sufficient time and energy for the mixing to be completed. Since this ideal situation seldom arises, it appears that there is no simple relationship between instantaneous concentration and the degree of mixing. A careful examination of the motion of the interface in mixing layers shows that intense mixing takes place within individual eddies, after they have been formed by the rollup of the interface. Mixing on a molecular level is more likely to take place within the vortex.

The life of a vortex and its interactions can be described by the following events. In the process of amalgamation, irrotational fluid is ingested and involved with the coalescing vortices. The resulting composite structure consists of the two or more coalesced vortices (plus all previous ones) and the ingested fluid. During its lifetime, the structure rotates and is subjected to stress. At the same time, internal mixing is occurring by the action of the small-scale turbulence and viscosity, and the new fluid is digested and incorporated into the structure. There may

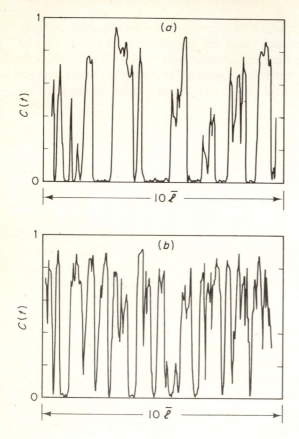

Figure 5.6 Concentration fluctuations in two-dimensional mixing layer. (*Roshko, 1976.*) (*a*) $y/x = -0.095$; $\bar{C} = 0.45$; UM $= 0.63$. (*b*) $y/x = -0.084$; $\bar{C} = 0.55$; UM $= 0.54$.

also be some small-scale turbulent diffusion laterally into the free stream fluid, resulting in growth of the structure before the next pairing of vortices.

Winant and Browand (1974) describe the pairing process when a wave develops into an S-shaped pattern, which evolves into two vortices that subsequently rotate around each other to form a new single vortex. This pairing process has been observed by a number of investigators, some of whom have been able to demonstrate the phenomena on a computer where calculations have been made without taking into account the effects of viscosity.

Unmixedness

The extent of molecular mixedness is defined by the unmixedness factor

$$\text{UM} = \frac{\int_{T_1}(C - \bar{C})\, dt_1 + \int_{T_2}(\bar{C} - C)\, dt_2}{(1 - \bar{C})T_1 + \bar{C}T_2} \tag{5.29}$$

where t_1 corresponds to time when $C > \bar{C}$ and t_2 to time when $C < \bar{C}$. C is the instantaneous concentration and $\bar{C}$ is the mean concentration. When the flow is completely mixed, there is no fluctuation of concentration about the mean and UM is zero. The flow is considered to be completely unmixed when $C(t)$ fluctuates between values of zero and one so that UM has a value of one. Variations of the unmixedness factor, as determined by Roshko, are shown in Fig. 5.7; (a) was for the case of a mixing layer between helium and nitrogen $(\rho_2/\rho_1 = 7)$, while (b) is for a mixing layer between nitrogen on one side and helium/argon on the other, so that $\rho_2/\rho_1 = 1$. The peaks of unmixedness are connected with the high intermittency at the edges of the mixing layer.

The unmixedness factor was originally measured in flames by Hawthorne et al. (1951). Probes inserted into hydrogen-oxygen flames provided measured average concentrations of oxygen and hydrogen from samples drawn from within the flame. Because of the high temperatures and high reaction rates between oxygen and hydrogen, Hawthorne et al. (1951) argued that hydrogen and oxygen could not be present simultaneously at any one instant of time. The turbulent fluctuations of the flame front across the probe resulted in samples being drawn from the flame which were rich, either in hydrogen or oxygen. The time-average concentrations of hydrogen and oxygen removed from the flame front region were directly dependent upon the magnitudes of the concentration fluctuations of flow across the probe. Hawthorne et al. assumed that the concentration fluctuations were directly related to the velocity fluctuations and utilized hot-wire anemometer measurements of velocity fluctuations in nonburning jets in order to assess the relationship between the velocity and concentration fluctuations. With the considerable advance in the degree of sophistication in experimental techniques, it is now possible to measure velocity fluctuations in flames and these are known to be different from the velocity fluctuations measured in corresponding nonburning jets. An experiment in which velocity fluctuations were measured by laser anemometer and gas concentration measurements, with a frequency response comparable to that obtained by Roshko (1976) in his helium-nitrogen experiments, would provide very useful information as regards the degree of unmixedness in turbulent diffusion flames.

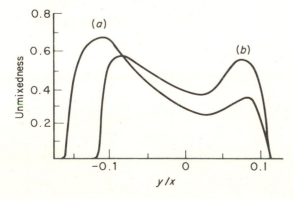

Figure 5.7 Distribution of unmixedness factor UM across two-dimensional mixing layers. (*Roshko, 1976.*) (a) U_1, $x/v_1 = 4 \times 10^4$; (b) U_1, $x/v_1 = 3 \times 10^5$.

Probability Density Distributions

Probability density distributions provide information as to the extent of the deviation from the mean and the degree of symmetry about the mean value. Most of the
information on probability density distributions in turbulent flows has come from
measurements of velocity but the small fast-response probe used by Roshko
(1976) to measure composition of a binary gas mixture has provided information
on the probability distribution for the concentration at a given point, as shown in
Fig. 5.8. Near either side of the mixing layer, there is a high probability, due to
intermittency, of detecting the pure gas from that particular side. In the middle of
the layer, there is a broad distribution about the peak. The deviations from the
mean are seen in Fig. 5.8 to be very extensive. At the center of the mixing layer, it
is still possible, at one time, to measure a concentration of pure gas from one side
of the layer, while, at another time, it is possible to measure a concentration of
pure gas from the other side of the layer. The distributions shown in Fig. 5.8
indicate the severe limitations of describing mixing in terms of the mean concentration at any one particular point in a mixing layer.

Reynolds Number Effects

The work of Roshko and coworkers has been considered to be restrictive on two
counts: firstly, the flow configuration which they used was designed to be two
dimensional, in that very special efforts were made in order to constrain the flow
in two dimensions, whereas turbulent flows are three dimensional; secondly, the
flow conditions and, in particular, the Reynolds numbers originally used by
Roshko and coworkers were such that the flow was not representative of fully
developed turbulence but more representative of transitional flow. There is little
doubt that almost all turbulent flows are three dimensional but there is no reason
to exclude the possibility of the existence of a two-dimensional turbulent flow or,
at least, that the large structures can be quasi two dimensional. If such a strictly
two-dimensional turbulent flow can be generated, it is extremely difficult to prevent this flow from becoming three dimensional. The contention that the flows
examined by Roshko and coworkers are not fully turbulent has been countered by

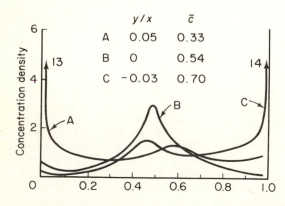

Figure 5.8 Probability-density distributions of concentration at three points
in two-dimensional mixing layer.
(*Roshko, 1976.*)

Roshko, who quotes Reynolds numbers based upon dimensions of the mixing layer as being well in excess of the critical Reynolds number for transition from laminar to turbulent flow. He also cites examples of flows in the atmosphere and in large rivers where the scale, and hence Reynolds numbers, are well above the critical Reynolds numbers for those particular flows.

In the sequence of pictures of a mixing layer, shown in Fig. 5.5, the Reynolds number was varied from 0.3 to 1.2×10^5. The large-eddy structure and the lateral extent of the mixing region is seen to remain unaltered by the increase in Reynolds number by a factor of 4. Measurements of the mean flow properties in these examples did not reveal any significant effects of changes in Reynolds number. As the Reynolds number is increased, the energy-transforming and energy-dissipating scales in a turbulent flow become separated. The photographs in Fig. 5.5 show that, as the Reynolds number is increased, the small-scale structure associated with viscous dissipation does change.

From an examination of the spreading rates of mixing layers measured by a series of experimentors over a range of Reynolds numbers, Roshko concluded that all the evidence suggested that any important effects of Reynolds number appear indirectly through the initial shear layer conditions and not through direct action of viscosity on the developing turbulent structure. The thickness of the initial mixing layer just after separation and the distribution of vorticity in it depends on the particular nozzle geometry, on the thickness of the splitter plate, and on the Reynolds number based on some nozzle dimension. Bradshaw (1966) has shown that at least one thousand initial momentum thicknesses are required before a mixing layer achieves a local similarity structure and becomes independent of the initial conditions of any particular experiment.

Transition Region

The term "transition region" is usually used to describe a flow regime between laminar and fully developed turbulent flow. In early studies of friction factor as a function of Reynolds number for flow in pipes, or around cylinders and spheres, the measurements of friction factor at a fixed Reynolds number within the transition region were found to be erratic and not reproducible. In later experiments, in which the propagation of disturbances in flows was studied, the flow was considered to be laminar when disturbances were damped. When disturbances were amplified, the flow was considered to be turbulent. Vortex shedding, as found in the wake of cylinders, is an example of an intermittent flow superimposed on a laminar flow.

Another criteria for distinguishing between laminar and turbulent flows is the extent to which the flow is dependent upon Reynolds number. In laminar flow, viscous effects dominate, whereas, in turbulent flows, the effects of viscosity, and hence Reynolds number, are insignificant. The transition region is a flow regime which commences at the lower Reynolds number range, where effects of viscosity are dominant; as the Reynolds number is increased, inertia and viscous effects become equal in magnitude; as the flow approaches the turbulent condition, inertia effects dominate the flow. In mixing layers between two laminar streams, the separated, laminar, free-shear layer thickness develops and rapidly amplifies downstream. This type of motion can be shown to be practically independent of viscous effects as shown by good agreement of measurements with predictions of the amplification and subsequent nonlinear development based on inviscid theory. The physical structure of mixing layers, as shown by the work of Roshko, and other work carried out in the intermittent region of jet flows, indicates that the distinction between laminar and turbulent flows is complex and cannot simply be described as being transitional.

5.6 COHERENT STRUCTURES IN AXISYMMETRIC JETS

The three-dimensionality of axisymmetric round jets makes observation by flow visualization more difficult than in the two-dimensional mixing layer. Whereas the two-dimensional mixing layer expands without restriction, the inner boundary of the round-jet mixing layer is restricted and can only extend to the centerline of the jet, where the conical surface converges to an apex, while the external boundary continues to diverge without restriction. The behavior of eddies within the initial region can thus be expected to be different from that in the fully developed region farther downstream, where there is a single interface between the jet and its surroundings. Yule (1977) designated the first seven diameters of the flow as the "transition region" in which streets of vortex rings were clearly visible. Farther downstream in the turbulent region, large eddies are more diffuse and less clearly visible. Visualization studies carried out in both water and air jets show that both transitional vortex rings and turbulent large eddies grow in scale by coalescing. Turbulent large eddies are formed initially by the coalescing of vortex rings; coalescing of two, and sometimes more, vortex rings frequently occurs without transition to turbulent flow. The transition and turbulent regions of jets are described schematically by Yule (1977) in Fig. 5.9, based upon visualization studies. Very clear demonstrations were obtained of the formation of turbulent large eddies by two vortex rings coalescing. From the analysis of movie films, the velocity and direction of movement of vortex cores during the coalescing process was determined. The velocity fields of vortex rings and local average passing frequencies of vortices were measured for a range of jet Reynolds numbers. The structures in the transition region were found to be repetitive, but wide variations were found in the strengths, velocities, and separations of individual vortices. Some of the photographs obtained in the transition region of a round jet, when viewed in cross section (Fig. 5.10), are remarkably similar to those observed by Roshko (1976) in a two-dimensional mixing layer.

In the turbulent region, downstream the transition region, Yule (1977) found traces of the transitional vortices within the large eddies of the turbulent region. These larger eddies were more clearly visible in the central and inner part of the turbulent mixing region. Entrainment into eddies was seen as an engulfment

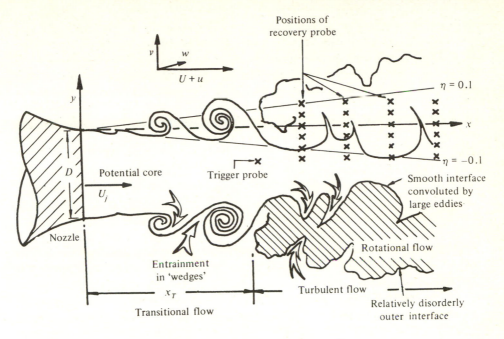

Figure 5.9 Mixing layer in axisymmetric round free jet. (*Yule, 1977.*)

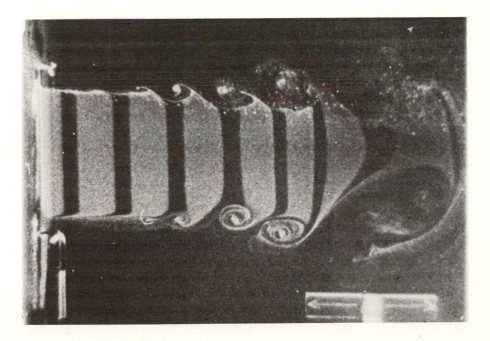

Figure 5.10 Vortex formation and coalescence in submerged water jet; visualization by pulsed formation of hydrogen bubbles. (*Yule, 1977.*)

process which was greatest during merging; during eddy interaction, fluid was ejected from the eddies to form an outer layer of diffuse slower moving fluid. Tracer fluid within the large eddies had a diffuse appearance, unlike the orderly spiral structure of the transitional vortices, indicating the presence of mixing on a scale smaller than the size of the eddies. The turbulent region was not symmetric across the jet centerline.

The separate experiments of Roshko (1976) on the two-dimensional mixing layer and those of Yule (1977) on the round jet have led to a reexamination of the relationship between two-dimensional and three-dimensional flows. It is extremely difficult to experimentally achieve a two-dimensional turbulent flow in which no changes occur in the third dimension. Whereas two-dimensional flows can be generated at low Reynolds numbers under laminar-flow conditions, the flow breaks up into three-dimensional eddies as the Reynolds number is increased, either by increasing velocity or by increasing distance downstream of the origin of a mixing layer. Fully developed turbulent flow has always been considered as being three-dimensional with a tendency of the turbulence toward isotropy, at smaller scales at least. In the initial region of a round jet, where the thickness of the mixing layer is small compared with the diameter of the jet, the flow has some aspects of two-dimensionality. As the mixing layer grows in size, there is a transition from this two-dimensional flow to a three-dimensional flow. Yule has contended that three-dimensionality is a critical factor distinguishing the turbulent and transitional flow regimes. On the basis of the similarity of the structures found in the initial region of the round jet with those obtained by Roshko, Yule argues that the flow conditions of Roshko were transitional and not turbulent.

Mixing Layers in Round Jets

The mixing region of a round jet is bounded on the inside by the potential core region and on the outside by stagnant surroundings or the mainstream flow. The origin of the mixing layer is a circle at the nozzle exit where contact between jet and surrounding fluid is initiated. The mixing layer spreads downstream with the inner boundary merging toward the axis of the jet at the end of the potential core. The outer boundary of the mixing layer is the outer boundary of the jet. The structure of the mixing layer is shown schematically in Fig. 5.9. In this figure, the interface separating jet and surrounding fluid is shown to roll up and form ring vortices, which subsequently become unstable and lead to the formation of large eddies. The flow in the mixing layer is initially laminar, subsequently becomes unstable, and, after passing through a transition region, becomes turbulent farther downstream. As the laminar shear layer near the nozzle becomes unstable, migration of vorticity takes place, resulting in the formation of a periodic "street" of circumferentially coherent vortex rings. Cinematographic studies of submerged water jets with visualization by hydrogen bubble pulsed time lines and dye visualization show the formation of individual ring vortices in the mixing region. Periodical ring vortices accelerate and coalesce, as shown in Fig. 5.10.

Transition

The transition distance in a round jet is defined as the distance from the nozzle exit to the beginning of the turbulent flow region. A number of different criteria can be used for designating the beginning of turbulent flow: (1) from flow visualization, the relative disorder and rapid rate of diffusion can be used as indicators of the onset of turbulent flow; (2) similarity of mean velocity profiles, generally of gaussian shape; (3) equality of the axial, radial, and circumferential turbulence intensity components $(\overline{u_i^2})^{1/2}$ on the centerline of the mixing layer, i.e., local isotropy; (4) transition from a periodic correlation near the nozzle to a characteristic fully developed form of turbulence covariance with time delay, that is, $\overline{u_i u_j}(\tau)$.

On the basis of measurements in a round turbulent jet, Yule (1977) has found that the asymptotic turbulent values of the three turbulence intensity components were the same to within 15 percent with a value of $(\overline{u^2})^{1/2}/U_j$, varying between 0.14 and 0.15. The jet mixing layer is considered to be locally turbulent when the peak values of $(\overline{v^2})^{1/2}$ and $(\overline{w^2})^{1/2}$ are within 15 percent of each other and the values are independent of x. It was also concluded that the attainment of similarity distributions of $(\overline{u^2})^{1/2}$ and U across the mixing layer could not be used as criteria for proving the local existence of fully developed turbulent flow. Within the range of Reynolds numbers up to 5×10^5, the experimental data provided the correlation $x_T U_j/v = 1.2 \times 10^5$ with transition distances being of the order $x_T = 4D$.

Vortex Rings in Transition Regions

A number of visualization studies have shown clearly the formation of distinct vortex rings in the initial laminar mixing region of round jets. From turbulence measurements in these regions, Bradshaw (1966) described how peaks in $\overline{u^2}$ and $\overline{v^2}$ intensity distributions near the nozzle can be related to the coalescing of these vortex rings. Yule (1977) has shown that the passage of vortex rings across fixed probes can be detected by peaks in velocity frequency spectra measured in the jet potential core. Direct comparison of the frequency spectrum measurements with high-speed movie films showed that individual vortex rings would be clearly seen in regions where frequency halving was found in the spectra. The average positions where coalescence of vortex rings could be seen was found to be the same as the positions where frequency halving was observed in the spectra.

Development of Turbulent Region

Visual observation of the cross section of a jet shows the ordered growth of wave deformations of the cores of the vortex rings. Jets of fluid are shed from regions between vortices. The interface of a jet has the appearance of various lobes, which are associated with the wave deformation of individual vortex rings. The instability of vortices in the transition region of the jet are remarkably similar to the instabilities of single vortex rings.

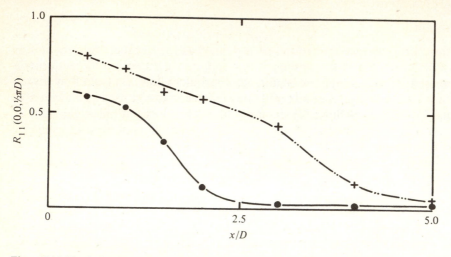

Figure 5.11 Physical structure of transitional axisymmetric round jet. (*Yule, 1977.*)

A general description of transitional jet structure, as given by Yule, is shown in Fig. 5.11. This schematic diagram shows the initial formation of a street of vortex rings, due to the natural instability of the laminar shear layer. As these vortex rings move downstream, they generally coalesce with neighboring rings so that the scale and separation of the vortex rings increase with distance from the nozzle. The movements and strengths of the coalescing vortices are random and the vortex rings lose their phase agreement across the jet as they move downstream.

Wave deformations of the cores of vortex rings grow almost linearly and this growth results in a decrease in the level of circumferential cross-correlations. As a result of growth and coalescence, vortex rings become deformed and develop into the large eddies of the turbulent region.

Large Eddies in Turbulent Region

Large eddies can be distinguished in the initial stages of the turbulent region. These large eddies have a wide range of sizes and trajectories at any position and there appears to be no obvious symmetry between structures within the jet. Turbulent eddies lack the circumferential coherence of transitional vortex rings. Some evidence has been found of coalescence of large eddies accompanied by an outward burst of jet fluid from the mixing layer. Because of the rapid diffusion, visualization of the flow is much more difficult in the turbulent region.

Conditional Sampling

Conditional sampling requires a triggering or conditioning signal, which is directly associated with the passage of a particular structure. A single probe, such as a hot-wire anemometer, can be used as a trigger, and a rake of probes farther downstream can be used for simultaneous measurement of signal variations with time. When this measuring system is coupled with movie photography, information such as that shown in Fig. 5.12 can be obtained. These measurements were made by Yule (1977) using a hot-wire anemometer, recording the u time histories at different radial positions and simultaneously filming the water jet with dye injection. A direct relationship can be seen between the features of these signals and the shape of the edges of the turbulent mixing layer, as indicated by the dye. The large peaks in the u signals could be most easily correlated with the shapes of dye patterns from visual scans of u time histories. At the outer edge of the mixing layer, a series of large positive u peaks, up to three times the local rms value of u, were found, and these corresponded to regions of dye projecting from the jet. At the high-velocity edge of the mixing layer, large negative u peaks corresponded to regions within the eddies. In the central part of the potential core, positive peaks in the u fluctuations corresponded to the passage of large eddies, and negative peaks corresponded to intervals between passing eddies. The positions of both the negative and positive u peaks advance and then retreat in phase with distance away from the center of the jet.

Vortex rings produce large negative u peaks at the outer edge of the mixing layer and large positive u peaks at the inner edge. Comparison of the vortex rings in the transition region with large eddies in the turbulent region show that, in the turbulent region, eddies produce higher-frequency, lower-amplitude u signals superimposed on the basic large-amplitude peaks. These higher frequencies are associated with the smaller scales of turbulence generated by interaction and

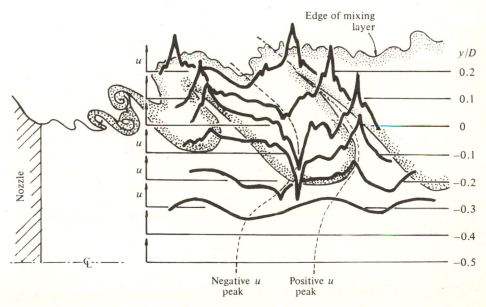

Figure 5.12 Velocity (u) time histories measured by array of hot wires in turbulent water jet. (*Yule,* *1977.*)

coalescence. Large positive and negative u peaks are found to occur at similar times as large peaks in small v fluctuations. The movements of fluid induced by the large eddies, which are responsible for these peaks, provide the major contribution to the shear stress uv.

5.7 COHERENT STRUCTURES IN COMBUSTION

Photographic and visual evidence of a wide range of turbulent flames indicate that flames are structured. Under certain conditions, flames are composed of groups of flamelets and there are indications that these flamelets are associated with the coherent large eddies in the flow system. In some cases, evidence of the important roles of coherent structures is more readily obtainable for burning flows than for nonburning flows. This is because luminous flames can provide very clear visualization of molecular scale fuel-oxidant mixing. Such regions form parts of the coherent large-eddy components of diffusion flames and will also be deformed and convected by these eddies. Observation of large-scale structured flamelets may provide evidence for the existence of large coherent eddies in the same flow. The coherent regions of flame associated with eddies in the flow are referred to as "flamelets." An eddy is defined as a vorticity-containing region of fluid which moves as a coherent structure. In practice, flamelets will form part of an eddy. However, the precise structure and position of a flamelet, relative to an eddy, varies according to the structure of the eddy and its position in the main flame. The visualization of a flamelet in a flow will not necessarily always provide a full indication of the shape of the eddy with which it is associated.

Yule, Chigier, and Thompson (1977) discuss the evidence from a series of photographs and movie films revealing large tongues of flame convected downstream, remaining coherent, and growing in scale until extinction occurs at the end of the visible flame. These flamelets have a substantial degree of coherence azimuthally, but they have a helical, rather than axisymmetric, structure. Helical large-scale structures have been predicted by stability analyses for the downstream regions of cold round jets.

Eddy structures in liquid spray flames are particularly important because of their roles in fuel droplet-turbulence interaction and, hence, in the determination of the local regimes of flow and combustion around droplets. The smaller droplets in the spray follow the gas flow, are carried along by the eddies, and vaporize in the eddies. Burning for these droplets is a cloud combustion process, which is basically similar to that found in gas diffusion flames. Larger droplets, however, do not follow the gas streamlines but can, instead, cross fuel-oxidant-product interfaces or even leave the large eddies completely. This process can improve the efficiency of burning and the stability of the spray flame, but it also affects pollutant levels. For example, soot particles can result from large droplets which leave the flame and burn individually but are quenched before burning has been completed.

Axisymmetric gas diffusion flames have a wide range of structures and, hence,

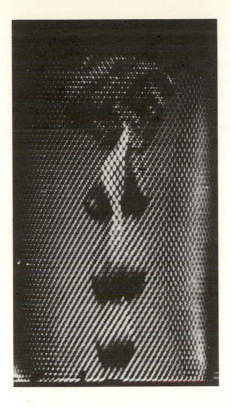

Figure 5.13 Schlieren photograph of propane diffusion flame. Burner diameter 5 mm; Re 10^3.

a wide range of visual appearances, depending on the Reynolds number of the fuel jet and the properties of the fuel. Schlieren visualization of gas diffusion flames (Fig. 5.13) shows that the flow contains vortex rings similar to those found in nonreacting flows. A combination of the laminarizing and expansion effects of combustion and also buoyancy effects, produces differences between the vortex ring structure for burning and nonburning flows of the same jet of gaseous fuel. Qualitatively, the length of the region in which vortex rings are found, and thus the length of the transition region, is greater for the burning than the nonburning case. As in cold jets, these vortex rings coalesce with their neighbors and this process results in a less orderly structure for the fuel-oxidant-product interfaces in the vortices, thus causing changes in the observed flame structure. As schlieren-type techniques indicate density gradients in the flow, they produce visualizations which differ from direct visualizations of the luminous flame regions. The primary reaction regions, or flamelets, associated with the vortex rings, are smooth tongues of flame which partially, or completely, encircle the outer interface between the vortex ring and the surrounding air.

As with nonreacting round jets, the length of the transition region, containing these vortex rings, decreases as the Reynolds number is increased, but coherent eddy structures can also be observed in the turbulent region of the flow, although with less clarity than the vortex rings. At higher Reynolds numbers, the flamelets associated with the turbulent eddies are observed as bulges of flame at the edges of the flow, which do not have the smooth, spiral shapes found in the transition region. Small regions of burning gas frequently detach themselves from the flame, and larger detached islands of burning gas separate at the end of the main flame.

Coherent flame and eddy structures are also observed in many large-scale industrial processes. Large eddies are produced by the combination of forced convection, buoyancy forces, and also wind-jet interaction. The gas has a significant content of higher hydrocarbons, which can form soot easily during combustion. In addition, the relatively long period between ignition and completion of combustion allows significant radiative heat transfer to the unburned gas, which can cause cracking. These factors account for the formation of considerable quantities of soot but it is interesting to note that soot-containing eddies do not disperse rapidly but maintain their coherence for a considerable distance. In addition to the coherent structures in the free flames, there is also strong evidence for the existence of coherent eddies and flamelets in combustion flows interacting with walls. In particular, coherent structures are clearly visible in the flow produced by a gas diffusion flame impinging on flat plates. Many practical combustors with swirling flow contain regions of recirculation which are similar in structure to the vortex ring-type eddies which are observed to be convected downstream in free flames.

Direct photographic evidence of coherent structures in flames is supported by quantitative point measurements. For example, velocity and temperature spectra and correlations, using laser anemometers and thermocouples, indicate the presence of quasi-periodic streets of vortex rings in the transitional regions of flames. For turbulent flames, the importance of the large eddies can be deduced from measurements which indicate unmixedness and also measurements of large temperature "spikes" (Ho et al., 1976), which can be related to the shapes, coherence, and convective properties of flamelets and eddies in the flow. This type of measurement indicates that coherent large eddies influence flame structure, even in the apparently complex burning of premixed flames downstream of turbulence grids. Other workers (Ballal and Lefebvre, 1973) have used schlieren techniques to show the deformation of flame regions by the eddies in this type of flow.

Figure 5.14 shows two frames from cine films of a diffusion flame on the lower surface of a flat plate, produced by an axisymmetric jet of propane impinging on the plate from below. In these experiments, the burner orifice diameter was 5 mm and the orifice was 100 mm vertically below the center of a 740-mm-diameter steel plate. Physically similar flame structures can also be observed for a wide range of these parameters. For example, the same phenomenon has been observed by Milson and Chigier (1973), using apparatus which was an order of magnitude larger. Figure 5.14 shows that, for $Re = 10^3$, the flame on the plate consists of a street of toroidal flamelets, which indicates a street of vortex rings. These rings of flame increase in diameter with time until burning is completed and they are no longer visible. At $Re = 3 \times 10^3$, evidence of instability can be seen in these rings. For $Re = 10^4$ a toroidal structure is still noticeable in the flow, the coherent structures now have the appearance of individual, three-dimensional cells of burning gas or flamelets. These structures are also visible at higher Reynolds numbers.

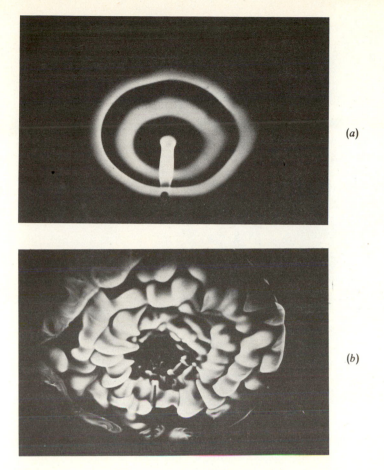

(a)

(b)

Figure 5.14 Diffusion flame impinging on flat plate. (a) Re 10^3; (b) Re 3×10^3.

It is not clear why these coherent flamelets are so clearly visible in this type of flow. Possibilities include the improvement of flame visualization produced by soot production and quenching at the flat plate or perhaps some stabilizing influence which is peculiar to the flow configuration. This type of flow is not completely unexpected, as it is known (Margarvey and MacLatchy, 1964) that non-burning smoke rings impinging on a flat plate break down into a similar three-dimensional cell structure, and a similar phenomenon is also found in the breakdown of transitional vortex rings in nonburning round jets (Yule, 1977). It thus appears that the coherent structures found in the impinging flame flow are formed by basically fluid mechanical processes and they do not arise because of effects peculiar to combustion processes. Thus, an examination of this flow can provide information on eddy-flamelet interactions and structures which may also be relevant to other types of combustion flow.

Figure 5.15 shows a sequence of cine film frames with the center of each frame showing a point 200 mm from the center of the plate. The individual flamelets are, on average, elongated in the flow (radial) direction and they grow in scale as they

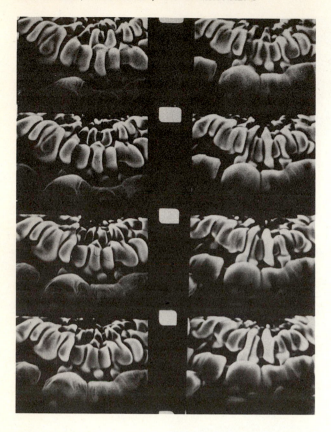

Figure 5.15 Sequence from cine film of diffusion flame impinging on flat plate. Re 10^4. Flamelets are moving from top to bottom in each flame. Interval between frames 0.016 s.

move downstream. A randomness in the movements and dimensions of the flamelets is evident in the films. The flamelets grow in scale as they move toward the edge of the plate, both individually and also by amalgamating with neighboring flamelets (both in the azimuthal and radial directions). In addition, flamelets are occasionally seen to move more quickly than average and they overtake and pass above other structures without merging with them. There is no visual evidence in these films of the eddies, which these flamelets delineate, breaking down into smaller scales, which invites comparison with the growth in eddy scale in two-dimensional nonreacting mixing layers, which was observed by Roshko (1976).

Transitional and Turbulent Flow in Flames

There is abundant evidence supporting the existence of coherent eddies and their associated flame structures in several types of flames. The photographic evidence indicates that the coherent eddy-flamelets found in combustion do not have a similar "universal" structure for all types of flow. In particular, for diffusion flames, one can identify a range of coherent structures, including: (1) unstable laminar flow, which contains an oscillating laminar diffusion flame; (2) streets of axisymmetric vortex rings with smooth tongues of flame at their interfaces; (3) other orderly vortex structures, including helical vortices, which also produce relatively smooth tongues of flame; (4) individual, coherent, three-dimensional eddies, which produce randomly moving cell-like flamelets (as in the flame imping-ing on a flat plate); (5) eddies containing coherent "ragged" regions of burning, which often form islands of burning that are separated from the main flame. This range of structures is due to the dependence of the eddy structure on the local rela-tive importance of inertial, viscous, and buoyancy forces in the flow. Eddy structure is also dependent on the history of an eddy from its point of formation. By compari-son with data (Yule, 1977, and Bradshaw, 1966) for nonreacting cold flows, the local existence of large eddies, which have an orderly vortex structure with asso-ciated smooth flamelets, is indicative of a local transitional structure for the combustion flow, for which viscous forces have an important stabilizing influence. As viscous forces become relatively less important, the orderliness of these vortex-eddies decreases as they become increasingly three dimensional and unstable. Thus, the existence of regions of ragged flame moving as coherent structures is indicative of a flow which is closer to the usually accepted conditions of fully developed turbulent flow. For this case, there are significantly energetic scales of motion present which are smaller than the large-eddy scale.

The accepted test for the local existence of fully developed turbulence is that the turbulence structure has Reynolds number similarity. This criterion cannot be simply applied to most combustion flows, because of the additional dependence of flame structure on variables such as the chemical kinetics, buoyancy forces, flame stabilization-ignition conditions, and a range of additional complicating factors which occur with spray flames. Further, flow in turbulent regions must be affected by any burning which occurs in Reynolds-number-dependent regions nearer to the burner nozzle. Most practical combustion flows do not achieve fully developed turbulence, at least for a significant length of the flame. Relationships for the length of the transition region x_T vary, probably because of differences in initial jet orifice conditions. For cold mixing layers, Bradshaw (1966) suggested $x_T = 7 \times 10^5 \, \mu/\rho U$, where ρ is the gas density and U is the jet orifice velocity. Fully developed turbulence occurs when the local ratio of an eddy diffusivity coefficient ε and the viscosity exceeds a certain value, i.e., the transition region ends when $\varepsilon/\mu = 7 \times 10^5$, where $\varepsilon = \rho U x$.

Burning in diffusion flames generally occurs at fuel-oxidant interfaces and these interfaces necessarily coincide with the vorticity-containing regions in the flow. Thus, regions where viscous forces act directly can coincide with the highest temperature regions of the flow. For propane and heavier gaseous fuels, viscosity increases approximately in proportion to absolute temperature. Temperature in the main reaction regions in diffusion flames may reach seven times atmospheric temperature, so that $\mu \simeq 7\mu_0$ where μ_0 is the viscosity at atmospheric conditions. The density and velocity terms, ρ and U, are representative of the flow structure responsible for inertial stresses. They are local scales for the flame cross section; for example, the mean velocity difference and an average density. In the case of pipe flow, mass conservation requires that ρU is the same for any section of the flow for both burning and nonburning conditions. However, for free shear flows, burning results in the expansion of gases in all directions and the rate of engulfment of cold fluid is modified so that mass flow rates cannot be readily calculated.

Experiments indicate that local mean velocities in burning and nonburning flows do not differ greatly under conditions where buoyancy forces may be neglected. However, thermal expansion decreases local mean densities, although to a much smaller extent than would occur if all the gas were at the flame temperature. Thus, as a first approximation, if it is assumed that the effect of combustion on inertial forces is much less than the effect on viscous forces, then

$$(x_T)_{\text{flame}} \simeq 7(x_T)_{\text{cold jet}} \qquad \text{or} \qquad (x_T/D)_{\text{flame}} \simeq 4.9 \times 10^6 \, \text{Re}^{-1}$$

where Re is the Reynolds number based on the burner diameter D. This relation predicts, for example, that the flow in the first 0.3 m of flame from a 5-mm-diameter burner is not fully turbulent when Re $= 10^4$. This implies that the impinging flame (Fig. 5.15) may not contain fully developed turbulent flow for half the flame length. In addition, if this criterion is applied to a variety of laboratory-scale diffusion flames and also to some large-scale industrial flames, it is found that significant lengths of the visible flame regions may not contain fully developed turbulence. These flames do, however, possess many of the important characteristics of turbulent flow, although fully developed turbulence has not been achieved.

Buoyancy forces in vertical free flames increase the convection velocities of vortices and thus decrease the residence time of the vortices at any position in the flow; the transition region can also be lengthened. In general, for diffusion flames, burning can increase the distance required to establish fully developed turbulent flow by at least one order of magnitude, especially where a fully formed cylindrical flame interface is established on and downstream of the jet nozzle.

Models of the Structures of Coherent Eddies and Flamelets

The mixing process at the interface between coflowing streams of air (above) and fuel gas (below) is shown in Fig. 5.16. The interface rolls up to form an eddy. The eddy represents a vortex ring, line vortex, or helical vortex. For nonburning conditions, the fuel-air interface is a double spiral around the vortex core. Figure 5.16 represents this eddy before it has interacted with other eddies or developed instabilities. The flame coincides with the fuel-air interface region, which has transverse concentration and temperature gradients and a local thickness which are dependent on the residence time of the vortex, the vortex strength, local diffusion coefficients, and the local chemical kinetics. The chemical kinetics of the burning in this interface region are the same as those of steady laminar diffusion flames. The stretching of this fuel-air interface, due to the interactions of the vorticity which it contains, enchances the molecular mixing. Preheating of fuel

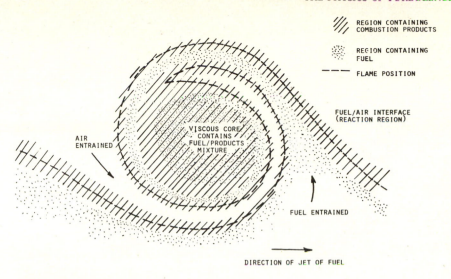

REGION CONTAINING
COMBUSTION PRODUCTS

REGION CONTAINING
FUEL

FLAME POSITION

FUEL/AIR INTERFACE
(REACTION REGION)

VISCOUS CORE
CONTAINS
FUEL/PRODUCTS
MIXTURE

AIR
ENTRAINED

FUEL ENTRAINED

DIRECTION OF JET OF FUEL

Figure 5.16 Rollup of interface between coflowing streams of air (above) and fuel gas (below). Formation of transitional vortex eddy in gas diffusion flame.

takes place mainly in the interface region, where mixing is on the molecular scale, so that this region has much higher temperatures than the unmixed fuel and air regions. As the vortex is convected along, the velocities induced by the vorticity within it produces a continuous "rolling up" of the interface, an increase in the vortex dimensions and the continuous engulfment of additional air and fuel. Under "cold" conditions, the central core of vorticity increases its dimensions as $(\mu t/\rho)^{1/2}$. Burning occurs near stoichiometric conditions and the vortex engulfs air and fuel in approximately equal quantities, so that, at a point along the interface, all of the air engulfed at some previous time is consumed, although there is still engulfed fuel remaining. This results in a region at the center of the vortex which contains both unmixed cold fuel and also a mixture of fuel and hot products. The volume of this fuel-products region increases with increasing residence time of the vortex.

Figure 5.16 indicates a continuous flame wrapped completely around the vortex and linked to adjacent vortices by the fuel-air interface. This may resemble a cross section of one of the eddies found in the impinging flame (Fig. 5.15). The length of the flamelet encircling the vortex varies depending on variables representing the vortex dynamics, the chemical kinetics, and molecular-scale mixing. For example, this flamelet is sensitive to the ratio of the vortex circulation and the molecular diffusivity.

One consequence of this orderly structure is that soot and other pollutants collect in the central region of the vortex. If such a vortex retained its organized structure until it reached a point in the flow where there was no longer a supply of fuel to be entrained and subsequently entrained only air, the possibility exists that the fuel-rich region in the vortex would never mix with air under conditions

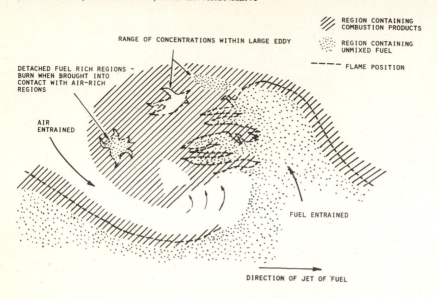

Figure 5.17 Structure of large turbulent eddy in gas diffusion flame.

suitable for complete burning of the remaining fuel. In general, these vortices do not remain coherent for the complete length of the flame, but instead the coalescing of neighboring vortices and the three-dimensional breakdown of the vortices (Yule, 1977) results in their fundamental restructuring. This restructuring is important in transitional flows as it distorts existing interfaces in the vortices, accelerates molecular mixing rates, and permits further mixing inside vortices between air and fuel-rich mixtures.

It is likely that the randomly moving eddies observed in flames impinging on flat plates are representative of many of the features of coherent large eddies in fully developed turbulent flames. Thus, one can consider three-dimensional eddies which are relatively elongated in the flow direction and which interact and coalesce in a similar manner to that observed for large eddies in cold-flow experiments. A cross section of such an eddy in a diffusion flame is sketched in Fig. 5.17. This eddy differs from the transitional eddy (Fig. 5.16) because of its three-dimensional structure, so that there is not complete coherence circumferentially around the flame, and also the existence of an irregular vorticity distribution within the eddy which can be considered as the existence of smaller eddy scales within the main eddy. Cold-flow visualizations show that these large eddies engulf wedges of unmixed fluid in a similar fashion to the transitional vortices. However, visualization experiments show that the subsequent molecular-scale mixing of the fluids engulfed from either side of the eddy is considerably more rapid for the case of the turbulent eddy than for the transitional eddy. Rapid mixing down to molecular scales is produced by vorticity stretching and interactions in the three-dimensional irregular vorticity field within the turbulent large eddy. The irregularity of the structure within the large eddy, whilst enhancing mixing, also introduces the possibility of a range of flamelet structures associated with the eddy. Thus, in general, a ragged flame front can be expected to exist behind the fuel entrainment wedge with a smooth flame at the relatively regular interface which separates eddies. This structure is observed in many diffusion flames. The likelihood exists of volumes of fuel detaching from the entrainment wedge and either existing, for some time, in a "sea" of combustion products or burning as small detached islands of burning within the eddy. This phenomenon is also observed in practical flames. Regions of air will be engulfed by the eddy and exist within the eddy for some time, perhaps mixing with products but not with fuel within flammability limits.

Although micromixing within large eddies is rapid in cold-flow experiments, there are conditions in the burning of fuel and oxidant which lead to significant differences from the cold-flow case, e.g., the presence of products, the need for molecular mixing within flammability limits, and the need for sufficient heat input. This requires mixing of fuel with very hot regions, as well as with air if a separated "island" of fuel in an eddy is to ignite and completely burn. Thus, it is possible for part of the fuel engulfed by a turbulent eddy to exist, unburned, in the eddy for a significant period before all of the conditions are met for combustion to be completed. As the eddies move downstream, a point is reached at which the fuel supply at the centre of the flow is depleted so that eddies on either side of the flow meet. After this point, the remaining unburned fuel in the entrainment wedges burn as large detached islands at the end of the main flame, as can be observed in practice.

SIX

TURBULENT COMBUSTION

6.1 TURBULENT-FLAME PROPAGATION IN PREMIXED GASES

Turbulent-flame propagation in premixed gases has been a subject of discussion in combustion literature for more than 30 years. There is clear evidence that flame propagation in premixed gases is faster under turbulent- than under laminar-flow conditions. In order to explain this increase in flame propagation speed, understanding is required of the changes in the flow field and the effect that this has on the chemical kinetics. In recent years, it has become possible to make detailed local measurements in turbulent premixed gaseous flames. These are beginning to show the influence of small-scale turbulence on flame propagation and structure. Also, the recognition of the dominating influence of large-eddy structures has led to the concept that turbulence can cause significant changes in the flow structure but that the fundamental chemical kinetic reaction mechanisms remain the same as in laminar flames. Consideration will first be given to the basic physics of flame propagation in premixed gases.

Laminar-Flame Instability

In turbulent flows with low levels of turbulence intensity and large turbulence scale, there is photographic evidence that flame fronts are laminar. Several forms of laminar-flame instability have been reported in the literature. Smithells and Ingle (1892) observed that laminar flames could break up into petallike segments, often having rapid rotation. Coward and Brimsley (1914) found combustion rings breaking up into small rising flame filaments. Böhm and Clusius (1948) observed as many as 100 flame filaments in a 5-cm-diameter tube. Smith and Pickering (1928) found pentahedral flames having three to seven surfaces rotating during transition from one shape to another. Di Piazza et al. (1951) obtained flame propagation only over a fraction of the cross section of a tube. All these instabilities were found in mixtures near the flame propagation limits and in which one of the reacting components had a very different molecular weight than the others. Preferential diffusion of lighter components toward the reaction zone can cause local shifts in stoichiometry. In addition to diffusional factors, hydrodynamic, thermal, and chemical phenomena influence the instability.

Landau (1944, 1953) made an analytical study of the fluid-dynamic stability of laminar flames. He considered the flame front to be a temperature discontinuity which propagates into the combustible mixture with a constant velocity. His analysis showed the flame to be unstable for disturbances of all wavelengths. Experiment showed that laminar flames can be stabilized and, in some cases, they are sufficiently stable to dampen artificially created disturbances. The influence of viscosity on flame stabilization has been analyzed by Einbinder (1951) and by Chu and Parlange (1962), who found viscosity had a destabilizing effect, whereas Yagodkin (1955) and Maxworthy (1962) showed that the influence of viscosity could be stabilizing.

Lewis and Von Elbe (1961) studied the thermodiffusional aspects of flame stability. For Lewis numbers $(Le = k/D)$ below unity, the enthalpy has a minimum near the flame front and the flame is stable, whereas for Le above unity, the flame is unstable. On the other hand, Barenblatt et al. (1962) came to the opposite conclusion, by arguing that, if the mass diffusivity is greater than the thermal diffusivity, the energy supplied with the mixture to the regions preceding the front exceeds the heat loss. This results in an increase in the velocity of flame propagation and a growth of perturbations leading to instability. Markstein (1951) extended Landau's theory by assuming that the flame speed S is a linear function of the radius of curvature of the disturbance R and a characteristic length of the order of the flame front thickness d, so that

$$S = S_0[1 + c(d/R)] \qquad (6.1)$$

where c is a constant, such that for $c < 0$ the effect is destabilizing, and for $c > 0$ it is stabilizing. From this theory, a critical wave number can be determined, above which the flame is stable. Also, the wave number for maximum growth rate of the disturbances can be calculated. For wavelengths of disturbances that are sufficiently large an instability of the flame occurs, irrespective of the constant c. The explanation for this is that stabilizing factors are proportional to the square of the wave number, while destabilizing factors are proportional to the wave number. Markstein (1951) showed experimentally that, for hydrocarbon fuels, a plane flame front separates into many small cells when the mixture is rich but does not separate when the mixture is lean. Transition from stable to unstable flame propagation was found to occur at exactly the stoichiometric composition. Methane flames differ from other hydrocarbon flames in that unstable flames occur for fuel-lean and not fuel-rich conditions. Of all the hydrocarbon fuels tested, only methane had a higher diffusivity than oxygen; hydrogen mixtures have also been found to behave similarly, so that it was concluded that the cellular instability was caused by preferential diffusion and occurred only for $c < 0$. Eckhaus (1961) examined the influence of flame perturbation on flame velocity. He showed that the perturbation of the flame propagation velocity should be proportional, not only to the flame curvature, but also to the rate of change of the tangential fluid velocities along the flame and to the relative acceleration of the flame. The flame stabilization process is governed by the convection of mass and heat, due to velocity perturbations tangential to the flame front surface.

Istratov and Librovich (1966) showed that differences in the molecular transport properties of the mixture result in the production of diffusive and thermal boundary layers of different thickness in the neighborhood of the flame front. The mass and heat transport along the flame is principally convective. Since this stabilizing effect is proportional to the deformation rate of the flame front, i.e., to the wave number, this provides an explanation for the existence of stable flames.

Hydrogen has a diffusivity about six times that of oxygen and the concentration of hydrogen can reach significant values in the reaction zone of hydrocarbon flames. Greater hydrogen loss from convex surfaces than from concave surfaces of the flame front results in stabilization of lean mixtures and destabilization of rich ones. The presence of highly reactive intermediate species and free radicals such as H, O, OH, CH, CH_3, etc., has a stabilizing influence.

Laminar flame instability is dominated by diffusional effects. These diffusional effects can only be of importance in flows with a low turbulence intensity, where molecular transport is of the same order of magnitude as turbulent transport. Flame instabilities do not appear to be capable of generating turbulence. Instabilities result in the growth of certain disturbances, leading to orderly three-dimensional flow structures. These structures may have complex forms but they are steady.

Large-Scale Turbulence–Flame Interaction for Stable Flames

Flame spread and flame propagation are accelerated by the action of turbulent diffusion. Turbulence causes the wrinkling of flames on a large scale relative to the flame thickness, and turbulence also causes small-scale changes in flame structure. Classical theories of flame structure in large-scale turbulent flow fields assume that the flame is confined to an interface between fresh and burned mixture. This interface is convected and strained by mean and fluctuating velocity components. Flame propagation is analyzed by considering the development of the flame surface in space and time under the influence of a random convective velocity field, starting from a given initial configuration. Simple relationships are determined between turbulent burning velocity and velocity fluctuation of the form

$$\frac{S_T}{S_L} = 1 + \left(\frac{u'}{S_L}\right) \tag{6.2}$$

where S_T is the turbulent burning velocity, S_L is the laminar burning velocity, and

u' is the velocity fluctuation component normal to the flame front. The basic assumption of classical theories, that the flame can be treated as a freely convected entrainment interface without affecting the turbulence, is incorrect. The classical theories only took into account the presence of turbulence in the approach stream. No account was taken of turbulence transformation within and beyond the flame and no account was taken of the influence of these transformations on flame propagation.

Karlovitz et al. (1951) proposed the concept of flame-generated turbulence and Scurlock and Grover (1953) attempted to show the validity of this concept. Flame-generated turbulence was considered to be additional to the approach stream turbulence and to have the same influence on flame propagation as turbulence in the approach stream. This concept has not been widely accepted but there is no definitive experimental evidence that clearly supports this theory. It is not, however, unreasonable to expect that the high temperature and density gradients and the consequent increases in volumetric flow rate within the flame could lead to significant changes in turbulence characteristics.

Chomiak (1973) has made a more detailed analysis of turbulent-flame propagation, based upon the equations of motion for unsteady flow in a stationary reference frame. His analysis followed the initial work of Tucker (1956), who obtained the boundary conditions at the flame front by perturbing the equations of conservation of momentum, energy, and mass flow rate. The flame was taken to be a surface of discontinuity and it was assumed that perturbed quantities of both sides of the flame were equal. The equations of motion were solved by Tucker, assuming that the initial velocity perturbation had the form of a sinusoidal shear wave. The general conclusions of Tucker have been shown to be incorrect, but he did indicate that, during the turbulence-flame interaction process, irrotational isentropic pressure waves can be generated, modifying the flow in the neighborhood of the flame and changing the "near-flame" turbulence level and structure. Chomiak (1973) also considered the interaction between the flame and a single oblique shear wave, representing a segment of the turbulent velocity field. It was postulated that there was no interference between the waves and, assuming low turbulence levels, small perturbation analysis was used. The reference system was chosen to move with the perturbation along the flame front so that the flow could be considered to be steady and, hence, time-dependent terms and instability did not need to be taken into account. A full description of the interaction between the flame and the turbulent flow field must take into account vorticity variations in the flame, caused by temperature variations. Chomiak (1973) obtained a solution for the case of a shear wave normal to a flame front by considering that the initial velocity disturbance had a sinusoidal variation. His analysis showed that velocity fluctuations tangent to the flame, or transverse to the flow, are produced in the neighborhood of the flame front. The ratio of the total intensity of fluctuations within the region adjacent to the flame front to their initial value is always less than unity. Beyond the flame front, the fluctuations decrease rapidly to values much lower than those before the flame. This results from the reduction in the vorticity of burned gas, due to an increase in volume of fluid elements crossing the flame front and rapid attenuation of potential disturbances. The ratio of the surface areas of the perturbed and unperturbed flame front, which is also the ratio of turbulent to laminar-flame velocity, was found to be

$$\frac{S_T}{S_L} = \left[1 + B \frac{1}{m^2} \left(\frac{u'}{S_L} \right)^2 \right]^{1/2} \tag{6.3}$$

where m is the density ratio and B is a constant of the order of unity.

Chomiak (1979) also considered the more complicated case of flame interaction with oblique shear waves. This takes into account the change in inclination and amplitude of the sinusoidal shear wave as it crosses the flame. An irrotational disturbance is generated in the form of an additional velocity field with associated pressure disturbances, which can be recognized as sound waves. Phase shifts in the disturbances also occur. From this analysis, Chomiak (1979) concluded that, as a result of an increase in volume of the flow crossing the flame front, there is a considerable reduction in the rotational component of the flow perturbations. The vorticity generated by the rippled front does not compensate for this reduction. The irrotational perturbation resulting from the flame-turbulence interaction decreases exponentially with increasing distance from the flame front. Within the flame, the generated perturbation is of considerable intensity and produces an intense acoustic field. The total intensity of turbulence near the flame never exceeds that of the unperturbed approach flow. However, in addition to perturbations in the main flow direction, perturbations are produced transverse to the flow, producing an apparent increase in turbulence within the region of the flame. These results are in agreement with experiments showing that increased turbulence occurs close to the flame (Galyun and Ivanov, 1970) and consists of intense transverse fluctuations (Durst and Kleine, 1974). Figure 6.1 shows the relative amplitude of the sinusoidal ripples on the flame front produced by the passage of a shear wave. The turbulent/laminar velocity ratio is proportional to this amplitude and is given by

$$\frac{S_T}{S_L} = (1 + F^2 i^2)^{1/2} \tag{6.4}$$

where i is equal to u'/w_1, which represents the turbulence intensity in the approach stream, and F is the relative amplitude of ripples in the flame front. The turbulent/laminar burning velocity ratio is, thus, dependent on (1) the amplitude of velocity fluctuations, (2) the inverse density ratio across the flame, and (3) the inclination of the flame front to the main flow direction. The strong competitive influence of the last two factors offers an explanation of the considerable discrepancies between results of measurements of turbulent burning velocity made by using different methods. Measurements made on unsteady spherical flames give turbulent burning velocities considerably lower than those measured with open tubes and still lower than those obtained by experiments with confined flames. The marked dependency of turbulent flame propagation on density ratios for small angles, shown by the Chomiak theory, may also explain the fact that measurements for spherical flames give burning velocities lower than theoretically predicted values.

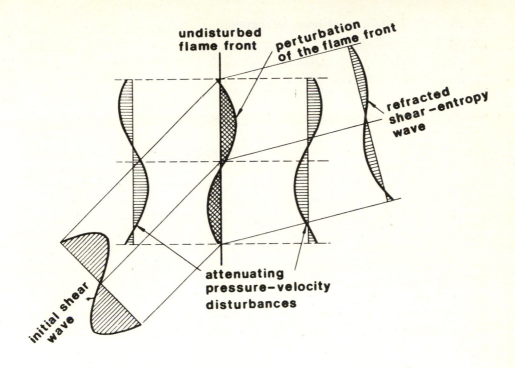

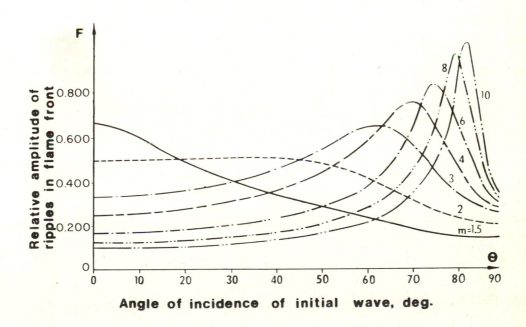

Figure 6.1 Interaction of a single elementary shear wave with a flame front. (*Chomiak, 1979.*)

Small-Scale Turbulence–Flame Interaction

Small-scale turbulence effectively increases the intensity of molecular transport processes. The flame propagation velocity can be represented by an analogy with thermal theories of laminar flame propagation, i.e.,

$$S_T \sim \frac{(K_1 + K_T)^{1/2}}{\tau_{r_T}} \tag{6.5}$$

where K_1 and K_T are the laminar and turbulent thermal diffusivities and τ_{r_T} is the reaction time in a turbulent flame. Transport processes inside a flame are enhanced by the relative motion of flame elements under the influence of small-scale turbulence. Kolmogorov's (1941) universal similarity theory shows that the energy spectrum of small-scale turbulent eddies in a high Reynolds number flow follows a universal equilibrium distribution. For very small scales, the properties of turbulence are determined by the viscous forces which govern the rate at which energy is transferred down the eddy cascade and, ultimately, dissipated. As the size of eddies increases, the effects of viscous forces are reduced so that, for large eddies, the properties of turbulence depend only on the rate of energy flow or dissipation. These two ranges are referred to as the viscous and inertial subranges of the Kolmogorov universal equilibrium range. The dividing length scale is the Kolmogorov microscale, defined as

$$\eta = (v^3/\varepsilon)^{1/4} \tag{6.6}$$

where v is the kinematic viscosity and ε is the rate of energy dissipation per unit mass of fluid. In the inertial subrange, the correlation between velocity fluctuations at two points, P_1 and P_2, has the form

$$\overline{(u_1' - u_2')^2} \sim (\varepsilon r)^{2/3} \tag{6.7}$$

where r is the distance between the points. The effective thermal diffusivity is given by

$$K_T \sim \varepsilon^{1/3} r^{4/3} \tag{6.8}$$

Richardson's (1926) empirical law for diffusivity is given by

$$\frac{dr^2}{dt} \sim (r^2)^{2/3} \tag{6.9}$$

This equation describes the relative dispersion of two particles by turbulence. Chomiak (1979) argues that, in flame analysis, the separation distance r should be taken as equal to the flame thickness d, because the principal contribution to relative dispersion comes from eddies of a scale of the same magnitude as the particle separation. The flame thickness d is known a priori only for the case when the turbulence levels are sufficiently low that the original laminar flame thickness and structure are unaffected by turbulence.

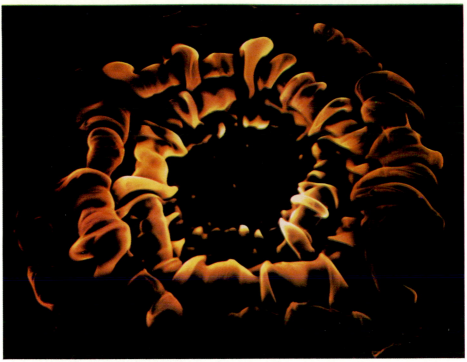

Spreading of propane jet flame after impinging on flat plate.

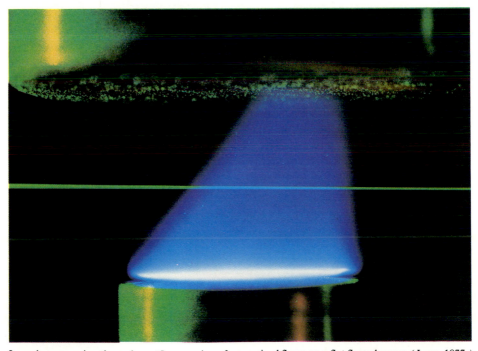

Laser beam passing through postflame region of a premixed flame on a flat flame burner. *(Lapp, 1977.)*

Structure of transitional jet flame.

6.2 INFLUENCE OF TURBULENCE ON COMBUSTION

It has long been recognized that turbulence can influence flame structure. Simple photographic studies have shown that changes in intensity and scale of turbulence can be visualized by the size and location of pockets of flame within the turbulent flow field. Damköhler (1947) put forward the hypothesis that turbulence augments transport rates within the reaction zone when the scale of eddies present in the flow is small compared with the laminar-flame thickness. He proposed a physical model of a turbulent-flame front as a wrinkled laminar-flame front in which the large eddies distort and increase the surface area of the flame front. Using the velocity fluctuations as a measure of the depth of indentation in the originally linear laminar-flame front, and the macroscale of the large eddies as a measure of the width of the indentation, Damköhler (1947) calculated the increase in flame surface area and, hence, showed that turbulent-flame speeds were greater than laminar-flame speeds. Very little experimental evidence has been produced to clearly demonstrate the presence of a wrinkled flame front. Turbulence was initially considered to be a random, unstructured phenomenon and, therefore, Damköhler's idealized model was not considered relevant to the problem of turbulent combustion. Later, turbulent flow came to be recognized as being dominated by large eddies and coherent structures, and this led to a revival in interest in some of the original concepts of Damköhler.

Ballal and Lefebvre (1973, 1975b) carried out a series of studies in which they systematically examined the influence of turbulence on flame structure and propagation. In their study of enclosed premixed propane-air, Ballal and Lefebvre (1973) showed that the influence of the integral scale of turbulence on flame speed was pronounced, and varied with the level of turbulence intensity. They recognized three distinct regions, each having different characteristics in regard to the effect of scale on turbulent burning velocity and, thereby, explained some of the anomalies that had arisen in the interpretation of previous studies on turbulent flames. In their subsequent study, Ballal and Lefebvre (1975b) examined the effects of the turbulent-flow parameters, such as Taylor microscale, Kolmogorov (dissipation) scale, and turbulent vorticity, on burning velocity and flame structure. Turbulence was generated at different levels by means of grids located at the entry to the combustion chamber. Ignition was achieved by means of a high-tension spark and the turbulent premixed flames in the combustion chamber were photographed, using a schlieren system. Turbulent burning velocity was calculated from the product of inlet velocity and the sine of the angle between the flow direction and surface of the flame, as measured from the schlieren photographs. The accuracy of measurements of turbulent burning velocity by this method is of the order of 10 percent.

The grids used by Lefebvre et al. generated near-isotropic turbulence. In the initial region, immediately downstream of the grids, the turbulence intensity is given by

$$u'/U \propto (x/b)^{n_1} \tag{6.10}$$

where u' is the rms value of fluctuating velocity, U is the mean velocity, x is the distance downstream of the turbulence grid, b is the bar size of the grid, and n_1 has values between 0.5 and 0.7.

The integral scale of turbulence L is given by the equation

$$L/b \propto (x/b)^{n_2} f(\text{Re}_b) \tag{6.11}$$

where Re_b is the grid Reynolds number (Ub/v) and n_2 has the value of 0.5.

The Taylor microscale λ is given by

$$\lambda^2 \propto v(x/U) \tag{6.12}$$

It can be seen from the preceding equations that the turbulence intensity and scales are simply related to the bar size and distance downstream of the grid, as well as to the mean velocity and kinematic viscosity of the fluid. In the experiments of Ballal and Lefebvre (1975b), the value of u' was fixed, and various combinations of x and b were selected so as to obtain a wide range of values of L and λ. These scales L and λ were calculated from energy spectra and autocorrelations measured by hot-wire anemometry.

The mean rate of dissipation ε is given by

$$\varepsilon = 15v(u'/\lambda)^2 \tag{6.13}$$

The Kolmogorov scale η is given by

$$\eta = (v^3/\varepsilon)^{1/4} \tag{6.14}$$

and the velocity parameter u_K is given by

$$u_K = (v\varepsilon)^{1/4} \tag{6.15}$$

Figure 6.2 shows the three regions found by Ballal and Lefebvre (1975b) which were subdivided as follows:

Region 1: $u' < 2S_L$ $\eta > \delta_L$
Region 2: $u' \gtrsim 2S_L$ $\eta \simeq \delta_L$
Region 3: $u' > 2S_L$ $\eta < \delta_L$

where S_L is the laminar burning velocity and δ_L is the laminar-flame thickness.

In region 1, turbulence and velocity levels are low and even the smallest values of η are larger than δ_L. Under these conditions, the flame front retains its smooth laminar appearance but the flame becomes wrinkled and, hence, the burning velocity is increased. The combined effects of intensity and scale on turbulent-flame speed conform with the relation

$$\left(\frac{S_T}{S_L}\right)^2 = 1 + 0.03\left(\frac{u'L}{S_L\delta_L}\right)^2 \tag{6.16}$$

The dimensionless group $(u'L/S_L\delta_L)$ corresponds to the ratio of the turbulent Reynolds number $(u'L/v)$ to the laminar Reynolds number $(S_L\delta_L/v)$. In region 1, the flame speed ratio (S_T/S_L) increases with increase in scale.

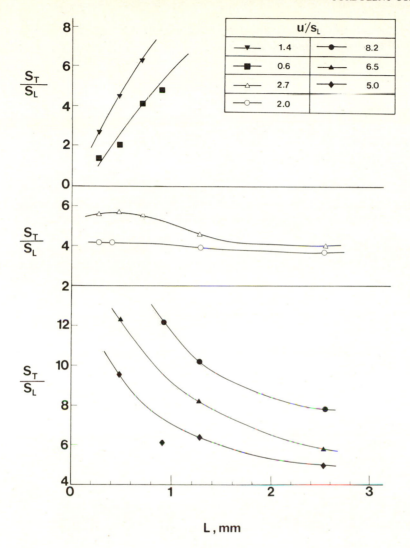

Figure 6.2 Influence of turbulence scale on the ratio of turbulent- to laminar-flame speed. (*Ballal and Lefebvre, 1975.*)

As the turbulence intensity increases, the scale decreases and, thus, at very high levels of turbulence, the eddies become too small to produce any noticeable wrinkling of the flame surface. The combustion zone takes on the form of a thick matrix of burned gases, interspersed with eddies of unburned mixture. As the turbulence intensity increases, a transformation occurs from a continuous, coherent laminar-flame surface to the conditions of region 3, where reaction takes place mainly at the interfaces formed between the combustion products and the eddies of fresh mixture.

If N is the number of eddies entrained per unit area of flame front, then the total volume of eddies contained within a unit area of flame front can be expressed by

$$V = N\tfrac{1}{6}\pi l^3 \tag{6.17}$$

and the total surface area of the eddies is

$$A_t = N\pi l^2 \tag{6.18}$$

where l is the effective mean eddy size. From the above two equations it can be shown that, for any given volume of fresh mixture, the total surface area of the eddies is inversely proportional to turbulence scale.

In region 2, the distribution of eddy sizes is such that the fresh mixture contains eddies which are both larger and smaller than the thickness of the flame. In this region, two different mechanisms for augmenting the surface area of the flame operate simultaneously:

1. The flame front is wrinkled by all eddies larger than its own thickness.
2. The area of interface between combustion products and fresh mixture is significantly increased by the eddies entrained between the inner and outer boundaries of the flame zone.

Both these mechanisms serve to increase the total surface area of the flame and, hence, the turbulent flame speed. Within region 2, which is a transition region between regions 1 and 3, the turbulent burning velocity is independent of both laminar-flame speed and turbulence scale, and is given by

$$S_T \simeq 2u' \tag{6.19}$$

The ratio of turbulent to laminar speeds in region 3 is given by

$$\frac{S_T}{S_L} = 0.5\left(\frac{u'\delta_L}{S_L\eta}\right) \tag{6.20}$$

Thus, in the highly turbulent mixtures of region 3, S_T diminishes with increase in scale.

The variation in turbulent energy distribution which occurs with change in turbulence intensity and its relation to the corresponding change in the physical structure of the flame is shown in Fig. 6.3. Proceeding from left to right it can be seen that, initially in region 1, turbulence intensity is low, and as both η and L are larger than δ, all the turbulence energy contributes to flame wrinkling. The top photograph in Fig. 6.4 shows these conditions with the flame having a smooth laminar appearance with the surface showing bulges which grow in size with distance downstream. The second diagram in Fig. 6.3 corresponds to maximum wrinkling of the flame, which occurs with $\eta = \delta_L$. The third diagram shows that, when $\eta < \delta_L$, part of the turbulence energy is contained in the eddies within the flame and a smaller amount is available for flame wrinkling. The flame surface appears to be more rough. In region 3, the wrinkling effect is small and all the

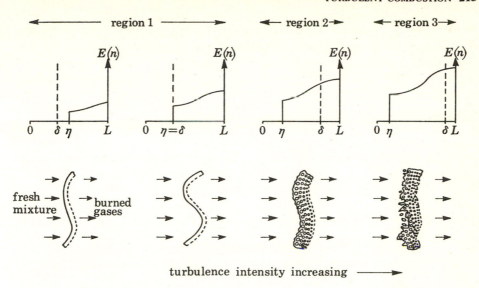

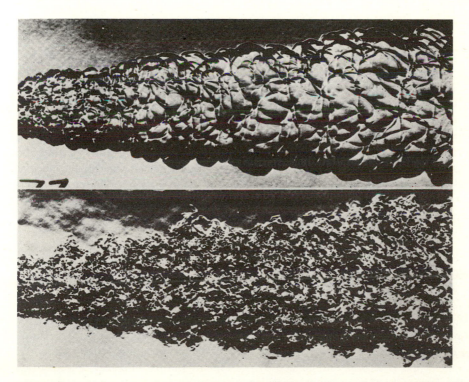

Figure 6.3 Diagrams illustrating the effect of turbulence energy distribution on flame structure. (*Ballal and Lefebvre, 1975.*)

Figure 6.4 Stoichiometric propane-air flames under conditions of low and high turbulence. Upper photograph $u' = 31$ cm/s; lower photograph $u' = 305$ cm/s. (*Ballal and Lefebvre, 1975.*)

available energy is distributed among the small eddies. Eddies of unburned mixture are gradually, and sometimes violently, consumed during their passage through an extended reaction zone. Within each eddy, the burning rate is enhanced by the flow of heat and active species from the enveloping flame front. In some instances, depending on the properties of the mixture and the turbulence scale, the acceleration of chemical reactions within an individual eddy may proceed to such an extent that eventually combustion occurs almost simultaneously throughout its volume, ahead of the advancing flame. Pressure pulsations are generated which, together with strain due to turbulence, lacerate and rupture the flame surface, as shown in the lower photograph of Fig. 6.4.

Influence of Vorticity on Flame Propagation

Ballal and Lefebvre (1975b) have considered the possible effects of vorticity on flame propagation and the nature of the flame front. Velocity gradients can cause translation and rotational movements of turbulent eddies. The stretching of vortex elements in any one dimension leads to reduction in length scales and increase in velocity components in the other two dimensions. This, in turn, leads to stretching of the fluid in the other two dimensions. The breakup of large eddies into smaller eddies can be explained in terms of vortex stretching, a process which terminates only when the eddies become so small that they are eliminated by viscous dissipation more rapidly than they multiply by disintegration. The production of vorticity due to stretching of vortex lines (leading to increased turbulent diffusion) occurs simultaneously with the destruction of vorticity due to viscous dissipation, leading to damping of the fluctuations. The production of vorticity due to stretching is related to the skewness factor, while the dissipation of vorticity due to viscosity is related to the flatness factor. The skewness and flatness factors can be measured by hot-wire or laser anemometry. Ballal and Lefebvre (1975b) calculated these functions from scale parameters and showed that the ratio of turbulent- to laminar-flame speeds reached a maximum value when the production of vorticity was about half the viscous dissipation. They concluded that this represented the ideal condition of turbulence for maximum flame propagation.

6.3 TURBULENT PREMIXED FLAMES

The speed of propagation of a laminar flame is a chemical kinetic and diffusive property which is independent of the flow field configuration and of the mechanism of flame stabilization. Turbulent-flame propagation is dependent upon the flow field and is strongly influenced by both mean and turbulence characteristics of the flow. Changes in experimental configuration can result in very different values of the turbulent flame speed. In grid-flow turbulence, nonstationary spherical turbulent flames have flame speeds of the order of, or in some cases less than, the laminar-flame speed (Mickelson and Ernstein, 1957). In grid-flow turbulence, the turbulent-flame speed tends to increase proportionally to the intensity of turbulence. In high-speed, ducted, premixed flows, where flames are stabilized in recirculation zones, Wright and Zukoski (1962) concluded that turbulent-flame speed grows, without apparent limit, in approximate proportion to the speed of the unburned gas flow. Flow in recirculation zones is at such a high level of turbulence intensity that variations in level of turbulence appear to have little influence on turbulent-flame speed.

Theories of turbulent-flame propagation relate turbulent-flame speed to increase in intensity of turbulence within the reaction zone. Scurlock and Grover (1953) proposed that flame-generated turbulence resulted from shear forces within the burning gas. Such a mechanism could explain the high propagation speed of turbulence in high-speed ducted flows. The existence of flame-generated turbulence is not universally accepted and, in unconfined flames, direct measurements of velocity indicate that there is no flame-generated turbulence. Bray (1975) has pointed out that the balance equation for turbulence kinetic energy in a reacting turbulent flow contains not only the usual terms which represent production due to mean flow shear and which can be influenced by combustion, but also mean flow dilatation terms, which can act to remove turbulent energy as a result of combustion. He suggested that some of the discrepancies between turbulent-flame propagation speeds observed in different flow situations might be explained in terms of the balance between these competing effects.

Theoretical analyses of premixed turbulent combustion include both the effects of combustion on the turbulence and the effects of turbulence on the average chemical reaction rates. Peak time-averaged reaction rates in a turbulent flame can be orders of magnitude smaller than the corresponding rates in a laminar flame because of turbulence-induced fluctuations in composition, temperature, density, and heat release rate within the flame. The presence of large-eddy structures and wrinkled laminar-flame fronts result in large fluctuations of these quantities. Bray and Moss (1974) proposed a unified statistical model of premixed turbulent combustion. The interacting effects of turbulence and of fluctuations in thermodynamic state variables are represented through balance equations for the kinetic energy of turbulence and for the mean square fluctuation. A "reaction progress" variable is used for relating the mass fraction of products. Fluctuations in the rate of reaction are incorporated through the introduction of a probability density function for the progress variable. The probability density function can be modeled using balance equations or it can be specified empirically. Bray and Libby (1976) applied the model of Bray and Moss (1974) to predict the speed of propagation and structure of plane turbulent combustion waves. They focused attention on the interaction between the dilatation and the shear-generated turbulence associated with heat release and the chemical reaction itself. They carried out their analysis for an idealized, oblique, planar, turbulent reaction zone.

The following main assumptions are made in analyses of turbulent premixed flames (Bray and Libby, 1976):

1. Combustion is controlled by a single-step, irreversible chemical reaction whose rate is specified by a global reaction rate expression.
2. Reactant and product species are treated as ideal gases.
3. The specific heat at constant pressure and the molecular weight of the reacting gas mixture are both constant and independent of the progress of the reactant.
4. The Lewis numbers of the reactant and product species are unity.
5. The flow is adiabatic, and occurs at a Mach number much less than unity, such that terms in the mean energy balance equation representing effects of pressure changes and viscous dissipation may be neglected.
6. Pressure fluctuations are assumed to be of small intensity and are neglected.

The global combustion reaction is characterized by a progress variable c which is defined as the mass fraction of the product of the reaction, normalized in such a way that $c = 1$ when combustion is complete, i.e., at least one reactant is exhausted, while $c = 0$ in the unburned mixture which has no product. The mixtures generally include either an inert diluent or an excess of one reactant.

Equations of State

The thermal equation of state of the gas is given by

$$\bar{p} = \rho R T \tag{6.21}$$

where $\bar{p}$ is the pressure, R is the gas constant for the mixture, while ρ and T are the instantaneous, time-varying values of density and temperature, respectively, and where the overbar denotes time averaging.

The caloric equation of state is

$$h = c_p T - cH \tag{6.22}$$

where h is the specific enthalpy, H is the heat of reaction, per unit mass of mixture (reactant plus diluent), and c_p is the specific heat at constant pressure. The specific enthalpy has been assumed to be constant throughout the flow, i.e.,

$$h = h_0 = c_p T_0 \tag{6.23}$$

where T_0 is the uniform temperature of the premixed reactants when $c = 0$. The parameter

$$\tau = H/h_0 = T_\infty/T_0 - 1 \tag{6.24}$$

has typical values ranging from four to nine.

Equations (6.21) and (6.22) then give

$$T = T(c) = (h_0/c_p)(1 + \tau c) = T_0(1 + \tau c) \tag{6.25}$$

$$\rho = \rho(c) = \rho_0/(1 + \tau c) \tag{6.26}$$

ρ_0 is the value of $\rho(c)$ when $c = 0$, and is used as a reference quantity.

Probability Density Function

The probability density function for the progress variable has the general form

$$P(c; \mathbf{r}) = \alpha(\mathbf{r})\delta(c) + \beta(\mathbf{r})\delta(1 - c)$$
$$+ [\eta(c) - \eta(c - 1)]\gamma(\mathbf{r})f(c; \mathbf{r}) \qquad (6.27)$$

where $\delta(c)$ and $\eta(c)$ are the dirac delta and Heavyside functions, respectively, the coefficients $\alpha(\mathbf{r})$, $\beta(\mathbf{r})$, and $\gamma(\mathbf{r})$ are nonnegative functions of position $(\mathbf{r})$, and $f(c; \mathbf{r})$ satisfies the condition

$$\int_0^1 f(c; \mathbf{r}) = 1 \qquad (6.28)$$

so that

$$\alpha(\mathbf{r}) + \beta(\mathbf{r}) + \gamma(\mathbf{r}) = 1 \qquad (6.29)$$

The delta functions at $c = 0$ and $c = 1$ are identified by Bray and Libby (1976) as "packets" of unburned and all-burned mixture, respectively, and $f(c; \mathbf{r})$ with product distributions associated with burning at $\mathbf{r}$; $\alpha(\mathbf{r})$, $\beta(\mathbf{r})$, and $\gamma(\mathbf{r})$ describe the partitioning among these three possible modes.

Thermodynamic variables are expressed as a function $g(c)$ whose time average is

$$\bar{g}(\mathbf{r}) = \int_0^1 g(c)P(c; \mathbf{r}) \, dc \qquad (6.30)$$

Favre Averaging

Bray (1973, 1975) and Bray and Libby (1976) have demonstrated the advantages of using Favre averaging in theoretical analyses and computations of turbulent flows with combustion. In Favre averaging, time-averaged quantities are mass-weighted by incorporating density terms. Thus, if $g(\mathbf{r}, t)$ denotes any variable, defined by

$$g = \tilde{g} + g'' \qquad (6.31)$$

where $\tilde{g} = \overline{\rho g}/\bar{\rho}$ so that

$$\overline{g''} = \bar{g} - \tilde{g} = -\overline{\rho'g'}/\bar{\rho} \qquad (6.32)$$

the mass-weighted mean of c is then

$$\tilde{c} \equiv 1/\bar{\rho} \int_0^1 c\rho(c)P(c) \, dc$$

$$= 1/\bar{\rho} \left[\beta\rho(1) + \gamma \int_0^1 c\rho(c)f(c) \, dc \right] \qquad (6.33)$$

while the mean square fluctuation in c is

$$\overline{\rho c'' c''} \equiv \int_0^1 (c - \tilde{c})^2 \rho(c) P(c)\, dc$$

$$= \bar{\rho}\tilde{c}^2 + \beta \rho(1)(1 - 2\tilde{c}) + \gamma \left(\int_0^1 c^2 \rho(c) f(c)\, dc - 2\tilde{c} \int_0^1 c \rho(c)\, dc \right) \quad (6.34)$$

If β is eliminated between Eqs. (6.13) and (6.14), then

$$\overline{\rho c'' c''}/\bar{\rho} = \tilde{c}(1 - \tilde{c}) - \gamma(1 + \tau\tilde{c})(M_1 - M_2) \quad (6.35)$$

where

$$M_k \equiv \int_0^1 \frac{c^k f(c)\, dc}{1 + \tau c} \qquad k = 0, 1, 2 \quad (6.36)$$

The mean density $\bar{\rho}$ is written

$$\bar{\rho}/\rho(0) = 1/(1 + \tau\tilde{c}) \quad (6.37)$$

The Mean Reaction Term

The instantaneous rate of reaction is, in general, a function which can be written as a function of c so that the mean reaction rate is given by

$$\bar{w} = \int_0^1 w(c) P(c)\, dc \quad (6.38)$$

Since $w = 0$ when $c = 0$ or $c = 1$, provided the reaction rate is finite, Eq. (6.38) can be written as

$$\bar{w} = w_{\max} \gamma I_3 \quad (6.39)$$

where

$$I_3 = \int_0^1 \frac{w(c)}{w_{\max}} f(c)\, dc \quad (6.40)$$

Also

$$\overline{\rho c'' w}/\bar{\rho} = \gamma w_{\max}(I_4 - \tilde{c} I_3) \quad (6.41)$$

where

$$I_4 = \int_0^1 c \frac{w(c)}{w_{\max}} f(c)\, dc \quad (6.42)$$

The mean value of c, weighted by the reaction rate expression, is given by

$$c_m \equiv \frac{I_4}{I_3} \left[\int_0^1 c \frac{w(c)}{w_{\max}} f(c)\, dc \right] \left[\int_0^1 \frac{w(c)}{w_{\max}} f(c)\, dc \right]^{-1} \quad (6.43)$$

c_m typically lies in the range $0.5 < c_m < 1.0$.

The Conservation Equations

The conservation equations generally used in analysis of turbulent flows with combustion include continuity and the equations of motion, energy, and the kinetic energy of turbulence. Favre averaging requires the introduction of density terms in each one of these equations, so as to provide mass-weighted time-averaged quantities. Equations for the conservation of the progress variable c and for its mean square fluctuation have also been shown to be important by Bray and Libby (1976). Molecular diffusion terms are neglected because they are of lower order of magnitude than turbulent diffusion terms. Nondissipative viscous effects are also generally neglected.

The conservation equations for stationary, adiabatic, two-dimensional, low Mach number turbulent flows, in terms of mass-weighted, time-averaged quantities, neglecting buoyancy forces, are given by

$$\frac{\partial}{\partial x}(\bar{\rho}\tilde{u}) + \frac{\partial}{\partial y}(\rho\tilde{v}) = 0 \tag{6.44}$$

$$\bar{\rho}\tilde{u}\frac{\partial \tilde{u}}{\partial x} + \bar{\rho}\tilde{v}\frac{\partial \tilde{u}}{\partial y} = -\frac{\partial \bar{p}}{\partial x} - \frac{\partial}{\partial x}(\overline{\rho u''u''}) - \frac{\partial}{\partial y}(\overline{\rho u''v''}) \tag{6.45}$$

$$\bar{\rho}\tilde{u}\frac{\partial \tilde{v}}{\partial x} + \bar{\rho}\tilde{v}\frac{\partial \tilde{v}}{\partial y} = -\frac{\partial \bar{p}}{\partial y} - \frac{\partial}{\partial x}(\overline{\rho u''v''}) - \frac{\partial}{\partial y}(\overline{\rho v''v''}) \tag{6.46}$$

$$\bar{\rho}\tilde{u}\frac{\partial \tilde{q}}{\partial x} + \bar{\rho}\tilde{v}\frac{\partial \tilde{q}}{\partial y} = -\overline{\rho u''u''}\frac{\partial \tilde{u}}{\partial x} - \overline{\rho u''v''}\frac{\partial \tilde{v}}{\partial x} - \overline{\rho u''v''}\frac{\partial \tilde{u}}{\partial y}$$

$$- \overline{\rho v''v''}\frac{\partial \tilde{v}}{\partial y} - \frac{\partial}{\partial x}(\overline{\rho u''q''}) - \frac{\partial}{\partial y}(\overline{\rho v''q''}) - \phi \tag{6.47}$$

$$\bar{\rho}\tilde{u}\frac{\partial \tilde{c}}{\partial x} + \bar{\rho}\tilde{v}\frac{\partial \tilde{c}}{\partial y} = -\frac{\partial}{\partial x}(\overline{\rho c''u''}) - \frac{\partial}{\partial y}(\overline{\rho c''v''}) = \bar{w} \tag{6.48}$$

$$\bar{\rho}\tilde{u}\frac{\partial}{\partial x}(\overline{\rho c''c''}/\bar{\rho}) + \bar{\rho}\tilde{v}\frac{\partial}{\partial y}(\overline{\rho c''c''}/\bar{\rho})$$

$$= -2\overline{\rho c''u''}\frac{\partial \tilde{c}}{\partial x} - 2\overline{\rho c''v''}\frac{\partial \tilde{c}}{\partial y}$$

$$- \frac{\partial}{\partial x}(\overline{\rho u''c''c''}) - \frac{\partial}{\partial y}(\overline{\rho v''c''c''}) - \bar{\chi} + 2\overline{\rho c''w}/\bar{\rho} \tag{6.49}$$

In these equations, $\tilde{u}$ and $\tilde{v}$ are the mass-weighted mean velocity components in directions x and y respectively; $\tilde{q} = (\frac{1}{2}\bar{\rho})\overline{\rho v''_\alpha v''_\alpha}$ is the mass-weighted turbulence kinetic energy, and $\bar{\phi}$ is the viscous dissipation term in the turbulent kinetic energy balance. The term $2\overline{\rho c''w''}$ represents the production of composition fluctuations due to chemical reaction, while $\bar{\chi}$ represents the annihilation of these fluctuations by molecular effects. The pressure fluctuation terms in the turbulent kinetic energy equation have been neglected.

The first four terms on the right-hand side of Eq. (6.47) represent energy exchanges among Reynolds stress components on the mean motion. Two of these terms contain the shear stress $\overline{\rho u'' v''}$, and are generally production terms for turbulent kinetic energy. The other two terms contain the normal stress components; in turbulent flames, these components can lead to removal of turbulent energy due to dilatation of the mean velocity field.

Modeling

The set of conservation equations for turbulent flows with combustion cannot generally be solved without the introduction of models which provide closure of the equations. The Prandtl-Kolmogorov model introduces an eddy kinematic viscosity v_T given by

$$v_T = a\tilde{q}^{1/2} l_1 \tag{6.50}$$

where a is a constant and l_1 is a turbulent length scale associated with the large eddies. Scalar flux terms are related to gradients by introducing v_T as follows

$$\overline{\rho u_i'' g''}/\bar{\rho} = -v_T(\partial \tilde{g}/\partial x_i) \tag{6.51}$$

The model for turbulent dissipation is

$$\bar{\phi} = b\bar{\rho}\tilde{q}^{3/2}(l_2)^{-1} \tag{6.52}$$

and for scalar dissipation is

$$\bar{\chi} = C\tilde{q}^{1/2} \overline{\rho c'' c''}(l_3)^{-1} \tag{6.53}$$

where b and C are constants and l_2 and l_3 are appropriate length scales, frequently but not necessarily equal to l_1.

Variable density effects are incorporated in Favre averaging. This is valid for low-speed flows, where the model accounts for variable density effects due to heterogeneities in composition; further models are required for high-speed flows.

As a result of the strain due to dilatation, there is a redistribution of the contributions to the turbulent kinetic energy. This requires the introduction of the further modeling assumption

$$\overline{\rho u'' u''}/\bar{\rho} = \varepsilon \tilde{q} \tag{6.54}$$

where $0 < \varepsilon < 1$.

Bray and Libby (1976) made a detailed analysis of premixed combustion in plane oblique combustion waves. Their analysis takes into account the influence of turbulence and inhomogeneity on the effective rate of heat release. They also take into account the effects of dilatation and turbulent production, due to shear in the turbulent kinetic energy balance. They relate their predictions to experiments on plane oblique flames stabilized in uniform turbulent flow of premixed reactants in the wake of cylindrical rods. The orientation of the reaction zone depends on the stream velocity, on the level of turbulence, and on the local thermochemistry.

In the case of normal flames, only dilatation associated with heat release occurs, so that the turbulent kinetic energy is diminished by chemical reaction. In the case of an oblique reaction layer, turbulent kinetic energy is generated by shear and this produces an effect opposing dilatation. Comparison of the theoretical analysis and numerical solutions of the equations with experimental studies of turbulent-flame propagation in premixed combustible gases led Bray and Libby to come to the following conclusions:

1. In the absence of strong shear, the turbulent-flame speed $\tilde{u}_0$ is proportional to $\tilde{q}_0^{1/2}$ and the constant of proportionality is of order unity.
2. The turbulent-flame speed is independent of the scale of turbulence l_0 of the approach flow. Some experimental evidence supports this theory.
3. The thickness of the turbulent-flame is proportional to the integral scale of the approach turbulence and the constant of proportionality is of order unity.
4. In the absence of shear, the turbulent-flame speed decreases as the mixture ratio approaches stoichiometric or as the mass fraction of inert diluent decreases. This prediction is not supported by experimental evidence.
5. In the absence of shear, the turbulent kinetic energy decreases on passage through the combustion zone because of the dilatation effect. No experimental evidence is available to test this prediction.
6. In flames which generate Reynolds shear stresses, production of turbulent kinetic energy competes with removal due to the dilatation effect. Turbulent energy initially decreases; later Reynolds stresses cause an increase in turbulence in the downstream region of the flame. No experimental evidence is available to test these predictions.
7. If the Reynolds stress becomes large, the flame angle approaches a constant value θ_c implying that the flame speed can increase without limit to match increases in the approach flow velocity. Experimental evidence confirms this prediction.
8. The limiting flame angle θ_c grows with increasing value of the heat release parameter τ. This is in agreement with some experimental evidence.
9. The limiting flame angle θ_c is independent of the initial turbulence $\tilde{q}_0$. This is supported by some observations.
10. In the strong interaction limit, the flame speed is predicted to be proportional to the final turbulent velocity, $\tilde{q}_\infty^{1/2}$.
11. The peak time averaged rate of heat release increases with increasing shear. This has been observed experimentally.
12. For nearly normal flames, the flame speed depends on the initial product concentration $\tilde{c}_0$ at the "cold boundary"; for highly oblique flames, flame properties are essentially independent of $\tilde{c}_0$.

Rod-Stabilized Premixed Turbulent Flames

Premixed turbulent flames can be effectively studied by inserting a rod stabilizer in a uniform premixed flow and examining the flame structure in the wake of the stabilizer. Such studies show clearly that combustion rates can be greatly increased by turbulence and that substantially larger volumetric heat release rates can be achieved relative to laminar combustion. Many attempts have also been made to measure flame speed in order to provide an indication of the overall reaction rates controlling the combustion process. Accurate determination of flame speed requires a precise measurement of the position of the flame front. Instantaneous determination of local orientation of flame surfaces, together with measurements of local mean velocity direction and magnitude would allow calculation of instantaneous local flame speeds. Since turbulence properties vary with time and space, correlations could be made between flame speed and turbulence properties. Examination could also be made of the effects of combustion-induced gas expansion of local flow direction. The use of high-speed photography and schlieren allows the determination of flame front location, and laser anemometry allows the determination of velocity and turbulence characteristics throughout the flow field.

Smith and Gouldin (1978) studied turbulent, unconfined methane-air flames in the wake of a cylindrical rod stabilizer. They determined local turbulent flame speeds as a function of local approach flow turbulence properties and fuel equivalence ratios. They used thermocouples to determine local time mean flame orientations; hot-film anemometry for mean velocity and turbulence measurements; and a laser anemometer to measure divergence effects due to combustion. Turbulence generating screens, located upstream of the flame stabilizer, were used to create turbulent flows of varying macroscale and intensity. Exit velocities were varied up to 30 m/s; turbulence intensity up to 7 percent; turbulence macroscales from 0.6 to 1.6 mm; and equivalence ratio from 0.75 to 1.00.

Local time mean flame positions were determined using fine-wire, silica-coated thermocouples to measure local flame isotherm inclinations. The laser anemometer was used to measure vertical and horizontal velocity components. Flame speed was defined as equal in magnitude to the component of the approach flow velocity normal to the $2T_0$ isotherm; T_0 is the cold approach flow temperature. The $2T_0$ criterion is an approximation to the requirement that flame speed should be determined at a surface of initial temperature increase. Weinberg (1955) has shown that the schlieren image of a one-dimensional laminar flame with a constant thermal conductivity corresponds to a flame temperature of $2T_0$. Smith and Gouldin (1978) used the $2T_0$ criterion to make a qualitative comparison of flame speed data derived from schlieren photography with their measurements made by thermocouples. In the studies of Smith and Gouldin, they found that intensity and macroscale were essentially independent of combustor velocity and they noted that the combustion process resulted in only a small change in direction of the approach flow. Intensities were determined at the points corresponding to the flame front locations but, since they were unable to obtain measurements in the flame zone with the laser anemometer, they used turbulence intensity measurements made in the cold flows for the correlation of their flame speed data.

Density gradients were determined from temperature data using a central difference approximation. The perfect gas law was used to relate gas temperature and density, assuming constant pressure and molecular weight. The variation of

SYMBOL	SCREEN	h,cm	Ū ,m/s
—○—	NONE	4.8	7.5
—●—	NONE	6.8	7.5
—□—	GRID	4.8	7.5
—■—	GRID	6.8	7.5
—△—	GRID	4.8	14
—▲—	GRID	6.8	14
—▽—	10 MESH	4.8	7.5
—▼—	10 MESH	6.8	7.5

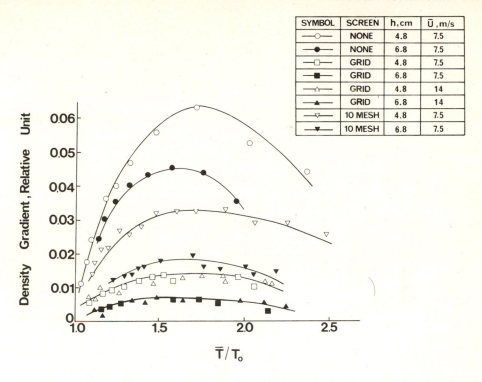

Figure 6.5 Variation of mean gas density gradient in low-temperature regions of turbulent flames. (*Smith and Gouldin, 1978.*)

mean gas density gradient in low temperature regions of a methane-air flame is shown in Fig. 6.5. As a function of $\bar{T}/T_0$, where $\bar{T}$ is the local mean temperature, maximum density gradients occur between the theoretically predicted limits for laminar and turbulent flames. Density gradients in more turbulent flames are smaller in magnitude because of the associated flatter temperature profiles. The variation of turbulent-flame speed with rms fluctuating velocity and equivalence ratio is shown in Fig. 6.6. At a constant macroscale, the turbulent-flame speed of lean methane-air mixtures increases with increasing velocity fluctuations and increasing equivalence ratio. At fixed values of equivalence ratio and rms velocity fluctuations, an increase in macroscale is accompanied by an increase in flame speed. The constant flame inclination angle suggests a linear relationship between flame speed and fluctuating velocity. It was also noted that flame thickness increased with distance downstream the stabilizer and also with increase in turbulence in the approach flow. Isotherms in flames are not parallel and this divergence of flame isotherms influences the accuracy of flame speed calculations. Temperature fluctuation measurements showed that maximum fluctuations occur in the central portion of the mean temperature profile and that peak rms temperature fluctuations increased in magnitude with increase in downstream distance. Temperature fluctuations also increase with increasing equivalence ratio and with increase in upstream turbulence.

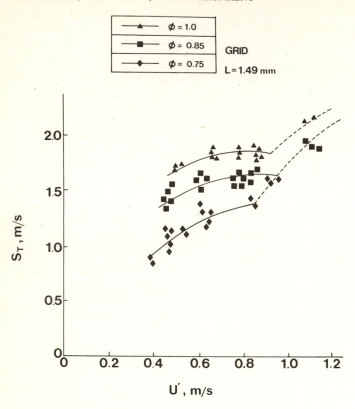

Figure 6.6 Variation of turbulent-flame speed with rms fluctuating velocity. (*Smith and Gouldin, 1978.*)

6.4 TURBULENT SHEAR FLOW WITH CHEMICAL REACTION

Bush and Fendell (1978) have developed a model for the prediction of the mean rate of reactant consumption, product formation, and heat generation for initially unmixed reactants in turbulent shear flows. They examine the unsteady conditions of a two-dimensional mixing layer, in the limit of large-scale-mixing-controlled combustion. This model is particularly suitable for systems where the reaction rate is large relative to the flow rate, so that the aerothermochemical system is in, or very close to, chemical equilibrium. In such cases, the rate-limiting step for the combustion process is macroscale, rather than microscale mixing. Molecular diffusion and chemical kinetics are not rate-limiting steps. Control of the system is dependent on the inviscidly controlled engulfing of fresh fluid from the non-vortical flow deep into the core of the shear flow and the subsequent breakdown of large-scale coherent organized structure. Once the large-scale structure is broken down, molecular-scale diffusion and chemical reaction proceed, followed rapidly by the formation of product gas and heat release from exothermic combustion. The rate-controlled processes are not dependent on Reynolds number, which influences the small-scale turbulence only.

Bush and Fendell have developed a simple model, based on a realistic appraisal of the limited amount of experimental data and the current physical understanding of turbulent transport. Design engineers require rapid and inexpensive means of explicitly characterizing changes in system performance as a function of alteration of controllable parameters.

Modeling of Two-Dimensional Mixing Layer with Chemical Reaction

Models have been formulated for a two-dimensional mixing layer of unmixed fuel and oxidant coflowing streams with chemical reaction. It is assumed that combustion is controlled by large-scale mixing and that flow conditions are unsteady. The aim of the study is to determine the mean rate of reaction consumption and product generation for the burning of the initially unmixed gaseous reactants. The large-scale process is inviscid and not dependent on Reynolds number; Reynolds number effects are significant for small-scale turbulence, where viscous effects are important. The separation of the mixing process into a large-scale structure, followed, after a time delay, by small-scale interaction, implies that the rate-limiting step for the process is macroscale mixing and not molecular diffusion or chemical kinetics.

The flowing fuel and air streams begin to mix within the mixing layer across the interface. Combustion occurs when mixture ratios reach the limits of flammability on the rich and lean sides of the mixing layer. The flame-sheet location coincides with the interface and can be described as a highly convoluted, rapidly translating, multiply connected sheet existing between initially unmixed, highly combustible reactants. Bush and Fendell (1978) put forward quantitative arguments suggesting that the principal effect introduced by unsteadiness was a finite, significant strain rate along the flame sheet. This strain rate enhances burning over steady-flow levels by stretching the flame sheet to increase the area of exposure of fuel to oxidant, and also by inducing additional fresh combustibles to the region of burning. The enhancement of burning by increasing the strain rate cannot extend indefinitely because of two counteracting phenomena: when the flame sheet becomes too extended, it begins to shorten due to local depletion of reactant and, secondly, when the strain rate becomes too great, it begins to exceed the rapid chemical kinetic rates. In steady mean situations, the two phenomena that cause increase in combustion rates are approximately balanced by the two phenomena causing decrease in combustion rates. The thin flame sheet expands into a layer of finite thickness and this leads to slower reaction rates and lower rates of product generation.

Cold-Flow Analogy

In order to circumvent the complexities of heat release, studies have been made of cold-flow analogies with dilute reactant concentrations and no, or virtually none, exothermic reactions. Toor (1962) developed a cold-flow analogy by using a one-step irreversible reaction between unmixed fuel and oxidant. Measurement of the probability density function (pdf) for a passive scalar permits the prediction of mean turbulent diffusion flame behavior in a flow with a Lewis number of unity. In such a flow, the dynamics alter the energetics but the energetics do not alter the dynamics. For certain specified conditions, predictions are made of the time-averaged mass-fraction profiles for a turbulent diffusion flame. Rates of product formation can be determined, provided that the pdf is known. Information on pdf distributions is still required to be obtained by experimental measurement. Bush and Fendell (1978) suggested the use of a mixed, clipped-type form pdf, with dirac delta functions representing the contributions from fluid from the bounding streams undiluted by molecular diffusion; a beta distribution is used to character-ize the continuous range of mixed fluid. This beta distribution permits the pdf to range naturally with parameter variation from (1) a bimodal form appropriate to typically larger deviations between the mean and instantaneous pictures (large turbulent intensity, pertinent to large-scale coherent structure) to (2) a trimodal form appropriate to typically smaller deviations between the mean and instantan-eous pictures (small turbulent intensity, pertinent to small-scale, well mixed structure).

Chemical Reaction with Heat Release

The exothermic chemical reactions in flames cause rapid heat release, with density decreases by a factor of five. Rapid chemical reaction must be associated with rapid local volumetric expansion of gases by a factor of the order of five. The cold-flow analogy requires to be modified by considering the effect of heat release on the system. Bush and Fendell (1978) analyzed the structure of a turbulent diffusion flame produced by the low-speed burning of initially unmixed, highly combustible gaseous streams of fuel and oxidant in an irreversible one-step reac-tion, close to equilibrium. Reaction takes place in a two-dimensional turbulent, isobaric mixing layer, formed by two parallel streams, downstream of a smooth flat splitter plate. For systems in which the constituent species have comparable molecular weights, heat capacities, and diffusion coefficients, the effect of heat release is to produce an increase in the mean temperature, which is inversely proportional to the decrease in the mean density for an isobaric flow. In the core of the mixing layer, local turbulent equilibrium is expected and use of mean fields in the equation of state is considered to be valid. Most of the combustion is expected to occur within the core of the mixing layer. This approximation permits the use of a coordinate transformation, involving the mean density, to correlate the compressible flow to an incompressible one. Under these conditions, the dynamics is decoupled from the energetics.

The mean rate of reactant consumption is taken to be proportional to the product of the local mean stoichiometrically adjusted mass-fractions for fuel and oxidant, and the appropriate local characterization of the mean rate of strain. It is argued that, in general, the local mean strain rate indicates the level of chemical activity in a turbulent diffusion flame, in which macroscopic mixing is rate-controlling. This explicit model isolates the dependencies of the mean rate of reactant consumption in the mixing-limited domain and incorporates the remaining dependencies in an empirical constant of proportionality. This formulation recognizes the role of mixing control, which tends to distribute the product species in a gaussian, symmetrical form about the mixing layer centerline, but also accounts for asymmetries in the product gas distribution owing to off-stoichiometric richness of the two bounding streams of the mixing layer. This model reflects the relative importance of enriching the faster, or the slower, stream of the mixing layer. This simple model enables both exact numerical and approximate analytical examination of the multiple parametric dependencies of mean consumption rate on turbulent Schmidt number, stoichiometry, velocity difference, and reaction order.

The Bush-Fendell model utilizes the concept of fully developed, self-similar flow and employs eddy viscosity-type exchange coefficients. These asymptotic conditions are not achieved in practical flames and turbulent diffusion flames normally consist of a large Reynolds number turbulent fuel jet exhausting into an ambient containing oxidant, for which the self-similar conditions do not apply.

Transitional Region of Turbulent Jet Diffusion Flame

The origin of the initial mixing layer is at the jet exit lip. The inner boundary of the mixing layers converges to merge on the axis of symmetry. This point of merger, the end of the potential core, is typically at a distance of five diameters from the jet exit. On the axis of a turbulent fuel jet, the principal strain vanishes, because of symmetry, and for a model in which mean consumption rate is proportional to the principal strain, this implies that mean consumption rate also vanishes. Bush and Fendell argue that there is reactant consumption immediately on the axis of symmetry, but the elliptical corrections to the parabolic form (in which one strain rate dominates the dissipation) are not considered to be essential to the description of the jet flame.

In a "thick" reaction zone, mean stoichiometrically adjusted reactant mass fractions coexist, even though the instantaneous reactant mass fractions do not, because the instantaneously thin flame fluctuates on the integral scale of the turbulence. Oxidant penetrates to the axis at increasingly higher levels with increasing downstream distance, ultimately achieving ambient levels on the axis. The flame locus, defined by the equality of the mean stoichiometrically adjusted mass fractions for fuel and oxidant lies initially off the axis but intersects the axis downstream, at a distance dependent mainly on the stoichiometry. The flame length increases with decreasing molecular weight of the fuel, all other parameters being held constant. Maximum-temperature levels never achieve the adiabatic flame temperature. The locus of the mean maximum temperature, in planes perpendicular to the axis, occurs initially off the axis but eventually converges to the axis. The "maximum-temperature" locus lies within the "flame" locus. The maximum temperature, in cross sections, increases with distance downstream until the fuel is depleted. It then gradually decays, because of thermal dilution, to the ambient level. The predictions of Bush and Fendell are consistent with experimental data. The centerline decay of the coupling function and that of the streamwise velocity component, with distance downstream, are compatible with experimental data. It was only necessary to adjust one empirical parameter to determine the reduced rate of jet spread in the initial region, the decreased rate of decay with downstream distance for the centerline value of the streamwise velocity component, and of any passive scalar. There is also agreement that the eddy diffusion coefficient initially decreases but subsequently increases with distance downstream.

Initial Conditions

The initial conditions for a jet flame are determined by the upstream flow patterns, thickness of boundary layers on surfaces, and turbulence intensities and scales. The walls of nozzles delivering fuel and oxidant will have finite thicknesses at the lip, leading to the generation of wake and recirculating flows. These regions can act as ignition sources. The walls of nozzles may become heated, due to back radiation from the flame, and act as heat sources to the flow. Nozzle material surfaces may also produce catalytic effects; deposition of liquid or solid fuel particles on wall surfaces can lead to coke formation and burning. Boundary layers formed on nozzle walls are generally thick, because of modest Reynolds numbers, resulting in nonuniform starting conditions for thermodynamic and dynamic variables. Both streamwise and transverse pressure gradients influence flow in the initial regions. Initial flow regions of jets are often dominated or influenced by wake flows, which decay as the initial region merges into the transitional zone.

6.5 TRANSITION FROM LAMINAR TO TURBULENT DIFFUSION FLAMES

The transition from laminar to turbulent jet flames is studied by progressively increasing the Reynolds number until the first signs of turbulence appear. As the Reynolds number is further increased, the flow regime becomes increasingly dominated by turbulence until, finally at high Reynolds numbers, fully turbulent jet flames prevail. Studies of nonburning gaseous jets show the onset of instabilities in the laminar jet, the gradual spreading of this turbulence throughout the jet, and the reduction of the length of the initial laminar regions. Under burning conditions, the effects of viscosity persist at much higher Reynolds numbers, and in the transition range of Reynolds numbers, turbulent nonburning jets can be almost completely laminarized after ignition.

Takeno and Kotani (1978) have made experimental studies on the transition and structure of turbulent jet diffusion flames developing in coflowing airstreams. By varying the type of fuel and the Reynolds number, they provided clear indications of the transition conditions of turbulent diffusion flames. These studies show that the flame structure changes are almost entirely governed by fluid mechanics and the physical transport processes. Many investigators in the past have noted that, as the velocity is increased, a location can be found in the laminar flame where the laminar flow breaks down and the flame is turbulent downstream from this break-point. The distance from the nozzle exit to this break-point is known as the "induction length." The induction length decreases with increase of injection velocity. This phenomenon is also seen in nonburning jets, where it has been found that, for some types of nozzles, there is a steady decrease of induction length with velocity, while, for other nozzles or injectors, sudden discontinuous decreases have occurred at critical velocities. Two possible mechanisms have been postulated for the generation of turbulence in jets: the turbulence in the pipe flow inside the injector and shear at the free jet boundary.

In the studies of Takeno and Kotani, the turbulence intensity of the coflowing airstream was maintained below 0.1 percent. The velocity of the coflowing airstream was 5 m/s at ambient temperature. A hot-wire anemometer was used to measure turbulence characteristics at the nozzle exit. Instantaneous (1 μs) schlieren photographs and high-speed (10,000 fps) motion schlieren photography was used to study nonburning and flame jets.

The jet Reynolds number is defined by

$$Re_0 = (\rho_0 u_0 d_0)/\mu_0 \qquad (6.55)$$

where ρ_0, u_0, d_0, and μ_0 are respectively, the fuel density, velocity, diameter, and viscosity at the nozzle exit. A flame Reynolds number, based upon nozzle exit velocity, but using flame density and viscosity is defined by

$$Re_f = (\rho_f u_0 d_0)/\mu_f = (v_0/v_f)\, Re_0 \qquad (6.56)$$

where ρ_f and μ_f are the density and viscosity at the flame surface and v is the kinematic viscosity.

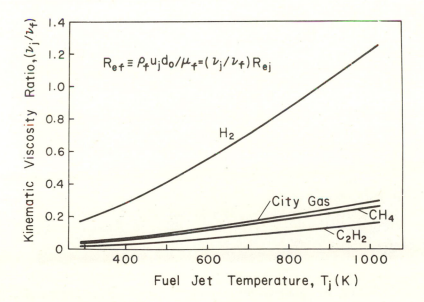

Figure 6.7 Kinematic viscosity ratio as a function of temperature. (*Takeno and Kotani, 1978.*)

Figure 6.7 shows the effect of variation of temperature on the kinematic viscosity ratio for a range of gases. For a hydrogen jet with a Reynolds number of 1240, the corresponding flame Reynolds number is 227, while, for an acetylene jet of Reynolds number 1240, the corresponding flame Reynolds number is 24. Throughout the transition region, the change from nonburning to burning conditions leads to a large-scale increase in kinematic viscosity, so that local Reynolds numbers are reduced by a factor of 10.

Schlieren photographs of hydrogen jets at $Re_0 = 1074$ show the cold jet to be turbulent beyond the transition point at eight diameters downstream. After ignition, the hydrogen flame was found to be laminar throughout the whole flow field. As Re_0 was increased from this value, the total flame length increased until, at $Re_0 = 1240$, some signs of transition were detected in the hydrogen flame. From this stage onward the flame had the characteristic configuration of an upstream laminar section, followed by a downstream turbulent section. The transition point in the flame jet moved gradually upstream with increase in Re_0. When Re_0 reached the critical value of 1950, where the pipe flow enters into the transition region, a drastic change in flame behavior occurred. The transition point began to fluctuate back and forth with a very high frequency. The fluctuations of the transition point were found to correspond with the intermittent flow fluctuations at the injector exit. Further increase of Re_0 increased the fraction of time during which the transition point was in the upstream position. At $Re_0 = 2500$, the transition point became fixed at an upstream position and this did not change with further increase of Re_0. A schlieren photograph of a fully turbulent hydrogen flame jet at $Re_0 = 3016$ for $u_0 = 299$ m/s is shown in Fig. 6.8. Figure 6.9 shows a comparison of the induction length (distance from nozzle exit to transition point), nondimensionalized by the nozzle exit diameter, as a function of the Reynolds number at the nozzle exit. For hydrogen, a striking difference is seen between the induction lengths of flame jets compared with cold jets, for $Re_0 < 2500$. This shows clearly that the flame has a strong dynamically stabilizing effect on the flow instability of a free jet.

Figure 6.8 Schlieren picture of H_2 flame jet ($Re_j = 3016$, $u_j = 299$ m/s). (*Takeno and Kotani, 1978.*)

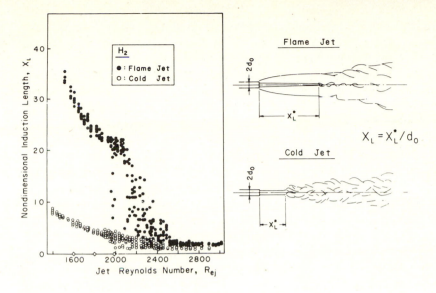

Figure 6.9 Nondimensional induction length of H_2 jet. (*Takeno and Kotani, 1978.*)

A similar study, made with acetylene, showed that at $Re_0 = 1853$, both the flame and the cold jet remained laminar. At $Re_0 = 1950$, the flame became turbulent suddenly with the whole flame fluctuating. The photographs suggested that the flame transition was triggered by intermittent disturbances from the ejected flow. Further increase in Re_0 led to an increase in the frequency of the intermittent fluctuations, until the flow in the flame jet was always turbulent. This interesting comparison between acetylene and hydrogen jets shows that there are important fluid-dynamic differences in the flames, which might lead one to suppose that these are due to differences in chemical composition of the two gases. Figure 6.7, however, shows that there are important physical property differences between the hydrogen and acetylene and, in particular, there is a large difference in the kinematic viscosity of the two gases. Since viscosity has such an important stabilizing influence on flow, these differences in kinematic viscosity affect the turbulence characteristics in the transition region. These studies also show that Reynolds number is not the sole parameter that determines the induction length.

When turbulent fluctuations inside the injector are small, they can be damped by a flame jet and the flame may remain laminar. As turbulence increases, flame stability can be affected, resulting in blowoff. Flame transition is thus dependent upon the turbulent characteristics of the pipe flow, which is affected by wall roughness, curvature, and length-to-diameter ratios. It has also been noted by several investigators that turbulent fluctuations and eddy structures originating inside the injector are ejected into the free jet and, subsequently, develop into large-scale eddy motion in the downstream region of the flames.

6.6 TURBULENT MIXING WITH CHEMICAL REACTION

The effect of chemical reaction on turbulent mixing processes has been a subject of discussion in many papers but there are very few facts which have been clearly established. In hydrocarbon flames, many complex chemical reactions take place within very short time periods, leading to the formation of combustion products and intermediates from the fuel and oxidant, coupled with heat release. In air-fuel systems, with the nitrogen remaining inert, it can be expected that the changes in chemical composition of the gases taking part in the reaction will, in themselves, have little effect on the turbulent mixing process. The heat release results in expansion and density change, which cause acceleration and expansion of the flow. The extent to which this heat release will affect mixing is dependent upon the relative energies of the turbulent mixing process and the heat release. The proportion of the total volumetric flow that is occupied by flame will determine whether the effects of reaction are local or sufficient to influence the whole flow pattern. It may be expected, as a first approximation, that the consideration of a flame as a local heat source may be sufficient to explain the differences between burning and nonburning flow systems.

Edelman and Harsha (1978) have reviewed recent advances in the prediction of turbulent chemically reacting flows, with emphasis on the detailed prediction of reacting jet flows coupling models of the turbulent mixing process with a detailed finite-rate kinetics scheme. Kinetics calculations are made using the quasi-global model for the higher hydrocarbons. Comparisons are made between a one-equation turbulence model, developed by Harsha (1974), and a two-equation turbulence model, described by Launder and Spalding (1972). The effects of heat release are approximated through the use of a fully reacted shifting-equilibrium chemistry model. The detailed prediction of the finite-rate chemical kinetics in hydrocarbon-air flow fields is made using the more general hydrocarbon finite-rate kinetics formulation, described by Boccio et al. (1973).

Turbulence Model Predictions in Reacting Flows

In the modeling of turbulent free shear layers, models that involve the solution of the turbulent kinetic energy equation as a means of obtaining the local turbulent shear stress distribution produce results that are far superior to those obtained with local eddy viscosity assumptions. These models are ranked according to the number of simultaneous differential equations which are solved to obtain the turbulent shear stress distribution. The one-equation model considers the turbulent kinetics energy equation alone, while the two-equation model considers the turbulent kinetic energy equation plus a differential equation for the turbulence length scale or, equivalently, the dissipation rate for turbulent kinetic energy. These equations are solved in addition to the other differential equations that describe the transport of mean momentum, energy, and species in a turbulent flow. The form of the kinetic energy equation used in the model is that for incompressible flows; the application of such models to flows with strong density variations assumes that the effect of these density variations on the turbulent kinetic energy is negligible. Bray (1973) has made a rigorous derivation for the turbulent kinetic energy equation for variable density flows and has shown that additional significant terms do arise in the equations.

One-Equation Model

The one-equation turbulence model of Harsha (1974) uses an algebraic formulation for the turbulence length scale. The first basic assumption of the one-equation model is that the relationship between the turbulent shear stress and the turbulent kinetic energy can be modeled by the equation

$$\tau = a_1 \rho k \tag{6.57}$$

where the parameter a_1 is a function of position. The turbulent kinetic energy is defined by

$$k = \tfrac{1}{2}(\overline{u'^2} + \overline{v'^2} + \overline{w'^2}) \tag{6.58}$$

The second basic assumption of the one-equation model is that the turbulence length scale can be approximated as a fraction of the local flow field scale. This length scale l_k is included in the definition of the turbulent kinetic energy dissipation rate

$$\varepsilon = a_2 k^{3/2}/l_k \tag{6.59}$$

The length scale is defined by a measure of the width of the shear layer in the potential core region of the jet and by a half width of the jet downstream of the potential core. The dissipation rate constant a_2 is given the nominal value of 1.69 for incompressible flows, while, for flows with strong density gradients, a_2 is allowed to vary as a function of both the initial jet- to free-stream-density ratio and the local turbulent Reynolds number

$$R_T = \Delta u l_k / v_{TM} \tag{6.60}$$

in which Δu is a local characteristic velocity difference and v_{TM} is the turbulent eddy viscosity evaluated at the radial location of the maximum shear stress at a given axial position

$$v_{TM} = \tau_M / \rho (\partial u / \partial r)_M \tag{6.61}$$

The turbulent kinetic energy transport equation is written as

$$\rho u \frac{\partial k}{\partial x} + \rho v \frac{\partial k}{\partial r} = \frac{1}{r} \frac{\partial}{\partial r} \left(\frac{r \rho v_T}{\mathrm{Pr}_k} \frac{\partial k}{\partial r} \right) + \rho v_T \left(\frac{\partial u}{\partial r} \right)^2 - \frac{a_2 \rho k^{3/2}}{l_k} \tag{6.62}$$

The first term on the right-hand side represents the diffusion of turbulent kinetic energy with a turbulent Prandtl number Pr_k taken to be 0.7; the second term is the exact expression for the turbulent kinetic energy production under the boundary-layer assumptions assuming that

$$\tau = \overline{\rho u' v'} = \rho v_T (\partial u / \partial r) \tag{6.63}$$

The third term represents the dissipation of turbulent kinetic energy.

Two-Equation Model

The two-equation model developed by Launder and Spalding (1972) is referred to as the $k - \varepsilon$ model. The eddy viscosity is given by

$$v_T = C_\mu l_k k^{1/2} \tag{6.64}$$

where C_μ is a function of the local axial velocity gradient with the basic value of 0.09. The turbulent shear stress then becomes

$$\tau = C_\mu \rho l_k^{1/2}(\partial u/\partial r) \tag{6.65}$$

The transport equation for the turbulent kinetic energy dissipation rate is derived from the Navier-Stokes equations. Modeling is used to reduce the higher-order turbulence correlation terms to a tractable form. The dissipation rate equation has the following form

$$\rho u \frac{\partial \varepsilon}{\partial x} + \rho v \frac{\partial \varepsilon}{\partial r} = \frac{1}{r}\frac{\partial}{\partial r}\left(\frac{r\rho v_T}{\text{Pr}_\varepsilon}\frac{\partial \varepsilon}{\partial r}\right) + C_{\varepsilon1}\frac{v_T\varepsilon}{k}\left(\frac{\partial u}{\partial r}\right)^2 - C_{\varepsilon2}\frac{\rho\varepsilon^2}{k} \tag{6.66}$$

The five constants have the following values

$$C_\mu = 0.09 \quad C_{\varepsilon1} = 1.45 \quad C_{\varepsilon2} = 1.96 \quad \text{Pr}_k = 1.0 \quad \text{Pr}_\varepsilon = 1.3 \tag{6.67}$$

Computations have been made by Edelman and Harsha (1978) using both the one- and two-equation models for the reacting hydrogen-air jet described by Kent and Bilger (1972). The initial boundary layers for the hydrogen jet and the coflowing external airstreams were assumed to follow a one-seventh power law profile, with an initial boundary layer thickness of 0.4 r_0 for the inner jet boundary layer. A step profile was used for the initial hydrogen concentration with $C = 1$ for $r < r_0$ and $C = 0$ for $r > r_0$. To obtain the initial kinetic energy, a constant eddy viscosity model was used with

$$v_T = 0.007(u_j - u_e)\delta \tag{6.68}$$

The initial turbulent energy dissipation rate was obtained from

$$\varepsilon = 1.69k^{3/2}/\delta \tag{6.69}$$

The Prandtl number was assumed to be 0.7 and the Lewis number was taken to be 1.0. A comparison of the centerline velocity decay predictions of the two turbulence models is made with experimental data in Fig. 6.10. Predictions of the centerline jet species mass-fraction variation are shown in Fig. 6.11. Comparison of the predictions made shows that there is little to choose between the two models. Provided that judicious choice is made of the turbulence length scale distribution, the one-equation model is sufficiently accurate for making these types of predictions. It should be noted that both the one- and two-equation models discussed in this section use constant density in reacting flows with significant heat release. Since there are large-scale variations in density in these systems, the assumption of constant density is not valid and this restricts the accuracy of the predictions.

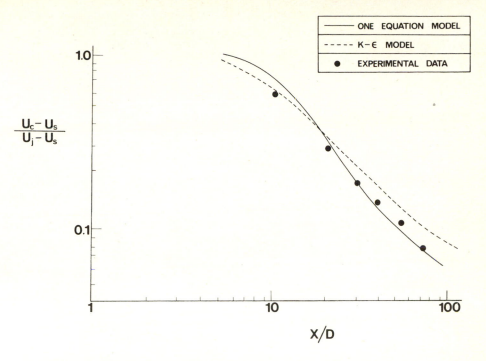

Figure 6.10 Comparison of one-equation model predictions for centerline velocity decay, reacting H_2-air jet. (*Edelman and Harsha, 1978.*)

Detailed Prediction of Finite-Rate Kinetics

In high-performance systems, the chemical equilibrium assumption is not valid for most species of interest. This is particularly true of NO_x, even for residence times well beyond the millisecond range. The oxidation of CO to CO_2 is halted, due to quenching at reduced temperatures, or dissociation at peak temperatures. Small reductions in combustion efficiency can lead to significant pollutant emission levels of CO and unburned hydrocarbons. Many of the intermediate atoms and free radicals that are an inherent part of any oxidation process play important roles in the formation and destruction of pollutants. Superequilibrium levels of such species occur during the early stages of combustion and they persist, because of the relatively slow three-body recombination reactions, even beyond what normally is considered the overall reaction time.

Several attempts have been made to develop a full finite-rate chemistry formulation for hydrocarbon reaction in turbulent flows. Insufficient information is available on the many reactions which are already known to take place and their respective rate constants. Most of the detailed predictions which have been made use simplified reaction schemes. Edelman and Fortune (1969) introduced the concept of quasi-global kinetics as a way of circumventing the lack of detail on the kinetics and mechanism of higher hydrocarbon oxidation while characterizing many of their important combustion characteristics. The concept of quasi-global

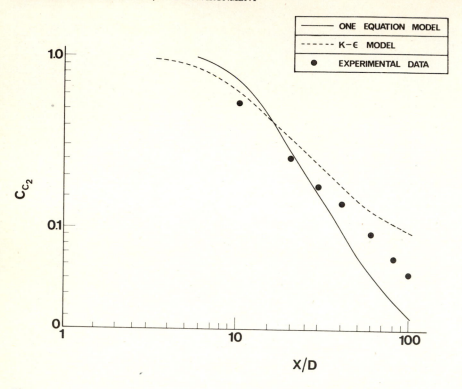

Figure 6.11 Comparison of one-equation and two-equation turbulence model predictions for centerline jet species concentration, reacting H_2-air jet. (*Edelman and Harsha, 1978.*)

kinetics involves the coupling of a set of subglobal reaction steps to a set or sets of detailed reversible reaction steps for those reaction chains where substantial information already exists for their kinetics and mechanisms.

The single most important consideration in the formulation of appropriate kinetic mechanisms is the recognition of the spectrum of local states that are encountered in various combustor flow fields. These include the distribution of fuel/air ratios, temperature, pressure ratios, and residence times. Edelman and Harsha (1978) have demonstrated that the rate of combustion of hydrocarbons can be predicted using the quasi-global kinetic scheme given in Table 6.1. Examination of experimental data shows, for example, that ignition delay times for many hydrocarbons are similar. This is particularly true for hydrocarbons in the presence of oxygen at initial temperatures above about 1000 K. Methane is an important example of a hydrocarbon in the paraffin (alkane) series with an ignition delay time that is significantly longer than the other hydrocarbons in the same series. This difference has been traced to the importance of the methyl radical (CH_3) formed during the decomposition of methane. For the higher hydrocarbons (propane and above) the similarity in oxidation behavior suggests that the prediction of significantly more detail of the oxidation process can be obtained by extending the notion of a single-step overall reaction to a scheme

Table 6.1 Extended C—H—O chemical kinetic reaction mechanism $k_f = AT^b \exp(-E/RT)$ (Edelman and Harsha, 1978)

Reaction	A		Forward b	E/R	
	Long-chain	Cyclic		Long-chain	Cyclic
(1) $C_nH_m + \dfrac{n}{2}O_2 \rightarrow \dfrac{m}{2}H_2 + n\,CO$†	6.0×10^4	2.08×10^7	1	12.2×10^3	19.65×10^3
(2) $CO + OH = H + CO_2$	5.6×10^{11}		0	0.543×10^3	
(3) $CO + O_2 = CO_2 + O$	3.0×10^{12}		0	25.0×10^3	
(4) $CO + O + M = CO_2 + M$	1.8×10^{19}		-1	2.0×10^3	
(5) $H_2 + O_2 = OH + OH$	1.7×10^{13}		0	24.7×10^3	
(6) $OH + H_2 = H_2O + H$	2.19×10^{13}		0	2.59×10^3	
(7) $OH + OH = O + H_2O$	5.75×10^{12}		0	0.393×10^3	
(8) $O + H_2 = H + OH$	1.74×10^{13}		0	4.75×10^3	
(9) $H + O_2 = O + OH$	2.24×10^{14}		0	8.45×10^3	
(10) $M + O + H = OH + M$	1.0×10^{16}		0	0	
(11) $M + O + O = O_2 + M$	9.38×10^{14}		0	0	
(12) $M + H + H = H_2 + M$	5.0×10^{15}		0	0	
(13) $M + H + OH = H_2O + M$	1.0×10^{17}		0	0	
(14) $O + N_2 = N + NO$	1.36×10^{14}		0	3.775×10^4	
(15) $N_2O_2 = N + NO_2$	2.7×10^{14}		-1.0	6.06×10^4	
(16) $N_2 + O_2 = NO + NO$	9.1×10^{24}		-2.5	6.46×10^4	
(17) $NO + NO = N + NO_2$	1.0×10^{10}		0	4.43×10^4	
(18) $NO + O = O_2 + N$	1.55×10^9		1.0	1.945×10^4	
(19) $M + NO = O + N + M$	2.27×10^{17}		-0.5	7.49×10^4	
(20) $M + NO_2 = O + NO + M$	1.1×10^{16}		0	3.30×10^4	
(21) $M + NO_2 = O_2 + N + M$	6.0×10^{14}		-1.5	5.26×10^4	
(22) $NO + O_2 = NO_2 + O$	1.0×10^{12}		0	2.29×10^4	
(23) $N + OH = NO + H$	4.0×10^{13}		0	0	
(24) $H + NO_2 = NO + OH$	3.0×10^{13}		0	0	
(25) $CO_2 + N = CO + NO$	2.0×10^{11}		-0.5	4.0×10^3	
(26) $CO + NO_2 = CO_2 + NO$	2.0×10^{11}		-0.5	2.5×10^3	

† $\dfrac{dC_{C_nH_m}}{dt} = -AT^b P^{0.3} C_{C_nH_m}^{1/2} C_{O_2} \exp - \left(\dfrac{E}{RT}\right)$, $[C]$ = g mol/cm^3, $[T]$ = degrees Kelvin, $[P]$ = atm, $[E]$ = kcal/mol.

which characterizes the reactions higher up in the chain, with one or more sub-global steps coupled to a set of reversible reactions to characterize the kinetics processes at the lower end of the chain.

Under fuel-rich conditions, hydrocarbon fragments, radicals, and a variety of oxygenated species are produced during the combustion process; account needs to be taken of these species in representing the fuel oxidation process, as well as the NO$_x$ formation process. In addition, soot formation is a potential problem under fuel-rich conditions and fuels with high aromatic content are particularly prolific in their sooting characteristics. Edelman (1978a) has developed an extended quasi-global model that accounts for soot formation and soot oxidation in fuel-rich combustion. For aromatics the net soot emission is found to depend most strongly upon the local unburned hydrocarbon concentration as well as on the oxygen concentration and temperature.

Table 6.2 CH$_4$—O—N chemical kinetic reaction mechanism system $k_f = AT^b \exp\left(-E/RT\right)$ (Edelman and Harsha, 1978)

Reaction	Forward		
	A	b	E/R
(1) $CH_4 + M = CH_3 + H + M$	2×10^{17}	0	44.5×10^3
(2) $CH_4 + OH = CH_3 + H_2O$	3.5×10^{14}	0	4.5×10^3
(3) $CH_4 + O = CH_3 + OH$	2×10^{13}	0	3.45×10^3
(4) $CH_4 + H = CH_3 + H_2$	2×10^{14}	0	5.95×10^3
(5) $CH_3 + O = HCHO + H$	1.9×10^{13}	0	0
(6) $CH_3 + O_2 = CHO + H_2O$	2×10^{10}	0	0
(7) $CH_3 + O_2 = HCHO + OH$	1×10^{14}	0	0.75×10^3
(8) $CH_3 + O = CHO + H_2$	1×10^{14}	0	0
(9) $HCO + OH = CO + H_2O$	3×10^{13}	0	0
(10) $HCHO + H = CHO + H_2$	1.7×10^{13}	0	1.5×10^3
(11) $HCHO + CH_3 = CHO + CH_4$	2.5×10^{10}	$\frac{1}{2}$	2.65×10^3
(12) $HCHO + O = CHO + OH$	3×10^{13}	0	0
(13) $HCHO + O_2 = CO_2 + H_2O$	7.3×10^{10}	$\frac{1}{2}$	0
(14) $CHO + O_2 = CO_2 + OH$	7.4×10^{11}	$\frac{1}{2}$	0
(15) $CHO + O = CO_2 + H$	5.4×10^{11}	$\frac{1}{2}$	0
(16) $CHO + O = CO + OH$	5.4×10^{11}	$\frac{1}{2}$	0
(17) $CHO + CH_3 = CH_4 + CO$	2.5×10^{10}	$\frac{1}{2}$	0
(18) $CHO + OH = CO + H_2O$	3×10^{13}	0	0
(19) $HCO + M = H + CO + M$	2×10^{12}	$\frac{1}{2}$	-14.4×10^3

Plus all reactions in Table 6.1.

One of the most exhaustively studied hydrocarbon oxidation mechanisms is that for the methane-air system, although substantial progress is being made on ethane (Gelinas, 1979). Agreement is found on the following aspects of the mechanism: (1) a basic H_2-air mechanism is crucial; (2) CO oxidation is controlled by the hydroxyl radical (OH) concentration level; and (3) the methyl radical (CH_3) dominates the initial phase of the oxidation process. Other important factors include the following: (1) the relevance of formaldehyde; (2) the importance of hydrogen peroxide (H_2O_2) and the hydroperoxyl radical (HO_2); (3) the appearance of higher hydrocarbons such as ethane (C_2H_6); and (4) the effect of other intermediates, including CN- and HCN-type species on NO_x formation. The methane oxidation mechanism used by Edelman and Harsha (1978) is shown in Table 6.2.

6.7 THE INTERACTION BETWEEN TURBULENCE AND COMBUSTION

Turbulence causes significant changes in turbulent flame speeds, minimum ignition energy, flame stabilization, and pollutant formation rates. The interaction between combustion and turbulence is described with the aid of mathematical and physical models. Models require to take into account the main controlling processes. These models are used to obtain closure of the exact equations of turbulent reacting flow.

Governing Equations

For any dependent variable $s(\mathbf{x}, t)$, the mass-weighted or Favre average is written $\tilde{s} \equiv \langle \rho s \rangle / \bar{\rho}$, where ρ is the mass density, and $\langle \rangle$ and $(^-)$ each denotes a time average. The fluctuation relative to the Favre average is $s''(\mathbf{x}, t) = s(\mathbf{x}, t) - \tilde{s}(\mathbf{x})$. Bray (1978) has demonstrated that Favre averages in flames differ significantly from the unweighted Reynolds averages of the same variables. The exact conservation equations are as follows:

Conservation of mass

$$\frac{\partial \rho}{\partial t} + \frac{\partial}{\partial x_\beta} (\bar{\rho} \tilde{v}_\beta) = 0 \tag{6.70}$$

where v_β is the velocity in the direction of x_β.

Conservation of species

$$\frac{\partial}{\partial t} (\bar{\rho} \tilde{c}_i) + \frac{\partial}{\partial x_\beta} (\bar{\rho} \tilde{v}_\beta \tilde{c}_i) = -\frac{\partial \bar{J}_{i\beta}}{\partial x_\beta} - \frac{\partial}{\partial x_\beta} \langle \rho c_i'' v_\beta'' \rangle + \bar{w}_i \tag{6.71}$$

where $J_{i\beta}$ is the mass flux of species i in the direction of x_β due to molecular diffusion, and w_i is the mass rate of production of i per unit volume due to chemical reactions.

The equation for fluctuating intensity for c_i is

$$\frac{\partial}{\partial t} \langle \rho c_i'' c_i'' \rangle + \frac{\partial}{\partial x_\beta} [\tilde{v}_\beta \langle \rho c_i'' c_i'' \rangle]$$

$$= -2 \left\langle c_i'' \frac{\partial J_{i\beta}}{\partial x_\beta} \right\rangle - 2 \langle \rho v_\beta'' c_i'' \rangle \frac{\partial \tilde{c}_i}{\partial x_\beta} - \frac{\partial}{\partial x_\beta} \langle \rho v_\beta'' c_i'' c_i'' \rangle + 2 \langle c_i'' w_i \rangle \tag{6.72}$$

The equation for turbulence kinetic energy is

$$\bar{\rho} \frac{\partial \tilde{q}}{\partial t} + \bar{\rho} \tilde{v} \frac{\partial \tilde{q}}{\partial x_\beta} = \left\langle f_{\alpha\beta} \frac{\partial v_\alpha''}{\partial x_\beta} \right\rangle - \langle \rho v_\alpha'' v_\beta'' \rangle \frac{\partial \tilde{v}_\alpha}{\partial x_\beta} - \frac{\partial}{\partial x_\beta} \langle \tfrac{1}{2} \rho v_\beta'' v_\alpha'' v_\alpha'' \rangle - \left\langle v_\alpha'' \frac{\partial p}{\partial x_\alpha} \right\rangle \tag{6.73}$$

where $\tilde{q} \equiv \tfrac{1}{2} \langle \rho v_\alpha'' v_\alpha'' \rangle / \bar{\rho}$ is the tubulence kinetic energy (TKE) per unit mass, and $f_{\alpha\beta}$ is the frictional part of the molecular stress tensor.

The above equations are exact. If further exact equations are introduced, to calculate the unknown covariances, new covariances, of higher order, are required, and this leads to the "closure problem." A closed set of equations can only be obtained by formulating models based upon empiricism. Empiricism is required to represent turbulent transport terms, for example, $\partial \langle \rho c_i'' v_\beta'' \rangle \partial x_\beta$ in

Eq. (6.71), terms representing dissipation of fluctuating quantities due to molecular action, e.g., the scalar dissipation

$$\overline{\chi_i} \equiv -2\left\langle J_\beta \frac{\partial c_i''}{\partial x_\beta} \right\rangle \tag{6.74}$$

term in Eq. (6.72) as well as chemical source terms such as $\overline{w_i}$ and $2\langle c_i'' w_i \rangle$.

The Effects of Combustion on Turbulence

Combustion affects the physics and the modeling of turbulent transport through (1) production of density variations, (2) buoyancy effects, (3) dilatation (expansion) due to heat release, (4) molecular transport, and (5) instability. Bray has shown that there is no justification in using the empirical closure and model equations for constant density unreacting flow conditions for flows with turbulent combustion. The exact, unclosed equations contain many extra terms which are not present in the constant density case, and additional dimensionless groups, such as density ratios, require to be introduced.

Combustion instabilities can lead to sharp increases in turbulent transport as a result of pressure fluctuations coupling with density and velocity fluctuations. Combustion can cause changes in noise levels and result in combustion-generated turbulence. In modeling, Bray (1978) utilizes a two-equation formulation with differential equations for the TKE $\tilde{q}$ and for the kinetic energy dissipation function $\bar{\phi} \equiv \langle f_{\alpha\beta}(\partial v_\alpha''/\partial x_\beta)\rangle$. In the TKE equation, effects must be included of any mechanism of flame-generated turbulence. Bray (1978) has shown that combustion can lead either to the generation of additional turbulence or to the removal of turbulence energy, depending upon the balance between positive and negative terms in the TKE equation. The following model procedure is adopted. The transport term $\langle \frac{1}{2}\rho v_\beta'' v_\alpha'' v_\alpha'' \rangle$ and the production terms due to Reynolds shear stresses $\langle \rho v_\alpha'' v_\beta'' \rangle \partial\tilde{v}_\alpha/\partial x_\beta$ ($\beta \neq \alpha$) are modeled by introducing an eddy viscosity ε related to $\tilde{q}$ and $\bar{\phi}$. The only experimental or theoretical support for these expressions comes from incompressible flow. Terms of the form $\langle \rho v_\alpha'' v_\alpha'' \rangle \partial\tilde{v}_\alpha/\partial x_\alpha$ describe removal of turbulence energy due to work done in dilatation by normal components of the Reynolds stress. This effect has been shown to be important in near-normal premixed flames where shear-generated turbulence does not predominate.

The modeling of the pressure term $\langle v_\alpha'' \partial p/\partial x_\alpha \rangle$ requires consideration of the buoyant production mechanism. This was considered by Bilger (1976) in the pressure term

$$\langle v_\alpha'' \partial p/\partial x_\alpha \rangle \overline{v_\alpha''} \partial\bar{p}/\partial x_\alpha + \partial\langle v_\alpha'' p' \rangle/\partial x_\alpha - \langle p' \partial v_\alpha''/\partial x_\alpha \rangle \tag{6.75}$$

where $\overline{v_\alpha''} \partial\bar{p}/x_\alpha$ occurs only with Favre averaging. The equation of motion is used to eliminate $\partial\bar{p}/\partial x_\alpha$ from this expression. The models for the dissipation function $\bar{\phi}$ require drastic modeling assumptions, such as assuming constant density flow without chemical reaction, and self-preserving flow. This also applies to the modeling of the scalar dissipation function, $\bar{\chi}$.

The Effects of Turbulence on Chemical Reaction Rates

The chemical source term w_i is highly nonlinear in the thermodynamic state variables, with the result that fluctuations in these variables, due to turbulence and incomplete mixing, can strongly influence the time average. An exact expression for $\overline{w}_i$ is

$$\overline{w}_i(x) = \iint \cdots \int w_i(\rho, T, c_j) P(\rho, T, c_j, x) \, d\rho \, dT \, dc_j \qquad (j = 1, 2, \ldots) \qquad (6.76)$$

where P is the joint probability density function (pdf) for the variables ρ, T, c_j at location x. The pdf may be specified a priori either empirically or from some insight into the flame structure. Alternatively, the pdf may be calculated from a modeled version of an exact balance equation.

Model of Diffusion Flame

The turbulent diffusion flame can also be treated in terms of a pdf, the shape of which is assumed to be known a priori. For the limiting case of fast chemistry, the chemical source term is eliminated by introducing a conserved scalar ξ which is defined as the mass fraction of all material originating in the fuel stream. The equation for the mass fraction, Eq. (6.71), is rewritten by inserting the Favre average $\tilde{\xi}$ in place of $\tilde{c}_i$ and removing the chemical source term. Models of this type involve specification of relationships between ξ and instantaneous values of all scalar variables, either from a simple flame-sheet description, or from an assumed chemical equilibrium. A pdf, $P_\xi(\xi)$, is specified, with $\tilde{\xi}$ and $\langle \rho \xi'' \xi'' \rangle$ as parameters, and transport equations are formulated for $\tilde{\xi}$ and $\langle \rho \xi'' \xi'' \rangle$. The mean value of a scalar $S(\xi)$ is given by

$$\bar{\rho}\tilde{S} = \int_0^1 \rho(\xi) S(\xi) P_\xi(\xi) \, d\xi \qquad (6.77)$$

Bilger (1976) has shown that, with $c_i = c_i(\xi)$, Eq. (6.71) yields an explicit expression for the instantaneous reaction rate w_i. After time averaging, this becomes

$$\overline{w}_i = -\left\langle \rho D (\nabla \xi)^2 \frac{d^2 c_i}{d\xi^2} \right\rangle \qquad (6.78)$$

Solution of this equation requires the specification of a joint pdf. When the assumption is made that ξ is uncorrelated with the scalar dissipation, Eq. (6.78) reduces to

$$\bar{w} = 0.5 \rho_s \tilde{\chi}_\xi C_B P_\xi(\xi_s) / \bar{\rho} \qquad (6.79)$$

For the one-trip irreversible reaction

$$F + r \cdot O \rightarrow (r + 1)P$$

where F is fuel, O is oxidant, P is product, and r is stoichiometric mass ratio of oxidant to fuel. In Eq. (6.79), $\bar{w} = \bar{w}_p / (r + 1)$ and $C_B = C_{F1} + C_{O2}/r$, where subscript 1 represents the fuel stream and 2 the oxidizer stream, $\tilde{\chi}_{\xi}$ is the scalar dissipation function for fluctuations in ξ, and subscript s denotes properties at the stoichiometric value of ξ. Equation (6.79) is not exact but it does contain significant features of the real process. In common with predictions for the premixed case, the reaction rate is directly proportional to the scalar dissipation function. However, unlike the premixed flame, the source term for the diffusion flame is also proportional to a particular value of the pdf, for $\xi = \xi_s$. Predictions are sensitive to the shape chosen, or determined, for the pdf. If the scalar dissipation function is assumed to have the form

$$\bar{\chi} = C\tilde{q}^{1/2}\langle \rho c''c'' \rangle/l \tag{6.80}$$

where C is a modeling constant and l is a length scale of the scalar field, then Eq. (6.79) becomes

$$W \equiv \frac{\bar{w}l}{\rho_s C_\xi \tilde{q}^{1/2} C_B} = 0.5P_\xi(\xi_s)\tilde{\xi}(1 - \tilde{\xi})g \tag{6.81}$$

where C_ξ is a modeling constant and $g \equiv \langle \rho \xi''\xi'' \rangle [\rho\tilde{\xi}(1 - \tilde{\xi})]^{-1}$ is the normalized fluctuation intensity.

Joint PDF

In cases where chemical reactions are coupled, and when partially premixed, nonadiabatic, or high Mach number flows are involved, the probability density function does not reduce to a function of a single scalar variable. Bray (1978) has developed a simple description of the premixed turbulent flame with arbitrarily complex chemistry, in terms of a single-variable pdf, with the assumption that combustion occurs only in laminar flamelets of known structure. This laminar structure is used to express all instantaneous species mass fractions in the form $c_i = c_i(T)$. The temperature pdf, $T_T(T)$, is specified as a function of the mean temperature and rms temperature fluctuation, so that the chemical source term is closed through solution of conservation equations for these two quantities alone. The validity of this description requires that the turbulence must not distort the laminar flamelet structure sufficiently to influence the assumed relationship between composition and temperature. This assumption was made by Williams (1975), who also used a similar approach for turbulent diffusion flames.

Bray and Moss (1977) assumed that chemical reactions occur sequentially. Hydrocarbon oxidation in a premixed turbulent flame is described by a two-step sequential process, in which CO is produced in the first (fast) reaction and then oxidized in a second (slower) reaction. The rate of hydrocarbon removal is given by

$$\bar{w} = K_2\bar{\chi} \tag{6.82}$$

and the rate of final product formation is given by Eq. (6.39). Two separate single-variable pdf's are required for the initial hydrocarbon and the final product mixture.

The preceding methods cannot be used when the chemical reactions cannot be described as a function of a single progress variable and when turbulent fluctuations are large. A joint pdf is required for the relevant concentrations and thermodynamic state variables. A general expression for a joint pdf, the shape of which depends upon a suitable number of parameters, requires to be postulated. The parameters are related to moments of the joint pdf representing mean concentrations, fluctuation intensities, and covariances, which are determined from balance equations. Donaldson (1975) has postulated a joint pdf consisting of dirac delta functions at fixed locations c_s $(s = 1, 2, \ldots)$ in the composition space, and the strengths α_s of these delta functions are parameters, variable in physical space, that require to be determined from appropriate moments. The mean reaction rate is given by

$$\bar{w}_i = \sum_s \alpha_s w_i(c_s) \tag{6.83}$$

The number of differential equations that require to be solved increases rapidly with the number of species, resulting in highly complex equations and computations.

Lagrangian Frame of Reference

The process of chemical reaction in turbulent combustion requires to be examined in a lagrangian frame of reference when attempts are made to follow the paths of molecules. Three flow field models, which combine lagrangian and eulerian features, and also claim to accommodate complex kinetics, have been proposed. Rhodes et al. (1974) introduced a diffusion flame model, in which all the fluid in the mixing layer is divided into "classes," each representing an instantaneous elemental composition within a specified range. Each class is treated as a one-dimensional transient stirred reactor, with mass addition, in which complex chemistry can be allowed to occur. In Spalding's ESCIMO concept (1979), one-dimensional "sandwiches" are formed by engulfment, at the edge of the mixing region, and are then stretched, folded, and convected by the turbulent flow. One-dimensional, time-dependent laminar diffusion and complex chemical reactions are allowed to take place in these sandwiches. Felton et al. (1978) combine a finite-difference prediction of the three-dimensional flow pattern in a gas turbine combustor with a network of interconnected stirred and plug-flow reactors. A multireaction chemical kinetic scheme is used and a simplified representation of the evaporation of a polydispersed liquid fuel spray is incorporated in the prediction method. The relationship between the lagrangian and eulerian descriptions is treated by Rhodes, Harsha, and Peters (1974), assuming an empirical connection between composition and velocity; Spalding (1979) uses an unknown, four-dimensional joint pdf, while Felton, Swithenbank, and Turan (1978) bypass the problem by sacrificing spatial information. In each case, an assumption is made about the decoupling of the flow field from the chemistry and the chemical description is given in terms of a pdf. Rhodes, Harsha, and Peters (1974) assumed an a priori pdf for mixture ratio, while Spalding's lagrangian sandwich formation of ESCIMO can be viewed as an intuitive derivation of a pdf balance equation.

Experimental Verification

Ballantyne and Bray (1977) and Ballantyne and Moss (1977) have carried out experiments to test simple premixed and diffusion flame models. They have used measured pdf's to test the assumption that instantaneous values of all relevant scalars are related to a single progress variable or conserved scalar. Scalar fluctuation intensities are deduced and compared with theoretical upper and lower bounds, providing a sensitive test of both model and experimental technique.

Experimental information is required on the form of the scalar dissipation function, which controls the reaction rate in many practical situations. Currently, this function is modeled on the basis of dimensional analysis and incompressible nonreacting flow data. Models of $\bar{\chi}$ involve a length scale of the scalar field, which is assumed to be proportional to a scale of the velocity field.

Kennedy and Kent (1978) used the light-scattering technique to measure distributions of a conserved scalar in a hydrogen jet diffusion flame. Mean rms fluctuations and pdf's were measured in the flame by seeding the fuel with an inert tracer. The vertical turbulent hydrogen jet diffusion flame issued into the still air and flames with Froude numbers of 0.6×10^6 and 0.75×10^5 were studied. An argon ion laser was used as the light source; light scattered at 90° to the incident beam was collected by a photomultiplier tube. The flow was seeded with titanium dioxide particles with a size range between 0.1 and 2.0 μm.

The scattered light intensity is proportional to the concentration of seeding material

$$I/I_j = (\rho \xi)/(\rho \xi)_j = (\rho \xi)/\rho_j \qquad (6.84)$$

where I is the light intensity, ξ is the mass fraction of nozzle fluid, and subscript j refers to conditions at the jet nozzle exit. Light-intensity signals can be converted to ξ by dividing by local density. If assumptions are made of equal diffusivities of all species, unity Lewis number and adiabatic equilibrium conditions, composition, temperature, and density can be related instantaneously to ξ. Measurement was also made of the following Favre quantities:

Favre mean

$$\tilde{\xi} = \overline{\rho(\xi)\xi}/\overline{\rho(\xi)} \qquad [\rho(\xi) \neq \rho(\xi)] \qquad (6.85)$$

Favre variance

$$\tilde{\xi}''^2 = \overline{\rho(\xi)\xi''^2}/\overline{\rho(\xi)}$$
$$= \overline{\rho(\xi)(\xi - \tilde{\xi})^2}/\overline{\rho(\xi)} \qquad (6.86)$$

Favre pdf

$$\tilde{p}(\xi) = \rho(\xi)p(\xi)/\overline{\rho(\xi)} \qquad (6.87)$$

Turbulence levels were found to be higher for the low Froude number flame and this was expected to be due to buoyancy exerting a greater influence and causing large-scale fluctuations. Up to x/D, pdf's were found to be almost gaussian. Farther downstream, the pdf's become very skewed as the signal becomes intermittent. Favre pdf's grow a prominent spike, whereas conventional pdf's have a rather unusual "shoulder." Intermittency found on the centerline is ascribed to large-scale buoyancy-induced turbulence. The more buoyant flame exhibited a more intermittent pdf. Radial variation of pdf shows that the signal becomes more intermittent and the intermittency spike grows with increasing distance from the centerline. No well defined separation between the turbulent and nonturbulent region was found, either on or off the flame axis. Comparison of the pdf shapes with isothermal results reveals a similarity between the isothermal pdf's and the Favre, rather than conventional flame pdf. This supports the hypothesis that Favre quantities behave like conventional quantities in isothermal flows.

SEVEN

ATOMIZATION AND VAPORIZATION OF LIQUID FUEL SPRAYS

7.1 THE EFFECT OF SPRAYS ON COMBUSTOR PERFORMANCE

Sprays play an important role in combustion performance and emission of pollutants. Local values of air/fuel ratio are governed by trajectories of individual droplets, rates of vaporization, and mixing of fuel vapor with air. Atomizers produce sprays with a wide range of droplet sizes, velocities, and initial direction of flight. These droplets interact with gas streams, which can deflect the particles from their trajectories. Vapor is released in the wakes of the droplets and these trails of vapor mix by convection and diffusion with the airstreams. The rates of vaporization are governed by the temperature and vapor pressure concentrations of the environment through which the droplets traverse. Since combustion efficiency and temperature distributions are directly dependent upon air/fuel ratio distributions, changes in spray characteristics can result in important changes in combustor performance. In most gas turbine combustors, droplet lifetimes are sufficiently long that sections of the combustor are dominated by the presence of droplets. For highly volatile fuels injected into high-temperature regions, vaporization of the complete spray may be rapid but, on the other hand, liquid droplets can survive long enough so that they impinge directly on combustor walls. The penetration of liquid droplets into airstreams and onto wall surfaces is undesirable and can, in the limit, result in the emission of unburned fuel. It is preferable to direct droplets into regions of high-turbulence intensity and high temperature, where rapid vaporization and mixing can take place.

Atomization of a liquid is most effectively accomplished by creating a high relative velocity between a liquid sheet and the surrounding air. In general, the higher the relative velocity, the smaller is the mean drop size. In an airblast atomizer, this high relative velocity is achieved by injecting fuel at low velocities into a high-velocity airstream. This type of fuel atomization is ideally suited for gas turbine applications, since high-velocity air is readily available as flow enters the combustion chamber from the compressor.

Airblast atomization has the additional advantage of providing airflow through the fuel nozzle and this avoids the very high fuel/air ratios that are found in the central regions of pressure atomizers and are mainly responsible for formation and emission of high levels of smoke. The improvement in fuel-air mixing can also lead to a reduction in the emission of nitric oxides.

Clouds of droplets lead to the formation of rich mixtures and low-temperature zones. Within these rich mixture regions, soot and CO can readily be formed in regions of low oxygen concentration. The soot and CO can be chemically frozen by coming into direct contact with low-temperature airstreams and may, subsequently, be emitted from the combustor. Soot may be formed, either directly from liquid droplets or in rich mixture zones in the wake of droplets. This, in turn, affects the radiative properties of the flame and the presence of large concentrations of soot can result in high radiative transfer rates from the flame to combustor walls, with the risk of damage as a result of overheating of the wall surfaces.

Ignition and engine relight are directly affected by the presence of liquid droplets. Liquid droplets absorb heat during vaporization and, if insufficient liquid has been vaporized, local equivalence ratios may be outside the limits of flammability. Impingement of liquid droplets on igniters results in coking and reduction in efficiency of ignition. Due to the gas inhomogeneity, the presence of large, rich, or lean eddies in the vicinity of the igniter can prevent ignition from taking place.

The temperature-pattern factor in the exhaust plane of the combustor exit can be radically altered by changing spray patterns. The location of heat release as a result of chemical reaction is governed by droplet trajectories and subsequent mixing. The design of a combustor, so as to provide a prescribed temperature distribution at the combustor exit, is dependent upon control of droplet trajectories and airflow patterns. A wide range of temperature pattern factors can be readily achieved by variation of spray angle and droplet size. Turbine blade life can be very adversely affected by the presence of particulate matter in the exhaust. This particulate matter originates in the fuel. Droplets require to have sufficient residence time in oxidizing high-temperature regions in order to prevent formation of soot particles.

7.2 ATOMIZATION

Liquid fuel requires to be broken up into small droplets in order that it can effectively burn in combustion chambers. Atomization of the liquid fuel is most commonly carried out by either injecting the fuel through small orifices at high pressure or by mixing the fuel with high-pressure air or gas. The most effective atomization is achieved when thin liquid sheets are formed which subsequently become unstable and then break up to form ligaments and large drops, which then break down further into small droplets.

The term *twin-fluid atomization* is used for systems in which a high-velocity gas stream is used to atomize fuel in a relatively low-velocity liquid fuel stream. Atomization is most effective when very thin liquid sheets are sandwiched between gas streams with either higher or lower velocities than the liquid stream, so that high rates of shear and velocity gradients are generated across the liquid gas interfaces. The atomizing fluid is generally high-pressure air, in gas turbine systems, and high-pressure steam in land-based and marine systems where steam is readily available from a boiler. The selection of steam or air is based upon its availability and cost, since, from the point of view of atomization, there appears to be no significant difference in using steam or air. From the point of view of combustion efficiency, air is preferable to steam. When mass flow rates of steam are less than 10 percent of that of the fuel, steam atomization has been found to cause no significant deterioration in combustion efficiency. In addition to the general overall reduction in sizes of droplets that can be achieved by twin-fluid atomization, as compared with pressure-jet atomization, production of large droplets can be avoided over a wide range of fuel flow rates. In practice, combustion systems are required to operate effectively at idle conditions as well as at maximum power, and difficulties have been experienced in achieving adequate atomization at both the very low and very high levels of fuel flow rate. The ratio of the maximum to minimum fuel flow rates is referred to as the turndown ratio. For marine applications, turndown ratios of 20 : 1 are required and, in gas turbines, ratios of 50 : 1 are needed. In the past, atomization systems were designed to cope with "cruise" or "normal" operating conditions and it was accepted that during startup and idle, as well as during rapid acceleration, atomization would be poor and high rates of emission of smoke and unburned hydrocarbons would occur for periods which are short, relative to the total period of combustion. In order to satisfy the strict limitation of total emission of pollutants, it is necessary to obtain good atomization over a wide range of operating conditions and the high rates of emission obtained during startup and idle require to be reduced. The most effective practical means of achieving this is by the use of twin-fluid atomization. The specific terms "air blast" and "steam blast" are used in the literature but the more general term is "twin-fluid atomization."

In gas turbine combustion chambers, the airblast atomizer is replacing the pressure-jet atomizer, which has been the most commonly used atomizer. An additional advantage of airblast atomizers over pressure atomizers is that small droplets are airborne in the high-velocity airstream so that their distribution throughout the combustion zone is dictated by the airflow pattern, which remains fairly constant under all operating conditions. In consequence, the temperature profile at the combustion chamber exhaust tends to remain constant, thereby extending the life of the turbine blades. Mixing of fuel and air is greatly improved, resulting in blue flames of low luminosity compared with the yellow flames of high luminosity, which can be found with pressure-jet atomization. Reduction in radiative heat transfer results in cooler flames and the reduction in quantity and dimensions of soot particles leads to overall reductions in exhaust smoke. The airblast atomizer has an important advantage over the gas turbine vaporizing tube atomizer, which is normally immersed in flame and very susceptible to overheating. The airblast atomizer is continuously cooled by high-velocity air flowing over it at compressor outlet temperatures.

Airblast Atomization

The physical process of airblast atomization is composed of the following steps (Lefebvre and Miller, 1966):

1. Formation of thin liquid sheets on a plate or along the inner walls of an internal-mix atomizer, or free sheets unattached to walls.
2. Disintegration of the liquid sheet by aerodynamic forces to form ligaments, large drops, and droplets.
3. Breakup of ligaments and large drops into droplets.
4. Acceleration of droplets by high-speed gas stream and/or deceleration of droplets by low-velocity and recirculation flows.
5. Formation of two-phase, liquid-gas spray, followed by spreading of spray jet and entrainment of gas from surroundings.
6. Evaporation of droplets as a result of temperature and vapor pressure differentials between droplet surface and surroundings.
7. Agglomeration of droplets by collision can occur but, except under conditions of rapid deceleration in the regions of a spray close to the nozzle, this mechanism is not considered to be significant.

Nukiyama and Tanasawa (1939) studied airblast atomization for application in piston engines. Fuel was injected through a fine orifice into a venturi of a carburetor, through which air was flowing (Fig. 7.1). They examined the effect of variation in fluid flow properties by using mixtures of water, alcohol, and glycerine and they also varied the relative velocity of air and liquid. They found that droplet size depended on the surface tension, density, and absolute viscosity of the liquid; on the ratio of liquid to air flow rate; and on the initial relative velocity of air and liquid. They derived the following empirical equation

$$d_0 = \frac{0.585\sqrt{\sigma}}{\Delta u \sqrt{\rho_l}} + 1.415 \times 10^6 \left(\frac{\mu_l}{\sqrt{\sigma \rho_l}}\right)^{0.45} \left(\frac{Q_l}{Q_a}\right)^{1.5} \tag{7.1}$$

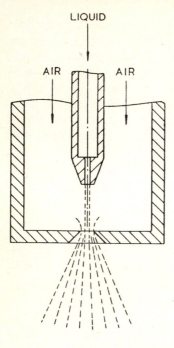

LIQUID

AIR   AIR

Figure 7.1 Airblast atomizer. (*Nukiyama and Tanasawa, 1939.*)

where d_0 = Sauter mean diameter of droplet, μm
    Δu = relative velocity of air and liquid, m/s
    σ = surface tension of liquid, N/m
    ρ_l = density of liquid, kg/m^3
    μ_l = viscosity of liquid, N s/m^2
    Q_l = liquid flow rate, m^3/s
    Q_a = airflow rate, m^3/s

For kerosine with specific gravity of 0.78 at 15°C, Eq. (7.1) reduces to

$$d_0 = \frac{10.57}{\Delta u} + 7\left(\frac{Q_l}{Q_a}\right)^{1.5} \tag{7.2}$$

Equation (7.2) is not dimensionally correct but Lewis et al. (1948) have shown that this simple equation can be used for airblast atomizers when the liquid density is between 700 and 1200 kg/m^3, the surface tension between 0.019 and 0.073 N/m, viscosity between 3×10^{-4} and 5×10^{-2} N s/m^2, and the air velocity is subsonic. In the derivation of Eq. (7.1) changes in air properties have not been taken into account so that its application is limited to conditions close to standard temperature and pressure. Nukiyama and Tanasawa (1939) concluded, on the basis of their results, that the effect of change in nozzle shape can be incorporated in changes in the discharge coefficient.

    Wigg (1964) measured size distributions by using the freezing wax method, in which molten wax is atomized and the solid particles formed in the air are then collected and measured. On the basis of his own measurements, and those of a number of other investigators, he concluded that Eq. (7.1) gives too high a droplet size at some velocities and overestimates the effect of a change in the ratio of liquid to airflow rates. On the basis of the energy required to effect atomization, Wigg developed an equation that was dimensionally correct and could be used for velocities greater than 100 m/s. The Wigg equation is

$$d_m = 0.004v^{0.5}W^{0.1}\left(1 + \frac{W}{A}\right)^{0.5} h^{0.1}\sigma^{0.2}\rho_a^{-0.3} \Delta u^{-1} \qquad (7.3)$$

where d_m = mass median diameter, μm
$\quad v$ = kinematic viscosity, m^2/s
$\quad W$ = liquid mass flow rate, kg/s
$\quad A$ = air mass flow rate, kg/s
$\quad h$ = height of air annulus, mm

This equation shows that the greatest changes in droplet diameter can be obtained by changing the relative velocity between liquid and air. In general, for a fixed nozzle, variation in the liquid/air mass flow rate ratio also affects the relative velocity. Among the physical properties kinematic viscosity has the exponent 0.5, while surface tension has the exponent 0.2. This is explained by the shear forces predominating over the surface tension forces, which is generally true for the high velocities used in twin-fluid atomization. Changes in dimensions of the atomizer do not necessarily change the thickness of the liquid film and, thus, atomizer dimensions have an exponent which is only 0.1.

Twin-Fluid Atomizers

Mullinger and Chigier (1974) made a comprehensive study of the effect of changing geometric variables in atomizer design as well as the variables which affect the distribution and the thickness of the fuel film in the mixing chamber of twin-fluid atomizers. They found that Wigg's equation for the determination of mean drop size agreed closely with their experimental results. They carried out tests with air/fuel mass ratios as low as 0.005, which is a factor of 10 smaller than that tested by Wigg (1964). They thus showed that Eq. (7.3) is still applicable at much lower values of atomizing air/fuel mass ratio than those for which it was originally developed. Furthermore, this agreement with Eq. (7.3) for a wide variety of different designs of twin-fluid atomizer, both internal and external mixing, confirmed that atomizer geometry has relatively little effect upon droplet size, except for the way in which if affects atomizing fluid velocity and density.

The type of atomizer tested by Wigg (1964) and Mullinger and Chigier (1974) is commonly used in boilers for both land and marine applications. The multijet, internal mixing, twin-fluid atomizer is used in oil burners, in boilers, and in industrial furnaces. Twin-fluid atomizers operate at high combustion efficiencies with low excess air requirements with a wide turndown ratio. For burners with large flow rates of fuel, multiple jets are used so as to provide flow rates of up to

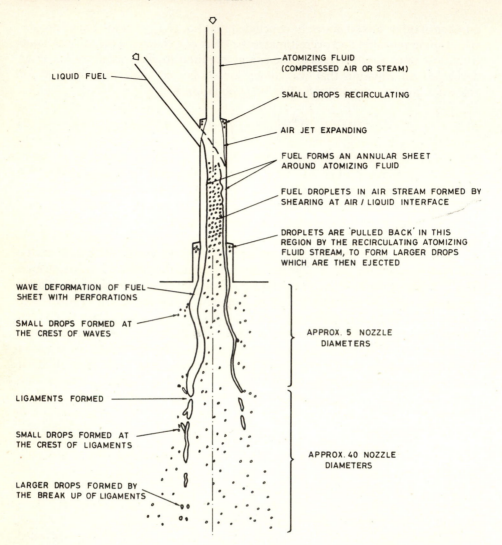

LIQUID FUEL

ATOMIZING FLUID
(COMPRESSED AIR OR STEAM)

SMALL DROPS RECIRCULATING

AIR JET EXPANDING

FUEL FORMS AN ANNULAR SHEET
AROUND ATOMIZING FLUID

FUEL DROPLETS IN AIR STREAM FORMED BY
SHEARING AT AIR / LIQUID INTERFACE

DROPLETS ARE 'PULLED BACK' IN THIS
REGION BY THE RECIRCULATING ATOMIZING
FLUID STREAM, TO FORM LARGER DROPS
WHICH ARE THEN EJECTED

WAVE DEFORMATION OF FUEL
SHEET WITH PERFORATIONS

SMALL DROPS FORMED AT
THE CREST OF WAVES

APPROX. 5 NOZZLE
DIAMETERS

LIGAMENTS FORMED

SMALL DROPS FORMED AT
THE CREST OF LIGAMENTS

APPROX. 40 NOZZLE
DIAMETERS

LARGER DROPS FORMED BY
THE BREAK UP OF LIGAMENTS

Figure 7.2 Liquid atomization in an internal mixing twin-fluid atomizer. (*Mullinger and Chigier, 1974.*)

15×10^3 kg/h in boilers for electricity generation. The physical processes of atomization are shown schematically in Fig. 7.2. Liquid fuel is injected into the mixing chamber at an angle while the atomizing fluid, compressed air, or steam, is introduced centrally to the mixing chamber with sufficient pressure to provide sonic conditions at the jet exit. The liquid fuel forms an annular film around the walls of the mixing chamber, with the high-speed atomizing jet passing centrally through the mixing chamber. Some atomization occurs within the mixing chamber, but the major portion of the liquid emerges from the atomizer in the form of a liquid sheet, which disintegrates into ligaments and, subsequently, into droplets. Secondary atomization occurring outside the atomizer continues for some 50 nozzle diameters downstream. Increasing the air/fuel ratio leads to a reduction in the breakup length.

The internal mixing within the atomizer (Fig. 7.2) was determined from photography of a transparent model and the effects of changes in the atomizer geometry were studied as they affected the size distribution of droplets in the spray. The variation of the Sauter mean diameter of droplets, as measured by the high-speed photographic technique, with changes in air/fuel mass ratio, is shown in Fig. 7.3 for a range of fuel flow rates. Figure 7.3 shows the significant reductions in droplet size that can be obtained by increasing the air/fuel mass ratio. For each fuel flow rate, however, there is a limit to the effectiveness of increased airflow rates, beyond which no further significant reduction in droplet diameter is achieved. Since energy considerations and the desirability of limiting the quantity of steam introduced through the atomizer are important, it is useful to establish the minimum air/fuel mass ratio for any given fuel flow rate. It can be seen in Fig. 7.3 that, for a fixed air/fuel mass ratio, the droplet diameter was reduced by increasing the fuel flow rate. This is explained as being due to the increase in air supply pressures required for the higher fuel flow rates and, hence, leading to an increase in the air density at the mixing point. Variation of atomizer geometry was shown to affect the atomizing fluid velocity and density, which had an important effect on droplet size distribution. Changes in geometry could also affect the distribution and thickness of the fuel film in the mixing chamber, which, in turn, affect the droplet size distribution. The design study showed that, in order to obtain the most effective atomization, the fuel should be supplied so as to form a

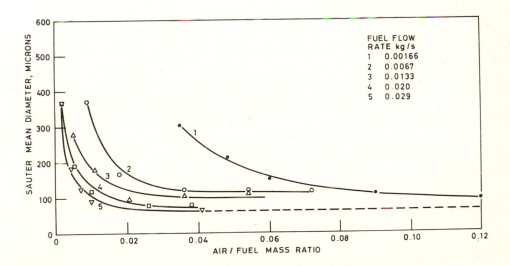

Figure 7.3 Variation of Sauter mean diameter with air/fuel mass ratio for various fuel-flow rates in a twin-fluid atomizer. (*Mullinger and Chigier, 1974.*)

thin, continuous, steady film in the mixing region. Increases in velocity and density of the atomizing fluid are the most effective means of reducing droplet diameters. A table of the effect of changes on atomizer geometry and performance, together with design recommendations, is given by Mullinger and Chigier (1974).

Airblast Atomizers for Gas Turbines

Airblast atomizers, as used for fuel injection in aircraft and industrial gas turbines, fall into one of two main categories: (1) *plain-jet*, in which the liquid is injected into a high-velocity airstream in the form of discrete jets, and (2) *prefilming*, in which the liquid is first spread out into a thin continuous sheet and then exposed on both sides to high-velocity air. Both types of atomizers exhibit very similar characteristics in regard to the influence on droplet size distribution of variations in airflow conditions and in the physical properties of the liquids. Lefebvre (1978) has shown that, under comparable operating conditions, the performance of the prefilming atomizer is superior to that of the plain-jet type.

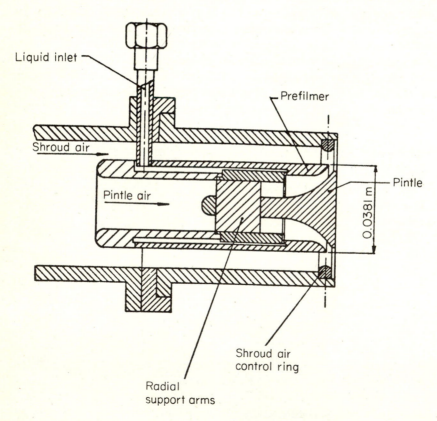

Figure 7.4 Prefilm airblast atomizer. (*Lefebvre, 1978.*)

Rizkalla and Lefebvre (1975*a* and 1975*b*) undertook a detailed experimental investigation of airblast atomization, using a specially designed form of airblast atomizer, which is representative of modern gas turbine practice. In this atomizer (Fig. 7.4), the liquid flows through six tangential ports into a weir, from which it spills over the prefilming surface before being discharged at the atomizing edge. The swirling motion of the liquid ensures that the liquid sheet attains a uniform thickness by the time it reaches the atomizing edge. One airstream flows through a central circular duct and is deflected radially outward by a pintle before striking the inner surface of the liquid sheet, while another airstream flows through an annular passage surrounding the main body of the atomizer. This passage has its minimum flow area in the plane of the atomizing lip in order to impart a high velocity to the air where it meets the outer surface of the liquid sheet.

In the first phase of their program, Rizkalla and Lefebvre (1975*a*) examined the effects of changes in liquid properties, namely, viscosity, surface tension, and density on mean droplet size. These studies were carried out using air supplied from a fan at atmospheric pressure. In the second phase of their work they examined the effect of changing airflow properties by varying the air temperature using a kerosine-fired preheater located upstream of the atomizer. Changes in air density were thus obtained by varying the air temperature between 23 and 151°C. Changes in air pressure were investigated by mounting the atomizer in a pipe connected to the outlet of a multistage compressor. Liquid flow rates were varied from 0.005 to 0.039 kg/s at pressure levels between 10^5 and 10^6 N/m². Mean droplet size was measured by using a light-scattering technique based on the forward scattering of a parallel beam of monochromatic light which had been passed through the spray.

Influence of Liquid Properties

The effects of variation of viscosity, surface tension, and density were examined by preparing special liquid solutions, in which each property could be separately varied. The effect of increasing viscosity is shown in Fig. 7.5, in which the Sauter mean diameter (SMD) is plotted against viscosity for various levels of air velocity at a constant liquid flow rate. High-speed photographs show that viscous forces tend to suppress the formation of waves on the liquid surface and lead to the formation of long ligaments. Viscous forces resist deformation of these ligaments into drops, so that atomization continues far downstream in regions of relatively low velocity. Increases in viscosity of liquids yields larger drop sizes, as shown in Fig. 7.5.

Surface tension forces tend to impede atomization by resisting disturbances or distortion of the liquid surface, and thereby oppose the creation of surface waves and delay the onset of ligament formation. Figure 7.6 shows that atomization quality deteriorates as surface tension is increased.

Increases in liquid density lead to increases in length of the coherent liquid sheet, so that ligament formation occurs under conditions of lower relative velocity between air and the liquid. Increase in liquid density also produces more

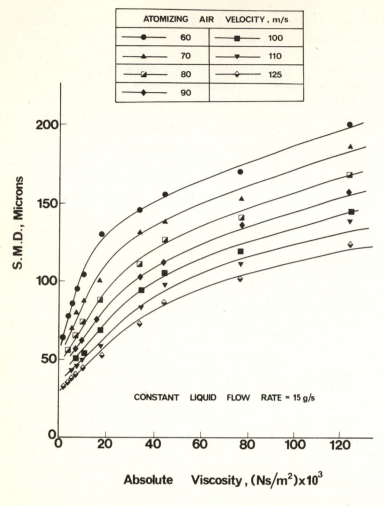

ATOMIZING AIR VELOCITY , m/s		
● 60	■ 100	
▲ 70	▼ 110	
◪ 80	◇ 125	
◆ 90		

CONSTANT LIQUID FLOW RATE = 15 g/s

S.M.D., Microns

Absolute Viscosity, $(Ns/m^2) \times 10^3$

Figure 7.5 Mean drop size in airblast atomizer as function of variation of liquid viscosity and air velocity. (*Lefebvre, 1978.*)

compact sprays that are less exposed to the atomizing action of the high-velocity air. Both these effects combine to increase droplet size. On the other hand, Lefebvre (1978) found that increase in density can lead to a reduction in the thickness of the liquid sheet at the atomizing edge and, hence, improve atomization. The net effect of these conflicting factors is that the influence of liquid density on SMD is fairly small.

Influence of Air Properties

The most important air property that influences mean drop size is air velocity. Drop size is found to be roughly inversely proportional to air velocity. In airblast

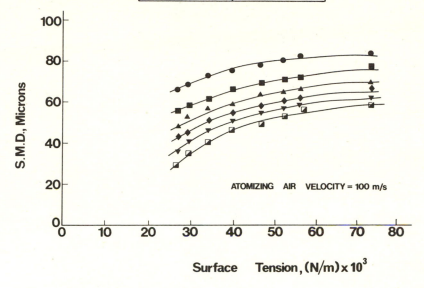

Figure 7.6 Mean drop size in airblast atomizer as function of variation of liquid flow rate and surface tension. (*Lefebvre, 1978.*)

atomizer design, the fuel sheets should be exposed to the highest possible air velocity consistent with the available pressure drop.

The effect of variation of mass flow rate of air and liquid is shown in Fig. 7.7. Atomization quality starts to decline when air/liquid mass ratios are below 4 and deteriorate quite rapidly when the ratio is below 2. At ratios above 5, only marginal reductions in SMD are gained by the utilization of more air in atomization. The optimum air/liquid ratio for design purposes is between 3 and 4. Lefebvre (1978) concluded that, if the air/fuel ratio is too small, the air momentum is insufficient to overcome the viscous and surface tension forces which act together to oppose drop formation, while an excess of air provides no additional advantage, since it is physically too remote from the liquid sheet to affect its disintegration. A lack of air quantity may be compensated to some extent by an increase in air velocity. Variation in air pressure and temperature indicated that SMD is roughly proportional to air density to the power -0.7.

Lefebvre (1978) has come to some general conclusions concerning the effects of air and liquid properties on mean drop size. For liquids of low viscosity, such as kerosine, the main factors governing SMD are air density and air velocity, whereas for liquids of high viscosity, the effects of air properties are less significant and SMD becomes more dependent upon the liquid properties, especially viscosity. Lefebvre derived a general equation for the effects of air and liquid properties

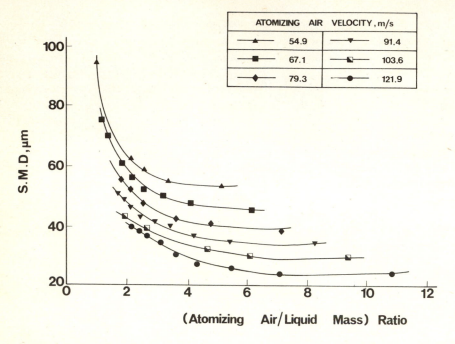

ATOMIZING AIR VELOCITY, m/s			
▲	54.9	▼	91.4
■	67.1	◣	103.6
◆	79.3	●	121.9

Figure 7.7 The effect of variation of atomizing air/liquid mass ratio on SMD. (*Rizkalla and Lefebvre, 1975b.*)

on mean drop size, expressed as the sum of two terms, one dominated by air velocity and air density, and the other by liquid viscosity. For prefilming airblast atomizers, the equation is

$$\text{SMD} = 0.073 \left(\frac{\sigma_L}{\rho_A U_A^2}\right)^{0.6} \left(\frac{\rho_L}{\rho_A}\right)^{0.1} D_p^{0.4}(1 + \dot{W}_L/\dot{W}_A)$$

$$+ 0.015 \left(\frac{\mu_L^2 D_p}{\sigma_L \rho_L}\right)^{0.5} (1 + \dot{W}_L/\dot{W}_A) \tag{7.4}$$

where SMD is Sauter mean diameter, σ is surface tension, ρ is density, μ is viscosity, U is flow velocity, D_p is diameter of prefilming lip, $\dot{W}$ is mass flow rate, and subscripts A and L are for air and liquid respectively.

Equation (7.4) is consistent with nearly all other published data. It is dimensionally correct and may be used with any consistent set of units. SMD values thus appear either in meters or feet. For liquids of low viscosity such as water and kerosine, the first term in Eq. (7.4) predominates and the SMD thus increases with increase in liquid surface tension, density, film thickness, and liquid/air ratio, and declines with increase in air velocity and air density. With liquids of high viscosity, the second term acquires greater significance and, in consequence, SMD becomes less sensitive to variations in air velocity and density. The range of variables tested by Rizkalla and Lefebvre (1975b) was: air velocities from 70 to 125 m/s; air/liquid ratios from 2 to 6; liquid viscosities from 0.001 to 0.044 N s/m^2; and surface tension between 0.026 and 0.073 N/m. For burning of kerosine, the second term on the right-hand side of Eq. (7.4) is negligibly small compared with the first term, and the SMD is inversely proportional to air pressure. Since gas turbines operate over a wide range of pressures, this finding can be usefully applied in practice.

Both the equations of Wigg (1964) and Rizkalla and Lefebvre (1975b) were obtained under atmospheric pressure conditions. The effect of increasing pressure was also examined. The studies of Lewis et al. (1948) and Ingebo and Foster (1957) suggest that droplet size falls with increasing pressure. In gas turbine combustion chambers, any increase in inlet air pressure is always accompanied by an increase in temperature, which increases the rate of evaporation. Airblast atomizers for gas turbine application should, therefore, be designed for the minimum pressure conditions and it can be assumed that, if combustion performance is satisfactory at atmospheric pressure, then atomization quality will be more than adequate at all higher levels of pressure.

Drop-Size Distribution Functions

Drop-size distributions are characterized by distribution functions of the number of droplets N with diameters greater than a specified value D. Tishkoff and Law (1977) have examined a number of distribution functions and determined a generalized function, which was used with digital computer programming. The general function for size distribution, as introduced by Nukiyama and Tanasawa (1939), has the form

$$\frac{dN}{dD} = aD^p \exp\left(-bD^n\right) \tag{7.5}$$

The four parameters a, p, b, and n are used for matching spray data to the generalized function. This generalization function has been simplified by assuming prescribed values for one or more of the four parameters.

1. Rosin-Rammler function (1933)

$$\frac{dN}{dD} = \frac{6bn}{\pi} D^{n-4} \exp\left(-bD^n\right) \tag{7.6}$$

2. Nukiyama-Tanasawa function (1940)

$$\frac{dN}{dD} = aD^2 \exp\left(-bD^n\right) \tag{7.7}$$

3. Griffith comminution function (1943)

$$\frac{dN}{dD} = aD^p \exp\left(-b/D\right) \tag{7.8}$$

For the three parameters of the Griffith comminution function, a logarithmic least-squares technique can be used. The Rosin-Rammler function contains only two parameters, b and n, and is thus the least accurate of the functions defined in Eqs. (7.5) to (7.8).

As an alternative approach to calculating mean diameters from Eqs. (7.5), (7.7), and (7.8), direct numerical integration can be used. The drop-size interval between the minimum and maximum drop diameters is divided into evenly spaced increments, and Simpson's rule is used for integration. Tishkoff and Law (1977), in their comparison of the size distribution functions, concluded that the generalized function, Eq. (7.5), was the most accurate expression but required digital computer programming to solve a nonlinear algebraic equation for one of the data approximation parameters. The Griffith function was almost as accurate as the generalized function and could be applied to data more easily by solving the linear algebraic equations.

7.3 VAPORIZATION OF DROPLETS IN HIGH-TEMPERATURE GAS STREAMS

Vaporization of liquid droplets is a subject which has been studied extensively for many years. A large number of industrial processes require the atomization of bulk liquid into droplets, vaporization, and subsequent mixing with surrounding gases. In the United States, nearly 50 percent of energy is derived from the burning of liquid petroleum products and, in some European countries, the fraction is even higher. Current requirements to improve simultaneously the fuel economy and limit the emission of pollutants from automobiles and other transportation systems have focused attention on the basic mechanisms of vaporization.

Understanding of vaporization phenomena requires the study of the combined effects of physical chemistry and fluid mechanics on heat and mass transfer across liquid-gas interfaces. The pioneering work in this field by Levich (1962) is described in his book *Physico-chemical Hydrodynamics*. During the period of the first development of gas turbine engines in the 1940s and 1950s, vaporization of liquid droplets was studied with the aim of increasing the efficiency of burning of liquid fuels—the first studies of Spalding (1955) were on the experimental investigation of vaporization and combustion of liquid fuel droplets. During the 1950s and 1960s, rocket propulsion became an important part of the development of space technology and the studies by Barrère (1960), Williams (1965), and Wise, Lorell, and Wood (1955) were focused on improving the combustion performance of liquid propellants. Penner (1957) examined the chemistry problems in jet propulsion, while detailed studies of injection and atomization characteristics of liquid fuels were made by Putnam (1957) and Lefebvre (1955 and 1975). Comprehensive reviews on liquid droplet combustion were written by Alan Williams (1968 and 1976) and Faeth (1977).

In the field of combustion, many experimental studies were made with single droplets of liquid fuel held in suspension in furnaces or allowed to fall freely through hot gas streams. The physical process was also simulated by using porous solid spheres, through which liquid was continuously injected so as to form a liquid film surrounding the sphere. In all of these cases, flames surrounded the droplets (or spheres). Under stationary conditions where free convection effects were small, flames enveloped the particles, while under conditions of convection, wake flames were formed in addition to the flame surrounding the fore region of the sphere. Rates of vaporization for these conditions could be calculated from solution of the equations governing the rates of heat and mass transfer between the flame and the droplet surface. Forced convection heat and mass transfer experimental results were correlated by Nusselt, Sherwood, Reynolds, and Prandtl number relations. Reasonably good agreement was found between prediction and experiment.

In recent years attention has turned to spray combustion. Under these conditions, large quantities of small droplets are injected from atomizers, to form liquid sprays. Initial studies by Chigier and coworkers, in which measurements were made within liquid spray flames, showed no evidence of individual flames surrounding single droplets. The spray region was found to be at a low temperature due to the quenching action of the large number of droplets present and, further, oxygen levels within the spray were extremely low, of the order of one percent. It was thus demonstrated that combustion would generally not take place within the body of the spray and the phenomena could be separated into a vaporization stage, with very little oxidation, followed by a mixing stage between the fuel vapor and surrounding air, leading to combustion in a flame surrounding the entire spray. Onuma and Ogasawara (1975) and Onuma et al. (1977) subsequently showed by a direct comparison of measurements made in liquid spray and gaseous flames that there was a close similarity between the temperature and species concentrations in these flames. A number of investigators have since confirmed these findings, which are in basic agreement with the initial model of spray burning as proposed by Chigier (1976). The work of Mellor (1976) and Tuttle et al. (1977), in which they studied sprays in simulated gas turbine combustor conditions, provided further evidence of the concept of initial vaporization, followed by subsequent gaseous combustion. Practical sprays can be shown to have the characteristics of clouds in the total group combustion region, while individual flame droplet burning may occur for large droplets with high momentum outside the main body of the spray and its surrounding flame.

The study of the physicochemical and fluid-dynamic aspects of liquid spray flames has been mainly based upon experimental evidence. Experimental techniques were specially developed for making studies in heterogeneous combustion systems with particle-laden flows. Developments, with particular emphasis on laser optical noninvasive probes, were reviewed by Chigier (1977). The dual-beam laser anemometer provides a noninvasive, nonperturbing means of making measurements within liquid sprays in high-temperature gas streams. The frequency shift of Doppler signals of individual particles crossing the interference fringes of two intersecting laser beams provides a measure of velocity, while the amplitude of the signal provides a measure of particle size. Digital electronic processing, directly interfaced with a computer, allows measurement of characteristics of individual droplets with a frequency response of the order of 10 kHz.

The physical processes of heat, mass, and momentum transfer for single droplets vaporizing in high-temperature gas streams will first be examined. Consideration is then given to effects of pressure and temperature change in the gas stream and the extent to which theoretical models are related to experiment. Droplet combustion, in which flames envelope individual droplets, is next examined and this is followed by discussion of spray combustion systems. The available experimental evidence, together with limited theoretical work on group and cloud combustion phenomena, is then reviewed in the light of the physical models formulated from experimental evidence. The use of laser optical techniques for measurement of particle size and velocity in spray flames is described in Chap. 10.

Models of Single Droplet Vaporization

The preparation of premixed, prevaporized systems for gas turbine and other transport engines is dependent upon achieving total vaporization of liquid droplets in airstreams without combustion taking place. In order to achieve this vaporization efficiently, the heat and mass transfer characteristics of droplets in gas streams with varying vapor pressure and temperature conditions needs to be considered. Further, since it has been demonstrated that, during spray combustion, droplets frequently evaporate in relatively cool gas mixtures with low oxygen concentrations, droplet evaporation is of primary importance in spray flames. In order that spray-flame structure can be modeled, it is important to be able to predict the physical characteristics of the vaporization of droplets with different fuel properties as functions of droplet diameter and the temperature, velocity, and species concentration of the gas environment.

Figure 7.8 shows the variations of temperature within a liquid droplet and in the surrounding gas stream for two conditions, early and later in the heating process. Also shown in Fig. 7.8 are the changes in mass fraction of the fuel vapor Y_F and the ambient air Y_A for the early and later conditions. When droplets are initially injected into combustors, temperatures and fuel concentrations at the liquid surface are low and there is little mass diffusion from the droplet early in the process. Under the influence of heat transfer from the gas to the droplet, the liquid temperature rises and the rate of mass transfer increases as a result of higher fuel vapor concentration at the droplet surface. The temperature within the liquid droplet increases from a minimum at the droplet center to a maximum at the surface. The latent heat of vaporization is supplied from the gas and a radial velocity is generated by the diffusion of vapor away from the droplet surface. This outward flow of fuel vapor reduces the rate of heat transfer to the droplet. The rate of increase of the liquid surface temperature diminishes and, later in the process, temperatures become more uniform within the liquid droplet. Eventually, a stage is reached where all the heat reaching the surface is utilized for the heat of vaporization. Faeth (1977) refers to the droplet stabilization temperature as the *wet bulb temperature*.

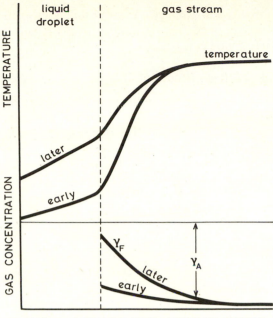

Figure 7.8 Variation of temperature and gas concentration during the process of droplet vaporization.

It has been shown both theoretically and experimentally that, after an initial heating up period, if forced and natural convection effects are small, the square of the droplet diameter decreases linearly with time, according to the equation

$$dD^2/dt = -K \qquad (7.9)$$

where K is the vaporization constant, which is a function of physical and chemical properties of liquid and ambient gas. Experiments show that, during the initial heatup period, there is little change in droplet diameter and, subsequently, the decrease of D^2 with time becomes linear. Figure 7.9 shows the variation of K with temperature, from the experiments of Masdin and Thring (1962), for single drop-

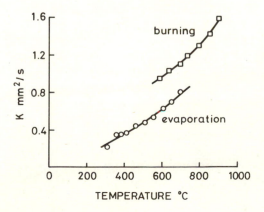

Figure 7.9 Comparison of vaporization rates (in stagnant nitrogen) and burning rates (in still air) for aviation kerosine. (*Masdin and Thring, 1962.*)

lets of kerosine evaporating in a nitrogen still atmosphere, and also for droplets burning in air. For ambient temperatures of 560°C, the burning rate constant is 75 percent higher than the vaporization rate constant.

Spherically Symmetric Quasi-Steady State Model of Evaporating Droplet

For volatile fuels at low pressures, the wet bulb temperature is close to typical injection temperatures, and liquid-phase properties can be assumed to be constant. At higher pressures, reduced liquid densities cause the droplet to swell and variation in physical properties needs to be taken into account. Until more detailed information becomes available, the assumption has been made, following Sparrow and Gregg (1958), that average properties can be evaluated at the following reference temperatures and conditions

$$T_r = T_s + 1/3(T_\infty - T_s) \tag{7.10}$$

$$Y_{F_r} = Y_{F_s} + 1/3(Y_{F_\infty} - Y_{F_s}) \tag{7.11}$$

where T is temperature, Y_F is mass fraction of fuel vapor, and subscripts r, s, and ∞ refer to reference, drop surface, and ambient conditions.

Analytical studies of droplet vaporization generally consider the droplets to be spherical. This assumption is valid for small droplets in low-velocity streams. The criteria for deformation and breakup of droplets is the critical Weber number

$$\text{We}_c = \rho u_\infty^2 r_s / \sigma \tag{7.12}$$

where u is velocity, r is droplet radius, and ρ and σ are density and surface tension. Critical Weber numbers are in the range 6–30 with specific values for breakup depending upon the Reynolds number, the rate of acceleration, and property ratios between the gas and liquid phase.

At moderate pressures, where the quasi-steady state gas-phase assumption is valid, liquid fuels have low solubilities for the nonfuel portion of the ambient gas. Spalding's (1953) insolubility condition is still considered to be valid so that the mass flux of the ambient gas is taken to be zero at the liquid surface. Gas-phase fuel concentrations at the liquid surface are obtained from the equilibrium vapor pressure for the pure fuel as a function of temperature. The assumption implied by the use of pure vapor pressure data is that the presence of other gases at the surface does not affect the equilibrium vapor pressure. It is also assumed that the presence of finite mass transfer rates at the surface does not appreciably influence the gas vapor pressure for a given liquid temperature. It may be concluded that effects of nonequilibrium are small for pressures greater than 1 atm and temperatures below the wet bulb temperature for droplets with diameters of the order of 200 μm. The effect of surface tension on vapor pressure relationships is small for pressures greater than 1 atm and droplet diameters greater than 1 μm. Faeth (1977) has also examined the evidence on continuum effects, from which it is concluded that droplets larger than 1 μm can be treated by continuum approximations for pressures greater than 1 atm. Examination of radial pressure gradients leads to the conclusion that pressure can be assumed to be constant.

Models of droplet vaporization have been derived for a spherically symmetric droplet with the quasi-steady state assumption that the gas phase always adjusts to the steady state structure for the imposed boundary conditions at each instant of time. With the assumption of a binary-diffusion law for mass transfer and neglecting terms which contribute little to the fluxes, the equations for a liquid fuel of a single pure component and a Lewis number of unity are:

Gas-Phase Equations

Conservation of mass
$$\frac{d}{dr}(\rho r^2 v_r) = 0 \tag{7.13}$$

Conservation of species
$$\frac{d}{dr}\left[r^2\left(\rho v_r Y_i - \rho D'\frac{dY_i}{dr}\right)\right] = 0 \tag{7.14}$$

Conservation of energy
$$\frac{d}{dr}\left[r^2\left(\rho v_r C_p(T - T_\infty) - k\frac{dT}{dr}\right)\right] = 0 \tag{7.15}$$

Since only the fuel is diffusing, D' is the binary diffusivity of the fuel with respect to other gas-phase species; similarly, C_p is the fuel specific heat. By definition

$$\sum Y_i = 1 \tag{7.16}$$

The analysis only considers two species, fuel and ambient gas. Therefore, through Eq. (7.16), only one conservation species equation must be solved. Integration of Eq. (7.13) yields

$$r^2 \rho v_r = \dot{m}_f/4\pi = \text{const} \tag{7.17}$$

The boundary conditions are

$$
\begin{array}{lll}
r = r_s & T = T_s & Y_F = Y_{F_s} \\
r = \infty & T = T_\infty & Y_F = Y_{F_\infty}
\end{array}
\tag{7.18}
$$

The fuel concentration and temperature at the liquid surface are related through the vapor pressure characteristics of the fuel

$$Y_{F_s} = f(T_s, p) \tag{7.19}$$

A final boundary condition is supplied by the insolubility assumption, which implies that the mass flux of ambient gas is zero at the liquid surface

$$r = r_s \qquad \rho v_r Y_a - \rho D'\frac{dY_A}{dr} = 0 \tag{7.20}$$

Specification of p, Y_{F_∞}, T_∞, and T_s provides Y_{F_s}, $\dot{m}_f$, and the heat transfer rate to the droplet. A heat transfer coefficient, which includes mass transfer effects, can be defined as follows

$$h = \left(k\frac{dT}{dr}\right)_s \bigg/ (T_\infty - T_s) \tag{7.21}$$

Solution of the equations then yields the following

$$\frac{\dot{m}_f}{4\pi\rho D'r_s} = \ln\left(1 + B_Y\right) \tag{7.22}$$

$$\mathrm{Nu} = \frac{2hr_s}{k} = 2[\ln\left(1 + B_Y\right)]/B_Y \tag{7.23}$$

where B_Y is Spalding's mass transfer number

$$B_Y = (Y_{A\infty} - Y_{A_s})/Y_{A_s} = (Y_{F_s} - Y_{F\infty})/(1 - Y_{F_s}) \tag{7.24}$$

For high transfer rates where $Y_{F_s} \to 1$, B_Y may be considered as the driving potential for the diffusion of one gas through a stagnant gas

$$\dot{m}_f = 4\pi r_s^2 K'(Y_{F_s} - Y_{F\infty})/(1 - Y_{F_s}) \tag{7.25}$$

where K' is the mass transfer coefficient. Substitution from Eq. (7.24) then yields the Sherwood number

$$\mathrm{Sh} = \frac{2K'r_s}{\rho D'} = 2[\ln\left(1 + B_Y\right)]/B_Y \tag{7.26}$$

At low mass transfer rates, $B_Y \to 0$, Eqs. (7.23) and (7.26) yield $\mathrm{Nu} = \mathrm{Sh} = 2$, which are the familiar values for a sphere at low mass transfer rates.

Liquid-Phase Equations

Conservation of mass of the droplet liquid yields

$$\frac{d}{dt}\left(\frac{4}{3}\pi r_s^3 \rho_f\right) = -\dot{m}_f \tag{7.27}$$

Employing Eqs. (7.22) and (7.27) yields an expression for the evaporation rate constant given in Eq. (7.9) as follows, where $\rho D' = kC_p$ for Lewis number of unity

$$K = \frac{8k}{C_p \rho_f} \ln\left(1 + B_Y\right) \tag{7.28}$$

where K is the vaporization constant, k is the thermal conductivity, and C_p is the specific heat of the gas surrounding the droplet.

For the case of rapid mixing within the drop, where the liquid temperature can be assumed to be uniform and the conservation of energy is given by

$$mC_{p_f}\frac{dT_s}{dt} = 4\pi r_s^2 h(T_\infty - T_s) - \dot{m}_f h_{fg} \tag{7.29}$$

where h_{fg} is the heat of vaporization. The last term in Eq. (7.29) represents the energy required to evaporate the fuel.

The initial conditions for Eqs. (7.27) and (7.29) are

$$t = 0 \qquad r_s = r_{s_0} \quad T_s = T_{s_0} \tag{7.30}$$

Equation (7.29) can be rearranged so that the wet bulb state can be interpreted as follows

$$\frac{dT_s}{dt} = \frac{\dot{m}_f h_{fg}}{mC_{pf}}\left(\frac{B_T}{B_Y} - 1\right) \tag{7.31}$$

where

$$B_T = C_p(T_\infty - T_s)/h_{fg} \tag{7.32}$$

Ignoring internal temperature gradients, Eq. (7.31) yields

$$B_{T_{WB}} = B_{Y_{WB}} \tag{7.33}$$

providing a relationship between concentration and temperature. A second relation is provided by the vapor pressure relation, Eq. (7.19), allowing solutions for $Y_{F_{WB}}$ and $T_{s_{WB}}$.

Alternatively, it has been assumed that there is no mixing within the droplet and thus temperature gradients occur within the droplet. For this situation, the energy equation must be solved numerically, in conjunction with the other equations. Faeth (1977) has compared measurements with predictions for the evaporation of droplets suspended within furnaces which are falling freely. Calculations taking into account internal temperature gradients were compared with results for the assumption of uniform drop temperature at each instant of time. The two methods yielded very similar results with respect to drop evaporation rates, therefore the approximate methods, ignoring internal temperature gradients, appear to be justified.

Transient Effects

Although many of the theoretical analyses in the past have considered steady-state conditions, recently more attention has turned to determining the extent to which variations with time are significant. Droplets are initially at a uniform cold temperature and, during the heating period, the surface temperature increases and temperature gradients develop within the droplet. Steady-state conditions may be achieved when the temperature at the droplet surface reaches a constant value. Temperatures within the droplet may reach a constant value equal to that of the surface temperature or a steady-state temperature gradient within the droplet may be achieved. Very little direct information is available on this subject and steady-state heat transfer may never be achieved within the droplet during its lifetime. The droplet radius changes very significantly. Some swelling may occur and, for total vaporization, the droplet ultimately disappears. The assumption, often made, that changes in droplet radius are not significant, cannot be valid, and it is even doubtful whether this approximation can be made for a short period of the droplet lifetime.

Droplets are injected into combustors with velocities which are generally higher than that of the gas stream. The relative velocity between droplet and gas, which is the basis for the Reynolds number calculation, is continuously changing during the trajectory of the droplet. The trajectory of the droplet is governed by the drag coefficient and the relative velocity. From the point of injection, where relative velocities are a maximum, to the position where droplet velocities have been reduced by drag to local gas velocities, the convection effects are continuously changing. The temperature and concentration ambient conditions vary over a wide range in sprays. Temperatures within the gas vary from the cold inlet conditions up to flame temperatures; the vapor pressure of the fuel varies from saturation at the droplet surface to zero in the airstream. An exact analysis of spray phenomena would, therefore, need to take into account both gas- and liquid-phase transient effects.

At low pressures, wet bulb temperatures are relatively low and heats of vaporization of the fuel are high. Under these conditions, the droplet heatup time is short in comparison with the lifetime of the droplet. At high pressures, the wet bulb temperatures are sufficiently high and the heat of vaporization is sufficiently small, so that the droplet heating time becomes more important. The heatup period always delays the start of significant evaporation but this effect becomes more important at high pressures. These effects have been examined in terms of a basic time scale, which is the liquid-phase transient time for the diffusion of heat within the droplet. At high Reynolds numbers and at low Reynolds numbers at low pressures, the gas-phase transient time is much smaller than the liquid-phase transient time. On this basis, models have been proposed assuming a quasi-steady state gas phase and the effect of radial velocity of the liquid surface is neglected.

Convection—Departures from Spherical Symmetry Model

The effect of forced or natural convection on heat transfer rates has been given in terms of the correlation proposed by Ranz and Marshall (1952). In this correlation, the Nusselt number for conditions without convection has the value 2 and the effects of convection are taken into account by adding a term which is a function of Reynolds and Prandtl numbers as follows

$$\text{Nu} = hD/k = 2 + 0.6\ \text{Re}^{1/2}\text{Pr}^{1/3} \tag{7.34}$$

For the case of mass transfer, the Nusselt number is replaced by the Sherwood number, and the Prandtl number is replaced by the Schmidt number to provide an expression similar to that for heat transfer, Eq. (7.34).

An alternative approach to treating convection has involved the assumption that the sphere is surrounded by a film or boundary layer of thickness δ. Faeth (1977) considered only the radial motion due to mass transfer and then selected a value of δ to yield the correct heat or mass transfer rate at a given Reynolds number, at the limit of low mass transfer rates, yielding the expression

$$\delta/D = (\text{Nu or Sh}) - 2 \tag{7.35}$$

where δ is the film thickness. When the Schmidt and Prandtl numbers are equal, the same film thickness is used for heat and mass transfer, otherwise two separate thicknesses are used for heat and mass transfer. The film theory approximation has been shown by Faeth (1977) to provide an adequate representation of more exact calculations and experimental results. For the simplified analysis considered above, with unity Lewis number, the general form of the equation is

$$(\text{Nu or Sh})B_Y/\ln(1 + B_Y) = 2 + f(\text{Re, Pr, or Sc}) \tag{7.36}$$

For most experiments, wet bulb conditions pertain and B_T can replace B_Y in Eq. (7.36).

Faeth (1977) has considered various data for Nu and has derived a synthesized correlation which has the correct limiting values for high and low Re, for $B_Y = 0$,

$$\text{Nu} = 2 + 0.555\ \text{Re}^{1/2}\text{Pr}^{1/3}/(1 + 1.232/(\text{Re, Pr}^{4/3}))^{1/2} \tag{7.37}$$

with an analogous expression for the Sherwood number.

Figure 7.10 shows that there is good agreement with available experimental data for Eq. (7.37). Combining Eqs. (7.37) and (7.36) gives a relation for Nu (and Sh) when B_Y is not small, which Faeth (1977) showed agreed reasonably well with available data.

High-Pressure Phenomena

At high ambient pressures, a higher vapor pressure at the liquid surface is required to provide a given fuel mass fraction. Therefore, the wet bulb temperature during steady evaporation increases as the ambient pressure is increased. Eventually, conditions are reached where a drop approaches or exceeds its thermodynamic critical point during its lifetime.

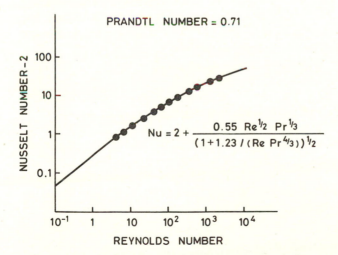

Figure 7.10 Correlation between Nusselt, Reynolds, and Prandtl numbers for heat and mass transfer from vaporizing droplets. (*Faeth, 1977.*)

Near the critical point, solubility increases (at the critical condition itself all characteristics of both phases are the same). The radial regression velocity of the liquid surface becomes comparable to the radial gas velocity since the densities of the two phases are the same. Gas-phase transient effects are as important as liquid-phase transients and, since the heatup period is a larger percentage of the total lifetime, both factors are important for the prediction of gasification rates. If the drop is moving, deformation, liquid circulation, and drop breakup become important due to reduced values of surface tension.

Manrique and Borman (1969) theoretically considered the effect of solubility and found that dissolved gas appreciably changed high-pressure vaporization characteristics. Kadota and Hiroyasu (1976) took into account solubility and compared their theoretical results with measurements for high-pressure drop life histories. Figure 7.11 illustrates their comparison between theory and experiment for n-heptane. At a fixed ambient temperature, the droplet lifetime decreases with increasing pressure, largely due to higher rates of natural convection, and the heatup period becomes a greater percentage of the lifetime.

Motion of a Vaporizing Droplet in a Cross-Flow Stream

Law (1977) has carried out an analysis for the motion of a single droplet undergoing evaporation in a constant cross flow. The droplet with initial diameter D_0 is

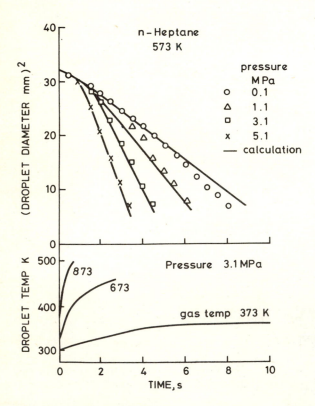

Figure 7.11 Variation of droplet diameter and droplet temperature with time for high-pressure vaporization. (*Kadota and Hiroyasu, 1976.*)

injected with velocity u_{po} and v_{po} in the x and y directions, respectively, into a gas stream with constant u_G in the x direction. It is assumed that the evaporation and motion of the droplet do not affect the properties of the gas stream. The rate of decrease of droplet size under forced convection is given by

$$d\hat{D} = -2kH(\text{Re, Sc}) \tag{7.38}$$

where $\hat{D} = D/D_0$, $k = 4M/(D_0^2 \rho_p)$, $M = \rho_G \delta_G \ln(1 + B)$ and $B = (C_G/L) \times (T_G - T_p)$; t is the time, ρ is the density, T is the temperature, μ is the viscosity coefficient, δ is the binary-diffusion coefficient, C is the specific heat, L is the specific latent heat of vaporization, and the subscripts 0, p, and G refer to the initial state, the droplet, and the gas, respectively.

The function H is given by

$$H(\text{Re, Sc}) = 1 + 0.276\ \text{Sc}^{1/3}\ \text{Re}^{1/2} \tag{7.39}$$

The Schmidt number is defined by

$$\text{Sc} = \mu_G/(\rho_G \delta_G) \tag{7.40}$$

and the Reynolds number is defined by

$$\text{Re} = \{D_0 \rho_G/\mu_G\}\{(u_G - u_p)^2 + v_p^2\}^{1/2} \hat{D} \tag{7.41}$$

The accelerations of the droplet in the x and y directions are, respectively,

$$du_p/dt = (3\mu_G/4\rho_p)(C_D\ \text{Re}/D^2)(u_G - u_p) \tag{7.42}$$

and

$$dv_p/dt = -(3\mu_G/4\rho_p)(C_D\ \text{Re}/D^2)v_p \tag{7.43}$$

The drag coefficient C_D is given by

$$C_D = K(\text{Sc})\quad H(\text{Re, Sc})\quad G(B)/\text{Re} \tag{7.44}$$

The function $G(B)$ accounts for the change in drag due to the outward mass transfer at the droplet surface.

Spalding (1959) used

$$G(B) = B^{-1} \ln(1 + B) \tag{7.45}$$

whereas Eisenklam et al. (1967) recommended

$$G(B) = (1 + B)^{-1} \tag{7.46}$$

Law (1977) found that C_D could be closely correlated to the standard experimental drag curve for solid spheres by using the function

$$K(\text{Sc}) = 23\ \text{Sc}^{-0.14} \tag{7.47}$$

over the range $0.5 < \text{Sc} < 10$ and $\text{Re} < 200$. For Reynolds numbers larger than 200, droplets tend to become unstable and break up. Dividing Eqs. (7.42) and

(7.43) by Eq. (7.38) eliminates the nonlinear term $H(\text{Re}, \text{Sc})$ and the following linear equations are obtained for u_p and v_p

$$du_p/d\hat{D}^2 = -(\alpha/2)(u_G - u_p)/\hat{D}^2 \tag{7.48}$$

$$dv_p/d\hat{D}^2 = (\alpha/2)v_p/\hat{D}^2 \tag{7.49}$$

where

$$\alpha = (3/16)\mu_G K(\text{Sc})G(B)/M$$

Integrating Eqs. (7.48) and (7.49) with the initial conditions that $u_p = u_{p0}$ and $v_p = v_{p0}$ at $\hat{D} = 1$, yields

$$u_p = u_G + (u_{p0} - u_G)\hat{D}^\alpha \tag{7.50}$$

and

$$v_p = v_{p0}\hat{D}^\alpha \tag{7.51}$$

The parameter α represents the ratio of the effects of drag to the rate of evaporation, that is, $\{\mu_G G(B)\}/M$. For $\alpha \ll 1$, the droplet vaporizes so fast that it retains its initial velocity, the drag force having little influence on its motion. During the final stages of its lifetime, the drag force begins to dominate and the droplet rapidly attains the local gas velocity. For $\alpha \gg 1$, the droplet rapidly loses its inertia after injection and attains local gas velocity before any significant amount of mass has evaporated. Vaporization occurs during the remainder of the droplet lifetime with only slight adjustments required to follow the gas motion. Thus, from Eqs. (7.50) and (7.51), for $\alpha \ll 1$, $\hat{D}^\alpha \to 1$ for $\hat{D} \not\ll 1$ such that $u_p \to u_{p0}$ and $v_p \to v_{p0}$; whereas for $\alpha \gg 1$, $\hat{D}^\alpha \to 0$ for $(1 - \hat{D}) \not\ll 1$ such that $u_p \to u_G$ and $v_p \to 0$.

Law (1977) has calculated the ratio of the penetration depths, $R = (y_{max})_{\text{Stokes}}/y_{max}$ for a heptane droplet with $D_0 = 10^{-1}$ mm and $v_{p0} = 1$ m/s injected perpendicularly into a 1-atm airstream with $\text{Sc} = 0.7$. At $T_G = 300$ K, vaporization is so slow that the droplet can be effectively treated as a solid sphere in computing its trajectory. At $T_G = 500$ K, the rapid reduction in droplet size tends to significantly increase the drag force per unit mass, but the drag reduction due to the enhanced mass transfer rate at the droplet surface more than compensates for this increase so that the penetration depth is larger than for the case of $D_G = 300$ K.

7.4 INDIVIDUAL DROPLET BURNING

When the oxygen concentration and the temperature of the gas surrounding an individual fuel droplet is sufficiently high to provide a mixture within the limits of flammability, the mixture, when ignited, forms a thin flame reaction zone surrounding the droplet. This phenomena is referred to as individual or single-droplet burning, but the physical process still involves vaporization of the liquid at the droplet surface and diffusion of fuel vapor away from the surface. When the

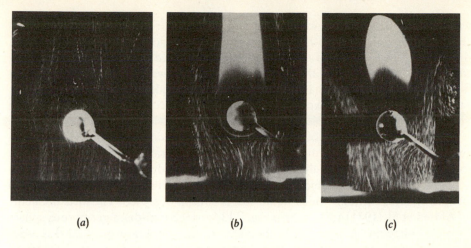

(a)   (b)   (c)

Figure 7.12 Particle-track photographs around porous spheres: (*a*) sphere without flame, Re = 92; (*b*) sphere with envelope flame, Re = 92; (*c*) sphere with wake flame, Re = 152. (*Gollahalli and Brzustowski, 1973.*)

flame is in close proximity to the droplet surface, within one droplet diameter, the gradients of temperature and fuel vapor pressure are high and, as shown in Fig. 7.9, the vaporization constant is significantly higher than for nonburning vaporization. These differences in values of the vaporization constant, K, become less as the ambient temperature is increased above 800°C.

Many experimental and analytical studies have been made on the burning of single liquid drops. These single drops have usually been held in suspension attached to the ends of wires, or the process has been simulated by the use of a porous sphere covered with a liquid film. These studies have been reviewed by Beér and Chigier (1972), Williams (1973), and Hedley et al. (1971). The more recent studies of Gollahalli and Brzustowski (1973) show that the flame around the porous sphere can either have an envelope flame or a wake flame (Fig. 7.12). This work follows that of Spalding (1953), who investigated the effect of relative air velocity on the combustion of liquid fuel spheres and observed a critical velocity above which the flame could not be supported at the upstream portion of the sphere. In low-velocity airstreams, a flame envelops the leading half of the sphere. Above a critical free-stream velocity (the extinction velocity) the flame on the leading half of the sphere is extinguished and a small flame is stabilized within the wake. A third regime of burning is also sometimes observed, in which the flame is stabilized in the boundary layer at the side of the liquid sphere. In the experiments of Gollahalli and Brzustowski (1973), *n*-pentane was supplied to a porous bronze sphere, 6 mm in diameter, which was suspended in a uniform airstream. Measurements were made of burning rate, temperature, and composition profiles in the wake, as determined from micro probe samples. They found that the envelope flame in free convection had a near-wake extending about 5 sphere diameters and a far-wake extending another 15 diameters. For the typical wake flame, these lengths were about 3–6 diameters respectively. The particle-track photographs in Fig. 7.12 show a recirculation zone extending for about 1 diameter in the wake of the sphere, without the presence of a flame. This recirculation zone is not visible in either the envelope or in the wake flame conditions. This is very clear evidence of changes in wake structure under combustion conditions and provides a partial explanation to the observation of reduction in drag coefficient of burning droplets.

The flame provides a source of heat and is also a sink for both fuel vapor and oxygen. Furthermore, combustion products are generated within the flame, and these diffuse both toward the droplet surface and radially outward. Several investigators have measured temperature and gas concentration distributions around porous spheres through which liquid fuel is continuously fed. The measurements of Aldred et al. (1971) are shown in Fig. 7.13 for a 9.2-mm-diameter porous sphere burning n-heptane in still air. At the flame front, temperatures of 2000 K were measured, and fuel and oxygen concentrations were close to zero. Traces of oxygen were detected between the flame and the droplet surface. It appears that dissociation and finite rate effects cause the oxygen concentration to be greater than zero in the flame zone and this oxygen diffuses toward the droplet surface. Intermediate hydrocarbons, C_2H_4, C_2H_2, and CH_4, were found with maximum

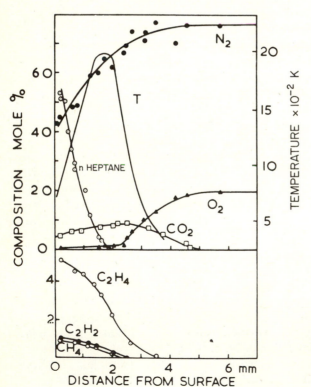

Figure 7.13 Temperature and concentration profiles at the lower stagnation point of a 9.2-mm-diameter porous sphere burning n-heptane in still air. (*Williams, 1976.*)

concentrations near the droplet surface. Thus, in the non-, or very low-oxidizing atmosphere between the flame and liquid surface, pyrolysis, cracking, and formation of soot particles occur.

The composition profiles measured by Gollahalli and Brzustowski (1973) suggested that the near-wake of the envelope flame is a pyrolysis zone in which n-pentane decomposes to produce lighter hydrocarbons, including acetylene. Combustion takes place at the edges of this zone, similar to a laminar diffusion flame. In the case of a flame totally in the wake of a sphere, the near-wake resembles the flame behind a flame holder. Most of the heat release occurs in this zone and only a small amount of soot burns in the far-wake. The burning rate was found to decrease by a factor of three when the envelope flame was transformed into a wake flame at the critical Reynolds number of 138 and envelope flames radiated much more than wake flames. The oxygen concentration of the ambient gas and the intensity of turbulence in the gas stream can have considerable effects on the extinction velocity.

Several attempts have been made to determine the reduction in drag coefficient of droplets under vaporizing and burning conditions. High-speed movie films show that the velocity with which vapor is emitted from the drop surface is comparable with that of gas flow surrounding the droplet, and this results in reduction of skin friction. The particle-track evidence of Gollahalli and Brzustowski (1973) shows substantial reductions in the size and strength of the recirculation zone in the wake of the droplet, which provides additional reduction to the drag coefficient. Khudyakov (1949) measured appreciable decreases in drag coefficient of metal spheres wetted with fuel in airstreams at high Reynolds numbers so as to produce wake flames. Spalding (1954) in similar work with kerosine-wetted cylinders also found reductions in drag coefficients in cylinders supporting wake flames.

In single-droplet combustion a heating up phase occurs prior to ignition, which is referred to as "ignition delay." This delay comprises the time interval for the combustible mixture to form around the drop (the physical delay time) and the time interval between the perceptible chemical reaction and self-ignition (the chemical delay time). The lengths of the chemical and physical delays are similar; the chemical delay is shorter than the physical delay at higher air temperatures and with more volatile fuels.

In almost all the experimental studies made with porous spheres, diameters are two orders of magnitude larger than average drop sizes in sprays, and in many of the single-drop burning studies diameters have been one order of magnitude larger than drop sizes in sprays. The extent to which these "large drop" results can be extrapolated to droplets of the order of 100 μm has not yet been determined. In most practical sprays, a large percentage of the droplets vaporize without burning. Isolated drops, particularly those with the largest diameters and highest velocities within the system, will cross the main flame and then burn in isolation. These isolated drops can make a major contribution to the emission of pollutants and lead to increases in overall flame length.

7.5 GROUP COMBUSTION OF DROPLET CLOUDS

The collective behavior of droplets in liquid sprays results in the formation of a fuel-rich nonflammable mixture in the spray core, due to insufficient air penetration. The radial transport of gaseous fuel by transverse convection and diffusion leads to the formation of flammable mixtures at some distance from the spray boundary. This flammable mixture burns as a gas-phase diffusion flame. The spray boundary surrounds the region containing vaporizing liquid droplets. As droplets move beyond the dense core region of the spray, droplet separation distance increases and drop size is reduced, so that more air can penetrate the spray boundary. Under these conditions, the flame may penetrate inside the spray boundary, while some liquid droplets may burn individually or as groups outside the flame. The inner region consists of droplets vaporizing in the absence of oxygen, while the outer region may contain droplets burning with multidrop flames. Turbulence and droplet dispersion will produce flame broadening.

Labowsky and Rosner (1976) have developed a model to predict the mode of burning of a spray of fuel droplets, i.e., cloud combustion, individual envelope flames, or a mixture of these two extremes. Their model is an extension of the quasi-steady burning model for a single fuel droplet to the case of a cloud of fuel droplets so that initial and final transients and convection effects are neglected. Both cloud and droplet radii are considered to vary slowly with time.

The *continuum total group combustion* criterion is established by determining when the cloud burns as a large pseudo-droplet with the flame located just outside the cloud droplet region. Under this condition, evaporating particles inside the cloud provide sufficient vapor so that fuel and oxidizer mix in stoichiometric amounts at the cloud boundary. Labowsy and Rosner (1976) also describe an alternative approach which predicts when the flames around individual adjacent droplets in the spray just meet, and this is defined as the *incipient group combustion* criterion.

Figure 7.14 shows flame locations calculated for dodecane fuel droplets in a cubical array. When the fuel particles are very far apart they burn as separate particles, each surrounded by its own *envelope* flame (Fig. 7.14a). When they are moved closer together (or as more particles are added to the cloud), oxygen has more difficulty in penetrating the cloud and, as a result, the flame radii of the interior particles increase. When the gaseous oxidizer has been sufficiently depleted, the flames around the particles in the center of the cloud will just touch and these particles are considered to burn as a group (Fig. 7.14b), while other particles continue to burn individually (incipient group combustion). With further reductions in particle separation, additional particles join the group until all fuel particles in the cloud ultimately burn with a common flame (Fig. 7.14d). This situation is called *total group combustion*.

Figure 7.15 shows the predicted regimes for individual droplet burning and cloud combustion as functions of the total number of particles N in the cloud and the dimensionless interparticle separation L. It is seen that the two modeling approaches produce essentially the same predictions. Labowsky and Rosner

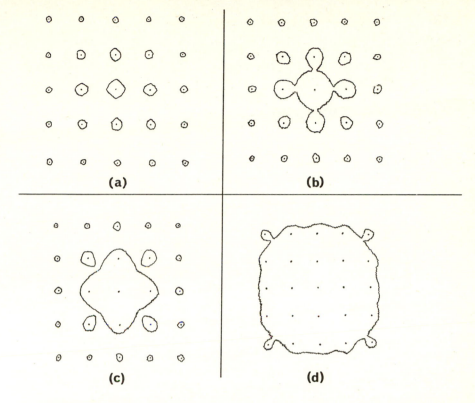

Figure 7.14 Superposition calculations of flame locations as a function of particle separation for a dodecane fuel cloud consisting of 125 spherical particles "burning" in air. "Incipient group combustion" is shown in (*b*) and "total group combustion" is shown in (*d*). (*Labowsky and Rosner, 1976.*)

(1976) conclude, in accordance with experimental evidence, that evaporating fuel particles are so efficient in preventing oxidizer penetration that, if a fuel cloud of practical interest were to approach a quasi-steady state, it would necessarily burn as a group.

Chiu et al. (1978) have developed a group combustion model for the structure and burning characteristics of liquid fuel sprays. Collective behavior of the droplets is accounted for by a simultaneous analysis of an inner heterogeneous region and an outer homogeneous gas-phase region. Spray combustion models are classified according to group combustion number.

The physical model of Chiu et al. (1978) is based on a two-phase flow description of a spray jet in an axisymmetric configuration. Chemical reaction between the gaseous fuel and oxidizer is assumed to be a one-step exothermic reaction with characteristic reaction times much smaller than the characteristic diffusion or flow residence time. Schmidt and Lewis numbers are assumed to be unity and the Schvab-Zeldovich transformation is used. Similarity solutions for velocity, temperature, and fuel vapor concentration are obtained in the spray core. The flow variables of the outer field, which extends outwardly from the spray bound-

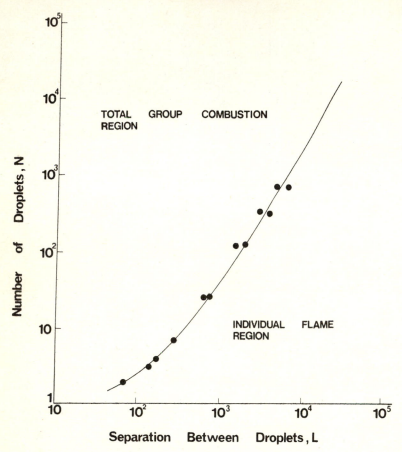

Figure 7.15 Criteria for separating regions of total group combustion and individual flames. Dimensionless numbers: N = number of droplets; L = separation between droplets. (*Labowsky and Rosner, 1976.*)

ary, are constructed by locally similar solutions that match the inner solutions at the spray boundary, and the uniform or nonuniform distributions of temperature and oxidant concentration in the free stream. Boundary perturbation techniques have been incorporated to accomplish matching at the spray boundary.

Numerical results have been obtained to predict the burning characteristics of sprays with a linearly decreasing axial velocity at the centerline. Sprays have been classified into "high G" and "low G" categories. The analysis shows that local and cumulative group burning rates at any location are dependent upon the group combustion number, mass ratio, and spray angle.

Chiu et al. (1978) define two group combustion numbers as follows

$$G = \frac{G_1 D_\infty}{R_\infty U_\infty} \tag{7.52}$$

and

$$G_1 = \dot{m}_0 n_0 R_\infty^2 / (\rho_\infty D_\infty) \tag{7.53}$$

where D is the diffusivity, R_∞ is the reference spray radius,

$$\dot{m}_0 = 4\lambda r_{l_0}(1 + 0.276 \, Re^{1/2} \, Sc^{1/3})/C_p \tag{7.54}$$

λ is thermal conductivity, r_{l_0} is initial droplet radius, C_p is specific heat, and n is the droplet number density; subscript 0 refers to initial conditions and subscript ∞ to reference quantities at the spray boundary.

The group combustion number G denotes the ratio of the mutual heat exchange between the two phases to the heat transfer by convection.

Chiu et al. (1978) came to the following general conclusions from their study:

1. Sprays with $G \gg 10^{-2}$ are classified as "high G sprays," in which external group combustion occurs with large flame standoff distances. Intermediate G sprays (10^{-3} to 10^{-2}) have external group flames in the vicinity of the spray boundary. Low G sprays ($G \ll 10^{-3}$) burn primarily with internal group combustion. Practical spray burners, as used in industrial and aerospace applications, operate in the "high G" spray regime. The transition to internal group combustion occurs where sufficient droplet dilution, reduction of droplet size, and enhanced mixing, favor gas-phase and single-droplet combustion.
2. Absence of combustion in the spray core is due to the high mixture equivalence ratio.
3. In sprays with large group combustion numbers, spray flames may be described by the "equivalent turbulent gas diffusion flame."

There is little experimental data on the burning of clouds or arrays of droplets which can be used to compare with the analyses of Labowsky and Rosner (1976) and Chiu et al. (1978). Experiments (Burgoyne and Cohen, 1959) with laminar and turbulent premixed droplet-air suspensions show that, under these conditions, both individual droplet burning, involving a network ignition process, and cloud combustion can be found. The experiments of Mizutani and Ogasawara (1965) have demonstrated that the burning velocity in laminar monosize sprays of tetralin is a function of the mass concentration and the droplet size. Mizutani and Nakajima (1973) showed experimentally that the presence of small numbers of fuel droplets in turbulent premixed gaseous fuel and air markedly accelerated the burning velocity and extended the region of stable combustion to the lean level. Polymeropoulas and Das (1975) found that for kerosine-air sprays there was an optimum size distribution which gave the highest burning velocity. It is likely that ultimately, if the burning of even these simple sprays is to be predicted reliably, an extension of the analytical techniques is required to include the effects of a range of droplet sizes and relative droplet-gas velocities.

7.6 PRESSURE-JET, SWIRL-ATOMIZED, HOLLOW-CONE SPRAY FLAMES

The pressure-jet, swirl-atomized, hollow-cone spray is used in many types of combustion chambers where pumping facilities are available for the high liquid pressures required and where high-pressure air or steam is not available. Liquid fuel is introduced tangentially at high pressure into a swirl chamber and passes through a diffuser to a circular orifice exit. The liquid is attached to the walls of the diffuser and leaves the atomizer in the form of an annular film with a central air core, generated by the pressure differences within the mixing chamber. This annular film spreads out to form a hollow conical spray, which becomes unstable and disintegrates into ligaments and large drops and, finally, into a spray. A description of the atomizer and the interaction of such sprays with uniform air-streams is given by Mellor et al. (1971). Following the nonburning study of Mellor, Chigier et al. (1974) studied pressure-jet spray flames in the wake of a stabilizer disk. The system is described schematically in Fig. 7.16. Kerosine is introduced at high pressure into a swirl atomizer, which produces a hollow-cone spray projected vertically upward. The spray emerges from the center of a stabilizer disk with a surrounding airflow. A recirculation zone is formed by the airflow in the wake of the disk, as shown by the streamline flow patterns and zero velocity boundary in Fig. 7.16. A primary reaction zone is formed near the edge of the disk by fuel vapor and small droplets which are transported by the recirculating gas flow toward the outer airflow. Since only a very small fraction of the fuel enters the primary reaction zone, the dimensions of this flame are restricted by the spray and surrounding airflow. This primary flame provides an ignition source, acting as a pilot flame, introducing hot products into the main airflow stream, which ignite the vapor mixture farther downstream in the secondary reaction zone.

Figure 7.17 shows the isotherms within the spray flame as measured by a fine-wire Pt/PtRh thermocouple. Temperature levels within the dense spray region are seen to be as low as 300°C. Temperature within the spray increases with downstream distance as a consequence of entrainment of hot products from the primary reaction zone, which can be seen outside the spray boundary near the edge of the stabilizer disk. The spray also entrains air from the surrounding cold airflow. Temperatures along the axis of the system are higher than within the spray, due to the transport of hot combustion products from both the primary and secondary reaction zones. The presence of a high concentration of droplets in the spray has a strong quenching action, which inhibits combustion but allows limited vaporization to take place. The spatial distributions of oxygen concentration, as determined by probe sampling, are shown in Fig. 7.18. Within the dense spray region measured oxygen concentrations were as low as one percent, showing that the mixture ratios are very rich and dominated by fuel vapor. The oxygen concentrations rise with distance downstream as a direct consequence of entrainment and penetration of oxygen from the surrounding airstream. The oxygen concentrations decrease in the secondary reaction zone due to burning. The experimental evidence in Figs. 7.17 and 7.18 shows that, in the initial dense-spray

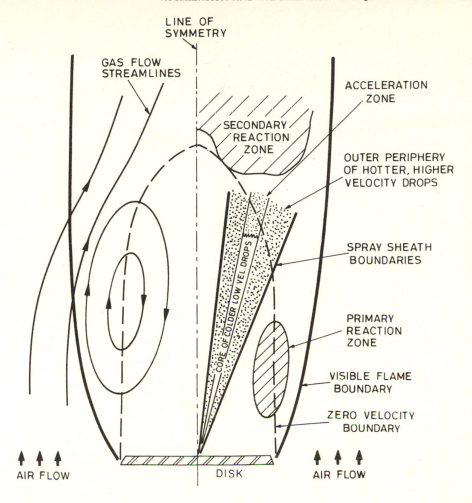

Figure 7.16 Hollow-cone pressure-jet spray burning in the wake of a stabilizer disk.

region, both the temperatures and the oxygen concentrations are too low to allow combustion to take place. Direct photography of this region also showed no evidence of burning. The double-flash high-speed photographic technique was used to measure droplet size and velocity and these measurements showed that large droplets were breaking up into smaller droplets in the initial regions of the spray. The size distributions change farther downstream as the smaller droplets vaporize rapidly and relatively small changes occur in the size of the larger droplets.

A study of the influence of the recirculating gas stream on the trajectories of large droplets (between 100 and 200 μm) showed that large droplets have almost linear trajectories and penetrate through the reverse flow zone with very little influence from the recirculating gas stream. The smaller droplets, less than 50 μm,

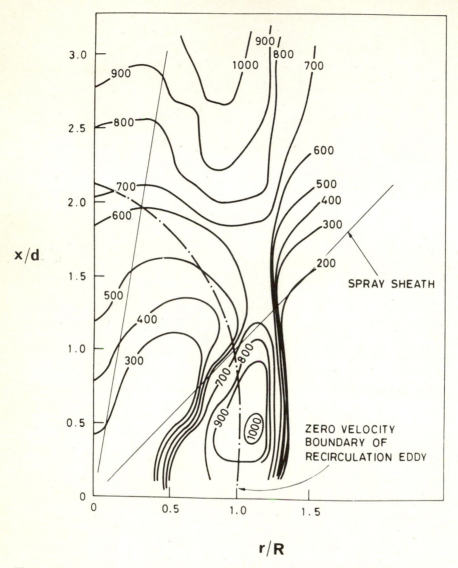

Figure 7.17 Isotherms in pressure-jet spray flame. (*McCreath and Chigier, 1973.*)

are deflected by the gas flow but there is insufficient time for droplets to be decelerated, and subsequently accelerated, in the opposite direction by the reverse flow, and little evidence was found of droplets moving in the reverse flow direction toward the stabilizer disk. The measurements of droplet velocities show comparatively small variations and the changes are explained as being due to the influence of acceleration and deceleration as droplets interact with gas streams of varying velocity.

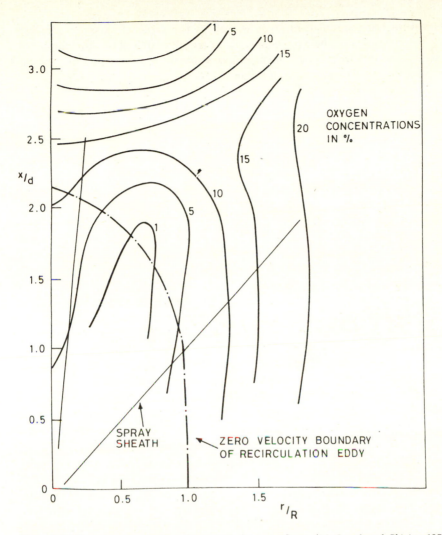

Figure 7.18 Oxygen concentrations in pressure-jet spray flame. (*McCreath and Chigier, 1973.*)

A comparison of trajectories and velocities of droplets, as measured by Chigier et al. (1974) in cold sprays and those of McCreath and Chigier (1973) in spray flames, shows that significant changes occur as a consequence of combustion. These differences are partly due to the reduction in drag coefficient of droplets vaporizing in the spray flame. The comparison showed that velocity of both small and large droplets was larger in the flame than in the cold spray. In addition to the effect of reduction in drag coefficient, which may be as high as 80 percent, the relative velocities between gas and droplets were lower and the density of the hot gas was also lower in the flame as compared with the cold spray. By increasing the velocity of the airstream from 8 to 40 m/s, the drag forces were increased by increasing the relative velocity between droplet and gas. The residence time of droplets within the recirculation zone as well as the rates of mass transfer are changed by increases of the air velocity. The smallest droplets were found to penetrate smaller distances as the reverse flow velocity was increased. For an air velocity of 40 m/s, droplets less than 50 μm were not found beyond 120 mm.

The physical models of spray burning show that sprays are initially dense, surrounded by vapor with rich mixture ratios at low temperatures in which no significant chemical reaction can take place. Reaction is forced to take place at the outer periphery of the sprays, where air/fuel ratios and temperatures are within the limits of flammability. No evidence was found of the classical model of droplets burning with surrounding individual flames. Within the spray core, the rate of evaporation of individual droplets plays no significant role in the combustion system, since evaporation is taking place in an atmosphere which is so rich that it is beyond the limits of flammability.

The three prime requirements for flame stabilization, i.e., mixture ratios within the limits of flammability, velocities low enough to match burning velocities, and sufficient supply of heat to retain reaction, are found in the primary reaction zone outside the spray boundary. The main combustion is deferred to distances farther downstream where the spray is more dispersed, more oxygen has been entrained from the surrounding airflow, and temperature levels and mixture ratios are within the limits of flammability. Liquid spray flames may generally be divided into a number of regions. Within the atomizer nozzle there is bulk liquid flow which interacts with the bulk air. Immediately after breakup of the bulk liquid flow there follows a dense-spray region with liquid particles becoming dispersed through the main airflow field. While the concentration of droplets remains high and the concentrations of oxygen within the spray sheath are low, no burning takes place within the spray and the main reaction zones occur at the spray periphery. This type of spray burning can persist for considerable distances from the nozzle and can, at times, be detected by visual observation of dark spray regions within the combustion chambers. Farther downstream, liquid particles become more dispersed and reaction zones converge toward the axis. Toward the end of the flame conditions can arise where droplets are in sufficient isolation, surrounded by air, that envelope flames can form around the droplets. The detailed measurements within spray flames have been made under laboratory conditions but evidence is coming forward that the physical models described above are applicable to combustion chambers in large power station boilers, gas turbines, diesel engines, and automobile engines. In each case it is necessary to determine the location of the flame front and establish either the presence of envelope flames around individual droplets or flames confined to the outer periphery of the spray sheath.

7.7 AIRBLAST SPRAY FLAMES

In airblast spray flames, the air which passes through the atomizer has the main function of breaking up the liquid sheet into droplets and providing sufficient momentum to transport the droplets. The oxygen in the atomizing air will react, after mixing with the fuel, provided that the temperature is sufficiently high and the mixture ratio is within the limits of flammability. Within the spray sheath, the mixture ratio changes from very lean conditions at the atomizer exit where almost all the fuel is still in liquid form, to the very rich conditions which occur when large proportions of the fuel have vaporized within the spray. Under burning conditions, the amount of oxygen entrained into the spray will be negligibly small because the flame surrounding the spray acts as an effective barrier to oxygen entrainment. In the initial regions of the spray, near the atomizer exit, the temperature levels are too low to allow combustion in the dense spray. As we proceed downstream, the spray temperature rises, due to heat transfer from the surrounding flame, and the vapor mixture ratio changes from lean to rich. The vapor mixture within the spray thus passes through the flammability limits, starting from conditions which are too lean and ending with conditions that are too rich. In sprays where the temperature is sufficiently high to allow ignition of mixtures within the limits of flammability, combustion can take place within the spray. When combustion does not take place within the spray, the characterization of the spray only requires consideration of the vaporization of droplets in a heated fuel-air gas stream.

Chigier and Roett (1972) carried out an experimental investigation of airblast kerosine spray flames using the same atomizer as Mullinger and Chigier (1974). The studies were also carried out under unconfined conditions with the spray directed vertically upward and initially ignited by a pilot gas flame but, subsequently, burning freely in the open atmosphere. Photographs of the open spray flame show that the central region of the spray is dark, due to the presence of the cloud of droplets, and the flame is seen to be initially confined to an annulus surrounding the spray. Farther downstream, flame is seen across the whole cross section of the jet but, at this stage, the spray region has ended since all the droplets have been vaporized. Chigier and Roett (1972) used the double-image high-speed photographic technique for measuring individual droplet sizes and velocities in the spray. Spray flames with air/fuel mass ratios of 0.2, 0.3, and 0.42, and total jet momenta of 0.061, 0.114, and 0.169 kg m/s^2 were studied. Series of photographs were taken at positions in radial and axial traverses and, following mass averaging, the changes in droplet size through the spray were plotted in terms of isomass fraction lines for droplets less than 100 μm (Fig. 7.19) and for droplets more than 200 μm (Fig. 7.20). The changes in droplet size within the spray are explained in terms of two separate and apparently distinct phases. In the initial phase, up to 150 mm, larger droplets break up into smaller droplets, leading to an increase in relative proportion of small droplets, with a corresponding decrease in proportion of large droplets. In the second phase, beyond 150 mm, atomization is complete and the effect of vaporization is dominant. The smaller droplets vaporize quickly with a net result that there is a decrease in the proportion of small droplets, as shown in Figs. 7.19 and 7.20. Radial distributions of droplet size show that the small droplets are concentrated near the center of the spray and that in the outer regions the proportion of small droplets decreases, due to the closer proximity to the flame. There is preferential vaporization of smaller droplets in

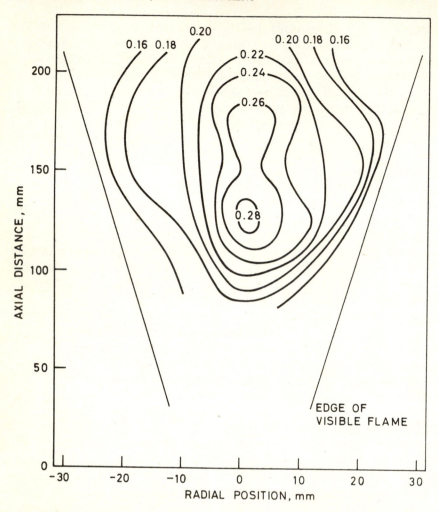

Figure 7.19 Isomass fraction lines for droplets less than 100 μm in airblast kerosine open spray flame (*Chigier and Roett, 1972.*)

regions of high temperature. Changes in size of droplets along the axis of the spray show that the percentage of droplets less than 100 μm initially increases. Photographs of individual droplets in this region showed some of the larger droplets to be nonspherical and in the process of breakup to smaller droplets.

Changes in the atomizer flow conditions were studied by maintaining the total jet momentum constant and increasing the air/fuel ratio, which led to the production of a finer spray. In studies where the air/fuel ratio was maintained constant but the total jet momentum increased, it was found that this again led to finer sprays, but a point is reached where further increases in momentum do not lead to reduction in average drop size. In the high-speed photographic techniques used by Mullinger and Chigier (1974) and Chigier and Roett (1972), volume

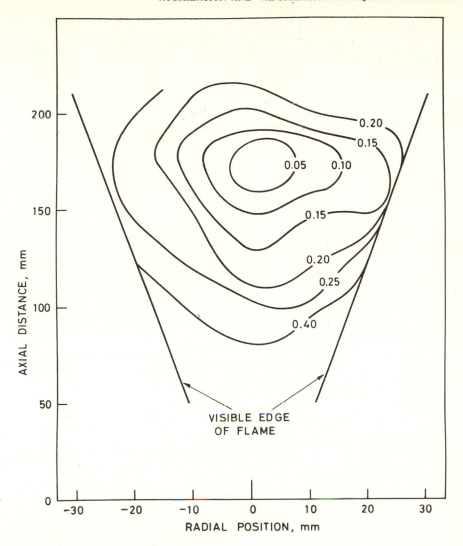

Figure 7.20 Isomass fraction lines for droplets more than 200 μm in airblast kerosine open spray flame. (*Chigier and Roett, 1972.*)

averaging was carried out at each " point " at which measurements were made in the axial and radial traverses. The changes in the droplet diameter distributions between one plane of the spray and another plane downstream are due to the breakup of large droplets, and vaporization, as well as spreading of the droplets within the spray jet by turbulence and droplet dynamics. These phenomena require to be separated in order to determine the vaporization constant from droplet size measurements.

Using the double-flash high-speed photographic technique, Chigier and Roett (1972) also measured velocity distributions of the droplets in the spray. Droplet velocity profiles were found to be gausssian in form with velocity maxima on the jet axis. Since the droplets are transported by the atomizing air jet, the droplet velocity profiles are governed by the air jet velocity profiles. The measured drop velocities were in the range 10–30 m/s and the drag-to-momentum ratio for droplets less than 30 μm was sufficiently high that droplet velocities did not differ significantly from the time-average air jet velocities. For the larger droplets, however, slip occurs so that velocities of large droplets are different from that of the surrounding gas stream. A direct measure of the velocity differential between droplet and gas streams is required in order to determine the drag forces.

At the atomizer exit, the airstream has a velocity just below sonic, while the large liquid droplets have relatively low momentum and velocity. Because of this initial velocity difference, the droplets are accelerated by the airstream. Farther downstream, gas velocities decrease, both due to the entrainment of air from the surroundings and as a consequence of exchange of momentum with the droplets being accelerated by the gas stream. Equilibrium conditions, neglecting gravitational forces, will be achieved where droplet and gas-stream velocities equalize. Beyond this position the droplets will retain their momentum to a greater extent than the surrounding gas, and Roett concluded that droplets were being decelerated by the gas stream in the tail region of the spray. Thermal expansion of the gases as a direct consequence of heating also leads to velocity changes and can result in local gas velocity increases.

Velocity distributions for a number of jet systems are compared in Fig. 7.21. In this figure the time-average velocity measured at several points in a radial traverse is normalized by dividing by the maximum velocity, which, in each case, is found on the jet axis at $r = 0$. This velocity ratio $\bar{U}/\bar{U}_m$ is plotted against $r/(x + a)$, where x is the distance along the axis from the nozzle exit and a is the distance from the nozzle exit to the effective origin of the jet system. In each of the experiments shown in Fig. 7.21 the profiles were found to be similar so that they had the same normalized velocity distribution at various cross-sectional planes along the jet. A velocity distribution such as shown in Fig. 7.21 gives a measure of the spread and the boundaries of the jet. The reference profile is that of the isothermal free jet, which is represented by an equation of the form

$$\bar{U}/\bar{U}_m = \exp\left(-k_u \xi^2\right)$$

where k_u is the velocity spread coefficient and $\xi = r/(x + a)$. The spray flame of Chigier and Roett (1972) is referred to in Fig. 7.21 as the spray with combustion. Since, in this spray flame, changes are expected in the velocity fields, both due to the presence of droplets and due to the presence of flame, profiles are also shown in Fig. 7.21 of an isothermal liquid spray from the measurements of Hestroni and Sokolov (1971). This profile, when compared with the reference isothermal free jet profile, shows that the presence of droplets reduces the spread of the jet. The droplets retain their momentum for longer periods of time than the equivalent particles of fluid, thus the exchange of momentum between a spray jet and its surroundings is less than that of a gas jet and, consequently, the spread and the rate of decay of velocity is reduced by the presence of droplets. In order to examine the separate influence of burning, the measurements of Chigier and Cher-

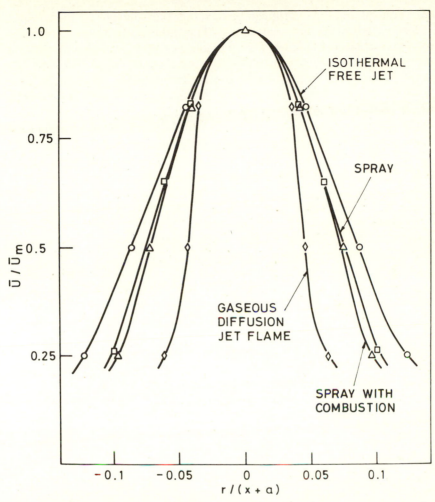

Figure 7.21 Radial velocity distributions for various jet systems. (*Chigier and Roett, 1972.*)

vinsky (1967) are shown in Fig. 7.21. For a turbulent diffusion flame with a cold core, the flame is concentrated in an annulus surrounding the cold core and the net result is that the spread and decay of the jet are reduced when compared with the equivalent nonburning jet (Beér and Chigier, 1972). The velocity spread coefficients, as determined from each one of these profiles, are given in Table 7.1.

Table 7.1 Velocity spread coefficient

	k_u
Isothermal free jet	92
Isothermal spray (Hestroni and Sokolov, 1971)	130
Spray with combustion (Chigier and Roett, 1972)	130
Gaseous diffusion flame (Chigier and Chervinsky, 1967)	360

From Table 7.1 and Fig. 7.21 it can be seen that the value of k_u in the burning spray is close to that of the isothermal spray. This indicates that, in the region of the spray where Chigier and Roett (1972) made measurements, the effect of the presence of droplets was dominant and more significant than the presence of flame. On the basis of the measurements of Chigier and Roett (1972), a physical model has been proposed and is shown schematically in Fig. 7.22. The twin-fluid atomized jet spray flame is seen to consist of a central core region with a high concentration of droplets and high droplet velocities. In this region no significant

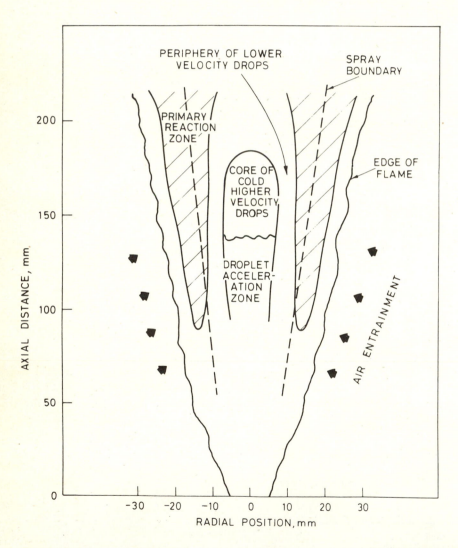

Figure 7.22 Schematic diagram of physical model of twin-fluid atomized jet spray flame. (*Chigier and McCreath, 1974.*)

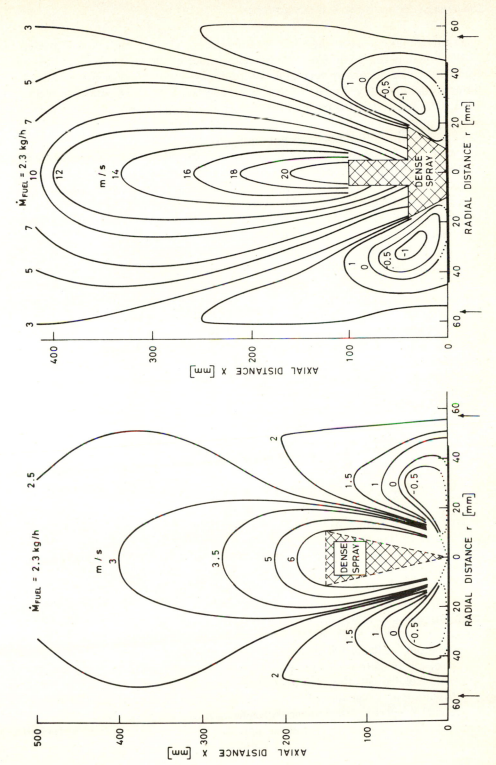

Figure 7.23 Isovelocity lines in (*a*) isothermal spray, and (*b*) spray flame as measured by laser anemometer with frequency shift for reverse flow measurement. (*Styles and Chigier, 1977.*)

reaction can take place as the mixture ratios are too rich and the quenching effect of the liquid is too great. Combustion takes place at the outer periphery of the spray, where, as a consequence of air entrainment, mixture ratios are within the limits of flammability. The flame acts as an effective boundary, confining fuel on the inside and restricting oxygen to the outside. Droplet diameters were seen to reduce rapidly on approaching the flame so that all droplets could be considered to vaporize within the spray and not enter into the flame.

A laser Doppler anemometer has been used for velocity measurements in an airblast spray flame. A comparison of velocity distributions measured in non-burning and burning sprays as measured by laser anemometer by Chigier and Styles (1975) is shown in Fig. 7.23. These results show the significant increase in velocities in sprays as a direct result of expansion due to combustion. Because of the wide spectrum of particle sizes with maximum droplet diameters up to 400 μm, some special problems arose in the use of the laser anemometer. For general applications in laser anemometry, the flow is seeded with particles of the order of 1 μm and the light scattered from these particles is used in order to determine the velocity of the fluid surrounding the particles. In order that the particles' movements will be representative of the fluid flow movements, the drag-to-momentum ratio requires to be large so that there will be no relative slip between particle and fluid. In the burning spray system differential velocities between particles and gas can be large and the majority of particles are in a state of acceleration of deceleration. The laser anemometer measures the velocity of all particles so that an average velocity at any one "point" does not differentiate between the variation in particle size and the associated variation in velocity. In order to take measurements which can be meaningful in such a system, it is necessary to simultaneously measure particle size and velocity.

7.8 BURNING OF LIQUID SPRAYS IN GAS-AIR MIXTURES

In many of the studies which have been made on single-droplet burning it has been assumed that the droplet is surrounded by an infinite quantity of air and that the diffusion of heat and mass can be explained in terms of a boundary layer surrounding the droplet. Mizutani and Nakajima (1973) have examined the case of mixing liquid fuel droplets with gaseous fuel and air. These systems can precipitate explosion disasters in industry, particularly in dual fuel-fired systems where both gaseous and liquid fuels are used. In industrial furnaces heavy fuel oil has been added to natural gas in burners as a means of increasing the radiative heat transfer from the flame and, in diesel engines, liquid petroleum gas has been added to diesel oil. In Otto-cycle engines it has been found useful to introduce easily ignitable fuel droplets to act as distributed ignition sources.

Mizutani and Nakajima (1973) have reported on the burning velocities and burning characteristics of propane–kerosine droplet–air systems as determined by using an inverted-cone-flame-burner apparatus. An air atomizer was used for atomizing the kerosine and the resultant droplets were added to a stream of propane-air mixture. Size distributions and average diameter of droplets were determined by collection of droplets on magnesium-oxide-coated glass slides. The burning velocity was determined from the local time-average flow velocity and the angle of the flame front. Local flow velocities and intensity of turbulence were measured by a hot-wire anemometer in the absence of flame and without injected kerosine. They found that addition of a small amount of kerosine droplets to a lean propane-air mixture increased the burning velocity for a fixed fuel/air ratio and also extended the region of stable burning toward leaner mixtures. They noticed that there was an optimum value of added kerosine droplets and, if kerosine droplets are added so as to exceed this optimal value, the burning velocity fell below that for the propane-air mixture. The effect of droplet addition was found to be greater for lower flow velocity or for weaker intensity of turbulence. The combustion-promoting effects of kerosine droplets were less prominent for higher flow velocities and for larger mean diameters of droplets. By making a separate set of measurements with kerosine mist, the combustion-promoting effects were observed to be directly due to the presence of the liquid droplets and not caused by adding a chemical substance of different thermal or chemical properties. Mizutani and Nakajima (1973) concluded that a small amount of added kerosine droplets results in acceleration of the combustion process for the following reasons:

1. Droplets interact with the originally smooth flame surface to make it wrinkled and to expand its surface area, resulting in an increased burning velocity.
2. Burning droplets act as high-temperature heat sources which accelerate the local burning velocities or act as stabilizers for the flame at extremely lean fuel/air ratios.
3. In the region surrounding evaporating droplets, regions of optimal fuel/air ratio are formed, thereby leading to increase in burning velocities.
4. Turbulence is generated due to local thermal expansion of gas associated with randomly located burning droplets, which lead to increase in burning velocities as a result of higher rates of diffusion.

Mizutani and Nakajima (1973) demonstrated that the contribution of radiative heat transfer from the flame and burned products to the kerosine droplets upstream of the flame was of little importance because of the low absorptivity of the droplets. They estimated that the radiative heat transfer would increase the temperature of a 60-μm kerosine droplet by only about 4 K above that of the surrounding air when the flame temperature is 1500 K and the temperature of air surrounding the droplet is 300 K.

Mizutani and Nakajima (1973) also studied burner flames in a cylindrical combustion vessel with a centrally located spark. This study has particular application to lean mixture operation of Otto-cycle engines. Flame speeds were measured using high-speed photography. In the cylindrical combustion vessel the conclusions they obtained were as follows:

1. A small amount of kerosine droplets added to a propane-air mixture intensifies the burning process, raises the maximum pressure for a given overall fuel/air ratio, and shortens the time between ignition and maximum pressure. In addition, both the burning and flame propagating velocities are significantly increased by addition of droplets.
2. Whereas a smooth flame sphere of low luminosity is observed in a propane-air mixture, addition of kerosine droplets results in a rough flame sphere with a luminous core region. The zone between the core region and the flame surface remains at low luminosity.
3. The optimal value of kerosine/air mass ratio was 0.0035, at which the burning and propagating velocities reached a maximum.
4. The effects of adding fuel droplets are less significant when the flame is turbulent.
5. There are some differences in the combustion-promoting effects of droplet addition between open burner flames and spherical flames in a vessel, since the acceleration processes of the unburned mixture is different for each case.

7.9 PHYSICAL MODELS OF SPRAY FLAMES

Most practical fuel sprays are burned in combustion chambers which have a wide range of sizes, from less than 200 mm in length in turbojet engines, up to tens of meters in length for heavy power plants. In addition, the flow patterns governing the mixing of spray and air within the combustion chamber have a wide range of complexity and generally there is strong three-dimensionality, swirl, and also reverse flow (to achieve flame stability). Considerable effort is being expended in developing techniques for modeling the burning of fuel sprays in these complex environments, with the ultimate aim of optimizing the design of combustion chambers to achieve efficient burning with minimal pollutant emissions. One problem at present is the lack of comprehensive and accurate data on the flows in combustion chambers, in particular point measurements of droplet sizes and velocities. However, available experimental information indicates broad features of the structures of burning sprays which are of use in deriving models.

Onuma and Ogasawara (1975) made a direct comparison between the structures of a spray combustion flame and a turbulent gas diffusion flame in a vertical cylindrical furnace. For the spray measurements, an air atomizing nozzle with kerosine fuel was used with a secondary air supply. Spatial distributions of droplet concentration and size were measured by inserting a probe containing a magnesium-oxide-coated slide, covered by a shutter. Measurements were also made of temperature, velocity, and gas concentrations, by using a thermocouple, pitot tube, and sampling tubes with gas-phase chromatography. Further measurements were made in a turbulent gas diffusion flame, using the same apparatus and replacing

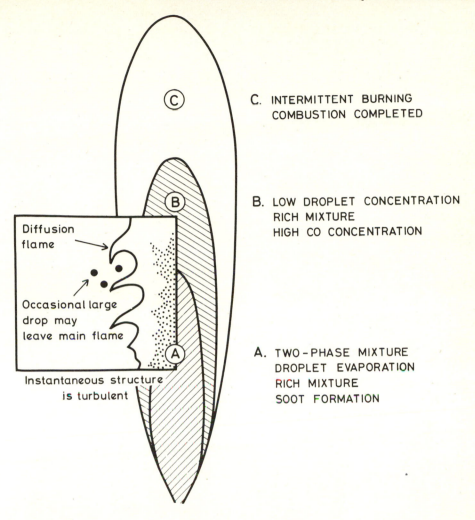

C. INTERMITTENT BURNING
   COMBUSTION COMPLETED

B. LOW DROPLET CONCENTRATION
   RICH MIXTURE
   HIGH CO CONCENTRATION

A. TWO-PHASE MIXTURE
   DROPLET EVAPORATION
   RICH MIXTURE
   SOOT FORMATION

Diffusion
flame

Occasional large
drop may
leave main flame

Instantaneous structure
is turbulent

Figure 7.24 Structure of liquid spray flame.

the liquid kerosine fuel by propane gas. This set of experiments serves as a test of the validity of the hypothesis regarding the structure of spray flames put forward by Chigier and Roett (1972) on the basis of their measurements made in an unconfined air-kerosine spray flame.

Spray flames can be subdivided into three main regions (Fig. 7.24): an initial region consisting of a two-phase mixture in which most of the droplets evaporate and where soot is formed, due to the very rich mixture ratios; an intermediate region where concentration and size of droplets is very low but concentrations of combustible gas are high, with particularly high concentrations of CO; and a final region where intermittent burning takes place as combustion is completed. Figure 7.25 shows the direct comparisons made by Onuma and Ogasawara (1975) be-

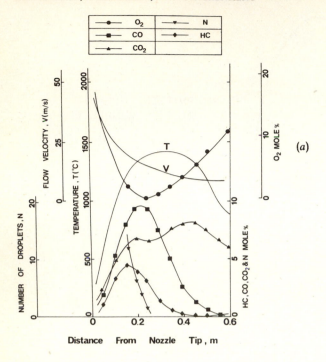

(*a*)

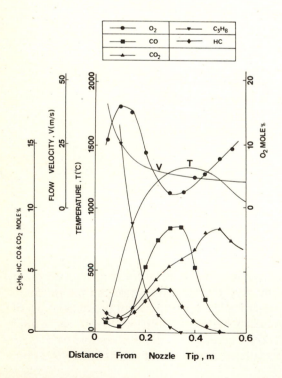

(*b*)

Figure 7.25 Temperature and gas concentration measurements along the axis of (*a*) liquid spray flame, and (*b*) gas diffusion flame. (*Onuma and Ogasawara, 1975.*)

tween temperature, velocity, O_2, CO, CO_2, and HC profiles in diffusion flames and spray flames; it is seen that they are very similar and the differences in the measured profiles were attributed to differences in emissivity and the initial conditions between the two flames. It was concluded that most droplets do not burn individually but that the vapor cloud from the evaporated droplets burned like a diffusion flame in the turbulent state. It was demonstrated that reduction in droplet size could be calculated on the basis of single droplets vaporizing in a hot gas environment.

Tuttle et al. (1977) made measurements in spray flames in a simulated gas turbine combustor and concluded that droplet burning does not take place. The same basic results were obtained when running propane and, subsequently, kerosine through the same atomizer. The following series of separate experiments have been carried out in which measurements were made in burning spray flames: the twin-fluid, air-assist kerosine unconfined spray flame of Chigier and Roett (1972); the hollow-cone pressure-jet unconfined spray flame in the wake of a stabilizer disk of McCreath and Chigier (1973); the confined spray flame of Onuma and Ogasawara (1975); and, finally, the spray flame in an experimental gas turbine combustor of Tuttle et al. (1977). In each of these experiments, it was concluded that there was no evidence of flames surrounding individual droplets; there was no evidence of burning within the dense spray region where temperatures were low and the mixture ratio was rich; the structure of the flames was found to be similar to that of gas diffusion flames. There is, thus, no justification to use results obtained from single-droplet burning tests in order to make predictions in spray flames. Application of the results from burning single drop experiments to spray flames will only be valid when evidence is found of envelope flames around individual droplets. From the tests which have been carried out on spray flames, droplets have been shown to be vaporizing in a low-oxygen, rich-fuel gas environment at temperatures substantially lower than the flame temperature. Heat transfer from the flame front to the drop surface takes place over distances of hundreds or tens of drop diameters by both radiation and turbulent convection. Mass transfer from the drop surface is restricted by the high concentrations of fuel in the gas surrounding the droplet and concentration gradients of fuel vapor can be expected to be much smaller than in the case of single-drop burning. Turbulence levels in the surrounding gas also need to be taken into account when predicting rates of vaporization. Most practical spray flames are thus, to a large extent, mixing-controlled, and the rate of burning is mainly dependent upon the rates of turbulent diffusion of fuel vapor and air to the flame front. In the initial regions of the flame, where mixtures are rich, vaporization of droplets will play no significant role in the rate of combustion.

In the outer regions and toward the end of certain sprays, a small proportion of droplets may be in relative isolation in a predominantly air surrounding, resulting in the formation of a flame around the individual drops. Further, in the burning of heavy fuel oils, droplets having a low volatility persist for long periods and distances in a combustor with the subsequent formation of envelope flames. More evidence is clearly required of the burning mechanisms and structure of spray flames but the balance of evidence reported in the literature indicates that individual droplet burning is generally not significant and that most spray flames burn similarly to gas diffusion flames, i.e., practical spray combustion of lighter fuels is basically a cloud combustion process, although allowance must be made for the possibility of occasional larger droplets leaving the main spray flame.

The characterization of a spray requires information on the initial size distribution, velocity, and direction of flight of droplets as they emerge from the atomizer, and also information on the sizes and trajectories of all droplets during the process of vaporization. Information needs to be provided on the turbulent structure of the air and gas flow patterns and their turbulence characteristics. The heat transfer to the droplets is a function of the temperature differences between droplet and surroundings and the convective and radiative heat transfer coefficients. Evaporation rates will be dependent upon the rate of heat transfer and gradients of vapor pressure in the droplet gas environment. A number of idealized situations are examined in order to focus attention on a particular aspect of the spray characteristics.

Interaction of Droplets with Uniform Airstream

For the case of droplets of different size, injected at a fixed angle to a uniform airstream, the particle trajectories will vary as shown in Fig. 7.26. For particle velocities of the order of 50 m/s and airstream velocities of the order of 20 m/s, droplets in the 300 μm range and above will have straight-line trajectories with no significant deviation by the airstream. Particles of the order of 10 μm will be deflected almost immediately by the airstream. Particles in the range between 10 and 300 μm will be partially deflected. The trajectories of particles can be accurately computed for this simple case using standard drag coefficients for spheres. The Reynolds number is based upon the initial velocity difference between the particle and the airstream.

If the airstream is considered to have a uniform temperature, computations can be made for the trajectory of the particles under conditions of evaporation as a function of the variation of the ambient air temperature. As particle size decreases, drag/momentum ratios decrease, both as a result of reduced mass and, hence, momentum of the droplet, as well as due to the reduced drag coefficients of droplets under conditions of evaporation. Trajectories of vaporizing droplets will deviate, as shown in Fig. 7.26, so that a particle of 300 μm could be deflected from its straight-line trajectory as the particle diameter decreases.

Sprays Injected into Recirculation Zones

Liquid droplets injected into recirculation zones will be deflected by air and gas streams, as shown in Fig. 7.27. Large drops (more than 200 μm) traverse almost undeflected through recirculation zones. Small droplets are partially deflected by the gas streams but, because of the high-temperature flame regions usually found within the recirculation zone, these small droplets are generally vaporized before

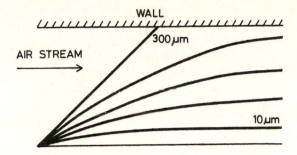

TRAJECTORIES OF DROPLETS WITH DIFFERENT
DIAMETERS INJECTED AT THE SAME ANGLE
INTO A UNIFORM AIR STREAM

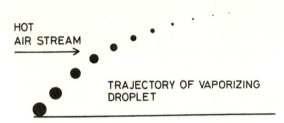

TRAJECTORY OF VAPORIZING
DROPLET

Figure 7.26 Droplet trajectories in airstreams.

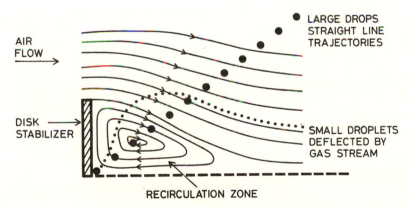

DEFLECTION OF LARGE & SMALL DROPLETS
INJECTED INTO RECIRCULATION ZONES

RECIRCULATED PARTICLES ARE ACCELERATED
OR DECELERATED BY GAS FLOW ACCORDING TO
SIGN OF RELATIVE VELOCITY $V_p - V_g$

Figure 7.27 Deflection of droplets injected into recirculation zone.

they are recirculated by the gas stream. Droplet exit velocities will be of the order of 50 m/s. Within the recirculation zone, gas velocities are lower than those in the main airstream, and negative values are found in the reverse flow regions. The recirculation zone is a region where drag forces will be low, due to low velocities and, hence, droplets will tend to be undeflected from their trajectories until they encounter the higher velocity main airstreams. The smallest droplets may either be accelerated or decelerated by the gas flow, according to the sign of the relative velocity between particle and gas.

Droplet Interaction with Turbulent Flow

Figure 7.28 indicates the type of fluctuation that a turbulent velocity field can cause to a particle. Large drops (more than 200 μm) will be unaffected by gas velocity fluctuations. Submicron particles have drag/momentum ratios sufficiently high that particles will follow the turbulent fluctuations, except for the very high frequencies. Very little information is available concerning the extent to which particle trajectories are influenced by the presence of turbulence in the gas flow but the net effect of velocity fluctuations can be expected to be of the form shown in Fig. 7.28.

Engulfment of Droplets in Coherent Large Gas Eddies

Turbulent flow is structured and many flows are dominated by large gaseous eddies which retain their coherence for considerable lengths of time during their passage through the combustor. The formation of these eddies is due to engulf-

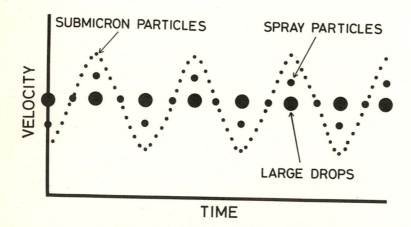

VARIATION OF VELOCITY OF PARTICLES OF
DIFFERENT DIAMETER IN FLUCTUATING
VELOCITY GAS FIELD

Figure 7.28 Variation of velocity of particles of different diameter in fluctuating velocity gas field.

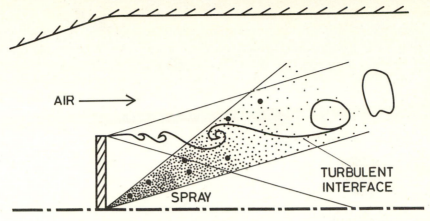

SPRAY INJECTED INTO SHEAR LAYER MIXING ZONE

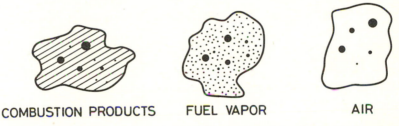

COMBUSTION PRODUCTS   FUEL VAPOR   AIR

DROPLETS ENGULFED IN EDDIES WITH DIFFERENT EQUIVALENCE RATIO

Figure 7.29 Interaction of droplets with airflow.

ment at interfaces in shear layer mixing zones. Gases may be considered to be well mixed within the eddies but mixing across eddy boundaries is slow, giving rise to the overall phenomena of unmixedness. The turbulent flow structure can be envisaged as being composed of eddies of three types: (1) air, (2) fuel vapor, and (3) combustion products, in which the designation is given to the dominant gas within the eddy. Figure 7.29 gives a schematic presentation of droplets engulfed in these types of eddies. Rates of vaporization of droplets within these eddies will be governed by fuel vapor concentrations. Within a fuel vapor eddy, vaporization rates can be expected to be reduced until, in the limit, vapor concentration reaches saturation levels.

Formation of Fuel Vapor Wakes

Particles crossing a uniform airstream with "straight-line" trajectories will leave vapor trails in their wakes, as depicted in Fig. 7.30a. The width of these wakes will be of the same order as the initial diameter of the droplets. When droplet trajec-

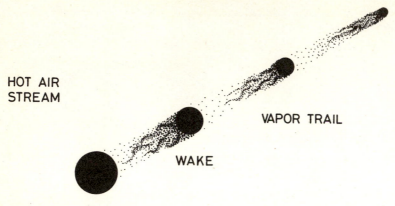

HOT AIR
STREAM

VAPOR TRAIL

WAKE

DROPLET INJECTION

LARGE DROP VAPORIZING IN LOW VELOCITY HOT AIR STREAM

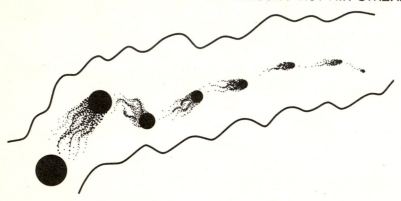

VAPORIZING DROPLET IN TURBULENT FLOW FIELD

Figure 7.30 Vapor trails in droplet wakes.

tories are affected by fluctuating components of the turbulent flow field, fuel vapor wakes can be expected to have the form shown in Fig. 7.30b. This shows a nonuniform vapor trail, centered around the irregular trajectory of the vaporizing droplets.

Effect of Environment on Vaporization Rate of Droplets

Vaporization rates of droplets are principally affected by the temperature and fuel vapor pressure of the gas in the immediate environment of the droplet. These effects are depicted in Fig. 7.31. For a droplet in relative isolation, so that it could be considered to be in an "infinite" air environment, the standard vaporization data for single droplets in hot gas environments can be used. Vaporization rates follow the "d^2 law." The number density and separation distance between droplets is sufficiently high in most particle sprays for the droplets to act as "clouds." Under these conditions, rich mixture ratios are formed within the cloud, and

EFFECT OF ENVIRONMENT ON VAPORIZATION RATE OF DROPLETS

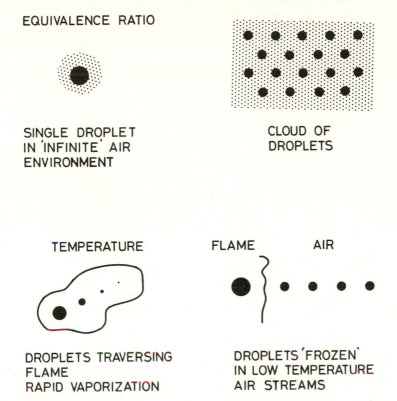

EQUIVALENCE RATIO

SINGLE DROPLET
IN 'INFINITE' AIR
ENVIRONMENT

CLOUD OF
DROPLETS

TEMPERATURE

FLAME   AIR

DROPLETS TRAVERSING
FLAME
RAPID VAPORIZATION

DROPLETS 'FROZEN'
IN LOW TEMPERATURE
AIR STREAMS

Figure 7.31 Effect of environment on vaporization rate of droplets.

vapor pressure concentrations will vary from a maximum, which may reach saturation near the center of the cloud, toward zero at the flame front. Temperatures within the cloud may also have relatively low levels, due to the quenching action of the presence of a high concentration of droplets. Figure 7.31 also depicts the situation of droplets traversing a flame zone, resulting in very rapid vaporization. Droplets entering low-temperature regions will be "frozen" and may continue without significant reduction in size until they impinge on wall surfaces or leave the combustor.

7.10 MATHEMATICAL MODELING OF SPRAY BURNING

A major objective of studies involving the vaporization and burning of fuel droplets, both individually and in clouds, is to aid the development of modeling techniques for the design of practical combustion systems. Simple one-dimensional models (Spalding, 1959) of sprays burning in combustion chambers have been developed which are based on the assumption that droplet vaporization or burning is the controlling rate, with fuel-air mixing and chemical kinetic rates being relatively fast. These techniques have been developed further, using computer modeling, to attempt to predict the vaporization rates and droplet histories for sprays injected into an airstream by including empirical relations for the drag coefficients of vaporizing droplets. Recent advances in computer modeling of spray combustion in practical combustors have been reviewed by Mellor (1976). One approach involves computing the turbulent flow field within the combustor, computing the structure and vaporization of the spray within the combustor, and then iterating for the complex interactions between the mixing and burning of the fuel vapor, the spray structure, and the gas flow. The formulation of models for turbulent mixing is a major problem in calculating combustor performance accurately. There is also a need for a fuller understanding of the regimes of droplet burning for sprays in different conditions. Knowledge of the local environments of droplets as they vaporize in sprays and a fuller and more accurate understanding of droplet characteristics as functions of their environment is required.

Idealized Spray Flame

On the basis of the measurements that have been made in spray flames, we can propose an idealized model of droplet vaporization in spray flames, as shown in Fig. 7.32. We use the hypothesis that spray flames and gas diffusion flames are similar and that both temperature and oxygen concentrations are low within the spray. In the idealized model, the flame acts as an interface, totally separating the inner fuel vapor from the outer air. All the fuel vapor originates from vaporization at droplet surfaces. All burning occurs as a consequence of fuel vapor diffusing outward and air diffusing inward to the flame front. This interface is convoluted by the turbulence so that, in the time mean, droplets and flame can occur at any one position but never at the same time. The droplet velocity can either be greater or smaller than the gas (fuel vapor) velocity. These velocities may typically have values of the order of 30 m/s with velocity differentials of the order of 5 m/s. The surrounding air velocity will vary from zero for stagnant air surroundings to velocities of the order of 100 m/s and thus higher than droplet velocities. The flame temperature may be assumed to be 1500 K, with both the droplet temperature and the ambient air temperature taken to be 300 K.

All droplets are injected into the spray from the atomizer and vaporization of all these droplets is completed within the flame volume. The major portion, say 99 percent, of all droplets is contained within the spray boundary so that there is a zone of fuel vapor, without droplets, between the spray boundary and the flame. Droplet sizes will vary from 1 to 500 μm, but particular attention is focused on the largest droplets. No interaction takes place between droplets and the separation between droplets is sufficiently large that each droplet can be considered in isolation. A spherical boundary layer, with a diameter of approximately twice the droplet diameter, envelops the droplet. We ascribe bulk gas temperatures,

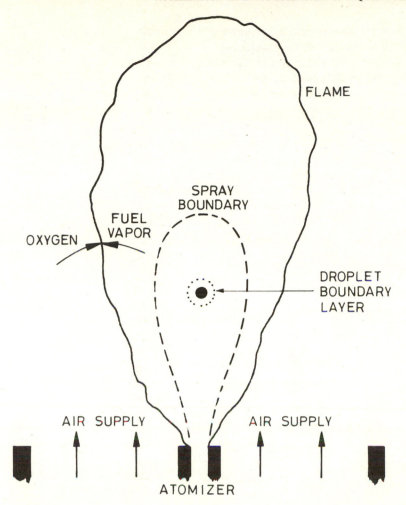

Figure 7.32 Vaporization of a typical droplet in an idealized spray flame.

velocities, and concentrations to the gas outside the boundary layer, while restricting the significant gradients of temperature, velocity, and concentration to within the boundary layer.

We subdivide the spray into an initial "cool" zone, where temperatures are maintained at low levels, due to the strong quenching action of the droplets, followed by a "hot" zone, where the temperature rises, due to turbulent convective transfer of heated products from the flame. Within the cool zone, heat transfer is restricted to radiation from the flame front to the droplet surface. In the hot zone, heat transfer takes place both by radiation from the flame front and by turbulent convection. Predictions of the rate of vaporization of droplets within the spray can be made by using the following equations:

The reduction in droplet diameter due to vaporization follows the "d^2 law"

$$-d(d^2)/dt = \lambda \tag{7.55}$$

where λ is the evaporation constant for forced convection.

The evaporation constant in a stagnant atmosphere λ_0 is given by

$$\lambda_0 = (8k/\rho_l c_p) \ln (1 + B) \tag{7.56}$$

where k is the thermal conductivity of the gas, c_p is the specific heat of the gas, ρ_l is the density of the fuel and

$$B = \text{the transfer number} = \frac{1}{L}\left[\bar{C}_p(T_\infty - T_l) + \frac{QY_{O\infty}}{i}\right]$$

$L = $ latent heat of vaporization per unit mass of fuel

$T_\infty = $ temperature of gas surrounding the droplet

$T_l = $ temperature of drop surface

$Q = $ heat of reaction

$Y_{O\infty} = $ mass fraction of oxidant in the surrounding atmosphere

$i = $ stoichiometric mixture ratio

The evaporation constant in forced convection λ is

$$\lambda = \lambda_0(1 + 0.276 \, \text{Re}^{1/2} \, \text{Sc}^{1/3}) \tag{7.57}$$

where Re is the Reynolds number and Sc is the Schmidt number.

Motion of droplet

$$m(du_a/dt) = F - mg \tag{7.58}$$

where m is droplet mass, u_a is droplet velocity, t is elapsed time, F is the drag force, and g is the gravitational constant.

Drag force

$$F = C_d\tfrac{1}{2}\rho_g(u_g - u_a)^2 A \tag{7.59}$$

where C_d is the drag coefficient of an evaporating droplet, u_g is the velocity of the gas surrounding the droplet, A is the surface area of the droplet, and ρ_g is the gas density.

The drag coefficient of a spherical droplet is

$$C_d = (24/\text{Re})(1 + 0.15 \, \text{Re}^{0.687}) \tag{7.60}$$

The drag coefficient for an evaporating droplet $\bar{C}_d$ is

$$\bar{C}_d/C_d = 1/(1 + B) \tag{7.61}$$

The heat transfer coefficient is obtained from the Nusselt number, which is given by the relationship

$$\text{Nu} = 2 + 0.6 \, \text{Pr}^{1/3} \, \text{Re}^{1/2} \tag{7.62}$$

where Pr is the Prandtl number.

Onuma and Ogasawara (1975) have used the above equations to calculate the reduction in diameter of droplets due to vaporization and also changes in the velocity of the droplets in a spray flame. In their calculations they have used the measured values of local mean gas velocity and temperature and the measured initial size distribution of droplets. They assume that the initial velocity of droplets was equal to the discharge velocity of kerosine from the atomizer. For the case of a total flame length of 0.53 m they found that evaporation of the spray was completed by a distance of 0.3 m from the atomizer. They also determined the distances to completion of evaporation of droplets with sizes from 15 to 155 μm.

The rate of vaporization of droplets is initially low because of the low temperature and high fuel-vapor pressure of the gas in the cool zone. This rate of vaporization increases in the hot zone so that, for droplets up to 150 μm in diameter, vaporization is just completed at the tip of the spray. For droplets greater than 200 μm, the possibility exists for isolated droplets to pass through the spray boundary and the flame. When these liquid particles come in contact with cooled chamber walls, deposition and coke formation arise. The rate of vaporization of droplets may, thus, play only a minor role in the combustion process where the rate determining step is the rate of turbulent diffusion of vapor and air to the flame front.

Before this idealized model of spray combustion can be applied to a practical combustor, it is necessary to verify that there is no evidence of envelope flames around individual droplets and that the flame front surrounds the spray. The flame front may not act as a complete separation interface between fuel vapor and air so that air may penetrate into the spray. When air-assist atomizers are used, a proportion of the total air is introduced directly into the spray and this will alter the flame structure. Given the choice between making predictions on the basis of single-droplet burning and that of the idealized spray flame model, the experimental studies which have been made on spray flames suggest that the above model is a good approximation to the practical conditions.

Spray Evaporation in Recirculating Flow

Boyson and Swithenbank (1978) have made predictions of sprays evaporating in recirculating flows, using a two-dimensional (axisymmetric) numerical computational method. The elliptic flow equations are evaluated using the $k\varepsilon$ model of turbulence. The simultaneous differential equations of droplet dynamics are evaluated in each cell of the flow field for each size group of the spray, using short time steps.

In order to make quantitative predictions of combustor characteristics, such as stability, efficiency, and pollutant production, calculations are required to be made of trajectories of droplets of various sizes and the distribution of fuel evaporated from these droplets. The droplets evaporate along their trajectories in the presence of forced convective motion of the gas, leaving behind pockets of fuel vapor, which later mix with the oxidizer and burn. Small droplets evaporate rapidly but larger droplets may pass through the flame front and burn as isolated droplets in the airstream.

The overall rate of evaporation is very strongly influenced by the relative velocity between droplet and gas, so that knowledge of the separate velocities of droplets and gas is required for prediction of evaporation rates. In practical combustors, the region in the vicinity of the fuel injector is of particular concern, since this is the region where soot may be formed from locally rich mixtures. Magnussen and Hertager (1977) have shown that the mechanism of soot formation can be described in two stages: the first stage representing the formation of radical nuclei and the second stage representing soot particle formation from these nuclei.

Boyson and Swithenbank (1978) have made predictions of the rate of evaporation of a fuel spray by first determining the aerodynamic flow field and then proceeding to calculate the trajectories of individual droplets, droplet velocities along these trajectories, the change of size distribution, and the rate of evaporation. They have concluded that, for conditions typical of gas turbine and industrial combustors, the influence of the droplets on the flow field is negligible, except in the initial region of spray breakup. Interaction only becomes significant when droplet velocities are an order of magnitude greater than gas velocities. This condition arises when fuel is injected into a recirculating flow of low velocity. For airblast atomizers, the interaction between droplets and flow field is considered to be negligible. Boyson and Swithenbank (1978) made their calculations for the case of injection of a spray into a recirculation zone in the wake of a circular baffle in a duct.

Prediction of Spray Characteristics

The equation of motion of a droplet in the absence of external forces is given by

$$\frac{du_p}{dt} = -\frac{3}{4}\frac{\rho_g}{\rho_l}\frac{C_D}{D}(u_p - u)|u_p - u| \tag{7.63}$$

where u_p and u are the droplet and gas velocities respectively, t is time, ρ_l and ρ_g are liquid and gas densities respectively, C_D is the drag coefficient, and D is the droplet diameter.

For a two-dimensional system, Eq. (7.63) is written in component form as

$$\frac{du_p}{dt} = -\frac{3}{4}\frac{\mu}{\rho_l D^2}C_D \,\text{Re}\,(u_p - u) \tag{7.64}$$

for the x direction and

$$\frac{dv_p}{dt} = -\frac{3}{4}\frac{\mu}{\rho_l D^2}C_D \,\text{Re}\,(v_p - v) \tag{7.65}$$

for the y direction, where μ is the gas viscosity and Re is the Reynolds number based on relative velocity defined by

$$\text{Re} = D\rho_g |u_p - u|/\mu \tag{7.66}$$

The rate of change of diameter with respect to time under conditions of forced convective evaporation is given by

$$\frac{dD}{dt} = -\frac{C_b}{2D}(1 + 0.23\ \text{Re}^{1/2}) \tag{7.67}$$

where C_b is the evaporation rate constant whose value depends on the type of fuel, as well as the temperature and properties of the surrounding gas. Quasi-steady state analysis of droplet combustion yields

$$C_b = \frac{8\lambda}{\rho_l c_p}\ \ln\left\{1 + \frac{c_p}{L}(T_\infty - T_l)\right\} \tag{7.68}$$

where λ and c_p are thermal conductivity and the specific heat at constant pressure of the surrounding gas, L is the latent heat of evaporation, and T_∞ and T_l are the ambient and drop surface temperatures respectively. Equations (7.65) to (7.67) constitute a set of simultaneous ordinary differential equations, which are supplemented with the auxiliary relations for the evaporation rate constant and the drag coefficient. The components of gas velocity required for the solution of these equations are determined from the aerodynamic predictions with the assumption that the presence of droplets does not change the aerodynamic flow field.

The initial conditions for the spray are normally obtained by measurement of the size distribution for a particular atomizer. The size distribution curve is divided into a number of size groups, each represented by a mean diameter D. The initial mass fraction of each group can then be expressed as

$$\Delta m = \rho_l\frac{\delta D^{\delta - 1}}{\bar{D}^\delta}\ \exp\{-(D/\bar{D})^\delta\}\Delta D \tag{7.69}$$

where Δm is the mass of drops of size $\pm\Delta D/2$ around D per unit volume of injected fuel, $\bar{D}$ is the size constant, and δ is the distribution parameter. Normally, 15 to 20 size intervals are required to represent the spray adequately. For each size range, the fuel calculations are performed at every time increment, according to the following procedure:

1. The relative Reynolds number is calculated using the conditions prevailing at the beginning of the increment, denoted by time t.
2. The drag coefficient C_D is obtained from the formula

$$C_D = \frac{K_1}{\text{Re}} + \frac{K_2}{\text{Re}^2} + K_3 \tag{7.70}$$

3. The drop diameter at time $t + \Delta t$ is calculated from

$$D^2 = D_0^2 - C_b(1 + 0.23\ \text{Re}^{1/2})(t - t_0) \tag{7.71}$$

4. The mass remaining at $t + \Delta t$ is found from

$$\Delta m_{t+\Delta t} = \Delta m_t(D_{t+\Delta t}/D_t)^3 \tag{7.72}$$

5. Values of droplet velocity components are computed.
6. The location of the droplet at time $t + \Delta t$ is computed.
7. The computational steps are repeated until the drop diameter falls to zero or impingement on all boundaries occurs.

From the evaporation histories of individual size groups, mass fractions are calculated and total evaporation time curves are plotted. These calculations yield drop trajectories, drop velocity histories along these trajectories, and size distribution curves at different times.

Computation of Spray Characteristics

Boyson and Swithenbank (1978) have made computations for a spray with a total cone angle of 45° injected into the wake of a baffle. All droplets were assumed to leave the nozzle with an initial velocity of 21 m/s. Calculations were performed for two different size constants of the Rosin-Rammler distribution, 50 and 100 μm, and the distribution parameter was set at 2. The values of the evaporation constant for $T_\infty = 1500$ K and $\lambda/c_p = 7.8 \times 10^{-4}$ g/cm s range from 0.002 to 0.01 cm^2/s, depending on the properties of the fuel. Trajectories of individual droplets of size between 50 and 130 μm, together with burnout locations, have been computed. The trajectories of the droplets show only slight deviations from the nominal cone angle in the recirculation region. In the main stream airflow, droplets are deflected according to the drag-to-momentum ratio. Droplets smaller than 50 μm evaporate within the recirculation zone, whereas droplets larger than 110 μm impinge on the wall of the duct. Calculations of rates of evaporation show that evaporation times of the order of 5 ms are required for evaporation of 50 μm droplets under conditions of forced convection. Computations have also been made of the change of distribution function as a function of time. These show an increase in mean size, with time, due to the preferential evaporation of smaller droplets.

Heat Transfer to Droplets

The evaporation lag is defined as the time required for the surface temperature of the droplet to reach the boiling point of the liquid. The time rate of change of droplet temperature is given by

$$\frac{dT}{dt} = \frac{6 \text{ Nu } k}{\rho_l D^2 c_{p,d}} (T_\infty - T) \tag{7.73}$$

where k is the thermal conductivity of the gas, Nu the Nusselt number, $c_{p,d}$ the specific heat of droplets, and T_∞ the temperature of the surrounding medium. The Nusselt number is given by the equation

$$\text{Nu} = 2 + 0.6 \text{ Re}^{0.5} \text{ Pr}^{0.33} \tag{7.74}$$

Equations (7.73) and (7.74) can be integrated to give

$$T = T_\infty - (T - T_0) \exp - (6 \text{ Nu } kt/\rho_l D^2 c_{p,d}) \qquad (7.75)$$

where T_0 is the initial droplet temperature. Calculations were made with $T_\infty = 1500$ K, $T_0 = 320$ K, $k = 10^{-3}$ W/cm K and $c_{p,d} = 3$ Ws/g K. The boiling point of fuel was taken to be 450 K.

Boyson and Swithenbank (1978) came to the following conclusions on the basis of their computations of spray characteristics. Droplets tend to travel along near-straight-line trajectories. Droplets smaller than 50 μm only travel distances of 20 to 30 mm before complete evaporation. Droplets greater than 100 μm may hit combustor walls. When droplets become small enough to be deflected by the gas flow, vaporization is completed soon after deflection. Wet surface spots can be detected on the walls of some gas turbine combustors, showing direct impingement of droplets on the walls.

7.11 SPRAY COMBUSTION IN GAS TURBINE ENGINES

Mellor (1973) has carried out a series of studies in a gas turbine test combustor with airflows up to 2.75 kg/s and pressures up to 15 atm. From the measurement of mass averaged exhaust emissions from the test combustor, using a water-cooled gas sampling probe with a side-mounted thermocouple, Mellor (1973) found that both NO and CO emissions decreased as the differential fuel injection pressure was increased to 15 atm. This demonstrated that the combustion characteristics could not be explained satisfactorily solely on the basis of homogeneous chemical kinetic effects, but rather that some aspect of injector performance was influencing the combustion.

Following Lefebvre (1968a), Mellor examined the parameters that could relate simplex injector performance to combustion characteristics and, in particular, to see how changes in fuel pressure could change flame characteristics. A characteristic time for droplet evaporation is given by $\tau = d_0^2/\lambda$, where d_0 is the initial droplet diameter and λ is a function of fuel and ambient gas properties and Reynolds number. It was shown that, since the mean droplet size d_m is proportional to $\Delta P^{-0.4}$ and the initial droplet relative velocity ΔU is proportional to $\Delta P^{0.5}$, the Reynolds number based on mean droplet size and initial droplet relative velocity is proportional to $\Delta P^{0.1}$. Taking λ to be approximately independent of ΔP, it was concluded that

$$\tau \simeq d_0^2 \simeq d_m^2 \sim \Delta P^{-0.8} \qquad (7.76)$$

On the basis of the above argument, the pressure drop across the atomizer is shown to be almost inversely proportional to the droplet evaporation time.

In the physical model of spray combustion, as postulated by Mellor, he considered three conditions: (1) $\Delta P < 15$ atm, (2) $\Delta P = 15$ atm, and (3) $\Delta P > 15$ atm. For the first case of low ΔP, the evaporation time is long compared to the residence time, resulting in low combustion efficiency, with long yellow flames.

Carbon monoxide emission levels are high, due to incomplete combustion, and the high NO levels are taken to be evidence of envelope flames around individual droplets. At $\Delta P = 15$ atm, Mellor found both a minimum in NO and CO emissions, and he concluded that this corresponds to a transition from droplet diffusion flames to wake flames or pure evaporation. For $\Delta P > 15$ atm, the flame was found to be blue and the flame length remained constant. The combustion phenomena resembled that of a gaseous turbulent diffusion flame. Nitric oxide levels were found to be unaffected by increase in ΔP, which is consistent with the evidence of no increase in flame length. The increase of CO emissions with increasing ΔP is due to the overall equivalence ratio increase with ΔP when the turbulent flame length remains unchanged.

Under conditions of poor atomization, such as those obtained by Azelborn et al. (1973), using an air-assist nozzle operating at low air flows, corresponding to idle conditions, it was concluded that the presence of a yellow flame, coupled with high emission levels of both NO and CO, could be ascribed to droplet burning. In a series of measurements made by Sanders et al. (1972) in a large utility boiler, they found no effect of increasing ΔP from 57.3 to 75.4 atm on NO emissions, and they concluded that droplet burning does not occur in a well atomized spray.

Characteristic Times for Spray Combustion

Mellor (1976) characterized the dominant physical processes in liquid spray combustion by time scales for fuel evaporation, turbulent mixing, and chemical reaction. On the basis of physical models of the flame structure in the wake of a disk with liquid fuel injected into the recirculation zone, the characteristic times are combined to form parameters correlating carbon monoxide and nitrogen exhaust emissions for different fuels over a range of burner operating conditions.

In liquid-fueled flames, three processes must occur before combustion of fuel is complete: (1) vaporization of liquid fuel droplets; (2) mixing of fuel vapor with oxidizer to form combustible mixtures; and (3) combustion of the resulting fuel-air mixture. The characteristic time for droplet evaporation is obtained from the "d^2 law" for droplet evaporation and is given by

$$\tau_{eb} = d_i^2/\lambda \tag{7.77}$$

where d_i is the initial droplet diameter and λ is the evaporation coefficient. The flame structure is dependent upon the magnitude of the fuel droplet lifetime τ_{eb}. When droplet lifetimes are short, liquid fuel is vaporized rapidly and the vapor is mixed within the recirculation zone and subsequently convected toward the shear layer flame zone. If the droplet lifetime is long, significant amounts of fuel penetrate through the shear layer into the free airstream. The flame zone becomes dependent upon the trajectory of droplets as they move through the main airstream. Tuttle et al. (1977) have reported on experiments for sprays of liquid propane with droplets ranging from 30 to 80 μm and Jet-A fuel with droplets between 100 and 200 μm injected into recirculation zones in the wake of a disk. The evaporation coefficient λ was found to be less dependent upon fuel type and ranged from 0.03 to 0.05 cm^2/s.

Exhaust emission measurements were made over a range of operating conditions for both the propane and Jet-A fuels. Calculated values of τ_{eb} for Jet-A fuel were an order of magnitude greater than those for propane, from which it was concluded that Jet-A fuel droplets penetrated into the free airstream. A change in emissions was measured as the fuel spray atomization was altered. As droplet lifetimes were increased, NO_x emissions decreased, while CO emissions increased. Both these effects were attributed to poorer combustion efficiency. The decrease in NO formation, with increasing τ_{eb}, precluded the existence of substantial droplet burning by envelope flames. The experimental and analytical studies of Mellor (1976) and Tuttle et al. (1977) demonstrated the importance of individual droplet lifetimes on the overall spray combustion process. In the experiments, the ambient gas temperature within the fuel vaporization zone was of the order of 1000 K. The average relative velocity between the droplets and gas flow was taken to be of the order of the initial fuel droplet velocity, 50 m/s. Under these conditions, droplet acceleration times were found to be considerably longer than droplet lifetimes. Hence, almost all of the liquid evaporation took place during the initial acceleration process. The calculations of Mellor required estimates to be made of the initial droplet diameter and evaporation coefficient. A controlled experiment, in which size distribution variation was measured during the process of evaporation, would allow a closer correlation to be developed between the characteristic time scales and the emission index for CO and NO_x.

7.12 SPRAYS IN DIESEL ENGINES

Diesel fuel is normally injected into diesel engines under pressures of the order of 10^8 N/m^2. The mean droplet diameter of a typical diesel spray is about 15 μm with maximum diameters of approximately 70 μm. Air temperatures in the combustion chamber approach the critical temperature of fuel so that droplets evaporate rapidly and the major portion of the entire spray is in the gaseous form. At low injection pressures, atomization is less effective than at higher pressures and both liquid ligaments and droplets with diameters greater than 100 μm have been found in diesel engines.

Because of the very high pressures and temperatures under which diesel engines operate, atomization and evaporation characteristics of diesel sprays can be substantially different from those of sprays burning under lower pressures and temperatures. Radcliffe (1955) and Hiroyasu and Kadota (1971) measured droplet sizes and distributions of diesel sprays under low air-temperature conditions. Several attempts have been made to study the specific effects of pressure and temperature on spray characteristics. Photographic observations show that at atmospheric conditions long ligaments of liquid fuel are observed in the central core of the spray. The breakup of the liquid can be seen to take place along the entire length of the spray. The breakup of the spray is not completed for some considerable distance downstream from the nozzle and the number density of droplets is smaller at the periphery of the spray compared with that occurring at a higher gas pressure. The high gas pressures result in higher gas densities, which significantly improve the gas entrainment and the atomization. At the high gas pressures used under normal diesel engine operating conditions, the spray usually breaks up into droplets rapidly, soon after the liquid emerges from the nozzle.

Gas temperatures at the high pressures in diesel engines are typically 200°C higher than the boiling point of diesel fuel and, under these conditions, it has been calculated that droplets would completely evaporate within 10 mm from the nozzle exit. The precise temperature and pressure conditions within diesel sprays have not yet been clearly established and, in the direct-photography studies which have been made, it has so far not been possible to clearly determine the extent to which the fuel spray remains in the liquid phase.

For a number of diesel injection systems, the injection pressure becomes very low toward the end of the injection period and, under these conditions, large droplets are formed. Poor atomization, due to inadequate injection pressure, results in the formation of long liquid ligaments and droplets of the order of 100 μm. This poor, low-pressure, atomization is one of the main factors resulting in the formation of smoke and unburned hydrocarbons in diesel engines. Development studies have shown that smoke levels can be reduced by increasing injection pressures and reducing the size of the injector holes—both of which lead to finer spray atomization.

The use of laser optical techniques in diesel engine research has made very substantial progress in recent years (Dougherty and Belz, 1971, and Osgerby, 1974). By inserting quartz windows in diesel combustion chambers, in order to allow optical access, measurements have been made using the following techniques: high-speed photography, holographic flow visualization, holographic fuel droplet sizing, resonance absorption spectroscopy, and laser velocimetry.

High-speed photography, up to 20,000 frames/s, has been used to follow the injection and combustion processes in diesel combustion chambers. The position, time, and propagation of combustion have been determined and smoke formation in the gas flow has been observed. Recording high-resolution images of small fuel droplets by direct photography has not been very successful, except when measurements are made of a very thin plane of droplets.

EIGHT

FORMATION AND CONTROL OF POLLUTION IN FLAMES

8.1 POLLUTION AND THE ENVIRONMENT

Pollution is defined as the contamination of human environment; all matter emitted from combustion chambers that alters or disturbs the natural equilibrium of the environment must be considered as a pollutant. Damage to the ecosystem by pollutants can be due to changes in the concentration of gas components of air and deposition of particulates and chemically reactive species. Deposition of particulate matter on the skin and inhalation of fine particles which settle in the lungs are health hazards. Deposition on plant and vegetable leaves can cause suffocation of plant life. Deposition of chemically reactive species is a general health hazard and affects the purity of drinking water. Corrosion of metallic surfaces and erosion of buildings are directly caused by deposition and subsequent chemical reaction of reactive species.

In addition to the above-mentioned direct consequences of pollution, indirect effects arise due, for example, to photochemical reaction, leading to the formation of smog. Particulate matter increases fog formation and rainfall. When heat is released in sufficient quantities, thermal pollution leads to changes in the local temperature gradients, causing inversion layers over cities; relatively small changes in temperature of water and air can be harmful to temperature-sensitive life systems. When water from rivers is diverted for cooling of power generation systems, the local increase in water temperature causes a reduction in oxygen concentration in the water, resulting in the killing of fish and other aquatic life. Heat pollution may ultimately limit power generation if the temperature of the biosphere as a whole is increased to any significant extent.

Noise generated by combustion can generally be characterized as one of two types: (1) combustion roar and (2) combustion-driven oscillations or pulsating combustion. Combustion roar has no specific frequency but consists of a broad spectrum of frequencies. Combustion-driven oscillations involve a feedback cycle that converts chemical energy to oscillatory energy, which is clearly distinguishable at specific frequencies. Legislation attaches greater importance to noise at a specific frequency than broad-band noise. In addition to the psychological and physiological damage of noise, pulsations can disrupt the combustion air supply. This may result in smoke and other pollutant formation and lower combustion efficiency. Pulsations can also produce large amplitude flexing of structural elements with consequent early fatigue failure and can locally increase heat transfer, causing a higher rate of material deterioration.

Smells are obnoxious and these very often result from incomplete combustion. Chemical plants and diesel engines are normally associated with distinctive smells; smells are generally included among the list of pollutants associated with combustion. Species that are sensitive to the olfactory glands are often at very low concentrations, fractions of parts per million, and are difficult to detect by instruments. Dihydrogen sulfide and mercaptans are well known substances with distinctive smells.

In large industrial complexes and urban areas, damage due to pollution has reached such high levels that the largest cities in the world have been threatened with suffocation. Under pressure from environmentalist lobbies, legislation has been introduced in the United States and other industrialized countries restricting the quantity of pollutants emitted from engines and industrial plant. The restrictions have been imposed with increasing severity and huge financial resources are being expended on changing engineering design, as well as monitoring and controlling the emission of pollutants. The climate in major cities in England has been radically changed, due to the elimination of domestic coal-fire burning, and by imposition of the Clean Air Act. Fog and smog levels have been reduced, sunshine has been increased, bird and plant life have reappeared, and the quality of life has improved. In the largest Californian cities, where automobile exhaust fumes are the main cause of pollution, strict control of emissions has led to reduction in the rates of increase of pollution levels and the aim is eventually to reduce levels by progressive reduction in emissions from automobiles, trucks, and power plants.

Energy conservation requires achievement of maximum combustion efficiencies. Burning of stoichiometric mixtures with long residence times leads to complete combustion and elimination of emission of CO, HC, and carbon particulates. Use of excess air leads to reduction in the overall thermodynamic and energy efficiency of systems. Burning at stoichiometric mixture ratios occurs close to the maximum theoretically attainable (adiabatic) temperatures. Formation of NO_x is directly related to temperature, so that attempts to reduce emissions of NO_x must involve reduction in flame temperatures, and this leads to reductions in both combustion and overall thermal efficiency. Hence, there is a basic conflict between attempts to increase energy conservation by increasing combustion efficiency and the requirement to restrict emission of NO_x. Removal of pollutants prior to emission by exhaust gas cleaning is also energy expensive. Reduction in consumption of energy can be achieved by making overall energy balances and seeking to improve the global energy efficiency of systems as a whole. Compromises need to be made between the economic necessity to maximize energy efficiency and the environmental need to minimize emission of pollutants. The major concern in pollution control is to prevent damage to human, animal, and plant life.

8.2 PRINCIPAL POLLUTANTS

Emissions from engine exhausts and chimney stacks are mainly in the gaseous phase with small quantities of solid particulate matter held in suspension. The gas constituents are made up of (1) inert gases which have passed through the combustion chamber unchanged, (2) products of combustion, and (3) unburned fuel and/or oxidant. The solid particulate matter arises from the fuel and consists of metallic compounds and other materials which cannot burn and hydrocarbons which were not completely burned in the flame. Under very poor combustion conditions, liquid fuel may be emitted, e.g., when full throttle is used during rapid acceleration.

The five principal classes of pollutant species emitted from combustion sources are particulates, sulfur oxides, carbon monoxide, nitrogen oxides, and organic compounds (largely unburned and partially burned hydrocarbons). In combustion systems, it is not necessary to restrict the formation of carbon monoxide, soot, and other pollutant species, which can be destroyed by chemical reaction within the flame prior to emission from the system. In systems where the radiative component of heat transfer is important, the formation of soot particles is actively promoted in order to increase the luminosity of the flame; these soot particles are subsequently burned. At the temperatures achieved in most flames, the presence of fuel and oxidant leads inevitably to the formation of some pollutants. In high-temperature flames, above 2000 K, atomic species, radicals, and ions are formed, and this can result in high concentrations of pollutants. These high concentration levels are reduced by further chemical reaction as the temperature is reduced, but rapid quenching, with insufficient time for equilibration, leads to the freezing of certain species, resulting in high emission rates. Pollutant formation, destruction, and control is thus intimately connected with the combustion process. The general aims of maximizing combustion efficiency and minimizing pollutants can be conflicting, since combustion efficiencies are maximized at, or close to, stoichiometric mixing conditions, where the highest temperatures are achieved. Carbon monoxide, unburned hydrocarbons, and particulates are thus minimized, but these high temperatures lead to a maximum formation of oxides of nitrogen. Optimization can only be achieved by very careful control of air/fuel ratio and temperature levels throughout the system.

The principal pollutants emitted from a 1000-MW electric power plant are

Table 8.1 Pollutants emitted from a 1000-MW power plant; comparison between gas, oil, and coal firing†

	Annual release, 10^6 kg		
	Gas‡	Oil§	Coal¶
Particulates	0.46	0.73	4.49
Sulfur oxides	0.012	52.66	139.00
Nitrogen oxides	12.08	21.70	20.88
Carbon monoxide	negligible	0.008	0.21
Hydrocarbons	negligible	0.67	0.52

† National Academy of Engineering, 1972.

‡ Based on the assumption that the plant burns 1.9×10^9 m³ gas/year.

§ Assuming that the plant burns 1.57×10^6 tons of oil/year which has 1.6% sulfur and 0.05% ash.

¶ Assuming that the plant burns 2.3×10^6 tons of coal/year with 3.5% sulfur content, of which 15% remains in the ash. The ash content of the coal is assumed to be 9% and the fly ash efficiency 97.5%.

shown in Table 8.1. The annual release of particulates, sulfur oxides, nitrogen oxides, carbon monoxide, and hydrocarbons is compared for plants burning gas, oil, and coal.

8.3 PARTICULATES

The particulate matter emitted from combustion chambers has three possible sources: (1) matter which was not combustible; (2) matter which was capable of being burned but was not burned; (3) matter formed during the process of combustion. Temperature conditions in most combustion chambers are sufficiently high for vaporization of liquids so that, except for conditions of very rich low-temperature burning, all emitted particulate matter will be in the solid phase. Particulate matter can be deposited on surfaces within the combustion chamber or may be emitted with the exhaust gases. Most pollution problems arise from particulates which are sufficiently small that they are held in suspension as they are transported by the exhaust gases. The particulate matter may be clearly visible as smoke or water vapor, but small concentrations of particles can be significant sources of pollution without being clearly visible.

Almost all hydrocarbon fuels contain traces of metals and other solid matter, which do not burn; they pass through the combustion chamber and are generally emitted in the form of metal oxides or salts. Carbon has a melting point of 3813 K, yet solid carbon burns easily in air, due to its low thermal conductivity. The particulate matter that is capable of being burned, and is emitted, consists mainly of unburned hydrocarbons, and these are referred to as "organic pollutants."

Most of the particulate matter emitted from practical combustion devices is solid carbon. Solid carbon particles formed in flames are known as *soot*. Soot can be formed from purely gaseous fuels, but is more commonly formed when liquid fuels are used. The presence of solid carbon plays an important role in the luminosity of flames and, for flames that rely on radiative heat transfer, the formation of soot particles is promoted. In automobile engines and other systems with water-cooled surfaces, coking, due to the deposition of hard carbon deposits, can readily arise when fuel is allowed to come into direct contact with the cooled surfaces. In gas turbine combustion chambers, formation of solid particles is not promoted so as to minimize radiative heat transfer from the flame to the combustor-can walls and reduce the requirement for additional air cooling. Particles cause damage to gas turbine blades, either by direct impingement of the particles, causing pitting of the blades, or by deposition on the blades, leading to subsequent corrosion. In industrial furnaces, flames are separated into two main regions: (1) a luminous region, where particles are present directly as original fuel, or have been formed by the combustion process, and (2) a less luminous region, farther downstream, where sufficient oxygen is supplied to allow reaction to be completed before the products leave the combustion chamber.

Soot particles are the main cause of luminosity of flames; the emission from the yellow luminous region of flames is mainly due to the presence of carbonaceous particles. In addition to the carbonaceous particles responsible for the main continuous radiation, some contribution from banded emission is made by large carbon and hydrocarbon molecules in regions close to the reaction zone. The mechanisms for the formation of soot involve the dehydrogenation of organic compounds and polymerization, leading to formation of large carbonaceous particles.

Both formation and destruction of soot are governed by residence time: if this is too short, soot will not form; and unless it is long enough, the soot may not be completely burned, even under favorable conditions of temperature and stoichiometry.

There are distinct differences in the soot-formation mechanisms in diffusion flames and premixed flames. For very small diffusion flames on circular burners, luminous regions appear as the mass flow is increased; further increase in mass flow leads to the formation of soot at the top of the flame. For paraffins, the tendency to smoke increases with molecular weight but the reverse is true for the olefine, diolefine, benzene, and naphthalene series. Primary alcohols form more soot as the molecular weight is increased and secondary alcohols produce more soot than primary alcohols. The C/H ratio is the most important parameter, but molecular structure also plays an important role. For the same C/H ratio, branched-chain paraffins produce more smoke than the corresponding normal isomers. Flame temperature affects soot formation in two ways: (1) higher flame temperatures provide greater temperature gradients, which favor the formation of soot particles, but (2) the higher flame temperatures also lead to faster burning. The low luminosity found in benzene and toluene flames has been explained as being due to the reduction in flame temperature as a result of soot liberation. For methane, carbon is formed in regions of relatively high temperature, where the particles tend to be consumed rapidly without formation of soot. Acetylene decomposes at a lower temperature and carbon particles form in cooler regions, where they are more likely to form soot because of the lower concentration of oxygen-containing substances. Turbulence leads to an increase in the formation of soot but also to a more rapid burning of the soot particles. Control of the mixing distribution in a turbulent flame allows control of soot formation and soot burning.

In premixed flames, the likelihood of soot formation is less than in diffusion flames. Soot particles will be burned, provided the temperature is sufficiently high, the residence time is sufficiently long, and there is sufficient oxygen. When there is insufficient oxygen to react with all the soot to form CO, then soot is found at the end of the flame. The formation and burning of soot is dependent upon local temperature and mixture ratio conditions and, unless these are clearly known or specified, it is not possible to predict the local concentrations of soot particles.

Soot luminosity occurs at mixtures which are much less rich than those required to liberate free carbon under equilibrium conditions. On an air/fuel ratio basis, the tendency to form carbon increases in the following order: aldehydes, ketones, ethers, alcohols, acetylene, light aromatic compounds, olefins, isoparaffins, paraffins, heavier monocyclic aromatic compounds, and then naphthalene derivatives. For methane, ethane, propane, and butane, soot formation has been observed well before the rich limit, although at equilibrium, soot should not be formed, even at the rich limit. The typical structure of a flat fuel-rich premixed flame consists of a blue reaction zone close to the burner surface, followed by a relatively dark zone which, in turn, is followed by a thick yellow carbon zone. The dark zone is composed of acetylene, methane, and traces of other hydrocarbons as well as water, hydrogen, and carbon monoxide.

The effect of increasing pressure is to increase soot formation. Smith (1940) found, for ethylene, that above atmospheric pressure, there was a marked increase in soot and decrease in the C_2 and CH radiation. Gaydon and Wolfhard (1970) concluded that, for premixed flames, the threshold for the first appearance of soot does not vary very much with pressure. However, once carbon is set free, the proportion of fuel so converted increases with increasing pressure. It has also been demonstrated by experiment that, for diffusion flames, there is a decrease in soot formation with decreasing pressure and that soot becomes more dense with increasing pressure, as in premixed flames.

The change of flame temperature has a complex effect on soot formation. Higher temperatures can lead to suppression of soot formation, whereas, for very rich flames, increases in the amount of soot deposit have been found. Preheating the gases of a premixed flame results in a shift in the threshold for soot formation to slightly richer mixtures; increasing the flow rate in premixed ethylene-air flames leads to an increase in flame temperature and the shifting of the threshold value to a higher carbon/oxygen ratio. Near the threshold, higher temperature decreases soot formation because it has a greater effect on the competing oxidation process than on the soot-formation process. In very rich mixtures, however, the competing oxidation is less important and the high temperature accelerates the dominant pyrolysis reactions.

Chemical Kinetics of Formation and Consumption of Soot

The chemical kinetics of the formation and consumption of soot in combustion processes is highly complex. There is no complete understanding of elementary soot-forming and soot-burning chemistry. Important advances have been made in understanding the mechanisms of decomposition of hydrocarbons, the nature of intermediates in sooting flames, the rates of growth of soot particles, the influences of additives on electric fields upon sooting, and the electric charges carried by sooting particles.

The formation of soot particles in flames is due to the thermal decomposition of the hydrocarbons. At high temperatures, these decompose largely into carbon and methane. Some pyrolysis processes occur in the preheating zones of flames. At high temperatures, pyrolysis occurs in a very short time.

Small hydrocarbon particles or organic molecules act as nuclei for the formation of soot, which grows into relatively large particles containing many thousands of atoms and a much higher carbon/hydrogen ratio. Both dehydrogenation and growth or condensation are necessary for soot to form. Dehydrogenation to atomic carbon or C_2 radicals, followed by condensation to solid carbon, has been found, as well as polymerization to very large hydrocarbon molecules, which then lose hydrogen and form graphite. Both polymerization and oxidation proceed by chain reactions, usually involving free radicals. Soot is formed from products of some of these chain reactions. Acetylene is the most common stable intermediate, observed in rich premixed flames of other hydrocarbons. Acetylene may form soot through the polyacetylene route or by surface decomposition on existing soot nuclei. Polymerization may also occur after formation of aromatics and polycyclic hydrocarbons.

During the burning of oil sprays, large droplets may crack to carbon, sometimes forming cenospheres. Condensation, in the form of mist formation, and graphitization have also been observed. From the observation that carbon forms on solid surfaces when a hydrocarbon is passed through a hot tube, it has been deduced that, at high temperatures, soot particles act as nuclei and result in growth. Small particles can also coagulate to form larger ones, and electron micrographs show the linking of particles together in clusters or chains. C_2 radicals, or carbon atoms, may also be important as nuclei and lead to liberation of carbon from a gas at relatively low temperature. Oxygen, in sufficient quantity, suppresses soot formation and carbon luminosity. In excess, oxygen may reduce soot formation by interfering with the polymerization process, or lead to direct burning of soot particles as they are formed. OH radicals are largely responsible for removing soot; they also play an important role in the dehydrogenation steps leading to soot formation. There are at least five major routes for the formation and growth processes of solid carbon in flames. For acetylene itself, the polyacetylene route is important. For diffusion flames of other fuels, polymerization appears, followed either by condensation to droplets or ring-closure to aromatics, leading to graphitization. In premixed flames, competition with oxidation may prevent the buildup of large polymers, and then growth of nuclei may occur.

Further understanding of the mechanisms of sooting has come from experimental studies using the following techniques.

Mass spectrometry sampling of hydrocarbon species in sooting flames (Homann and Wagner, 1967) has allowed comparison of mass spectra of species evaporated from soot, early and late in the soot-forming region of a flame. Early soots contain a vast array of compounds, including aromatic molecules or radicals, many with side chains. These compounds are intermediates which are not present in late soots. Species evaporated from late soots contain fused-ring aromatics having molecular weights as high as 646. These species are by-products of the process that leads predominantly to formation of a graphitic structure.

Electron microscopy reveals that soot particles contain graphitic layer planes in an onion-like structure, but with distortions and numerous dislocations. The high resolution of electron microscopy also provides information on the clustering of soot particles and on the interactions of growing particles with various gaseous species (Ban and Hess, 1969).

Multireflection spectroscopy allows measurement of the concentrations of the radicals C_2, C_3, and CH in fuel-rich C_2H_2–O_2 flames (Jessen and Gaydon, 1969). This experimental evidence suggests that C_3 is formed from soot and that C_2 serves as a starting point for the development of species that ultimately become soot particles.

Well stirred flame reactors have been used to observe C/O thresholds for sooting. Critical values were found to be slightly different from those found in premixed flames (Wright, 1969). *Isotopically labeled ethanol* (^{14}C) has been used in a diffusion flame to establish that the nonhydroxylated carbon atom in the molecule contributed twice as much carbon to the soot as the hydroxylated one (Lieb and Roblee, 1970).

Electric Fields

Weinberg (1968) has shown the strong effects of electric fields in controlling soot deposition and in modifying the size of particles. Electrical effects have also been shown to play some role in nucleation, growth, and coagulation in flames.

Electric fields have been used to show that carbon particles carry one unit of charge (Mayo and Weinberg, 1970). The study indicated that carbon growth occurs both on neutral and on charged species; some particles acquire a charge after substantial growth has occurred. Natural flame ions and certain ionizable additives, such as cesium, can serve as nuclei for carbon growth. Soot can, however, be formed under purely pyrolytic conditions, in the absence of flame generated ions.

Additives

Additives have, in some cases, produced striking changes to soot formation in flames. The addition of 0.1 percent of SO_3 to a town's gas flame causes it to show strong soot luminosity. 0.2 mole percent of SO_3 added to isobutane-air flames increases soot by 40 percent. This effect is explained by the strong oxidizing action of SO_3. The extremely fast reaction between SO_3 and O leads to the formation of the less reactive species SO_2 and O_2. Thus, a small amount of SO_3 prevents the removal of incipient soot particles by O atoms and, hence, permits rapid growth of soot particles. Both SO_2 and H_2S tend to decrease soot formation in premixed flames, and these additives also lead to decreased soot concentration in diffusion flames. Addition of CO or H_2 to premixed flames results in increased soot formation, and nitrogen dilution will slightly increase soot formation in flames of benzene or kerosine with air.

The majority of substances, when added to premixed flames in small quantities, have little effect, whereas, in diffusion flames, there is a general tendency for inactive diluents to reduce soot formation if added in sufficient quantity. An addition of 45 percent CO_2 can stop soot formation in methane diffusion flames. Soot formation has been found to be reduced in diffusion flames by addition of nitrogen and by recirculating part of the flame gases.

Reduction in sooting tendencies of flames can be achieved by introducing *metallic additives*. Many metals have been found to be effective (Cotton et al., 1971). Alkaline earths (Ba, Sr, Ca) and molybdenum act as catalysts for radical production. Metal atoms can also provide an exchange of charge with flame ions. Barium addition increases the rate of NO formation at the same time that it inhibits sooting—this provides further support for the radical-production mechanism for soot suppression by Ba. Manganese and iron, in contrast, appear to operate by being incorporated into soot particles and catalyzing soot burnout. The efficiency of Ba is approximately independent of overall air/fuel ratio, whereas Mn and Fe function more effectively as the air/fuel ratio moves toward stoichiometric.

Removal of Particulates

Particulates can be removed from flue gas emissions by wet collection devices, electrostatic separators, and a number of other devices. It has been estimated that wider use and further improvement of these devices could lead to reduction in particulates to about one-third of their present level by the year 2000. Most of the present devices remove particles down to 1 μm only. Finer particles remain suspended and are emitted with the gas stream. These fine particles reduce visibility and can lodge in the deep recesses of the lung. Small particles also interact with sulfur dioxide in the air and become more of a health hazard than the SO_2 or particulate pollution when they are separated.

8.4 NITROGEN OXIDES

Oxides of nitrogen are formed during the combustion process, mainly as a result of chemical reactions of atomic oxygen and nitrogen. The oxides of nitrogen are referred to as NO_x and the two major oxides of nitrogen emitted from combustion systems are nitric oxide, NO, and nitrogen dioxide, NO_2.

Effect of NO_x on the Atmosphere and Environment

Nitrogen oxide emissions initiate atmospheric reactions which lead to the production of photochemical smog. Emission of NO_x from supersonic transport in the stratosphere can reduce the ozone levels, leading to increased penetration of ultraviolet radiation with a possibility of increase in the incidence of skin cancer.

Nitrogen oxides have a severe impact on plants. Plants develop acute leaf injury when exposed to concentrations of NO_2 greater than 25 ppm for one hour. Continuous exposure of citrus trees to NO_2 for eight months at levels of 0.25 ppm or less increases the leaf drop, while the yield of tomatoes exposed to NO_2 of 0.25 ppm can be reduced by 22 percent. Peroxyacetyl nitrate, a photochemical smog ingredient, causes acute damage at levels as low as 0.03 ppm.

Nitrogen dioxide is recognized as a toxic gas in industrial environments. If inhaled in high enough concentrations, it produces pulmonary edema after a latent period. Laboratory experiments with animals have shown that concentrations of NO_2, only slightly higher than those occasionally found in polluted air environments, can produce cell changes, some of which resemble those seen in some human lung diseases. Nitric oxide is a very active molecule, capable of forming compounds with hemoglobin, as does carbon monoxide. Other effects of photochemical smog products include damage to fabrics, cracking of rubber, eye irritation, loss of atmospheric visibility, and a related significant increase in automobile accidents. A recent estimate of the cost of chemically inhibiting ozone attack on automobile tyres in Los Angeles alone was $15 million per year.

Nitrogen oxide concentrations in flue gases from large furnaces and boilers vary between 100 and 1500 ppm. World total emissions of nitrogen oxides are of the order of 20×10^6 tons from stationary sources with an equal amount emitted from transport vehicles. Natural sources produce about 500×10^6 tons of NO per year. Nevertheless, the NO_x concentrations in urban atmospheres have been measured to be as high as one hundred times those in rural atmospheres. Emissions of NO_x are compared with other pollutants in Table 8.2.

In large cities such as Los Angeles, the principal pollutant is the NO emitted from automobile exhausts. Concentrations of nitrogen dioxide are generally considerably lower than those of NO but significant concentrations of NO_2 have been measured in the exhaust from gas turbines. Pollution arising from NO leads to physical discomfort, eye smarting, and feelings of suffocation in locations with high smog concentration. In Tokyo, oxygen masks have been issued to police on

Table 8.2 Comparison of natural and manufactured emission of pollutants

	10^6 tons				
	CO	Particulates	SO_x	HC	NO_x
Manufactured					
United States (1970)	147	25	34	35	23
% change (1969–1970)	−4.5	−7.4	0	0	+4.5
Worldwide	304	70	150	90	53
Natural					
Worldwide	3000	1000	220	1000	800

traffic control, and in United States' cities, school classes are cancelled for children when smog levels exceed a threshold value.

Nitric oxide, the major component of NO_x, does not appear to cause damage directly, but only as a result of conversion to NO_2. The mean lifetime of NO in the atmosphere is estimated at about four days, so that large concentrations of NO_2 are not usually found, except in photochemical smog. Smog is formed in conditions where there is a large concentration of automobiles exhausting into an environment with a limited air volume due to the presence of surrounding hills creating a stable atmospheric inversion layer. In the atmosphere, NO_x enters a photolytic cycle, which would normally seek to reach steady-state concentrations of NO, NO_2, and O_3 were it not for the presence of unburned hydrocarbons. The hydrocarbons yield peroxy radicals which provide an additional pathway for converting NO to NO_2 with concomitant accumulation of ozone. The peroxy radicals also react with NO_2 to form peroxyacetyl nitrate (PAN), a plant poison and eye irritant. Aerosol formation is abetted by smog oxidants converting SO_2 to SO_3.

Nitric oxide can be formed in one of the following three ways: (1) at the high temperatures found in flames, N_2 reacts with oxygen to form *thermal NO*; (2) when the fuel has nitrogen-containing compounds the nitrogen is released at comparatively low temperatures to form *fuel NO*; alternatively, if NO is formed by mechanisms other than those for the formation of fuel and thermal NO, it is referred to as *prompt NO*. Prompt NO arises principally from reactions of fuel-derived radicals with N_2 which ultimately lead to NO. In most combustion devices, thermal NO is the dominant source of oxides of nitrogen. Crude oil and coal often contain significant amounts of organic nitrogen compounds and fuel nitrogen can be an important source of NO. Under the relatively low temperature conditions, approximately 1300 K, of the combustion of coal in fluidized beds, fuel NO is the dominant source. Prompt NO is formed in turbulent diffusion gaseous flames where maximum temperature levels may be as low as 1600 K.

Thermal NO

The principal reactions governing the formation of NO from molecular nitrogen during the combustion of fuel-air mixtures are given by the Zeldovich equations:

$$O + N_2 \leftrightarrows NO + N \tag{8.1}$$

$$N + O_2 \leftrightarrows NO + O \tag{8.2}$$

$$N + OH \rightarrow H \tag{8.3}$$

The forward part of reaction (8.1) is rate determining. Reaction becomes important only in near stoichiometric and rich flames that are held at high temperature long enough to produce significant amounts of nitric oxide. The atomic species arise from the decomposition of O_2 and N_2 during the chain reactions in which concentrations of atomic species can reach several times the equilibrium values; this phenomenon is referred to as *atom overshoot*. The rate constant expressions for the Zeldovich equations are given in Table 8.3.

Invoking a steady-state approximation for the N-atom concentration and assuming that the reaction

$$O + OH \rightarrow O_2 + H \tag{8.4}$$

is equilibrated, the NO formation rate is given by

$$\frac{d}{dt}(NO) = 2k_1(O)(N_2)\left(\frac{1 - (NO)^2/K(O_2)(N_2)}{1 + k_{-1}(NO)/[k_2(O_2) + k_3(OH)]}\right) \tag{8.5}$$

where k_1, k_2, and k_3 are the rate constants for the forward reactions (8.1), (8.2), and (8.3) respectively; k_{-1}, k_{-2}, and k_{-3} are the reverse rate constants for reactions (8.1), (8.2), and (8.3) respectively, and $K = (k_1/k_{-1})(k_2/k_{-2}) =$ equilibrium constant for the reaction $N_2 + O_2 \leftrightarrows 2NO$. Nitrogen oxide formation rates are calculated by inserting the rate constants from Table 8.3 and values of local temperature and local concentrations of O_2, N_2, O, and OH in Eq. (8.5).

The NO-formation rate is much slower than the combustion rate and most of the NO is formed after completion of combustion. By designating a *postflame zone* (Fig. 8.1) which is downstream of the main reaction zone, the NO-formation process is decoupled from the combustion process and NO-formation rates can be calculated, by assuming equilibration of the combustion reactions. The calcula-

Table 8.3 Rate constants for NO formation (*Bowman, 1975*)

Reaction	Rate constant $(m^3/mol \cdot s)$	Temperature range (K)
$O + N_2 \rightarrow NO + N$	$7.6 \times 10^7 \exp(-38{,}000/T)$	2000–5000
$N + NO \rightarrow N_2 + O$	1.6×10^7	300–5000
$N + O_2 \rightarrow NO + O$	$6.4 \times 10^3 T \exp(-3150/T)$	300–3000
$O + NO \rightarrow O_2 + N$	$1.5 \times 10^3 T \exp(-19{,}500/T)$	1000–3000
$N + OH \rightarrow NO + H$	3.3×10^7	300–2500

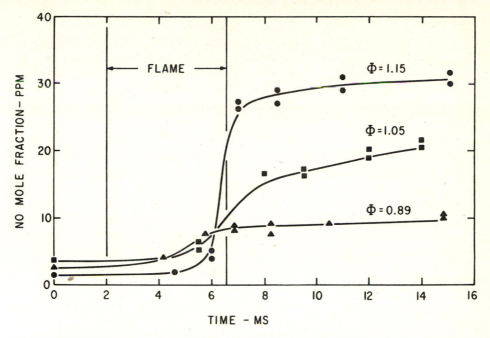

Figure 8.1 Measured nitric oxide profiles in methane-air flames demonstrating the importance of NO formation in the postflame zone. Nitrogen oxide formation increases as the fuel/air equivalence ratio, ϕ, increases. (*Bowman 1975.*)

tion of NO-formation rates is, therefore, greatly simplified by introducing the equilibrium values of temperature and concentrations of O_2, N_2, O, and OH in Eq. (8.5).

For the combustion of lean fuel-air mixtures, $k_3(OH)_{eq} \ll k_2(O_2)_{eq}$, and for conditions where $(NO) \ll (NO)_{eq}$, Eq. (8.5) can be approximated by

$$\frac{d}{dt}(NO) = 2k_1(O)_{eq}(N_2)_{eq} \tag{8.6}$$

The equilibrium O-atom concentration is given by

$$(O)_{eq} = \frac{K_O}{(RT)^{1/2}}(O_2)_{eq}^{1/2} \tag{8.7}$$

where K_O is the equilibrium constant for the reaction $O \leftrightharpoons \frac{1}{2}O_2$. Introducing the value of $K_O = 3.6 \times 10^3 \exp[-31{,}090/T]$ atm$^{1/2}$, used by Westenberg (1971), and the value for k_1 given in Table 8.3, leads to the expression

$$\frac{d}{dt}(NO) = 6 \times 10^{10} T_{eq}^{-1/2} \exp[-69090/T_{eq}](O_2)_{eq}^{1/2}(N_2)_{eq} \qquad \text{m}^3/\text{mol} \cdot \text{s} \tag{8.8}$$

Equation (8.8) shows the strong dependence of the NO-formation rate on temperature. Maximum NO-formation rates can be expected to be found under conditions of high temperature and high O_2 concentrations.

From a survey of experimental studies, Bowman (1975) showed that, in the postcombustion zone of some laboratory flames, measurements of NO are in good agreement with predictions made using the simplified Eq. (8.8). When, however, measurements have been made in the combustion zone, NO-formation rates are found to be significantly larger than predicted by Eq. (8.8). Nitrogen oxide-formation rates increase as the fuel/air equivalence ratio increases and the largest discrepancies between measured NO-formation rates and rates predicted by Eq. (8.8) have been observed in the combustion of fuel-rich hydrocarbon-air mixtures. For the majority of gaseous flames, and liquid fuel flames with negligible concentrations of fuel-bound nitrogen, since combusion takes place under lean or near stoichiometric fuel-air conditions, NO is formed principally in the postflame region after completion of combustion. For predictions of the final emission level, based upon integration of rates of formation throughout the system, steady-state equilibrium approximations can be used. These partial equilibrium models improve in accuracy as temperature or pressure increase, for two reasons: (1) prompt and fuel NO become much less important than thermal NO and (2) radical concentrations are nearer their equilibrium values.

In diffusion flames, NO formation is confined primarily to the fuel-lean side of the reaction zone. Almost all the experimental investigations that have been carried out have shown that the levels of emissions of NO are controlled primarily by the rate of NO formation in the high-temperature paths of the flame.

Fuel NO

Fuel NO is formed during combustion of fuels with chemically bound nitrogen. The nitrogen content of fossil fuels varies considerably. The nitrogen content of distillate fuels reaches the highest levels in the asphaltines fraction, 2.30 wt percent, and an average value for heavy distillates is 1.40 wt percent. The average nitrogen content of crude oil is 0.65 wt percent. The nitrogen content of most coals ranges from one to two percent by weight. Organic nitrogen compounds undergo thermal decomposition in the preheating zone.

The oxidation of low-molecular-weight nitrogen-containing compounds (NH_3, HCN, CN) is rapid, occurring on a time-scale comparable to that of the combustion reaction. Nitrogen oxide concentrations exceed calculated equilibrium values in the combustion zone, and in the postflame zone the NO concentration decreases, relatively slowly for fuel-lean mixtures and more rapidly for fuel-rich mixtures. The amount of fuel-nitrogen converted to NO is referred to as the *NO yield*.

The yield of fuel NO is defined by

$$Y = \frac{\text{mol NO}_{\text{exhaust}}}{\text{mol N}_{\text{combined in fuel}}} \times 100\% \qquad (8.9)$$

The yield is mainly affected by (1) the amount of fuel N present—the higher the fuel N, the lower Y—and (2) the stoichiometry—Y is much higher in lean flames than in rich flames.

High NO yields are obtained for lean and stoichiometric mixtures, whereas rich mixtures provide relatively low yields. Nitrogen oxide yields are only slightly dependent on temperature. Under cool flame conditions, as occur in fluidized-bed combustion of coal, fuel NO is the dominant source and, as temperatures rise, thermal NO concentrations increase until, at high temperatures, thermal NO will generally be the dominant source. Reactions of fuel nitrogen produce reactive radical species which contain nitrogen and these species react rapidly with oxygen-containing species. Nitric oxide is formed from nitrogen-containing free radicals through some of the following mechanisms:

$$NH + O \leftrightarrows N + OH \tag{8.10}$$

$$NH + O \leftrightarrows NO + H \tag{8.11}$$

$$HCN + O \leftrightarrows NCO + H \tag{8.12}$$

$$CN + O_2 \leftrightarrows NCO + O \tag{8.13}$$

$$NCO + O \leftrightarrows NO + CO \tag{8.14}$$

$$NCO + O_2 \leftrightarrows NO + CO + O \tag{8.15}$$

Fenimore (1972) and De Soete (1975) have proposed mechanisms by which the primary fuel nitrogen compounds react to form intermediate nitrogen compounds by pyrolysis or reaction with the fuel. The nitrogen intermediates react via two competitive reaction paths—reaction with oxygen-containing species to form NO or reaction with NO to form N_2. Since NO formation and combustion processes occur on a similar time-scale, the reactions involving nitrogen-containing species (with the exception of N_2) can be assumed to be sufficiently rapid so that the concentrations of these species are in equilibrium relative to one another from early in the reaction. Nitrogen oxide yields are calculated by coupling the pool of partially equilibrated nitrogen species to the combustion process and to a kinetic scheme for formation and removal of NO. The simplest calculation procedure assumes that the combustion reactions are equilibrated at the adiabatic combustion temperature, that NO is included in the pool of partially equilibrated nitrogen species, and that decay of the nitrogen pool, and hence NO, is kinetically controlled by the single reaction $N + NO \rightarrow N_2 + O$.

Flagan et al. (1974) showed that the above-mentioned simple model overestimates NO yields in very fuel-rich and fuel-lean flames due to a breakdown in some aspects of the partial equilibrium assumption. Superequilibrium concentrations of O, OH and H occurring near the combustion zone result in accelerated production of NO by increasing the rate of production of N atoms. These superequilibrium concentrations are referred to as *radical concentration overshoots*. De Soete (1975) has determined kinetic coefficients by experimental observations on NO yields from hydrocarbon flames with fuel-nitrogen additives. De Soete (1975) obtains an expression for the net rate of formation of nitric oxide from fuel nitrogen which contains four terms:

1. Gross formation of NO from fuel N

$$k_a[\text{fuel-N}][O_2] \exp\left(-E_a/RT\right)$$

2. Rate of reaction of NO with fuel nitrogen to form N_2

$$-k_b[\text{fuel-N}][\text{NO}] \exp(-E_b/RT)$$

3 and 4. Formation and removal of NO by reactions not involving fuel nitrogen.

The kinetic coefficients in the above mechanisms depend on the nature of the fuel nitrogen and the mixture strength. Some similarities can be expected between the NO-formation mechanisms on the basis of experimental observations that nitrogen intermediates found during conversion of fuel-nitrogen to NO have also been found during NO formation in fuel-rich hydrocarbon-air flames.

Prompt NO

Several investigators have shown experimentally that substantial concentrations of NO can be formed which cannot be ascribed either to thermal NO or to fuel NO. The term prompt NO was initially used by Fenimore (1971) because he found a rapid and appreciable formation of NO in the flame front. The prompt NO mechanism is the specific mechanism whereby NO is formed from molecular nitrogen upon reaction with hydrocarbon radicals (for example, CH, C_2, C, ...), the existence of which is limited to the early flame stages. Nonequilibrium radical concentrations can be calculated using the partial equilibrium approximation, by assuming local equilibrium of the rapid bimolecular reactions

$$O + H_2 \leftrightharpoons OH + H \qquad (8.16)$$

$$H + O_2 \leftrightharpoons OH + O \qquad (8.17)$$

$$H_2 + OH \leftrightharpoons H + H_2O \qquad (8.18)$$

Concentrations of O and OH are related to the concentration of stable species, which are readily measured. If radical concentrations and temperature during combustion are evaluated using Eqs. (8.16) to (8.18), the Zeldovich equations (8.1) to (8.3) may still be used for the calculation of NO-formation rates. The second possible source of prompt NO is via the CN-group containing intermediates. Fenimore (1971) and Iverach et al. (1973) observed that NO-formation rates in fuel-rich hydrocarbon flames were significantly larger than partial equilibrium values, and this could not be explained by superequilibrium concentrations of O and OH. Reactions involving the hydrocarbon species are

$$CH + N_2 \leftrightharpoons HCN + N \qquad (8.19)$$

$$C + N_2 \leftrightharpoons CN + N \qquad (8.20)$$

Several experiments have shown relatively large concentrations of HCN near the reaction zone, and the rapid decay of HCN is associated with a rapid formation of NO, as shown in Fig. 8.2. De Soete (1975) has suggested that, in the flame front, molecular nitrogen acts as a fuel-nitrogen compound, reacting to form NO via a HCN intermediate. This theory of de Soete links together the NO-formation mechanisms for fuel and prompt NO.

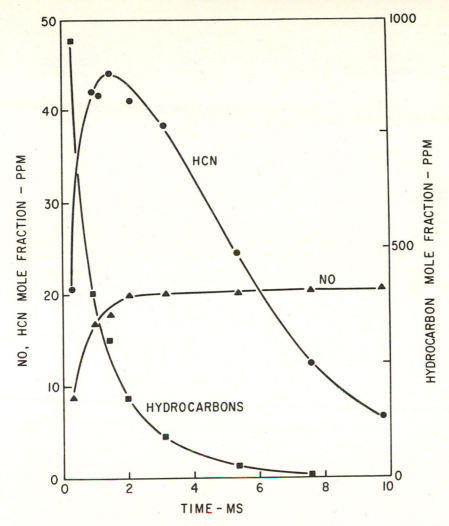

Figure 8.2 Large concentrations of HCN are measured near the flame reaction zone of fuel-rich flames. Rapid formation of NO corresponds with the appearance of HCN. (*Bowman 1975.*)

The products of Eq. (8.19) are presumed to yield NO by a sequence such as:

$$
\begin{array}{l}
\boxed{\text{N}} + \text{O/OH} \longrightarrow \text{NO} + \text{O/H} \\[1mm]
\text{HCN} + \text{OH} \longrightarrow \text{NCO} \\[1mm]
\qquad\qquad\qquad\qquad \downarrow \\[1mm]
\text{NH} + \text{CO} \longleftarrow \text{HNCO} \\[1mm]
\Big\downarrow {\scriptstyle \pm\,\text{H}_i} \\[1mm]
\text{NH}_{i\,\text{pool}} \xrightarrow{\text{OH}} \text{NO} + \text{H}_i
\end{array}
$$

The main characteristics of prompt NO are:

1. Absolute dependence on presence of hydrocarbons (CO and H_2 flames yield no prompt NO).
2. Relative independence of temperature, fuel type, or mixture ratio. Hayhurst and Vince (1980) have shown that, in stoichiometric flames, prompt NO levels are of the order of 50–90 ppm, yielding 1.5–1.9 kg NO_2/tonne fuel. Change of temperature from 1900 to 2350 K did not significantly affect prompt NO levels. Change of equivalence ratio of 0.9 to 2.0 at 1900 K also produced no significant change in prompt NO levels. In leaner flames, prompt NO levels were found to be negligible.
3. Independence of residence time.

In very rich relatively cool flames where superequilibrium NO occurs, allowing sufficient residence time can lead to destruction of virtually all the NO through reactions such as

$$CH + NO \longrightarrow products$$

and

$$NH_i + NO \longrightarrow N_2 + \cdots$$

The third possible cause for augmentation of NO formation is due to temperature fluctuations. In turbulent flames, instantaneous temperature levels exceed the time-average temperature levels and, during these periods, NO formation will be increased above the level predicted on the basis of the time-average temperature. Thompson et al. (1975) examined the effect of temperature fluctuations on NO formation in turbulent diffusion flames. They showed that a fluctuating temperature of $\Delta T = 100$ K can increase the NO yield by 150 percent. The temperature fluctuations occur, due to both variations in the mixture ratio and bulk movement of the flame. These fluctuations affect the time-mean NO-formation rate in particular regions, since the Zeldovich NO formation is strongly and nonlinearly temperature dependent. In turbulent diffusion flames, large eddies of fuel move through airstreams and large eddies of air are entrained in fuel streams. Coherent structures persist for relatively long periods before breaking up into smaller eddies. Conditions favorable for the formation of prompt NO can readily arise at the interface between these eddies and the surrounding medium. Figure 8.3 shows the effect of adding organic nitrogen compounds to the fuel for an oil-fired furnace. Nitrogen oxide concentrations measured in the exhaust flue gases show clear increases according to the percentage of nitrogen added to the fuel.

In gaseous hydrocarbon flames, and in engines burning higher distillate liquid fuels, the principal mechanism for NO formation is thermal NO via the Zeldovich mechanisms. In furnaces fired by heavy fuel oil or pulverized coal, the fuel-NO mechanism is responsible for at least 80 percent of the total emission of NO.

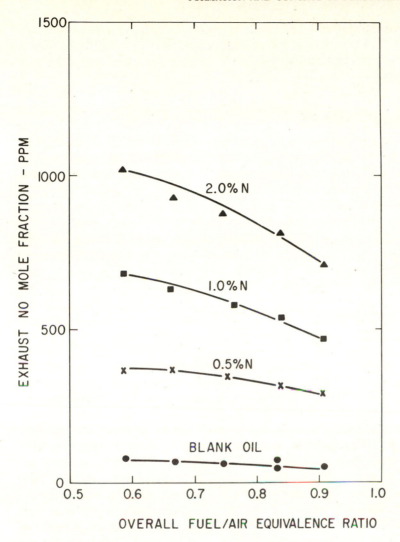

Figure 8.3 Nitrogen oxide concentrations in the flue gas of an oil-fired furnace show the effect of adding organic nitrogen compounds to the fuel. (*Bowman 1975.*)

NO$_2$

Nitrogen dioxide concentrations in exhaust gases are generally negligibly small compared with NO concentrations. Relatively large concentrations of NO$_2$ can be formed in the combustion zone, followed by subsequent conversion of the NO$_2$ back to NO in the postflame region. Nitrogen dioxide is formed from NO upon reaction with HO$_2$ radicals. Rapid mixing of hot and cold regions of the flow in turbulent flames can result in a rapid quenching of the NO$_2$, resulting in relatively large NO_2 concentrations in the cooler regions of the flow, followed by subsequent emission in the exhaust gases. Nitrogen dioxide can readily be formed from NO during sampling, so that NO_2 concentrations measured by probe sampling may be significantly different from NO_2 levels in the flame.

Abatement of NO_x Pollution

Oxides of nitrogen have only been identified as a pollutant in the United States and Japan during the last decade. Legislation controlling emissions of NO_x was first introduced in California and now covers all of the United States. Control by legislation has not yet been introduced in Europe. Because of the lack of control, NO_x has experienced a higher growth rate than any other major pollutant.

In practical combustion systems where thermal NO is the main source of NO_x emissions, the routine steps in emission control, taken to reduce CO, HC, and smoke levels, tend to maximize emission of thermal NO. Further, since SO_x standards have been tightened, the problem of NO_x has become more intractable as a result of SO_x reaction with oxygen atoms in flames.

Once NO is formed it remains relatively stable. Attempts to use heterogeneous catalysis for NO reduction have not succeeded in providing satisfactory reduction. Gas scrubbing has been shown to be impractical as NO is insoluble in water.

The principal means of reducing emissions of thermal NO are by reducing the peak temperature levels in the flame. This is achieved by using off-stoichiometric or staged combustion. Flue gas recirculation, water injection, and damping of temperature oscillations are methods which have been used to reduce peak temperature levels. Reducing residence time at the peak temperature is achieved by altering combustion chamber configurations and burner design. Changes in aerodynamic flow patterns produce important changes in fuel-air mixing; flow patterns can be designed for minimum formation and maximum destruction of NO_x. When temperatures are maintained below 1800 K, thermal NO formation and emissions are at very low levels.

The presence of fuel-bound nitrogen results in the formation of substantial quantities of fuel NO in most practical combustion chambers. Control of fuel NO is thus directed toward destruction in the flame or removal in the exhaust. In fluidized beds, dolomite removes a substantial proportion of NO_x and the inclusion of dolomite in fluidized-bed combustors is generally practiced. The burning of rich hydrocarbon jets in the exhaust of a polluting flame has been shown to lead to an order of magnitude reduction in NO_x emissions but the concept of burning fuel as a means of reducing pollution is not acceptable when so much effort is being made to conserve fuel.

Prompt NO emissions can be reduced by (1) replacement of hydrocarbons by other fuels, (2) very lean combustion, or (3) very rich ($\phi > 3$) combustion in the primary zone, followed by rapid change to lean combustion in secondary and dilution zones.

The practical methods required to minimize prompt NO are at variance with those required to minimize thermal NO. The general requirement to simultaneously minimize all three NO_x-formation rates without producing unacceptable levels of other pollutants still remains a formidable task for which there is currently no simple solution.

Recent evidence is suggesting that solid particles such as char, coal, soot, and ashes can lead to a heterogeneous reduction of NO into N_2. The presence of particles is thought to play an important role in the low levels of emission found in combustors where the dispersed solid phase is present in large concentrations as, for example, in fluidized-bed combustors.

8.5 ORGANIC POLLUTANTS

Organic pollutants, often referred to as unburned hydrocarbons, are the consequence of incomplete combustion of hydrocarbons in the fuel. The inability to complete the combustion of the fuel lowers the combustion efficiency and the primary aim of combustion engineering is to maximize the combustion efficiency by minimizing the emission of hydrocarbons. The most common cause for incomplete burning of fuel is insufficient mixing between the fuel, air, and combustion products. If the air and fuel could be completely mixed on the macroscale and sufficient time given for reaction to take place, after mixing on the microscale, overall burning could take place at stoichiometric fuel/air mixture ratios. Temperatures require to be sufficiently high for reaction to be completed during the period of contact between fuel and air; combustion products provide the heat source for raising the air and fuel mixtures above the ignition temperatures. When insufficient mixing takes place, excess air (above the stoichiometric ratio) is required, and the poorer the mixing the more excess air is required in order to complete combustion. Excess air reduces the combustion efficiency due to the lowering of temperatures and the increase in heat losses arising from heat being convected out of the combustion system via the increased mass flow rate of exhaust gases. The general aim of minimizing pollutants and maximizing combustion efficiency is achieved in practice by increasing the efficiency of mixing between fuel, air, and combustion products, increasing the residence time and minimizing the amount of excess air used.

Some hydrocarbon species, e.g., paraffins, are not considered to provide a serious health hazard, being almost inert at low concentrations, from a physiological point of view. Other hydrocarbons, such as the polynuclear organic compounds, have been shown to cause cancer when deposition is above certain threshold concentrations. The more highly reactive hydrocarbons can be the direct cause of production of smog, whereas other organic compounds may be virtually unreactive.

The relative concentrations of hydrocarbon emissions are greatly influenced by the composition of the fuel. For fuels containing large concentrations of olefins and aromatics, exhaust gases contain relatively high concentrations of reactive hydrocarbons and polynuclear organic compounds (POM). Correlations have been found between the concentration of high molecular weight hydrocarbons in the fuel with high levels of POM emission. The complete understanding of the history of organic compounds in the combustion process requires detailed knowledge of individual pyrolysis and oxidation reactions of individual hydrocarbon species. Hydrocarbons decompose thermally via chain reactions, which are initiated by unimolecular decomposition of the parent hydrocarbons. Synthesis reactions occur during combustion, leading to the formation of complex hydrocarbon molecules in the postcombustion region of flames with simple hydrocarbon fuels. The basic reaction paths for hydrocarbon species in flames are (Bowman, 1975):

1. Oxidative dehydrogenation of an aliphatic hydrocarbon molecule to form ethylene and acetylene
2. Chain lengthening of acetylene to form various unsaturated radicals
3. Dehydrogenation of these radicals to form polyacetylenes
4. Reaction of these radicals via cyclization to form C_6–C_2 aromatic compounds
5. Stepwise synthesis of C_6–C_2 radicals to form polycyclic organic compounds

The steps 1 to 5 are closely related to the soot-forming steps. Soot is formed in the presence of unburned hydrocarbons but it is possible for unburned hydrocarbons to be present without necessarily leading to the formation of soot.

Oxidation mechanisms of hydrocarbon species are divided into two principal regions: low-temperature $(T < 1000 \text{ K})$ and high-temperature regions $(T > 1000 \text{ K})$. Most of the hydrocarbon species found in exhaust gases are formed in the low-temperature regions. Initiation of the reaction chain occurs via

$$RH + O_2 \rightarrow R + HO_2 \tag{8.21}$$

where R is the hydrocarbon radical. The high-temperature oxidation mechanisms differ from those at low temperature because of the decreased importance of several intermediate species, such as peroxides, hydroperoxides, aldehydes, and peracids. The dominant chain centers are H, O, and OH and the dominant chain-branching reactions are of the H_2–O_2 system.

The models proposed for kinetic reactions of fuel can be divided into four categories: (1) detailed kinetics; (2) quasi-global, finite rate; (3) quasi-global, infinite rate; (4) C—H—O equilibrium. With the exception of the simplest hydrocarbons, such as methane, detailed kinetic models are not available and some approximations need to be made. Many of the approximate models have been found to provide closer agreement with the detailed kinetics models as the pressure, temperature, or air/fuel ratios increase.

Before oxides of nitrogen were considered to be a major pollutant, the emission of hydrocarbons from many combustion systems was reduced to very low levels by the use of efficient mixing, and high combustion efficiency was achieved with low excess air. In these systems, temperature levels were high—favoring the formation of oxides of nitrogen. The focus of attention on oxides of nitrogen as a major pollutant led, initially, to the use of off-stoichiometric mixture ratios, which resulted in the lowering of temperatures and, consequently, decrease in emission of NO_x, but with reductions in combustion efficiency. Satisfying the legislative requirement of maintaining both emissions of hydrocarbons and NO_x below statutory limits has only been achieved by careful control of mixing patterns, temperature levels, and residence time distributions throughout the system.

8.6 CARBON MONOXIDE

Carbon monoxide was one of the first combustion products to be recognized as a pollutant. Because of its hazard to health and its capability of causing death, careful monitoring and control of CO emissions have been introduced in most industrial plants. Dangerous levels of CO can be formed in buildings where combustion chambers and exhaust ducts are not securely sealed. The principal source of carbon monoxide is motor transport vehicles. Carbon monoxide concentrations in the atmosphere have been found to be 0.14 ppm in the northern hemisphere and 0.06 ppm in the southern hemisphere. The worldwide emissions of CO are of the order of 200×10^6 tons per year. Levels of concentration of CO in the atmosphere appear to remain constant. CO is removed by natural processes, such as the metabolic conversion of CO to CO_2 and CH_4 by soil microorganisms. When inhaled, carbon monoxide combines with hemoglobin and leads to a decrease in the capacity of the blood to transport oxygen from the lungs to tissues. Carbon monoxide is formed as an intermediate species in the oxidation of carbon-containing fuels. The reaction of CO to CO_2 is almost exclusively due to the elementary reaction

$$CO + OH \leftrightarrows CO_2 + H \tag{8.22}$$

Since this is the sole major mechanism for the conversion of CO to CO_2, it has been concluded (Caretto, 1975) that all carbon initially present in the fuel will form CO. All efforts in controlling emissions of CO are, therefore, concentrated on the completion of oxidation of CO rather than attempting to inhibit its formation. The concentration distribution of gases in premixed hydrocarbon flames is shown in Fig. 4.19. These measurements show that, when CH_4 is burned, CO is formed as an intermediate, which reaches a maximum concentration just prior to the completion of combustion of the CH_4. If sufficient oxygen and residence time is available at flame temperatures, CO concentrations fall to very low levels after reaction, to form CO_2. Maximum CO concentrations in flames are generally larger than the equilibrium values for adiabatic combustion of the reactant mixture. The levels of CO detected in exhaust gases are lower than the maximum values found within the flame but are significantly larger than equilibrium values for the exhaust conditions.

Both the formation and destruction of CO in combustors is kinetically controlled. Carbon monoxide formation is one of the principal reaction paths in the hydrocarbon combustion mechanism, which can be described schematically by

$$RH \rightarrow R \rightarrow RCHO \rightarrow RCO \rightarrow CO \tag{8.23}$$

where R is the hydrocarbon radical. The principal CO formation reaction is due to thermal decomposition of the RCO radical. Bowman (1975), in his examination of CO kinetics, showed that it is possible to use a quasi-global model for CO formation by a one-step reaction in which the hydrocarbon fuel reacts with molecular oxygen to form CO and H_2:

$$C_nH_m + \frac{n}{2}O_2 \rightarrow nCO + \frac{m}{2}H_2 \qquad (8.24)$$

The rate of oxidation of CO to CO_2 is relatively slow compared with the CO-formation rate. In hydrocarbon flames, which generally have relatively large OH concentrations, the oxidation of CO via the reaction

$$CO + O_2 \rightarrow CO_2 + O \qquad (8.25)$$

is very slow and may, in many cases, be neglected. The various elementary steps for the oxidation of CO are known. The steps by which CO is formed are not known, except for simple fuels such as methane.

8.7 SULFUR OXIDES

Sulfur is an impurity found in most forms of coal and oil. During the combustion process, sulfur reacts with oxygen to form, primarily, sulfur dioxide, SO_2, and a small quantity of sulfur trioxide, SO_3. These oxides of sulfur are emitted from the combustor and, unless they are removed, they will be emitted to the atmosphere. It has been estimated that the total emission of sulfur oxides from combustion systems is 93×10^6 tons per year, of which about 70 percent is emitted directly by fossil-fueled electricity generating stations.

Sulfur dioxide is converted into sulfates and sulfuric acid in the air soon after emission, and there is no evidence of global increase of sulfur oxides in the atmosphere. The rate of oxidation of sulfur dioxide to sulfuric acid depends upon the degree of sunlight, the concentrations of moisture, hydrocarbons, catalysts, and other reactive materials in the atmosphere. The sulfuric acid formed in the atmosphere is highly reactive and leads to corrosion of metallic structures, as well as of building materials. The acidity of rainfall has been found to be higher in large industrial cities where pH values of 4 have been reported and ascribed directly to emissions of sulfur oxides.

Sulfur dioxide absorbed by plant life can cause acute injury associated with high concentrations over short intervals, resulting in drying of injured tissues to a dark brown colour. Chronic injury leads to chlorosis, in which the chlorophyll-making mechanism is impeded and leads to a gradual yellowing of the plants.

Rainfall with high acid concentrations has been shown to reduce populations of microorganisms and lead to the leaching of nutrients from plant leaves. Sulfuric acid mist causes spotted injury to leaves at concentrations of 0.1 mg/m^3. Low levels of sulfur dioxide can also cause reductions in growth rates.

Sulfur dioxide at concentrations found in the ambient air is innocuous until combined with other substances. Sulfur dioxide is only a mild respiratory irritant.

Particle size and mass concentration are important parameters in determining the toxicity of sulfur dioxide mist. In the London fog of 1952, concentration of sulfur dioxide rose abruptly to levels of 0.715 mg/m^3. When daily concentrations of sulfur dioxide have exceeded 1.5 mg/m^3, increases in the death rate of 20 percent, or more, over baseline levels have been reported in several major industrial cities.

Chimney stacks up to 300 m high are used in large electricity power stations and these have led to very substantial reductions in local ground-level concentrations of sulfur oxides. Concern is, therefore, focused on emission of sulfur oxides from domestic chimneys and small factories with low chimneys. Sulfur dioxide emitted at high level can produce harmful effects at ground level at distances of several hundred kilometers from its source. Scandinavian countries have claimed notable increases in high-level emissions from the United Kingdom as a direct consequence of increase in the height of chimney stacks. The deposition of sulfur oxides on metallic surfaces causes severe corrosion problems, both inside and outside combustion equipment. Corrosion is usually associated with conversion of SO_2 to SO_3, which combines with water vapor and then condenses as a liquid on cooled surfaces. Sulfur trioxide concentration in flames can greatly exceed thermodynamic equilibrium concentrations, based on the molecular reaction between SO_2 and O_2. Rapid cooling of combustion gases can result in the freezing of SO_3. When residence times are short, as occurs in some boilers, only five percent of SO_2 is oxidized to SO_3. In automobile engines, over 40 percent conversion to SO_3 is possible.

Reduction of Emission of Sulfur Oxides

Four possible means are available for reductions of the emissions of sulfur oxides:

1. Use of low-sulfur fuels, such as natural gas or desulfurized liquid and solid fuels. These low-sulfur fuels have become premium fuels but there is widespread acceptance of the necessity to burn medium and high sulfur-content fuels.
2. Desulfurization of fuels—the sulfur content of solid fuels can be reduced to approximately 1.5 percent by normal washing processes. During the processes of distillation of petroleum fuels sulfur is concentrated in the heavier fuel oils, thus the higher distillates normally have lower sulfur concentrations than the residuals. Liquefaction of solid fuels and gasification of both solid and liquid fuels lead to substantial reductions in sulfur content. This becomes particularly important for very high sulfur-containing fuels (above four percent).
3. Removal of sulfur oxides from combustion gases before they are released to the atmosphere. Both wet and dry processes are available for the removal of sulfur oxides from flue gases. The sulfur is converted to sulfuric acid or some other sulfate and removal efficiencies are of the order of 90 percent.
4. Removal of sulfur during combustion. In the fluidized-bed combustion of coal and heavy fuel oils, alkaline earth oxides such as lime or dolomite are added to the bed. These earth oxides absorb the sulfur, which is retained in the solid residue which is subsequently removed from the combustion chamber.

NINE

ENGINE COMBUSTION CHAMBERS

9.1 GAS TURBINE COMBUSTORS

The various stages in the evolution of the conventional aircraft combustion chamber have been described by Lefebvre (1968). The simplest combustor is one in which liquid fuel is sprayed into an air duct of constant cross-sectional area. This geometry is impractical because the velocity of the air coming directly from the compressor is so high that the pressure loss associated with combustion would be excessive. Introduction of a diffuser reduces the combustor inlet air velocity to about 30 m/s, but this is still too large to allow flame stabilization on the fuel injector. Recirculation zones are generated in the wake of bluff-body stabilizers or baffles or, alternatively, by swirl generation using vaned swirlers. Fuel is sprayed directly into these recirculation zones, which provide intense mixing and high volumetric heat release rates. Temperatures are generally too high for acceptance by the blades of the turbine. The air supply from the compressor is, therefore, divided into one portion, which is introduced directly into the primary zone, and a second portion, which bypasses the primary zone on the outside of a liner or flame tube, and, subsequently, mixes with the products from the primary zone. Combustion efficiencies in excess of 99 percent can readily be achieved at maximum power conditions but combustion efficiency is reduced during engine idle, when both fuel and air flows are reduced.

Gas turbine combustors require to operate over a wide range of conditions. For an aircraft, before takeoff, there is the *idle* condition, where only sufficient power is required just to keep the engine turning, followed by *takeoff*, where maximum power is required by the aircraft in order to takeoff from the ground when the aircraft load is maximum with fuel tanks full. These are the two extreme power requirements since, during cruise and landing, power requirements of the aircraft are very substantially lower than those required during takeoff. Pressure conditions within the combustor vary because of differences in pressure of the intake air at various altitudes and also because pressure varies with the power output of the engine. One of the major considerations in design of gas turbine combustors is to ensure flame stabilization, high combustion efficiencies, and low emission of pollutants over the wide range of fuel and air flows used between idle and full power conditions. The overall fuel/air ratio in the primary and secondary zones varies very considerably with power output, and these have significant effects upon combustion efficiency and emission of pollutants.

Gas turbines are used predominantly in aircraft but there is a steady increase in the number of gas turbines being used for marine applications and for land-based electricity generation. Intensive development has taken place within the aircraft industry but very little change has been made in the basic design of the combustion chamber. A typical design is shown in Fig. 9.1. The chamber is made up of: (1) *a primary zone* where liquid fuel is injected into a region of high turbulence and recirculation. The overall mixture ratio is near to stoichiometric and the aim is to complete the evaporation of the fuel under conditions of maximum heat release rate with combustion efficiency of about 70 percent. Provision is also made for flame stabilization in this zone. (2) *a secondary zone* in which additional air is added in order to complete the burning of the fuel. As mixture ratios are weakened, heat release and maximum temperatures are reduced but conditions are more favorable for the completion of combustion; and (3) *a dilution zone* where air is added and mixed with the burned gases so as to reduce the temperature and provide a temperature profile at the combustion chamber exit which is compatible with the turbine and nozzle guide vanes. Table 9.1 shows the equivalence ratio ϕ, average gas temperatures T, and combustion efficiency η, for two air inlet conditions, (1) 800 K at 25 atm and (2) 300 K at 1.5 atm.

Table 9.1 Characteristics of gas turbines

	Air inlet conditions		Zones		
	Temp. K	Press. atm	Primary	Secondary	Dilution
	800	25			
ϕ			1.0	0.5	0.3
T			2200	1900	1500
η			75	98	100
	300	1.5			
ϕ			0.5	0.25	0.15
T			1100	950	700
η			40	96	97

ϕ = equivalence ratio within a given zone.
T = gas temperature—average in zone, K.
η = combustion efficiency, %.

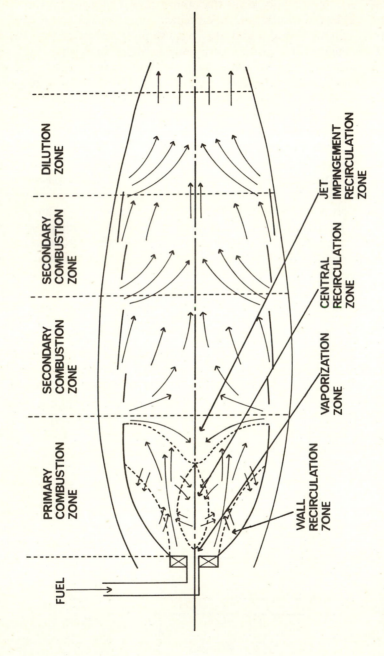

Figure 9.1 Gas turbine combustion chamber. (*Mellor, 1976.*)

DILUTION ZONE

SECONDARY COMBUSTION ZONE

SECONDARY COMBUSTION ZONE

PRIMARY COMBUSTION ZONE

JET IMPINGEMENT RECIRCULATION ZONE

CENTRAL RECIRCULATION ZONE

VAPORIZATION ZONE

WALL RECIRCULATION ZONE

FUEL

Combustion efficiency, as fuel loading is increased and aircraft combustor performance is improved, is illustrated by plotting combustion efficiency as a function of fuel loading where

$$\text{Fuel loading} = \frac{m_f}{VP^2} \tag{9.1}$$

where m_f is the mass flow rate of fuel, V is the volume of the combustion chamber, and P is the pressure. Table 9.1 shows that efficiencies of 100 percent can be achieved in the dilution zone, while efficiencies in the primary zone can be as low as 40 percent for pressures of 1.5 atm.

Combustion in the Primary Zone

Within the primary zone, the total time required for combustion is given by

$$t_c = t_e + t_d + t_r \tag{9.2}$$

where t_e is time for evaporation of droplets, t_d is time for diffusion or mixing, and t_r is time for reaction. Soon after initial evaporation, diffusion and reaction take place simultaneously with evaporation so that the overall combustion time is dictated by the longest of the above times. For any particular system, attempts are made to estimate the limiting process and thereby predict the combustion efficiency. Brzustowski (1965) has shown how physical and chemical control of combustion processes varies as a function of pressure for a single droplet (Fig. 9.2). If we follow the combustion process of a 50-μm droplet (as shown in Fig. 9.2) as a function of pressure, we note that initially, at pressures below 10^{-1} atm, reaction rates and evaporation are slower than diffusion. In the region of 0.5 atm the chemical reaction rates dominate and, as we proceed to pressures of 1, 5, and 12 atm, the combustion process is limited by diffusion, non-steady state effects, and, finally, supercritical phenomena, respectively. Mellor (1976) came to the conclusion that the vaporization of droplets has only a small effect on combustor performance when droplets are less than 80 μm but they could be responsible for the generation of pollutants.

Almost all the attempts which have been made to consider the effects of droplet evaporation have been based on the very extensive work carried out on single-droplet burning, which has been surveyed comprehensively by Williams (1973). Odgers (1975) has used a best-fit-line technique, fitted to experimental data, in order to arrive at the following equation for the rate of reduction in mass of an evaporating droplet

$$-dm/dt = 6.3 \times 10^6 d(T_g - T_f)(H_2/H_1)(1 + 0.38\ \text{Re}^{1/2}) \tag{9.3}$$

where m is the mass of liquid, in grams, in the droplet; d is the diameter of the droplet, in micrometers; T_g, in kelvin, is the gas temperature; T_f, in kelvin, is the film temperature of the droplet at boiling point; H_2 is the heat required, in joules, to evaporate 1 gram of fuel at pressure in the system; and H_1 is the heat required, in joules, to evaporate 1 gram of fuel at 1 atm. In order to use the

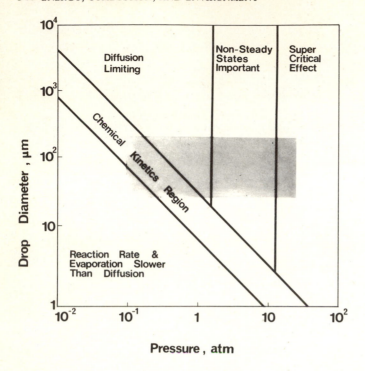

Figure 9.2 Physical and chemical control of combustion processes. The heavy lines denote the boundaries of the regions controlled by the phenomena indicated. The shaded area denotes operating region of combustion chambers. (*Brzustowski, 1965.*)

above equation, some estimate is made of the gas and liquid film temperatures as well as the Reynolds number, which is based upon the relative velocity between droplet and surrounding gas. The inaccuracies which occur in practical prediction procedures are largely associated with the inability to determine the variations in temperature and velocity as the droplet moves through the heated gas. The Reynolds number will be in the range from zero to 25 and a typical dwell period within the primary zone is about 5 ms. Flame temperatures are based upon stoichiometric mixing, of the order of 2330 K. The maximum size of droplet which will just evaporate within the primary zone is then calculated by inserting appropriate values into Eq. (9.3) and obtaining a solution in the integrated form.

Odgers (1975) has calculated the size of droplets which would just evaporate for residence times of 5, 10, and 15 ms in combustion chambers for engines having a range of compression ratios from 5 to 25/1 and expressed the results by the following equation

$$d = [48.5 + 5.6t - (0.67t + 10.05) \log P_c][1 + \mathrm{Re}^{1/2}] \qquad (9.4)$$

where d is the initial droplet diameter, t is time, and P_c is the combustion pressure in atmospheres. Re refers to the initial conditions. The equation takes into account the temperature rise due to compression from ambient pressures and also the change of fuel boiling point with pressure. Evaporation of the fuel (aviation kerosine) is assumed to take place under the influence of a flame with stoichiometric temperatures. Equation (9.4) can only be used for obtaining a rough estimate of the time required to evaporate a droplet injected into the primary zone with an initial velocity and diameter d. Some measurements have been made of temperature and mixture ratios within gas turbine combustion chambers (Mellor, 1976), from which it is now possible to make a better estimate of the properties of the environment within the various zones, but insufficient information is available at the present time to calculate droplet evaporation times with a high degree of confidence.

Mixing and Chemical Reaction

The performance of a combustor is dependent upon the rates of mixing and chemical reaction in each of the three zones of the gas turbine. Mixing is characterized by changes in the concentrations of gases due to turbulence dissipation and convection, which, in turn, are governed by the aerodynamic flow fields. A great deal of effort has been expended on making predictions of combustor performance by computing changes occurring through the system as a function of the initial and boundary conditions. Chemical reaction is generally incorporated within the mixing to take into account both the effects of chemical reaction on the flow and the influence of flow—and hence local air/fuel ratio—on the chemical reaction rate.

Spalding (1971) has developed a complex set of computer programs in order to solve the equations of conservation of energy, mass, momentum, and species. The specific heats of the gases are assumed to be equal and constant and combustion is only presumed to occur in those regions where the temperatures exceeds 400 K and where the fuel and oxygen concentrations are in excess of 0.0001. Free-stream ratios, free-boundary layers, and wall-boundary layers are considered separately, and a series of alternative hypotheses have been used to describe the mixing. The technique has been applied to the prediction of flow, composition, and temperature patterns in premixed turbulent flames. The application to real combustors is still hampered by lack of experimental information required to test the validity of the mixing and kinetic models.

Another approach has been to consider the primary zone as a stirred reactor and the secondary zone as a plug flow reactor. A perfectly stirred reactor is one in which the stirring is so vigorous that the time required for mixing between reactants and products can be considered negligible in comparison with the reaction time. Such vigorous stirring has been achieved in spherical combustors where fuel and oxidant are introduced under high pressure through tiny orifices, and conditions within the combustor have been shown to be close to homogeneous. Within the primary zone, intense recirculation zones are formed as a result of flow through swirlers in the burner and by impinging jets of air through the side walls of the combustor can. Turbulence intensities are high but there is evidence to suggest that conditions are not homogeneous and that there are considerable variations in temperature and concentration within the primary zone. High-speed movies show periodic passage of bursts of flame, indicating the presence of poorly mixed, large-scale eddies. Thus, despite the large amount of effort which has been expended on stirred and plug flow reactor concepts, there is no evidence that the assumptions of homogeneity and perfect mixing in the primary zone, or the absence of mixing in the secondary zone, are valid for gas turbine combustion chambers.

Chemical reaction takes place between a large number of molecular and atomic species, as well as radicals under the high temperature conditions existing in gas turbine combustion chambers. Hammond and Mellor (1971) have drawn up a table of elementary reactions which is claimed to be applicable to any hydrocarbon combustion process using air as oxidant. Fourteen reactions are listed involving the reactants C_aH_b, O_2, CO, OH, H_2, H, O, H_2O, N, and N_2. More complex and comprehensive multistep kinetic systems have been used by other investigators, making predictions of combustion performance and pollutant formation. These kinetic schemes are used in conjunction with a model of flow conditions, and computations are carried out with computers having large memory banks. There is still uncertainty as to the total number of reactions which should be included, as well as the magnitude of the constants in the reaction rate equations. Some of the models of flow conditions have been shown to be not representative of real conditions within combustors but, nevertheless, some models succeed, after adjustment of constants in the equations, to correlate temperature, velocity, and concentration conditions in the exhaust plane of the combustor. Very few of the models have proved to be successful when attempts have been made to predict conditions for which there is no experimental information. The failure of these models can be due to incomplete or inadequate modeling of kinetic, mixing, or droplet evaporation characteristics of the system.

There does appear to be some lack of sensitivity to changes in reaction rate, as can be seen by comparing results of predictions made by quite different reaction schemes. This relative insensitivity to kinetics suggests that the systems are mixing-controlled, in which case the use of a simple kinetic system would be all that is required in order to make predictions.

Secondary Zone

In the secondary zone, air is added in the form of jets through the combustor side walls with the aim of achieving rapid mixing with the products from the primary zone. If there are only a few admission points for air at the beginning of the secondary zone and mixing takes place rapidly, then it is assumed that the mixture moves as a plug flow through the secondary zone. In practice, most combustion chambers have several admission points along the length of the secondary zone and, thus, it is difficult to imagine that the actual flow structure is anywhere near that pictured by plug flow. Several investigators have found it, nevertheless, convenient to use the plug flow concept for making predictions. Combustion is usually completed in the secondary zone and there seems to be some justification in the assumption that, except for the small fractions of one percent of burnout of carbon monoxide and hydrocarbon, no significant amount of combustion takes place in the dilution zone.

The requirement to meet increasingly stringent restrictions on emission of pollutants, coupled with the increasing cost of fuel, is causing many manufacturers of gas turbine chambers to examine the combustion mechanisms and performance in much greater detail. For the majority of systems currently being used, in which there is no premixing, it appears that variations in reaction rate do not lead to any significant changes in combustor performance. When changes have been made to atomizers which are known to vary the spray characteristics, and from the use of vaporizing units in which there is very little atomization, it appears that, provided droplets are less than 100 μm, the presence and evaporation of droplets has little effect upon combustion efficiency. Premixing of the fuel and air, prior to injection into the combustion chamber, has been avoided in practice, mainly due to the safety hazard and risk of flashback. The experiments which have been carried out with premixed combustors have shown that combustion efficiency is greatly superior to that of a spray-type combustor and approaches the performance of spherical combustors. It has also been demonstrated that the flow pattern and pressure loss characteristics of the premixed chamber are very similar to those of conventional chambers. The use of airblast atomizers, resulting in partial premixing, has led to improvements in combustion efficiency as well as reduction in the emission of pollutants. Further improvements in combustion performance will be mainly dependent on improving the mixing processes within the combustor, and the highest efficiencies will be obtained in premixed combustors.

Pollution Emissions from Gas Turbine Engines

The primary pollutants emitted by aircraft gas turbine engines are carbon monoxide, hydrocarbons, aldehydes, smoke particulates, and nitric oxide. Measurements of aircraft engine exhaust compositions from several different turbojet, turbofan, and turboprop engines, reveal striking similarities in their pollutant emission characteristics (Sawyer et al., 1973). When aircraft are cruising at high altitude, the contribution to pollution is not normally considered significant and legislation concerning limitation of emission of pollutants is mainly concerned with emissions below 1 km altitude. It is estimated that 20 to 25 percent of all fuel consumed by U.S. civil aircraft is consumed during operations near air terminals, including takeoff and landing. Sawyer et al. (1973) have plotted aircraft gas turbine exhaust emissions of carbon monoxide, hydrocarbons, aldehydes, smoke-density, and NO_x (expressed as g/kg fuel) as a function of engine power output. At low power settings (idle) CO, HC, and aldehydes emission levels are high due to poor mixing and rapid cooling by an excess of air. At higher power levels, emissions of these species drop to very low levels. Smoke density increases with power level, due to a combination of increased combustor pressure and increased sizes of fuel-rich zones in the combustor. Oxides of nitrogen emissions increase with power and oxides of nitrogen have also been shown to be related directly to the combustor inlet air temperature.

Exhaust Smoke

Exhaust smoke results from the production of finely divided soot particles in fuel-rich regions of the flame. The conditions most favorable for the formation of soot are high temperature and fuel-rich regions. Pressure-jet atomizers produce hollow-cone sprays in recirculation zones which are at high temperatures but with low oxygen concentrations. Local pockets of fuel and fuel vapor can be enveloped in recirculating burned products moving upstream toward the burner, leading to the production of large quantities of soot. The use of airblast atomization results in substantial reductions in soot formation because of the improved atomization as well as the supply of oxygen into the spray.

The physical process of atomization, evaporation, and mixing between fuel and air are more important factors governing the formation of soot than reaction kinetics. In the dense region of the spray, droplets evaporate, forming a rich mixture which initially interacts with hot recirculation products and only subsequently mixes with the main airstream. The rate of soot formation will be governed by the local temperature and mixture ratios. The momentum of fuel droplets is reduced, both by reduction of mass as a result of evaporation, as well as by exchange of momentum with the gas surrounding the droplets due to drag forces. There is some evidence (Sjögren, 1973) to suggest that fuel droplets reach a certain critical velocity, dependent on droplet size, below which they become completely surrounded by attached diffusion flames. Envelope flames around individual droplets provide conditions which are most conducive to the formation of soot and it would be preferable to avoid the formation of envelope flames. Reduction of droplet size by improved atomization leads to more rapid evaporation, and increasing the relative velocity—either by increasing the droplet or gas velocities—leads to more rapid evaporation and the formation of wake flames rather than envelope flames. Reductions in soot formation can also be achieved by improving the rate of mixing between fuel vapor and air. The spray requires to have sufficient momentum to penetrate into the main airstream. Lefebvre (1975) has shown that, if improved atomization is accompanied by lower spray penetration, the smoke output may actually increase as was found by operating duplex and dual-orifice atomizers at high pressures.

The introduction of unmixed liquid fuel and air flows into the primary zones results in the formation of large eddies with mixture ratios considerably richer or leaner than the overall mixture ratio within the primary zone. The size of these eddies is dependent upon the dimensions of the orifices through which air is introduced into the combustor as well as the size and strength of the recirculation zones. Increasing the flow of air into the primary zone lowers the temperature, improves the mixing, and raises the oxygen level with the net result of drastically reducing soot formation and smoke. Since ignition and flame stabilization can be adversely affected by the introduction of additional air to the primary zone, an acceptable compromise must be reached.

Variations in fuel properties can affect smoke emissions in a number of ways. Reductions in viscosity and, to a lesser extent, surface tension lead to reductions in droplet size. Fuels with higher volatility will evaporate more quickly. Increases in the aromatic content, boiling point, and carbon/hydrogen ratio have all been shown to increase the soot-forming propensity of a fuel.

Additives for Smoke Reduction

Exhaust smoke from gas turbine combustors consists of small, graphite-like carbon particles. The visibility of smoke is high when the mean particle diameters are of the order of the wavelengths of visible radiation. Electron micrographs of gas turbine smoke from distillate fuels indicate that smoke particles are irregularly shaped agglomerates with dimensions of the order of 1 μm, composed of approximately spherical particles of the order of 0.01 μm.

There are two competing processes which determine the amount of smoke in the exhaust: (1) the formation of carbon particles in the fuel-rich regions of the primary zone and (2) the combustion of these particles in the fuel-lean secondary zone. The reduction of smoke levels is best achieved by reducing the formation of soot but, as often happens in practice, if soot is allowed to be formed in the primary zone, substantial reductions can be achieved by oxidation of the soot particles prior to emission. Despite the many studies which have been carried out, the detailed kinetics of carbon particulate formation has not been established, but there is ample evidence that the carbon particles originate in fuel-rich regions of the combustion zone, due to local nonhomogeneity of fuel-air mixtures. Experiments suggest that the formation process is either dominated by agglomeration of small particles nucleated on hydrocarbon ions or by a polymerization process. Extensive testing of gas turbine engines has shown that the variables most strongly affecting the formation of solid carbon are operating pressure, fuel composition, fuel/air ratio, and the fuel-air mixing process.

Temperatures in the secondary zone are sufficiently high for significant oxidation of carbon particles resulting in the elimination of the smaller particles and diminishing the size of the initially larger agglomerates. Two methods of controlling exhaust smoke have evolved: (1) combustion chamber modifications, leading either to reduction of soot formation or to increasing the oxidation of soot, and (2) adding small amounts of metallic-organic additives to the fuel.

Pagni et al. (1973) experimentally examined the effects of manganese-based additives on the mass, size distribution, and chemical composition of particulate emissions from gas turbine combustors. Various additives containing soluble metallic compounds have been developed, both for reduction of smoke exhaust as well as for reduction of carbon deposits on the inner walls of the combustion chambers. Additives with a composition of approximately 25 percent manganese by weight have proved to be the most effective with the least adverse effects on long-range engine performance. The metallic compounds can inhibit the agglomeration of small pure carbon particles and they also act as surface catalysts, which increase the rate of oxidation of soot. A number of claims have been made that additives eliminate smoke but these have generally been based upon observation of visible exhaust smoke. Pagni et al. (1973) added a manganese-based fuel additive to jet fuel, JP-4, in a model gas turbine combustor. Smoke particles were collected on the filter of a gas sampling probe introduced into the combustor. Visual examination of the interior of the combustion chamber showed heavy carbon deposition after ten hours of testing when fuel was used without additive, whereas, after twenty hours of testing a blend of fuel and additive, no carbon deposition was seen on the walls of the combustor. Electron micrographs showed that the average particle size for fuel with 0.06 percent by volume of additive was approximately 0.03 μm. A comparison of the size distribution of particles measured with the pure fuel, as compared with that including the manganese additive, showed that the net effect of the additive was to redistribute the mass emitted to small particle sizes where it was no longer visible. They concluded that deposition or agglomeration inhibition occurred in addition to any catalytic effects on the oxidation process. The net result of using the additive was found to alter the emitted size distribution to produce fewer particles larger than 0.2 μm and many more particles, smaller than 0.2 μm. Almost all the manganese added to the fuel was emitted with negligible manganese deposition within the combustor. When small concentrations, less than 0.01 percent by volume, of manganese additive were used, the mass of emitted carbon (g per kg fuel) was found to decrease. However, at concentrations greater than 0.01 percent, the mass of carbon emitted increased substantially with the increase of additive. The addition of very small quantities of additive acts as an inhibitor for agglomeration as well as a catalyst for oxidation, leading to a reduction in total mass emitted. The increase in mass of carbon emitted at higher concentrations of additive is due to the inhibition of carbon deposition on the inner walls of the combustion chamber, resulting in an increase in the total mass of carbon emitted.

There have been many advocates for the use of additives to fuels who have claimed to achieve complete elimination of smoke while, at the same time, achieving 100 percent combustion efficiency. The disappearance of visible smoke was used as a demonstration of the effectiveness of the additives. The careful analysis of investigators such as Pagni has shown that the disappearance of visible smoke is no proof of the disappearance or even the reduction of the quantity of particulate matter emitted from a combustor. Even worse, instead of reducing pollution, additives have, in fact, increased the health hazard for the following reasons: (1) the metal oxides added to the fuel are emitted from the combustor and, at sufficiently high concentrations, oxides such as manganese oxide act as additional pollutants; (2) small carbon particles provide adsorption and nucleation sites, which result in the promotion of smog-producing heterogeneous reactions; (3) the reduction of particle size to approximately 0.03 μm means that particles of carbon and metal oxide exist as invisible dust, which can be inhaled and ingested into the lungs and, hence, contribute toward cancer.

Carbon Monoxide

Carbon monoxide is an intermediate product of hydrocarbon oxidation and appreciable amounts are formed during combustion in the gas turbine combustor. This CO can be burned and not emitted from the combustor, provided that sufficient oxygen is available at a temperature high enough for reaction to be completed during the time of residence between the primary zone and the combustor exit. Carbon monoxide in the exhaust gas is a measure of combustion inefficiency and is caused by insufficient burning in the primary zone, combined with quenching of the CO in airstreams near the walls of the combustor or in the cooler zones downstream. Increases in inlet air temperature and pressure both accelerate burning rates and thereby reduce CO.

In practice, reductions in CO and hydrocarbons are achieved as part of the general attempt to maintain combustion efficiency as high as possible. The following methods have been found to be effective in practical combustors:

1. Improved fuel atomization. This requires elimination of large droplets and this can be most effectively done by using airblast atomizers. At low fuel-flow rates, the air pressures are not sufficient for effective atomization and additional supplies of compressed air can be supplied directly to the air atomizer from a compressor. This system is known as air assistance.
2. Increasing the airflow to the primary zone so as to give overall air/fuel ratios of of approximately 0.85.
3. Increasing the primary zone volume so as to increase the residence time.
4. Reduction of film-cooling air issuing from the primary zone. This air usually contains high concentrations of CO and HC. These species can readily be carried along the cool wall layer regions and be emitted through the exhaust. This problem can be overcome by directing the air from the primary zone, containing high concentrations of CO and HC, into central high temperature zones where they can react.
5. Compressor air bleed. This is particularly useful for low-power operation when air is bled directly from the compressor. This leads to an increase in primary zone fuel/air ratio, which leads to improvements in combustion efficiency.
6. Fuel staging. When fuel is introduced in various stages, so as to create separate zones of combustion, the systems are referred to as either fuel staging or staged combustion.

Usually there are two stages involved: in the first stage the major portion of the fuel is introduced with air below the stoichiometric requirements, so as to produce an overall fuel-rich mixture; this then is followed by a second stage, in which more fuel is added with air in excess of the stoichiometric requirements, so as to produce an overall lean mixture. Alternatively, the second stage is one in which the fuel which was not burned in the first, rich zone is burned to completion in the second, lean zone. Thus, a standard gas turbine combustion chamber, which operates with a rich mixture in the primary zone, is a form of staged combustion. The term "fuel staging" is reserved for designs of gas turbine combustors in which the fuel-flow rate is controlled and introduced separately at various locations in the chamber.

In annular combustion chambers, a number of fuel injection nozzles are located at various points on the circumference. In circumferential fuel staging, the fuel supply to alternate nozzles may either be stopped or reduced significantly below that of adjacent nozzles. This system produces segments of the combustion chamber which are alternately fuel-rich and fuel-lean. Another method of circumferential fuel staging is to inject fuel through all the nozzles in one half of the chamber.

Fuel can be introduced at various radial locations so as to produce annular zones which alternate between fuel-rich and -lean. This technique of radial fuel staging has been used in double-banked annular combustors. Lefebvre and Fletcher (1973) have suggested the introduction of a separate nozzle into the secondary zone, through which extra fuel is added when required at higher power levels at one or more locations downstream. Such a system is referred to as axial fuel staging.

Influence of Temperature on CO and NO$_x$ Emissions

The maximum temperature attained in any one zone of a combustor has a major effect on the emissions (Fig. 9.3). Carbon monoxide levels can be drastically

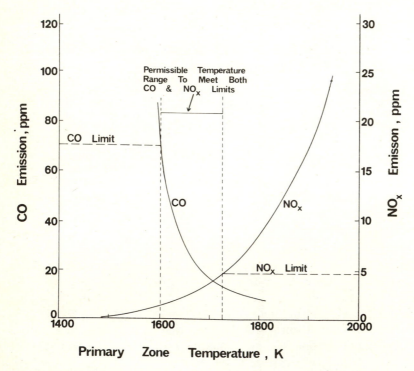

Figure 9.3 Influence of primary zone temperature on CO and NO$_x$ emissions. (*Lefebvre, 1975.*)

reduced by increasing temperatures from 1600 to 1800 K. Emissions of NO_x, on the other hand, begin to become significant at 1600 K and increase exponentially to values well above the acceptable limits. Figure 9.3 shows that, if 50 ppm is taken as the upper limit for CO and 5 ppm as the upper limit for NO_x, then it becomes possible to satisfy both these emission limits by maintaining the temperature in the reaction zone to within the very narrow range of 1600 to 1700 K. Many of the advanced designs of gas turbine combustors, which are required to simultaneously satisfy NO_x, CO, and HC limits, are based upon attempts to control and maintain reaction temperatures within these narrow limits. In conventional combustion chambers relying on turbulent diffusion between fuel and air for mixing, it is extremely difficult to avoid reaction taking place outside these limits. In premixed systems there is little difficulty in guaranteeing that all the combustion will take place at a desired temperature level. The assessment of the performance of a combustor is based on the emissions of both CO and NO_x. Conventional combustors have CO emissions of the order of 100 g CO/kg fuel at idle conditions and this reduces to a value of five at full power. The corresponding values of NO_x emissions are 2 g NO_x/kg fuel at idle, increasing to a value of 40 at full power. A design of a combustor which can claim to provide reduced pollution must result in reductions in both the NO_x and CO over the full operating range between idle and full power. Such reductions have been achieved by using airblast atomization and by controlling mixture ratios within the combustion chamber. Lefebvre (1975) has tabled a list of methods of pollutant reduction from gas turbines in which he shows the advantages and drawbacks of various proposed techniques.

Characteristic Times in Combustion Processes

The mixing and combustion processes in the primary zone of a conventional combustor can be described in terms of characteristic times required for completion of the processes. Mellor (1976) has summarized these characteristic times as shown in Table 9.2. The fuel droplet lifetime is dependent upon the initial diameter of the droplet as it leaves the atomizer and the rate of evaporation as it

Table 9.2 Characteristic times for combustion and pollutant formation in two-phase turbulent flow (Mellor, 1976)

Time	Symbol	Physical or chemical process
Fuel droplet lifetime	τ_{eb}	Droplet evaporation and/or combustion
Eddy dissipation time for injected fluid	τ_{fi}	Small-scale turbulent mixing near the fuel injector in the recirculation zone
Eddy dissipation time in the shear layer	τ_{sl}	Large-scale turbulent mixing between fresh air and the recirculating burned gas-fuel mixture
Fuel ignition delay and burning time	τ_{hc}	Homogeneous combustion of the fuel to CO_2
Nitric oxide formation time	τ_{no}	Homogeneous kinetics for NO formation

passes through regions varying in temperature and concentration. Mixing processes are characterized by τ_{fi}, which is referred to as the eddy dissipation time for the injected fluid. This is based upon small-scale mixing between eddies of fuel vapor and eddies of hot recirculating burned gases near the fuel injector in the recirculation zone. The mixing of fresh air from the surrounding shear layer with fuel-burned gas mixtures is referred to as the eddy dissipation time in the shear layer, τ_{sl}. This is a large-scale turbulent mixing phenomena. The chemical composition of the fuel and the rate of combustion determine the fuel ignition delay and burning time, τ_{hc}. Because of the particularly long time associated with NO formation, this species is labeled separately, τ_{no}, the NO-formation time. The separation of the mixing and combustion process into these separate characteristic times is an oversimplification, but Mellor has shown that, by making order of magnitude estimates of the separate times, it becomes feasible to explain the success achieved by adopting various pollution reduction techniques. Improvements in performance and reduction in emissions have been achieved by:

1. Improving fuel atomization (decreasing τ_{eb})
2. Increasing velocity of air injected with fuel (decreasing τ_{fi})
3. Increasing air flow rate through the primary zone (decreasing τ_{sl})
4. Increasing local reaction temperature (decreasing τ_{hc})
5. Decreasing local reaction temperature (increasing τ_{no})

Fuel ignition delay and burning time τ_{hc} are generally short compared with the time required for NO formation, τ_{no}. Flame blowoff occurs when the approach velocity is just increased past the point where

$$\tau_{sl} = \tau_{hc,\,sl}$$

where $\tau_{hc,\,sl}$ is the fuel ignition delay and burning time under the conditions in the shear layer. Examination of the influence in changes to the characteristic times shows that improved mixing leads to faster rates of combustion and, subsequently, to improved combustion efficiency and lower NO. Very significant improvements can be obtained in performance by improving the fuel atomization, and these improvements are maximum when premixing is used.

Characteristic times are calculated as follows (Vranos, 1974):

1. τ_{eb} from the "d^2 law" corrected for forced convection and aerodynamic drag. $\tau_{eb} = d_0^2/\beta$, where d_0 is the initial droplet diameter and β is the evaporation coefficient.
2. τ_{fi} is related to a fuel penetration length scale L_{fp} divided by an appropriate velocity in the free stream V_a. The appropriate length scale is assumed proportional to the product of the droplet evaporation time τ_{eb} and the initial fuel velocity V_f (Plee and Mellor, 1979)

$$\tau_{fi} = (V_f/V_a)\tau_{eb}$$

Table 9.3 Estimated characteristic times for primary zone conditions (Vranos, 1974)

Average primary zone equivalence ratio, ϕ_p†	τ_{sl}, ms	τ_{hc}, ms	τ_{no}, ms
0.7	10	0.20	330
0.8	10	0.17	53
0.9	10	0.15	15
1.0	10	0.13	7
1.1	10	0.16	23
1.2	10	0.20	109
1.4	10	0.40	> 1000

† Varied by changing fuel flow at constant airflow rate.

3. $\tau_{sl} = c(mL^2/P)^{1/3}$, where c is a proportionality constant evaluated from stirred reactor experiments, m is the fluid mass in the primary zone, L is a characteristic combustor dimension, and P is the power input calculated from the liner pressure drop and airflow rate.
4. τ_{hc} is obtained from laminar-flame speed and deflagration thickness.
5. τ_{no} from the initial kinetic rate of NO formation compared with NO concentration at chemical equilibrium.

Vranos (1974) has calculated characteristic times for a combustor with fixed geometry and flow rate with a combustor inlet pressure of 14.67 atm and inlet temperature of 700 K. Table 9.3 shows the calculated characteristic times for primary zone conditions when the average primary zone equivalence ratio is varied from 0.7 to 1.4. The airflow rate is maintained constant and, therefore, τ_{sl} is constant. In Table 9.4, characteristic times for liquid fuel injection are given when the mean droplet diameter is varied from 150 to 10 μm.

These calculations allow order of magnitude estimates to be made. The values in Tables 9.3 and 9.4 show that

1. $\tau_{sl} \gg \tau_{hc}$. The times for chemical reaction are negligibly small compared with turbulent mixing times.

Table 9.4 Estimated characteristic times for liquid fuel injection (Vranos, 1974)

Mean drop diameter, μm	τ_{eb}, ms	τ_{sl}, ms	τ_{hc},† ms
150	9.8	10	0.13
100	3.6	10	0.13
50	1.2	10	0.13
25	0.4	10	0.13
10	0.2	10	0.13

† Taken for stoichiometric combustion.

2. $\tau_{no} \geq \tau_{sl}$. The time required for formation of oxides of nitrogen is only comparable to the mixing time near stoichiometric conditions. For rich mixtures ($\phi < 0.7$), the times required for formation of NO are so long, compared with mixing times, that no appreciable amounts of NO have time to form.
3. $\tau_{no} \gg \tau_{hc}$. The general conclusion that chemical reaction may be considered very fast, compared with mixing times, does not apply to the very slow chemical kinetic rate of NO formation.
4. For large droplets, $\tau_{eb} \simeq \tau_{sl}$, whereas for small droplets, $\tau_{eb} < \tau_{sl}$. For very fine atomization, droplets evaporate so rapidly that they do not influence the combustion process, while for droplets above 100 μm there may not be sufficient time to complete the evaporation of the larger droplets.

Combustion Control Techniques

The efficiency of combustion and the level of emissions are determined by the combustion chamber geometry, the flow rates of fuel and air, and degree of atomization. Rapid and thorough mixing of fuel and air tends to minimize emissions of smoke, HC, and CO but, if the mixing is too rapid, so that there is insufficient time to complete the reaction in high-temperature regions, the reactions become quenched and result in emissions of HC and CO. Following Mellor (1976), characteristic times are controlled by variation of:

1 Fuel droplet lifetime (τ_{eb}) Small droplets in high-temperature, lean mixtures diameters decrease with fuel flow rate but airblast atomization allows good atomization over a much wider range of fuel-flow rates. Atomizers are generally size, reduced heat transfer to the droplet, and increase in vapor pressure of fuel surrounding the droplet. With pressure atomizing nozzles, mean droplet diameters decrease with fuel flow rate but airblast atomization allows good atomization over a much wider range of fuel flow rates. Atomizers are generally designed for full power conditions, and combustion efficiencies deteriorate at lower power ratings, due to deterioration in fuel atomization. The liquid spray needs to be directed toward the main reaction zone and, since the location of this main reaction zone will vary with fuel flow-rate, spray angles and droplet momenta need to be altered accordingly. When large, high-momentum droplets bypass or pass through the main reaction zone and enter the main airstream, an envelope flame may form around the droplet, providing favorable conditions for soot formation. Alternatively, the droplet vaporization may not be completed, as a result of quenching in the cool airflow.

For airblast atomization, some of the air discharged from the compressor is directed through the fuel injector. The term "air-assist atomization" is used when an auxiliary pump is provided to pressurize the air above the compressor discharge pressure. Atomization characteristics improve and mean droplet diameters are reduced progressively by using pressure-atomizing, airblast, and air-assist nozzles. Decrease in droplet diameter does not always result in improvement of combustion performance but, if mixing is controlled so as to minimize quenching, reduction in mean droplet size leads to both improved combustion and reduced emissions.

2 Fuel mixing time (τ_{fi}) Fuel vapor is mixed with air introduced directly into the chamber or through the airblast or air-assist nozzle. Combustion products are added to this mixture by the recirculation gas stream. Increase in air velocities provides increased momentum to both the air and recirculated combustion product streams, resulting in faster mixing. When air is directed through a swirler in the burner, mixing rates are greatly increased. The regions of high mixing are those where velocity gradients, and hence shear, are greatest and very considerable improvements in performance have been obtained by vigorous and rapid mixing of fuel vapor, air, and recirculated products in the central region of the primary zone. When fuel vapor fails to enter or succeeds in escaping from the vigorous mixing region, chemical reactions are quenched, resulting in increase in emissions.

Concentration measurements within conventional combustors (Mellor, 1976) show that most of the NO emitted in the engine exhaust is initially formed around the central recirculation zone of the primary zone. Intense mixing with rapid droplet evaporation avoids the formation of envelope flames around droplets and, hence, decreases NO emissions. On the other hand, it has been shown that mixing of air and fuel at high temperatures leads to locally high concentrations of NO. Minimization of NO formation thus requires careful control of the mixing process since, in practice, it has been found that increase in mixing has resulted in some cases in increase of NO emissions, while in other cases it has led to decrease in NO emissions. In order to obtain reductions both in HC and NO emissions some form of staged combustion is required; otherwise, the improved mixing leads to reduction of HC while, at the same time, resulting in increase in NO emissions because of the high temperatures associated with improved mixing.

Chemical reaction between molecules of fuel and air can only take place after mixing has been completed at the molecular level. In turbulent flows large-scale transport of fuel-rich mixtures into air-rich mixtures takes place. Subsequently, mixing is completed by molecular diffusion but turbulence can increase the rate of diffusion by increasing the concentration gradients. Measurements of mean concentrations in flames with gaseous fuel have indicated that fuel and oxygen can both be present at a particular location but, due to turbulent fluctuations, they may be present at the location at different times. This results in the phenomenon of unmixedness (Hawthorne et al., 1951) in which additional time is required to complete mixing on the molecular level, after mixing appears to be complete on the basis of mean concentration measurements. The studies of Tuttle et al. (1975) have shown that gas turbine primary zones have a combustion mechanism which is characteristic of diffusion flames. The rates of chemical reaction are limited by the rate of mixing of fuel and air. Chemical reaction occurs mainly at or near the regions of stoichiometric mixing. Because of the fluctuations of the flame front it appears that, instead of chemical reaction taking place in a large volume, the instantaneous reaction may be taking place in a much smaller volume where mixtures are close to stoichiometric. Because temperatures are highest within the flame front, the characteristic times associated with turbulent fluctuations of the flame front are important for nitric oxide formation.

Soot formation mainly occurs under very fuel-rich conditions and is characterized by time scales associated with the rate of mixing of fuel-rich regions near the fuel nozzle. Carbon monoxide is found in large concentrations at flame temperatures. Completion of combustion of this CO takes place in the relatively fuel-lean regions on the air side of the flame and is characterized by mixing times in this region.

3 Shear layer mixing time (τ_{sl}) The primary zone is separated schematically into various regions, as shown in Fig. 9.4. Each subdivision is associated with a separate characteristic time. Within the shear layer, mixing takes place between fuel, combustion products, intermediates, and air. The magnitude of the air velocity affects the mixing time as a result of convection, as well as determining the velocity gradients and, hence, shear. Improvements in combustion efficiency are achieved by increasing τ_{sl}, since slower mixing allows more complete combustion of all of the fuel. This also, simultaneously, allows more time for NO formation. If fuel should pass through the shear layer without being completely burned, it can result in significant levels of emissions. Wade and Cornelius (1972) showed that by changing the total airflow to the primary zone, so as to provide rich, lean, or early

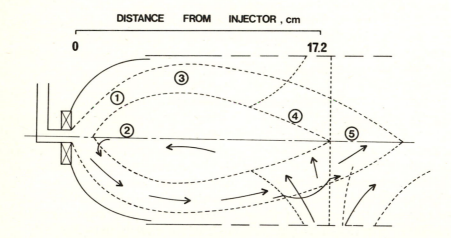

REGION : 1. Very Fuel Rich, Much CO Formation, Fuel Drops Continue Downstream
2. CO Quenched By Cold Recirculating Air, NO Entrained From Wall Flow
3. Measured Temperatures Low Due to Drop Impingement, Some CO Oxidation And NO Formation
4. Penetration Air Feeds Recirculation Zone
5. NO Formation Throughout This High Temperature Region 3—5

Figure 9.4 Regions within primary zone associated with characteristic times. (*Mellor, 1976.*)

quenching conditions, NO emissions were lowered over the entire engine design range with as much as a 40 percent reduction at full power. These reductions in NO were achieved by merely changing the location and/or diameter of holes in the liner walls for injecting air into the primary zone. When power is reduced by reducing fuel-flow rate, while maintaining the airflow pattern and mixing rates, HC and CO emissions increase due to quenching of combustion products. This can be overcome by bleeding air away from the compressor and thereby decreasing airflow to the combustor, leading to an increase in τ_{sl}. Under these conditions the gain in average combustor residence time is more important than the decrease in mixing with the net result that combustion efficiency can increase for idle conditions. Since the usual spray combustion process resembles a turbulent diffusion flame, the burning rate is equal to the mixing rate, the equivalence ratio at which combustion occurs is on or close to stoichiometric, and thus the concept of a primary zone equivalence ratio only reflects differences in combustion volume or flame length (Mellor, 1976).

4 Chemical kinetic times (τ_{hc} and τ_{no}) The time required to complete the oxidation of fuel and CO (τ_{hc}) and to form NO (τ_{no}) are governed by chemical kinetic rates. These chemical kinetic rates have a strong temperature dependence so that, under conditions of stoichiometric mixing, combustion is completed rapidly by the high rates of oxidation while, at the same time, leading to high NO formation. When reaction occurs at off-stoichiometric conditions, the lower temperatures are associated with both rich and lean mixtures. Control techniques are based on attempting to maintain temperatures within a narrow temperature band, which is just sufficient to allow completion of combustion of the hydrocarbons, while minimizing the formation of NO. In prevaporizing premixed combustors, the kinetic times τ_{hc} and τ_{no} dominate, though account must still be taken of local dilution effects as the premixed mixture mixes with combustion products.

NASA Swirl-Can Combustor

In a standard combustor all the fuel is injected into the primary zone, which produces a flame occupying a major portion of the cross-sectional area of the combustor. Temperatures in regions of stoichiometric burning are 1000 K in excess of the required exit temperatures and, because of the associated heat transfer by convection and radiation, provision needs to be made for film cooling to the walls of the combustor. In the NASA swirl-can combustor between 100 and 300 individual swirl-can modules are arranged in concentric rows with separated supply of fuel to each module. Burning is completed in small flames attached to the individual swirl-can modules with air flowing between the modules, resulting in high rates of mixing in reverse flow zones in the wake of the swirler. Flame lengths can be of the order of 100 mm and mixing of the hot products with the surrounding airstreams can be achieved in much shorter distances than required in conventional combustors. The combustor with typical swirl-can modules is

Figure 9.5 NASA swirl can combustor and swirl module.

shown in Fig. 9.5. Staging can easily be arranged by varying the fuel-flow rates to groups of individual swirl can modules. Overall mixture ratios are controlled aerodynamically by changing the geometry of the module and the angles of the vanes in the swirlers. Little attention was initially paid to atomization of the fuel but the more advanced designs include film atomizers for reducing the droplet size. Substantial reductions in NO_x emissions have been achieved at full power conditions. For low power conditions, fuel flow is restricted or stopped to a number of modules and, under these conditions, the CO and HC emissions are not excessive.

Premixed Combustors

In a premixed combustor, the fuel is introduced into a premix chamber in which the fuel is vaporized and then mixed with primary air. Air velocities are maintained high and care is taken to avoid any ignition source or obstacles to the flow in which a flame could be stabilized. This premixture is then introduced into a primary combustion chamber in which ignition takes place, followed by subsequent mixing with dilution air. An example of such a system is shown in Fig. 9.6, which is for use in small gas turbine engines, suitable for automobiles and light duty vehicles (Azelborn et al., 1973). Sufficient heat must be supplied to the primary air in order to complete vaporization of the fuel and sufficient time is needed in order to complete the mixing between the fuel and the air before the mixture enters

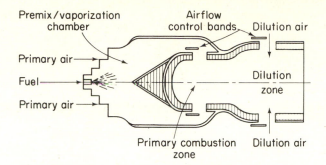

Figure 9.6 A prevaporized premixed gas turbine combustor. (*Azelborn et. al., 1973.*)

the primary combustion zone. With premixing, the characteristics times for fuel vaporization, τ_{eb}, and for fuel mixing, τ_{fi}, are eliminated. Within the primary combustion zone, mixing still takes place between the premixed mixture and recirculated burned gases as well as by mixing with secondary air. The equivalence ratio of the premixed mixture is the maximum possible ratio anywhere in the combustor. By decreasing the inlet equivalence ratio below stoichiometric, the time for NO formation τ_{no} can be significantly increased without too large an increase in hydrocarbon-formation time τ_{hc}. This leads to substantial decreases in NO and very small increases in HC and CO emissions (Mellor, 1976).

In premixed systems, the possibility of flashback arises in which a flame could propagate from the primary zone upstream into the premixing chamber. This will occur if velocities are allowed to decrease below the flame speed of the premixed mixture. The flame speeds of these mixtures are considerably higher than the laminar-flame speed for such mixtures, due to the high turbulence levels and elevated temperatures and pressures of air coming from the compressor discharge.

Premixing eliminates the uncertainties and lack of homogeneity which arise in the nonpremixed diffusion systems. If the mixture is made lean, then throughout the combustion chamber there is no possibility of burning taking place under stoichiometric conditions. Thus, the "hot spots" where stoichiometric mixtures burn at the maximum temperatures are eliminated. It also becomes possible to operate close to the limits of flammability without facing the possibility of partial flashback or blowoff. The pyrolysis and formation of soot associated with liquid droplets in the primary zone are also eliminated under premixed conditions. All these advantages have been known for a considerable time, but, in practice, premixing has been avoided because of the potential hazard of explosions in the premixed gas under uncontrolled conditions. More experience is, however, being gained on the prevention of explosions, and a number of gas turbines have been designed using premixing.

9.2 SPARK IGNITION ENGINES

Spark ignition engines in automobiles and trucks are a major source of urban air pollution. Unburned hydrocarbons (HC) are emitted from the exhaust and can also diffuse from the crankcase, the fuel tank, and the carburetor. Oxides of nitrogen (NO and NO_2) and carbon monoxide are the major pollutants emitted through the exhaust. In addition to mobile spark ignition engines, there are also stationary engines which, in certain localities, can be a dominant source of pollution.

Legislation is continuously being reviewed concerning the maximum permitted levels of pollution emission. Initially the prime concern was in reduction of levels of emissions but, more recently, these have been coupled with required increases in fuel efficiency. This legislation has initiated large-scale research by engine manufacturers and national research organizations, which has led to significant reductions in the level of emissions and is leading to some radical changes in engine design. In addition to the conventional spark ignition (SI) engine, the Wankel and stratified charge engines, which were previously considered unsuitable for mass production, are being reexamined. Heywood (1976) has reviewed the status of engine technology with special relation to pollutant formation and control. He has examined the mechanisms involved in the combustion process and also discussed the utilization of exhaust reactors and catalytic converters for reducing the emission of pollutants after the exhaust gases have left the combustion chamber.

In a four-stroke engine, the effects of each individual stroke on the combustion and mixing processes are examined. During the intake stroke, fuel and air from the carburetor pass through the inlet valve. Usually the fuel has been vaporized before entering the cylinder but, under conditions of high fuel-flow rate, liquid droplets can also enter the cylinder. Swirling motions are generated in the airstream and continue to prevail within the combustion chamber. Flow past the inlet valve will result in the formation of an annular diagonal jet which, subsequently, impinges on the cylinder walls. Residual gases left behind in the cylinder from the previous outlet stroke are entrained into the fresh mixture jet.

It is generally assumed that, during the completion of the inlet stroke and the major portion of completion of the compression stroke, a homogeneous mixture is formed with negligible residual motion. There are indications that this may not always be true and it is possible for nonhomogeneities to persist within the cylinder and for swirling and turbulent fluctuating motions to be present at the time of ignition. The various stages of the combustion process can be seen from the graphs of pressure, temperature, and mass fraction of charge burned, as shown in Fig. 9.7. Ignition takes place toward the end of the compression stroke by a spark formed across the spark plug. The flame propagates radially outward from the spark plug toward the cylinder walls and piston crown. The rate of flame propagation is dependent upon the physical properties of the fuel-air mixture, the degree of uniformity of mixture, and the turbulence intensity. Mixture ratios are within the limits of flammability and it is generally assumed that the mixture ratio is uniform. Turbulence intensity will be dependent upon the inlet flow conditions during the inlet stroke and any dampening which will have occurred during the completion of the inlet stroke and the compression stroke.

The shape of the flame front will be distorted according to the shape of the volume containing the gas mixture. As the flame front approaches the cool solid

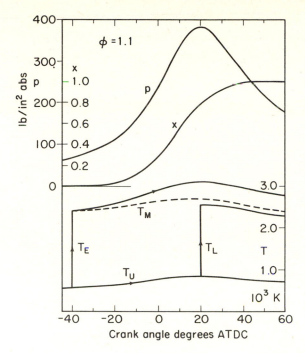

Figure 9.7 Graphs of cylinder pressure p, mass fraction of the charge burned x, unburned gas temperature T_u, and burned gas temperature T_b as function of crank angle. Zero degrees is crank angle at top-dead-center position. Lower graph is for two elements, one which burns early in combustion process and one which burns late. Dashed line shows the average burned gas temperature. (*Heywood, 1976.*)

surfaces, quenching takes place so that the flame is arrested before coming in contact with the cool walls. Quench layers of unburned gas form on all these surfaces and are typically of the order of 0.1 mm thick. Crevices between the piston and cylinder walls are regions where the flame does not generally propagate and gases remain unburned. Almost all of the combustion is completed during the few milliseconds that the flame front propagates from the spark to the walls but some residual combustion may take place as unburned gases come in contact with hot combustion products.

Temperatures are sufficiently high for large-scale formation of nitric oxide, due to the high temperature nonequilibrium reactions involving nitrogen and oxygen. Carbon monoxide is formed as an intermediate combustion product.

During the expansion stroke the volume of gases increases and temperature decreases and the rapid cooling leads to a freezing of NO and CO reactions. During the exhaust stroke the exhaust valve is opened. The piston scrapes the thickened quench layers on the cylinder walls during the exhaust stroke so that the emissions are made up of frozen NO and CO and unburned hydrocarbons from the quench layers.

The emission concentrations are largely determined by the fuel/air equivalence ratio, as shown in Fig. 9.8. With lean mixture ratios CO levels are low and HC levels are also lower than for rich mixtures. As mixture ratios become richer, both the CO and HC levels increase. Nitric oxide concentrations are a maximum just on the lean side of stoichiometric and NO levels decrease when either the mixture ratio is made more lean or more rich than stoichiometric. During starting,

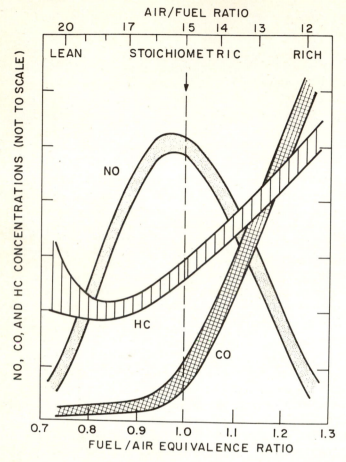

Figure 9.8 Variation of HC, CO, and NO concentration in the exhaust of a conventional SI engine. with fuel/air equivalence ratio. (*Heywood, 1976.*)

a choke is used to provide a rich mixture near the spark plug with the resultant high emission of CO and HC. As the engine warms up, mixture ratios are moved from rich to lean by releasing the choke so that, under normal operating conditions, overall mixtures are lean, tending to minimize HC, CO, and NO emissions. During periods of acceleration, when maximum power is required, attempts are made to control mixtures to be close to stoichiometric.

The processes responsible for the production of NO and CO are distinctly different from those for HC. Nitric oxide and CO formation occur in the bulk gases and are primarily dependent upon the kinetics of combustion. Unburned HC is a result of flame quenching at the walls and is related to the aerodynamic boundary layers, mixing processes, and oxidation near the wall surfaces. It is generally assumed that the state of the burned gas in an SI engine is very close to thermodynamic equilibrium and that this can be used as a basis for predicting temperature, density, and concentration of major species distributions in the burned gases.

The Engine Combustion Process

The combustion process is dependent upon the state of the fuel-air mixture prior to ignition. This state is determined by the mixing history of fuel and air from the time at which they are initially separate. Carburetors are designed to mix fuel and air and, to do this effectively, the fuel requires to be vaporized and then, subsequently, well stirred with the air. Many carburetors, particularly under high fuel-flow conditions, do not suceed in completing the vaporization and mixing process, so that inlet mixtures may contain liquid fuel droplets and pockets of fuel-rich and fuel-lean mixtures. In recent years, stricter control has been placed upon the design of the intake manifold and carburetor in order to guarantee more uniform mixture ratios. These designs mainly concentrate on normal operating conditions and it still appears to be necessary to accept that, under conditions of starting, cold weather, and high-power acceleration, some liquid fuel enters the cylinder. The mixing process between incoming fresh mixture and the residual gases may not have sufficient time for completion by the time ignition takes place. In multicylinder engines, when there is no separate metering and control to each individual cylinder, mixture ratios may vary within ± 10 percent from one cylinder to another.

The various stages of the combustion process are shown in Fig. 9.7 as a function of crank-angle position in degrees with $0°$ corresponding to the top-dead-center position. Pressures rise, initially as a consequence of the piston movement in the compression stroke and, subsequently, as a consequence of an expansion of the gases after combustion. Maximum pressures are found at $20°$ after top dead center. Thereafter, pressure decreases as a result of the increase in volume during the completion of the expansion stroke. The curve x shows the mass fraction of the charge which has been burned. From the temperature curve, ignition can be seen to take place at $-40°$, resulting in an almost instantaneous increase in temperature. Burning continues until $30°$ after top dead center.

There seems little possibility that all the burning takes place simultaneously and there is every indication that the burning process takes place in several separate stages. In order to make a complete prediction or calculation of the combustion process, the extent of homogeneity requires to be known both prior and subsequent to ignition. Overall calculations based upon thermodynamics can be made from variations of pressure and volume as a function of time. Lavoie et al. (1970) have developed a simple model by making the following assumptions: equilibrium conditions prevail; the original charge is homogeneous; pressure is independent of position; the volume occupied by the gas in a state of partial combustion is negligible; the unburned gas is "frozen" at its original position and undergoes an isotropic compression; both burned and unburned gases have constant local specific heats. Using the equations for conservation of mass and energy and equations of state for the burned and unburned gases, the mass fraction of the charge which has been burned, x, is calculated as a function of the observed pressure history.

Temperature distributions are dependent upon the rates of chemical reaction which, in turn, are dependent upon the initial mixture ratio distributions and, subsequently, upon the extent of mixing between burned and unburned gases. The simplest assumption to make is that each element of the charge which burns mixes instantaneously with the previously burned gas, resulting in a uniform temperature throughout the burned gases in the cylinder.

Heywood (1976) shows that this completely mixed model is incompatible with the observations of a relatively thin turbulent flame front, as well as substantial temperature gradients in the burned gas. Heywood and coworkers have made a more realistic approximation that there is no mixing during the early part of the combustion process, and that each element of gas which burns is isentropically compressed from its state just behind the flame front as the pressure rises. The unburned mixture is also compressed isentropically. Each element of the charge burns at constant but different pressure and enthalpy, while compression and expansion takes place isentropically.

Figure 9.7 shows the temperature variation as a function of crank-angle degree for two elements of gas; one which burned early in the cycle, denoted T_E, and one which burned later, denoted T_L. For each of these two gas elements the temperature of burned gas T_M and for the unburned gas T_U. In a practical engine, The separate factors, due to isentropic compression, are shown for the mean temperature of burned gas T_M and for the unburned gas T. In a practical engine, elements of gas will burn between the two extremes of early and late burning with a net result that there will be substantial temperature variations during the overall combustion process. The calculations of Heywood (1976) show that differences of 500 K in peak temperature can occur between elements of gas which burn early and those which burn late. These temperature variations can have an important effect on NO formation in the burned gas.

If the assumption is made that the burned gases are close to thermodynamic equilibrium, the cylinder pressure, temperature distribution in the unburned and burned gases, and mass fraction of the mixture burned can all be calculated. When this information is coupled with the kinetic measurements, the combustion characteristics can be determined, and computer programs have been developed for engine design and control of operating variables so as to minimize emissions and maximize efficiency.

Flame Propagation in Spark Ignition Engines

In spark ignition engines the compression ratio is governed by the requirement that the temperature of the gas mixture at the end of the compression stroke (TDC) is below the autoignition temperature. Combustion is initiated between the electrodes of the spark plug, and flame propagation from the initial gas kernel to the next layer of gas proceeds as a result of heat transmission between gas layers. As the combustion wave propagates toward the walls, the unburned mixture is compressed; if the temperature of this unburned gas rises above the autoignition temperature, the remainder of the mixture burns spontaneously and instantaneously. The higher the octane number of the fuel, the higher is the autoignition temperature. Therefore, the higher the compression ratio of the engine, the higher the octane number of the fuel that is required to avoid spontaneous ignition.

In addition to ignition by the spark, ignition may also occur from hot deposits on the walls or, if there is insufficient cooling, on the cylinder head. Under these conditions, wave propagation is initiated from several sources and the pressure rise within the chamber is much more rapid, resulting in pressures of 20 or 30 bars above the normal maximum pressure. This leads to greater heat losses and increase in wear of piston rings. This effect is cumulative, leading to further increases in wall temperatures and, hence, providing more ignition sources. In general, engine designers aim to raise wall and cylinder-head temperatures in order to reduce the thickness of quench layers on the walls and also to reduce heat losses from the engine. In order to optimize these requirements and avoid the risk of multiple ignition, designers aim to maintain a very uniform wall temperature, close to the spontaneous ignition temperature but sufficiently below this temperature to avoid ignition at the walls.

When the spark advance timing is too great, ignition and combustion take place during compression, resulting in large increases in peak pressures (20–40 bars). If the mixture is not homogeneous, the ignition delay and the flame propagation speed varies across the mixture. If the mixture is stoichiometric, the ignition delay is short and the cycle is normal. As the mixture between the spark plugs becomes rich or lean, the ignition delay increases, resulting in a cycle with poor efficiency. Nonhomogeneity of the mixture leads to severe vibrations due to nonuniform cycles and lower average efficiencies. This phenomenon is most severe in single-cylinder engines.

Nitric Oxide Formation in Spark Ignition Engines

The only oxide of nitrogen produced in significant quantities inside spark ignition engines is nitric oxide which forms in both the flame front and the postflame gases. The formation in the postflame gases is more significant, due to the higher temperatures occurring as a result of compression of postcombustion gases. It is assumed that the species O, O_2, OH, H, and N_2 are in local equilibrium in the burned gases and that N is in steady state. Taking into account the Zeldovich mechanism, the mass fraction of NO as a function of time for any element of burned gas can be obtained by numerical integration.

Figure 9.9 shows NO concentrations as functions of crank angle in two elements of gas which burn at -30 and 15 crank-angle degrees. The much higher rates of NO formation associated with a gas which burns early, as compared with that which burns late, are due to the high reaction rates associated with the higher temperature of gases which burn early in the cycle.

Experiments have been carried out by Alperstein and Bradow (1967) and Starkman et al. (1969) using gas-sampling techniques and by Lavoie (1970) using chemiluminescent radiation in order to measure NO concentrations. Figure 9.10 shows measurements and calculations of Lavoie (1970) of NO mole fraction as a function of crank-angle degrees in single-cylinder engine experiments. Nitric oxide mole fractions were measured at two distances from the spark plug and the flame front was shown to reach window W_2 earlier ($-5°$) than window W_3 ($10°$). The

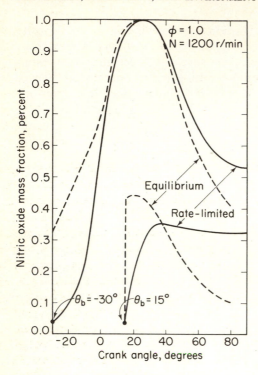

Figure 9.9 Calculated NO concentrations as functions of crank angle in two elements which burn at different times ($-30°$ and $15°$). Dashed lines are NO mass fractions if NO were in equilibrium: solid lines are rate-limited NO concentrations. (*Heywood, 1976.*)

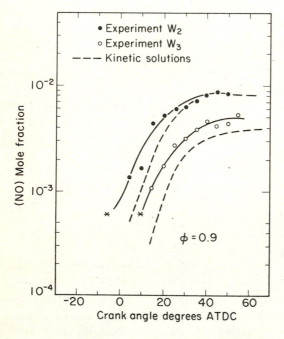

Figure 9.10 Measured and calculated NO concentrations in single-cylinder engine experiments. NO concentrations were measured by monitoring radiation from burned gases through quartz windows in cylinder head. Flame front reaches window W_2 earlier ($-5°$) than window W_3 ($10°$). (*Heywood, 1976.*)

NO levels observed at window W_2 closest to the spark were found to be substantially higher than those observed at window W_3. These experiments and calculations demonstrated that the formation process of NO is rate limited, NO chemistry is frozen during expansion, and NO concentration gradients exist across the combustion chamber.

The most important design and operating variables which affect NO emissions are compression ratio, combustion chamber shape as it affects flame speed and length of flame travel, engine speed, fuel/air equivalence ratio, inlet mixture pressure and temperature, spark-timing, and charge dilution with residual gas and recycled exhaust. Changes in the time history of temperature and oxygen concentration in the burned gases during the combustion process and early part of the expansion stroke are the most important factors influencing the NO formation.

In lean mixtures, NO concentrations freeze early in the expansion stroke and little NO decomposition occurs. In rich mixtures, substantial NO decomposition occurs from peak concentrations present when the cylinder pressure is a maximum, before freezing of the NO chemistry takes place.

Exhaust gas recycle (EGR) has been found to be an effective means of reducing NO emission. A fraction of the engine exhaust is recycled and mixed with the intake gases. When exhaust gases are mixed with the intake mixture, this results in a dilution with inert gases, which is equivalent to a reduction in the energy density distribution. As a consequence, burned gas temperatures as well as flame speeds are reduced. Experiments have shown that exhaust gas recycling can lead to substantial reductions in NO emissions, particularly for lean mixtures.

Many of the efforts to control NO emissions directly affect the specific fuel consumption. Changes in compression ratio, spark timing, burning rate, fuel/air equivalence ratio, and charge dilution with exhaust gas recycle change both NO emissions and fuel consumption. Calculations made by Heywood (1976) indicate that slower burning engines should give slightly better specific fuel consumption at low NO emission levels. At high NO emission levels, faster burning engines should give slightly better specific fuel consumption.

Carbon Monoxide Emissions

Carbon monoxide emissions from conventional SI engines are controlled primarily by the fuel/air equivalence ratio. For fuel-rich mixtures, CO concentrations in the exhaust increase steadily as the amount of excess fuel increases. For fuel-lean mixtures, CO concentrations in the exhaust vary little with equivalence ratio and are of the order 10^{-3} mole fraction.

In the postflame combustion products, at conditions close to peak cycle temperatures (2800 K) and pressures (15–40 atm), equilibrium conditions prevail. At peak cylinder pressures and temperatures, the time required for CO to reach equilibrium is faster than times associated with changes of burned gas conditions due to compression or expansion. The CO concentration rapidly equilibrates in the burned gases just downstream of the reaction zone following combustion of the hydrocarbon fuel. The burned gases remain close to equilibrium until about 60 crank-angle degrees after top dead center.

From an examination of a series of experimental and theoretical studies, Heywood (1976) has come to the following conclusions: the measured average exhaust CO concentrations for fuel-rich mixtures are close to equilibrium concentrations in the exhaust gases; for close to stoichiometric mixtures, partial equilibrium CO predictions are in agreement with measurements and are orders of magnitude above CO equilibrium concentrations corresponding to exhaust conditions; for fuel-lean mixtures, measured CO emissions are substantially higher than partial equilibrium predictions. Carbon monoxide formed in the quench layer at the combustion chamber walls makes a significant contribution to total exhaust CO for lean engine conditions. Carbon monoxide can also be formed as a result of partial oxidation of HC during the exhaust process.

The most effective control of CO emissions has been found by working with lean intake mixtures. Carbon monoxide emissions during engine warmup are much higher than emissions in the fully warmed-up state, due to the rich mixture resulting from the use of a choke. Also during periods of rapid acceleration, the rich mixtures result in increased CO emissions.

Unburned Hydrocarbon Emissions

Minimum hydrocarbon emissions occur at air/fuel ratios of approximately 18/1. Hydrocarbon emissions rise rapidly as the mixture becomes very lean due to engine misfire. Flame propagation rates become so slow under very lean conditions that combustion is not completed before the expansion process cools the gases and quenches the flame reactions. For fuel-rich conditions, the excess fuel is responsible for the emission of unburned hydrocarbons. Under normal engine operating conditions, the bulk of the gas mixture is completely burned with the exception of the quench layers and crevices at the cylinder walls.

Figure 9.11 shows schematically the quench layers formed on the cylinder walls and on the piston head as well as the crevice region between the piston and cylinder wall above the piston ring. The cool walls arrest the propagation of the flame, due both to heat transfer and destruction of radicals in the unburned mixture. Flame reactions are quenched so that a layer of the order of 100 μm thick is formed along all the cool surfaces. These quench layers contain hydrocarbon compounds formed by heating but not completely burning the fuel. Gas chromatographic analysis of the engine exhaust has shown that over 200 organic compounds have been identified. The reactivity of these different compounds with the atmosphere varies widely and only some compounds make a major contribution to the formation of photochemical smog. Exhaust hydrocarbon composition is influenced by fuel composition—fuels with higher olefin and aromatic content resulting in exhaust hydrocarbons with higher reactivity. Pyrolysis and partial oxidation of fuel components result in the formation of many hydrocarbon compounds that were not initially present in the fuel.

The thickness of the quench layer primarily depends on the pressure and temperature of the unburned mixture at the time quenching occurs and also on the fuel/air ratio. Quench-layer thickness is least for the rich mixtures and in-

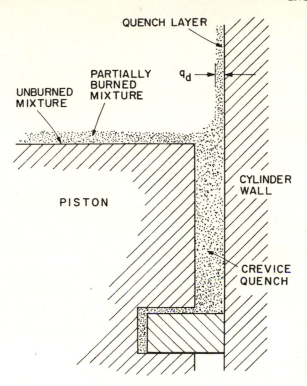

Figure 9.11 Schematic section of part of the piston and cylinder wall showing quench layers and the piston crown-first ring-cylinder wall quench volume. (*Heywood, 1976.*)

creases substantially for lean mixtures, due to the reduction in flame speed as mixtures are made more lean. Quench-layer thicknesses have been estimated to vary from 75 μm at full throttle to 400 μm at idle conditions. Since the flame reaches different parts of the cylinder wall at different times, quench distances are not uniform over the cylinder head and piston face.

Quench crevices around the valves and between the piston face and cylinder head have now been largely eliminated so that the most important remaining crevice is the volume between the piston and cylinder wall above the first piston ring. Wentworth (1971) showed the importance of the piston crown-cylinder wall crevice and was able to reduce exhaust HC levels by as much as 74 percent by changing the design of the piston. He also showed that wall temperatures affect HC emissions. Doubling of the surface temperature from 70 to 150°C led to a decrease of exhaust HC by a factor of two. Wentworth (1971) concluded that about half the decrease in exhaust HC measured, as load was increased, was due to increasing combustion chamber wall temperature. The remainder of the decrease was due to postquench oxidation of the quench HC. Increase in the surface roughness of walls or deposition on the surfaces also increases exhaust HC emissions.

Unburned gas in the quench layers is emitted through the exhaust valve as a result of pressure differences inside and outside the engine and also due to the gas flow generated by the piston movement during the exhaust stroke. The extent to

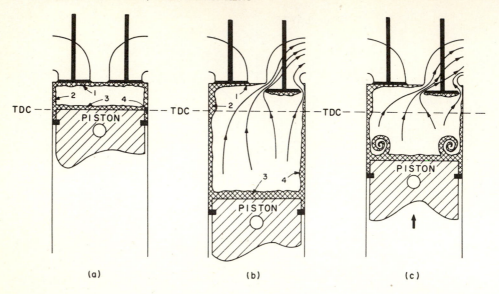

Figure 9.12 Schematic summarizing processes important in hydrocarbon emissions. (*a*) Formation of quench layers 1, 2, 3 and crevice quench 4 as flame is extinguished at cool walls. Not to scale, quench layers are about 0.003 in thick. (*b*) Gas in quench volume between piston crown and cylinder wall above the first ring, 4, expands as cylinder pressure falls and is laid along cylinder walls. When exhaust valve opens, head quench layers 1 and 2 exit cylinder. (*c*) Rollup of hydrocarbon-rich cylinder wall boundary layer into a vortex as piston moves up cylinder during exhaust stroke. (*Heywood, 1976.*)

which the unburned hydrocarbon is removed from the quench layers is dependent upon the aerodynamics of the gas flow. Some of the unburned HC which mixes with the bulk burned gas will be oxidized, dependent upon the temperature and fuel/air ratio of the bulk mixture. Measurements of HC concentration in the engine exhaust port with a rapid acting sampling valve show that high HC concentrations are measured at the end of the exhaust stroke.

Tabaczynski et al. (1972) have explained their measurements of exhaust hydrocarbons by means of a physical model of quench-layer formation and development, as shown in Fig. 9.12. The quench and crevice layers are formed as the flame is extinguished at the combustion chamber walls. These layers expand during the expansion stroke. When the exhaust valve opens, bulk gas is blown through the valve due to pressure differences between inside and outside the engine, and part of the quench layers are entrained and exit from the cylinder. As the piston moves up during the exhaust stroke, the boundary layer on the cylinder wall is scraped off the wall and rolled into a vortex. Toward the end of the exhaust stroke, a major portion of this HC rich vortex is expelled from the cylinder.

The results of experiments carried out by Tabaczynski et al. (1972) on a single-cylinder engine are shown in Figs. 9.13 and 9.14. Figure 9.13 shows the instantaneous mass flow rate out of the exhaust valve as a function of crank-angle degrees. Soon after the exhaust valve opens the blowdown process commences with choked flow through the exhaust valve. At bottom dead center, flow changes from supersonic to subsonic flow and the mass flow rate decreases. Up to this stage

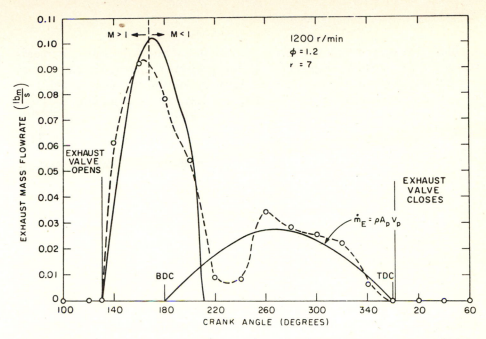

Figure 9.13 Measured instantaneous mass flow rate (dashed line) out of the exhaust valve. Solid lines show an isentropic compressible flow model of the blowdown process, and an incompressible model of piston displacement process. (*Heywood, 1976.*)

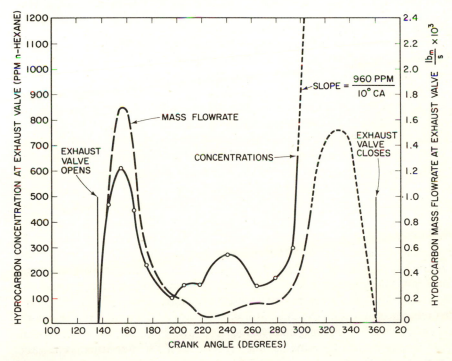

Figure 9.14 Variation of HC concentration and HC mass flow rate at the exhaust valve during the exhaust process. (*Heywood, 1976.*)

the blowdown process dominates the exhaust mass flow rate and piston movements play a negligible role. As the piston moves upward from bottom dead center, gases are forced out of the exhaust valve by the piston movement. The solid lines of Fig. 9.13 are calculations of the blowdown process based on an isentropic compressible model according to the equation

$$\dot{m}_E = \rho A_p V_p \tag{9.5}$$

where A_p is the cross-sectional area of the piston head and V_p is the velocity of the piston.

The variation of HC concentration and HC mass flow rate at the exhaust valve during the exhaust process is shown in Fig. 9.14. This figure shows clearly that HC emissions are initially associated with the blowdown process, during which part of the quench layers are entrained into the bulk gas escaping from the cylinder. The high concentration and mass flow rate of HC at the end of this stroke is explained by Heywood (1976) as being due directly to emission of part of the vortex formed as the piston scrapes the quench layers from the cylinder walls. Heywood has calculated both the exhaust mass flow rate and HC concentration and HC mass flow rates on the basis of a physical model. The results are shown as solid lines in Figs. 9.13 and 9.14, which are in close agreement with the measurements.

The effects of engine design and operating variables on exhaust HC emissions are given in Table 9.5. The effects of any variable are considered in four steps:

1. Formation of quench regions in the combustion chambers
2. Postquench oxidation in the combustion chamber
3. Fraction of HC leaving the combustion chamber
4. Oxidation in the exhaust system

Control of engine HC emissions is achieved by combining design and operating parameters so as to reduce the amount of HC formed by quenching and to increase the exhaust temperature and oxygen concentration so as to promote some reaction by afterburning. Increasing the exhaust gas temperature does, however, lead to an increase in the specific fuel consumption and is thus generally not acceptable.

Exhaust Reactors

The previous discussion has concentrated on methods of minimizing the emission of pollutants by control of the combustion process within the cylinder. Until these methods have been perfected, it is accepted that all engines emit some pollutants and these are treated by the use of exhaust reactors. Two principal types of exhaust reactor have been developed and introduced in automobiles. The catalytic convertors contain reducing catalysts for NO and oxidizing catalysts for HC and CO. The reactors complete the combustion of unburned HC and CO.

Table 9.5 Engine design and operating variables affecting HC emissions (Heywood, 1976)

1. Formation of HC quench regions	2. Postquench oxidation in combustion chamber	3. Fraction HC leaving combustion chamber	4. Oxidation in exhaust system
a. Effective surface/volume ratio i. compression ratio ii. combustion chamber design iii. spark timing	a. Mixing with bulk gas i. speed ii. induction sys. design iii. combustion chamber design	a. Engine geometry i. compression ratio ii. combustion chamber design iii. valve overlap	a. Oxygen concentration i. mixture ratio ii. air injection in exhaust port
b. Quench thickness i. mixture ratio/EGR ii. load iii. wall temperature iv. wall roughness v. deposits	b. Oxygen concentration i. mixture ratio	b. Operating param. i. load ii. exhaust pressure iii. speed	b. Exhaust gas temp. i. compression ratio ii. spark timing iii. load iv. mixture ratio/EGR v. burn rate vi. heat losses
c. Quench crevice i. combustion chamber design ii. mixture ratio/EGR iii. load iv. deposits	c. Bulk gas temp. during expansion i. compression ratio ii. spark timing iii. load iv. mixture ratio/EGR v. burn rate		c. Residence time i. volume of exhaust system ii. exhaust flow rate
	d. Wall temperature		d. Exhaust reactors i. thermal reactor ii. catalytic converter

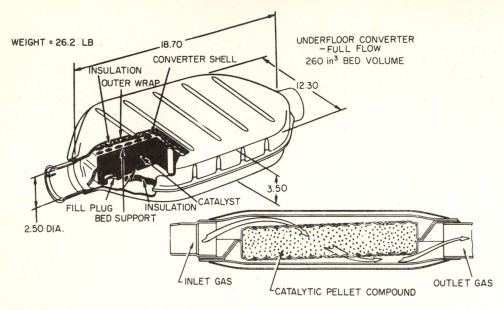

Figure 9.15 Schematic of pelleted catalytic convertors. (*Heywood, 1976.*)

Catalytic Convertors

A catalytic convertor developed for automobile use is shown in Fig. 9.15. Exhaust gas from the engine is forced through a bed of catalyst pellets. The catalyst is usually made up of a small mass of active material such as noble metal, or a combination of transition and nontransition metals, deposited on thermally stable support materials, such as aluminum. Catalysts suitable for HC and CO control include the noble metals Pt, Pd, and Rh. These materials have a high resistance to poisoning, react very little with support materials, and have better lightoff characteristics. These catalysts do, however, become poisoned by lead, phosphorus, and sulfur and require the use of fuel containing very low levels of these catalyst poisons.

Reduction of nitric oxide in the presence of CO and H_2 can also be achieved by catalytic conversion. Ruthenium is a favored catalyst material for NO reduction. The principal requirements of an NO-reduction catalyst are resistance to oxidation, selectivity of N_2 rather than NH_3 as a reaction product, resistance to poisoning, thermal stability of catalyst surface structure, and its performance as a function of temperature and fuel/air ratio of the gases. Figure 9.16 shows the degree of conversion of CO and HC in catalytic and thermal reactors as a function of reactor temperature.

Because of the high sensitivity of catalytic convertors to air/fuel ratio, oxygen sensors have been introduced into exhausts, which are coupled with a feedback loop to control the fuel flow. A special problem that has arisen with catalytic convertors is that sulfur contained in the fuel is emitted as sulfuric acid and there is concern at the high sulfate concentrations which can arise near motorways.

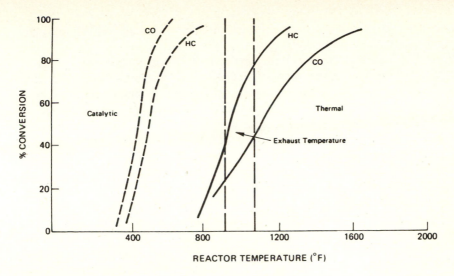

Figure 9.16 Comparison of catalytic convertor and thermal reactor for oxidation of CO and HC. (*Heywood, 1976.*)

Thermal Reactors

In a thermal reactor the hot exhaust gases are mixed with air at a temperature sufficiently high to allow reaction of unburned HC and CO. Thermal reactors are designed for low heat losses, long residence time, and low thermal inertia so as to achieve rapid warmup after engine start.

The operating temperature depends on the reactor inlet gas temperature, heat losses, and the amount of HC, CO, and H_2 burned up in the reactor. Fuel-rich cylinder exhaust gases require secondary air and, as a result of the burning of the excess fuel, temperatures may rise by 200°C. For lean engine exhaust gases, temperatures are considerably lower within a thermal reactor, and it is more difficult to achieve substantial reductions in CO and HC emissions. The principal disadvantage of thermal reactors is that the burning of fuel within the reactor makes no contribution to propulsion and, thus, results in a lowering of fuel efficiency.

Unconventional Spark Ignition Engines

Many alternative designs to the conventional spark ignition engine have been proposed. The main aim of alternative designs has been to improve the fuel economy and reduce the emission of pollutants. The developments which have taken place in the design of the conventional spark ignition engine have been such that it is very diffucult for any alternative design to compete with the reliability and comparatively low cost of current mass-produced conventional engines. The tendency has, therefore, been to make minor modifications to conventional spark ignition engines as a means of improving fuel economy and level of emission of pollutants. Two principal unconventional engines, the Wankel and stratified charge engines, have reached mass production stage and may possibly play an important role in automobile transport.

In the Wankel rotary engine, unburned mixture is introduced into the space between the rotor head and the rotor housing. As the rotor moves the shape of this combustion chamber space changes and the gases burn, expand, and are finally expelled from the engine so that the operating cycle has the same sequence of processes as the reciprocating spark ignition four-stroke cycle. Very significant reductions in weight of the engine can be achieved by the rotary mechanism but problems of sealing between the rotor and the rotor housing have led to less satisfactory performance than was initially desired. Currently, HC emissions are higher, CO about the same, and NO_x lower than for equivalent piston engines. Variation in engine NO emissions as a function of change of equivalence ratio is qualitatively similar to that of a piston engine with maximum NO concentrations at fuel/air ratios slightly lean of the stoichiometric. The effects of exhaust gas recirculation, spark timing, and engine load on NO emissions have also been found to be qualitatively similar to those of conventional piston engines. Seal leakage into the exhausting chamber is an additional source of unburned hydrocarbons which, at low engine speeds, can be greater than that due to the quench and crevice layers in the engine.

Figure 9.17 illustrates the motion of quench HC and leakage HC during the exhaust process in a Wankel engine. Combustion has been completed in chamber (1) and quench layers of unburned mixture are along the walls of the rotor and rotor housing. As the rotor rotates, it begins to scrape these quench layers into a vortex with burned gas at the trailing end of the chamber. At stage (2), the exhaust port opens and the burned gas is blown down, entraining some of the quench layer on the lead portion of the rotor. Unburned fuel and air begin to leak into the trailing zone from the compression chamber, and this leakage continues until the flame front in the combustion chamber reaches the leading apex seal. Further rotor movement toward position (3) removes the quench layer from the vicinity of the port so that only burned gas is exhausted. The vortex continues to grow until, at position (4), it is of sufficient size to be exhausted. A high peak in HC concentration is found in the Wankel engine, similar to that in the conventional spark ignition engine.

The effective surface-to-volume ratio is higher in a Wankel engine and this, together with equivalence ratio, engine load, engine speed, and ignition timing, influences HC emissions. The combustion chamber recess profile, seal design, and effective leakage area can also be important factors influencing emissions. Both thermal reactors and catalytic convertors have been incorporated in engines in order to make further reductions in emissions.

In stratified charge engines, the mixture ratio is purposely varied in different parts of the combustion chamber. Generally, a fuel-rich region is created, either in a separate precombustion chamber or in a certain portion of the volume of the combustion chamber. Combustion then takes place in two stages—initially rich, followed by a rapid mixing so that combustion is completed under lean conditions. By avoiding burning at stoichiometric mixtures, gas temperatures are lower

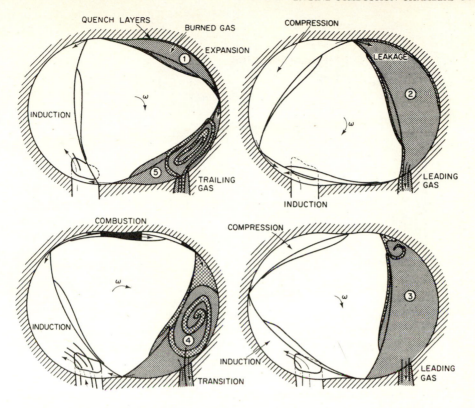

Figure 9.17 Schematic illustrating the motion of quench HC and leakage HC during the exhaust process in a Wankel. (*Heywood, 1976.*)

during combustion and also during expansion. Dissociation is reduced and increased compression ratios can be used since the detonation possibilities are reduced. Substantial improvements in fuel economy, as well as reduction of overall emissions, have been achieved by using the stratified charge. It has been found that engines can be operated at much leaner overall mixtures and much larger dilution recycled exhaust gases than in conventional engines.

Figure 9.18 shows an open chamber stratified charge engine with a high pressure injector. The piston head is recessed and the fuel spray is directed into this recess, where the airflow is swirled. Fuel injection is timed so as to provide a rich mixture near the spark plug, which is easily ignitable, while the overall fuel/air ratio in the engine is maintained lean. This type of open chamber stratified charge engine has certain similarities to diesel engines. These engines provide low levels of NO_x, due to the avoidance of the high temperatures associated with stoichiometric mixtures. Hydrocarbon and CO emission levels are comparable to or better than current conventional SI engine levels. Significant improvements in fuel economy have also been achieved.

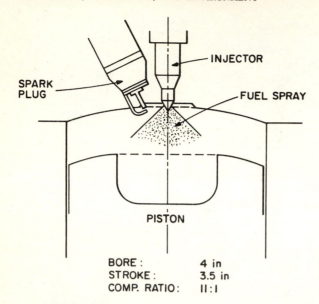

BORE :	4 in
STROKE :	3.5 in
COMP. RATIO :	11:1

Figure 9.18 Open-chamber stratified-charge engine with injector. (*Heywood, 1976.*)

Figure 9.19 shows a divided chamber engine, where a small prechamber has been added to the cylinder head. Fuel is injected through the injection nozzle directly into the prechamber and the rich mixture is ignited. The expanding gases pass through the orifice between the prechamber and the main combustion chamber to provide a burning jet, which ignites the very lean mixture which has been fed to the main combustion chamber. In some designs all the fuel is added in the prechamber and only air is in the main combustion chamber.

Nitric oxide emissions have been found to be low and relatively insensitive to overall fuel/air equivalence ratio. Carbon monoxide emissions are also low for lean overall engine operation and some very low HC exhaust concentrations have been obtained by careful optimization of fuel injection rate, pressure, and timing.

Particulate Emissions from SI Engines

The main particulate emissions affecting air pollution from spark ignition engines are due to lead, soot, and sulfates. Engines operated with lead additives have solids emissions up to 0.5 g/km. Solids emissions from automobiles operated with unleaded fuel are substantially lower. Particulate emission rates are considerably higher for cold engines than for hot engines. The particle size distribution with leaded fuel is about 80 percent by mass below 2 μm and about 40 percent below 0.2 μm.

Most of these particles are formed in the exhaust system, due to vapor-phase condensation, and they grow in size, due to coagulation. Particles are emitted both directly and after deposition on the walls where agglomeration may occur. During acceleration, when the exhaust flow is suddenly increased, particles, rust, and scale on the walls are suddenly emitted. Lead compounds constitute between 20 and 80 percent of the total particle mass emitted. Soot emissions can be reduced to insignificant levels by proper adjustment and control of the engine.

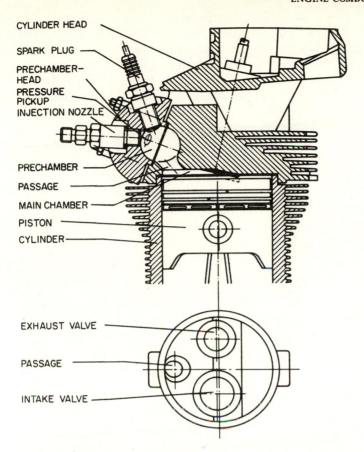

CYLINDER HEAD
SPARK PLUG
PRECHAMBER–HEAD
PRESSURE PICKUP
INJECTION NOZZLE
PRECHAMBER
PASSAGE
MAIN CHAMBER
PISTON
CYLINDER
EXHAUST VALVE
PASSAGE
INTAKE VALVE

Figure 9.19 Prechamber combustion system for stratified charge engine with 42 percent of clearance volume in prechamber. (*Pischinger and Klocker, 1974.*)

9.3 DIESEL ENGINES

Diesel engines are used extensively in transport systems and also as a source of power for stationary systems. Buses and trucks almost exclusively use diesel engines and, in many countries, the diesel engine has replaced the steam engine for railway traffic. For electricity generation in isolated areas and as a standby for failure of the main grid system, the diesel engine is widely used. It is beginning to gradually replace the regular spark ignition engine in private passenger cars and its advantages have been proved for public automobiles such as taxis and limousines.

The special advantage of diesel engines is the greater fuel economy when compared with the spark ignition engine. The thermal efficiency of diesel engines is considerably higher than that of spark ignition engines, mainly due to the high compression ratios, the autoignition process, the lower pumping losses due to the absence of the throttle valve, and the overall lean mixture ratios used in the combustion process. Diesel engines have, however, always been easily recognized by the emission of black smoke with a pungent odor, coupled with high noise levels. As a result of extensive research in order to meet legislation on emission, very substantial reductions have already been achieved in the emission characteristics of diesel engines. Most diesel engines emit less CO and unburned hydrocarbons than comparable SI engines. Since many of the special problems in diesel engines are associated with the direct ignition of liquid fuel into the combustion chamber, much attention has been devoted to the vaporization and burning characteristics of the liquid fuel sprays under high compression conditions. Since there is no spark, autoignition can only take place where the mixture ratio, temperature, and heat balance requirements for flame propagation are satisfied. Because the system is not homogeneous, there are substantial variations in temperature and mixture ratio throughout the chamber. The main emphasis in research has been to determine the principal mechanism of combustion and then further to optimize the system for maximum combustion efficiency, coupled with minimum emission of pollutants.

Diesel engines are subdivided into two principal groups; the direct injection (DI), in which the fuel is injected directly into the combustion chamber, and the indirect injection (IDI), in which the fuel is injected into a separate prechamber and, after mixing with a portion of the total air, the mixture passes into the main combustion chamber. The direct injection engine is the one most commonly used and most of the combustion problems are associated with the interaction of the fuel spray with the airflow. In the indirect injection engine, the aim is to complete all vaporization in the prechamber so that no liquid droplets enter the main combustion chamber. Developments which are being made in spark ignition engines are also incorporating the prechamber and direct fuel injection concepts so that the principal distinguishing feature between these engines is the high compression ratios in the diesel engine, which allow autoignition without the requirement of a spark.

Direct Injection Engines

In direct injection (DI) or open chamber (OC) diesel engines, the kinetic energy for the mixing of fuel and air is contributed partly by the liquid jet injected through the nozzle and partly by the air. The specific kinetic energy of the fuel is given by the equation

$$E_f = C\,\frac{\Delta P_f}{\rho_f} \tag{9.6}$$

where C is a velocity coefficient, ΔP_f is the pressure drop across the orifice, and ρ_f is the fuel density. The pressure drop across the orifice varies during the time of injection, starting from zero, rising to a maximum and then falling to zero at the end of injection. This pressure difference affects both the size of droplet and the rate of mixing of the fuel and the air. In a photographic study of diesel-engine combustion, Rife and Heywood (1974) showed the effects of reduction in pressure drop across the orifice near the end of injection, resulting in poor mixing.

During the intake stroke, air flows through the intake ports and generates air motion. Intake valves or ports are sometimes masked so as to direct the airflow tangentially into the cylinder, thereby generating a swirling motion. After the close of the intake ports and during the compression stroke, the nonswirling components of flow are severely damped but the swirling motion is retained. Fuel is injected radially as the swirl approaches top dead center and air is entrained into the spray. The main flow pattern of air is, thus, made up of a tangential component, due to initial swirl, and a radial component, due to fuel injection. Dent and Durham (1974) measured velocity distributions by hot-wire anemometry and concluded that a forced vortex solid-body rotation occurred over the whole of the compression period and that an inward radial movement takes place near the end of the compression stroke. This radial movement is caused by the flow into the combustion bowl.

Photographs by Watts and Scott (1970) showed that the core of the spray is not significantly deflected by the air swirl, as the spray tip passes across the combustion chamber toward the walls. These photographs mainly represent the large droplets which are concentrated in the core of the spray. Smaller droplets and fuel vapor are carried away from the core by the swirling air.

The degree of atomization and the magnitude of air swirl require to be carefully balanced in a diesel engine. Excessive fuel atomization results in poor penetration while excessive air swirl can also result in less penetration and overlap of the sprays. Both of these factors result in poor combustion, due to poor mixing. The optimum swirl and degree of atomization varies from one engine to another, depending upon the geometry of the combustion chamber, the number and size of injection holes, the droplet size distribution, and the relative momenta of air swirl and fuel injection.

The variation of cylinder pressure with crank-angle degrees is shown in Fig. 9.20. This shows that maximum pressure of the order of 6 MPa is reached at approximately 10° after TDC. Figure 9.20 shows the variation in cylinder pressure history for moderate fuel injection pressures (108 MPa) compared with that for high injection pressure conditions (176 MPa) as determined by Shahed et al. (1978). The cylinder pressure history is used as a measure of cycle efficiency. The use of very high injection pressures causes more rapid mixing and results in a faster pressure evolution and higher cycle efficiency. Figure 9.21 shows the variation of mixing temperature, equivalence ratio, and nitric oxide formation with crank-angle degrees for moderate injection pressure conditions (108 MPa) as determined by Shaded et al. (1978). Temperatures rise as equivalence ratio decreases. Nearly all the NO is formed within a narrow band of equivalence ratio close to stoichiometric conditions where the temperature reaches a peak. Use of very high injection pressure leads to higher local temperature earlier in the expansion process, resulting in a far more rapid NO-formation rate.

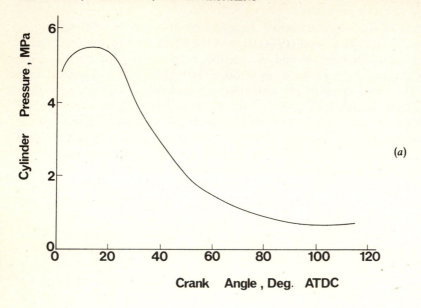

(a)

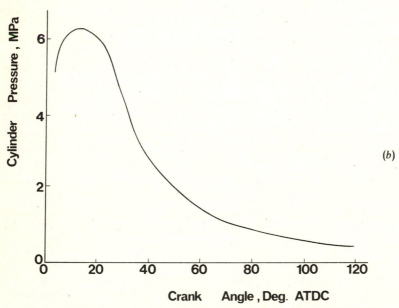

(b)

Figure 9.20(*a*) Cylinder pressure as function of crank angle for moderate injection pressure (108 MPa). (*b*) Cylinder pressure as function of crank angle for high injection pressure (176 MPa).

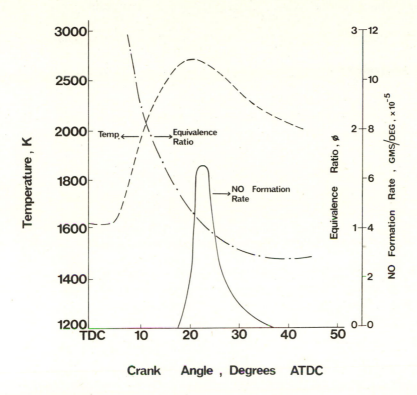

Typical Zone History, 108 MPa

Figure 9.21 Mixing temperature and nitric oxide formation history of typical combustion zone at moderate injection pressure conditions. (*Shahed et al.*, 1978.)

Prechamber Engines

In prechamber engines, the combustion chamber is subdivided into a prechamber and a main chamber. Fuel is injected into the air in the prechamber and, on expansion, gases pass into the main chamber, generating turbulence as they pass through the orifice between the chambers. These engines are sometimes referred to as indirect injection (IDI) or divided chamber engines.

In prechamber engines, mixing is achieved by airflow into the prechamber during the compression stroke and, subsequently, by the flow of products of combustion from the prechamber to the main chamber. Since mixing is no longer dependent upon the kinetic energy of the fuel jet, injection pressures in the prechamber engines can be lower than those in open chamber engines. The maximum pressures, rates of pressure increase, and noise levels in the prechamber engines are also lower than in the open chamber engines. The increased surface-to-volume ratio, as well as the higher levels of turbulence, lead to higher cooling losses, lower thermal efficiencies, and higher local thermal stress, caused by the impingement of high-velocity gases discharged from the prechamber into the main chamber.

In the early stages of combustion in the prechamber, carbon formation is high because of the relatively rich overall mixture. During the later stages of combustion, the high velocity and turbulence of the gases discharged from the prechamber lead to effective mixing with the air in the main chamber. Oxidation reactions are, thus, very effective, with the net result that smoke emission is lower than in open chamber engines. Nitric oxide, HC, and CO emissions have, in general, been found to be less in the prechamber engines than in open chamber engines.

The M-System

In the M-system, a small amount of fuel is allowed to evaporate and mix with the air before the start of autoignition. The majority of the fuel is injected at an acute angle on the walls of a spherical chamber, so as to form a thin liquid film. Swirling air passes over the liquid film and there are several ignition sources formed by injecting a small percentage of fuel into the chamber. In this film-type mixture formation, the maximum pressures, the rates of pressure increase, and noise are all lower than those found in DI engines.

Fuel Injection

Fuel injection in diesel engines is effected by the use of high-pressure pumps and nozzles with very small dimensions. The principal types of nozzles are: throttle, pintle, single orifice, or multiorifice. In the single-orifice nozzle a hole is drilled through the chamber wall so that the liquid jet passes through a short cylindrical tube to form a round liquid jet at the orifice exit. Because of the high pressure differential across the nozzle, liquid velocities are of the order of several hundred meters per second and initial liquid-jet diameters are less than 1 mm. Breakup into droplets occurs rapidly so that, within a very short distance from the discharge, a spray of liquid droplets is formed. The high velocity liquid particles entrain air from the surroundings so as to form a two-phase jet with the droplets as the source of momentum.

The liquid disintegration process is due to the instability of the liquid jet as well as the shear between the high-velocity liquid and lower-velocity surrounding air. This leads to a distribution of size of droplets. The drag deceleration forces on the droplets are a function of the size of the droplets and the differential velocity between the droplets and the air. Because of their high momentum, larger droplets will penetrate farther into the chamber than smaller droplets. In addition, the total momentum of the jet increases continuously as fuel is injected so that the environment which droplets encounter on emerging from the orifice varies with time. At the beginning of injection droplets encounter relatively stagnant air, whereas, subsequently, droplets enter a jet which was generated by preceding droplets.

Because of the short duration and periodicity of fuel injection, the diesel spray must be considered as a nonsteady process with characteristics distinctly different from those of continuous spray injection, as in gas turbines and furnaces. Differences in relative retardation of droplets lead to coagulation and changes in size distribution patterns along the length of the spray.

Spray Atomization

In diesel sprays, as in other sprays, liquid breakup begins with the formation of ligaments, followed by large drops, which subsequently break up into small droplets. Droplet breakup continues whenever the Weber number exceeds a critical value. The Weber number is defined as the ratio of the inertia body forces to the surface tension forces

$$\text{We} = \frac{\rho D V^2}{\sigma} \tag{9.7}$$

where ρ is mass density, D is droplet diameter, V is differential velocity between droplet and airstreams, and σ is surface tension. Critical Weber numbers are of the order of 10.

The quality of spray atomization is of great importance in diesel engines. The liquid requires to be fully vaporized and combustion must be completed within a restricted period of time. All droplets require to be completely vaporized before coming into contact with chamber walls. If this is not achieved, carbon deposition on cooled surfaces and emission of unburned hydrocarbons can occur. Sufficient time is required after vaporization to allow mixing with the air and completion of combustion before gases are cooled and emitted through the exhaust valve. In order to achieve this, it is particularly important that no large drops are formed in the spray and that droplet size should be of the order of 20 μm and below.

The shape of the jet and the droplet size distribution depend mainly upon the nozzle pressure differential, the geometry of the nozzle hole, the air density, and liquid properties such as viscosity and surface tension. In an experimental study under conditions which simulated diesel injection, Sass (1929) showed the effect of variation in nozzle diameter on droplet size distribution. Using liquid injection pressures of 280 bar and with air pressure of 10 bar in the chamber, droplet sizes ranged from a diameter of 2 to 40 μm when the nozzle diameter was varied from 0.4 to 0.8 mm (Fig. 9.22). The effect of variation of nozzle diameter on droplet size distribution is shown in Fig. 9.23. The effect of variation in chamber pressure on droplet diameter, as well as the effects of variation of injection pressure, were measured by Sass (1929) and are shown in Fig. 9.24. It should be noted that the measurements of Sass were made in 1929. More sophisticated optical methods have been developed and these are currently being used for more accurate determination of size distribution in diesel sprays.

Natarajan and Brzustowski (1970), Natarajan (1976), and Newman and Brzustowski (1971) have studied evaporation and combustion of single droplets and droplet clusters under sub- and supercritical conditions. From studies made

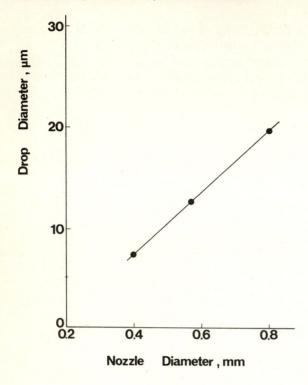

Figure 9.22 Average droplet diameter as a function of variation in nozzle diameter in diesel sprays. Injection pressure, 280 bar; chamber pressure, 10 bar. (*Sass, 1929.*)

at pressures and temperatures slightly higher than critical, they concluded that there was no substantial difference from the behavior at subcritical conditions. In the major part of their investigations, however, they concluded that a liquid jet injected into a supercritical atmosphere can be treated as a continuum because of droplet shattering at vanishing surface tension as the critical point is approached.

Spray Penetration

The extent of penetration of the liquid spray into the diesel chamber is determined by the distance from the injection nozzle, where evaporation of all droplets has been completed. This penetration distance cannot be rigorously defined, since it is dependent upon the lower limit of droplet size that is detectable. Since spray penetration is normally determined by photography, the extent of penetration that can be detected is dependent upon the fineness resolution of the optical system. The spray penetration distance was determined by Dent (1971) in the form of the correlation

$$x = 56.1 \left[\left(\frac{\Delta p}{\rho_a} \right)^{1/2} td \right]^{1/2} \left(\frac{530}{T_g} \right)^{1/4} \tag{9.8}$$

where x is the spray penetration distance (mm), Δp is the effective injection pres-

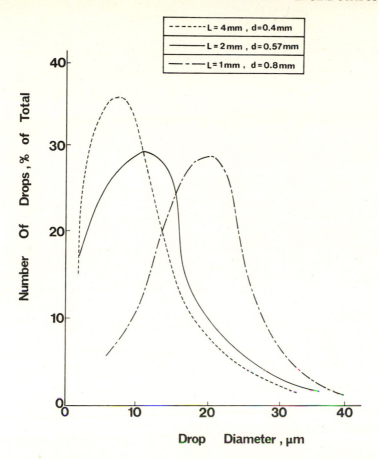

Figure 9.23 Effect of nozzle-hole diameter and length on drop-size distribution. Injection pressure, 280 bar; chamber pressure, 10 bar. (*Sass, 1929.*)

sure (bar), ρ_a is the chamber air density (kg/m³), t is the time from start of injection (s), d is the nozzle-hole diameter (mm), and T_g is the chamber air temperature (K). This correlation shows that penetration distance can be reduced by reducing the diameter of the nozzle hole and the injection time. The chamber air density and temperature are predetermined by overall engine design considerations. The Dent correlation indicates that penetration distance is proportional to $\Delta p^{0.25}$, which is based on momentum considerations. Since, however, average droplet size decreases with increase in injection pressure, the usual diesel practice is to have high injection pressures, so as to achieve droplet sizes of the order of 10 μm.

As a result of friction, the droplets exchange their momentum with the air so that, at a short distance from the nozzle exit, the spray can be considered as a two-phase jet. The spread of this jet is dependent upon the rate of entrainment of air into the jet with distance downstream and on the ratio of the density of fluid

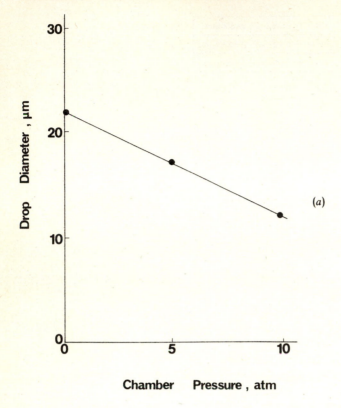

(a)

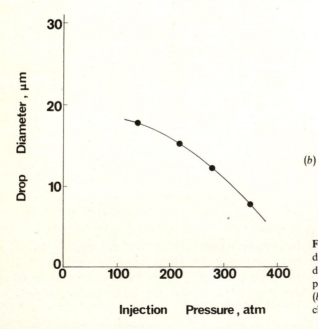

(b)

Figure 9.24 Average drop size in diesel spray with nozzle 0.57-mm diameter. (a) Variation of chamber pressure; injection pressure 280 bar. (b) Variation of injection pressure; chamber pressure 10 bar. (*Sass, 1929.*)

within the jet to that surrounding the spray. Considering diesel fuel injection as a jet phenomena, Melton (1971) derived the expression for jet penetration as

$$x = \frac{d}{2\alpha}\left[\sqrt{\frac{4\alpha V_f}{d}\sqrt{\frac{\rho_f}{\rho_j}}\, t + 1} - 1\right] \qquad (9.9)$$

where d is the nozzle hole diameter, α is a jet spreading factor, assumed to be a constant equal to 0.085, V_f is the injection velocity of the liquid fuel, ρ_f and ρ_j are the density of fuel and jet, respectively, and t is time.

Fuel Evaporation

Fuel is injected into diesel engines at very high pressures, which range from 200 to 2000 bar. These pressures usually exceed the critical pressures of the majority of the fuel constituents, so that flash evaporation is caused by the sudden drop in the fuel pressure as it is injected through the nozzle. Cylinder air pressures are usually in the range between 20 and 60 bar, although higher pressures may be reached in supercharged engines.

The amount of fuel that is evaporated due to flash evaporation, compared with that due to heat transfer from the gases or wall surfaces, is not known but it is generally assumed that the amount of flashed vapor is small in comparison with the total amount of fuel injected. This small amount, however, can play an important role in initiating the preignition reactions. Fuel evaporation in diesel engines, particularly those with turbocharging, may occur under supercritical conditions. Since diesel fuels are made up of compounds with considerable differences in thermodynamic properties, the degree of vaporization under supercritical conditions will be dependent upon the chemical composition of the fuel.

The rates of vaporization of liquid droplets will be a function of their size, trajectory, and temperature, and gas concentrations in the gas surrounding the droplets. As droplets decrease in size, the surface-to-volume ratio of the droplets is increased, leading to higher rates of heat transfer and, subsequently, higher rates of vaporization. In addition, the large droplets are concentrated in the core of the spray, resulting in a quenching of the surrounding gases; the reduced temperature difference between gas and droplet further reduces the rate of vaporization of large droplets.

Watts and Scott (1970) estimated that the fuel sprays take about four crank-shaft degrees to reach the wall. This is equivalent to about 0.4 ms at a speed of 1800 r/min. During this time, some of the small droplets are completely evaporated and the larger droplets are partially evaporated. Sprays have been seen to directly impinge on to the solid surfaces in the combustion chamber.

Henein (1976) has calculated that, under typical air conditions for naturally aspirated diesel engines, droplets of N-hexadecane with diameters less than 11 μm are completely evaporated. Most droplets of this size are carried away from the core of the spray by the swirling air.

In multicomponent liquid sprays, the concentration of the higher-boiling components in the liquid droplets increases as the spray evaporates. As air temperatures increase, this increased concentration of higher-boiling components becomes more pronounced. This leads to a partial separation of fuel vapor; fuel droplets in the core have a larger percentage of the higher-boiling compounds while the fuel vapor carried away by the swirling air has a large percentage of lighter compounds.

Physical Models of Fuel Spray Injection

On the basis of cine-film records, Henein (1976) has developed a physical model for a spray at the beginning of combustion, as shown in Fig. 9.25. The initial size distribution of droplets will be gaussian in form with the larger droplets concentrated in a core along the centerline of the spray. The swirling air carries fine droplets from the outer periphery of the spray, across the core, causing a skewed air/fuel ratio distribution, as shown in Fig. 9.25. This air/fuel ratio profile changes with distance downstream the spray. The smaller droplets will be concentrated toward the downstream edge of the spray. Most of the droplets carried away from the core will be completely evaporated before the start of ignition.

Photographic studies made through windows of diesel engine combustion chambers (Scott, 1968; Rife and Heywood, 1974; and Watts and Scott, 1970), showed that ignition starts in the spray envelope near the downstream edge of the spray. Once ignition starts, small independent nonluminous flame fronts propagate from the ignition nuclei and ignite the combustible mixture around them. The

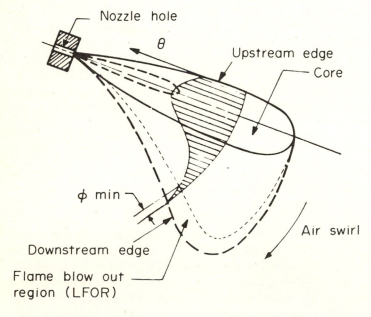

Figure 9.25 Physical model of fuel spray injected in swirling air for diesel engines. (*Henein, 1976.*)

overall mixture ratio, on a mass average basis, is lean, though pockets of richer mixture exist where the burning takes place preferentially. Near the far downstream edge of the spray, the mixture becomes too lean for ignition but temperatures are sufficiently high for some fuel decomposition and partial oxidation to take place. The partial oxidation products contain aldehydes and other oxygenates, while the decomposition products consist of lighter hydrocarbon molecules. Fuel entering this lean edge of the spray will not be completely burned and is considered as the major source of emitted unburned hydrocarbons. It corresponds with the quench zone in spark ignition engines. The influence of this lean edge of the spray is reduced by increasing the overall temperature and pressure within the combustion chamber. Substantial reductions in the emission of unburned hydrocarbons can be achieved by controlling the momentum, size, and direction of injection of liquid droplets, so that vaporization, mixing, and combustion can be completed within the bulk spray. If this is not achieved, liquid droplets or fuel vapor pass into regions of low temperature, where mixture ratios are too lean for flame propagation.

Following ignition and combustion in the lean region at the edge of the spray, the flame propagates toward the core. Heat is transferred mainly by radiation, since the main convection flow is from the spray core toward the outer edge. Droplets within the spray core are carried toward the flame by the air motion. There appears to be little evidence of individual flames surrounding individual droplets and, as in other sprays, combustion takes place in the vapor phase subsequent to vaporization.

Under part-load operation, there is an adequate supply of oxygen to the core of the jet, leading to completion of combustion with consequent high temperatures and high NO_x formation. Near full-load conditions, incomplete combustion occurs in many locations in the fuel-rich core. In fuel-rich saturated hydrocarbon flames, H-atom stripping occurs and recombination between hydrocarbon radicals leads to the formation of hydrocarbons containing molecules with more or less carbon atoms than the original fuel. Near full-load, unburned hydrocarbons, carbon monoxide, oxygenated compounds, and carbon are formed and emitted.

Spray Tail

During the short period of liquid fuel injection, the fuel pressure rises from zero, reaches a maximum, and then falls back to zero. The fuel-flow rate follows the same cycle and the atomization of the spray will vary with the variations in pressure and fuel-flow rate. Toward the end of injection, large droplets have been seen to form, caused by a combination of decreased fuel injection pressure and increased cylinder gas pressure. The low momentum of this "spray tail" results in poor penetration. Under high-load conditions, the spray tail has insufficient momentum to enter regions with adequate oxygen concentration. The temperature of the surrounding gases is, at this time, close to the maximum cycle temperature. The rate of heat transfer to the droplets is very high, so that droplets tend to evaporate quickly and decompose. The decomposed products contain unburned hydrocarbons and a high percentage of carbon monoxide. Partial oxidation products include carbon monoxide and aldehydes.

Under medium and high loads, many injection systems produce "afterinjection," due to the injector valve opening for a short time after the end of the main injection. In general, the amount of fuel delivered during afterinjection is very small but, due to the relatively small pressure differential, both atomization and penetration are poor. Emissions of carbon particles, unburned hydrocarbons, and CO can be substantially reduced by using a check valve and throttle orifice, so as to reduce this afterinjection.

Fuel Deposition of Walls

Fuel sprays can impinge directly on the cylinder walls. Any droplet, which has sufficient momentum and is not deflected by the airstream or vaporized, will reach the cylinder walls. In small, high-speed engines, it is not uncommon for liquid to reach the walls and form a liquid film. The rate of evaporation of this liquid film is dependent upon the temperatures of the gas and wall, gas velocity, gas pressure, and properties of the fuel. If air is sweeping past the liquid film, the flame can propagate to within a small distance from the wall. Combustion of the remainder of the fuel on the walls depends upon the rate of evaporation and mixing of fuel and oxygen. If the surrounding gas has a low oxygen concentration, or there is insufficient mixing, evaporation can occur without complete combustion.

As the piston moves on the expansion stroke, gases flow outward radially to fill the space between the piston top and the cylinder head. In shallow-bowl combustion chambers, most of the combustion process takes place in the bowl. In deep-bowl types, the outward radial flow is more significant than in shallow-bowl chambers. The swirl motion continues at a lower rotational velocity than that during the compression stroke. The combination of radially outward flow and swirl draw the fuel vapor and the partial oxidation products from the wall toward the cylinder walls. Cine films show eddies and heavy smoke clouds outside the bowl near the areas where sprays impinge on the wall. These eddies burn with a luminous flame, due to the presence of carbon particles.

Heat Release Rates

Henein (1976) has determined the heat release rates from pressure traces obtained in open-chamber diesel engines. Figure 9.26 shows the plot of gas pressure, rate of heat release, and cumulative heat release as a function of crank-angle degrees, as reported by Lyn (1963). Injection takes place at 22° before top dead center (TDC), as the pressure in the cylinder is rising. During the ignition delay of approximately 1.2 ms, heat is transferred from the hot air to the liquid fuel, resulting in a negative heat release. Droplets at the leading edge of the spray are the first to evaporate, forming a nonhomogeneous fuel vapor-air mixture. The complete evaporation of these droplets takes place in much less time than the ignition delay period. Igni-

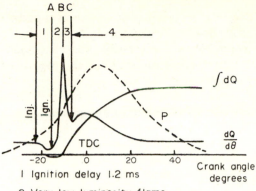

1 Ignition delay 1.2 ms

2 Very low luminosity flame

3 Orange flame

4 Highly luminous flame

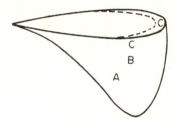

Figure 9.26 Heat release in a DI engine as related to pressure trace and spray penetration. (*Henein, 1976.*)

tion occurs at 13° BTDC and the rate of burning reaches its peak at about 10° BTDC. At the end of the ignition delay, combustion commences near the outer downstream edge of the spray. The flame observed by Lyn was of very low luminosity up to 10° BTDC, indicating that the burning was essentially confined to the premixed part of the jet.

At the point of maximum heat release rate, point B, Fig. 9.26, the cumulative heat release is about five percent of the total computed heat release. This reflects the approximate amount of fuel burned soon after ignition as a percentage of the total amount of fuel. At 10° BTDC, the first appearance of the orange-coloured luminous flame was observed but this did not spread to surround the tip of the jet until 7° BTDC. During the remainder of the combustion process, the flame was observed to have high luminosity, due to the presence of carbon particles. This high luminosity is characteristic of diffusion-type flames and is a further indication that the liquid initially vaporizes and, subsequently, diffuses toward the air, where it burns as in a gaseous diffusion flame.

Heat transfer losses from the cylinder are correlated with fundamental engine-operating parameters such as piston speed and engine load. Radiation can contribute as much as 40 percent of the total heat transfer in the central portion of the head of a diesel engine with a small amount of air movement and a large piston bowl. Instantaneous radiation heat transfer may rise to 50 percent of the total instantaneous heat transfer rates. The source of the infrared radiation is carbon particles within the flame which attain particle temperatures of the order of 2000 K. Radiant emission is very strong in the early vigorous stages of the combustion process and falls off rapidly to low levels after 30° ATDC, when the intensity of the energy release process decreases.

Turbulence intensity determines the level of velocity and temperature fluctuations. Radiation heat transfer has an extreme sensitivity to local temperatures, so that fluctuations of temperature of 100 K or more within the flame can very significantly increase the radiative and total heat transfer. Convective heat transfer at the walls can also be significantly increased by turbulence bursts and lead to periodic surges in heat transfer. The experiments being conducted on use of high-temperature ceramic linings and increased insulation are resulting in "hot" components including piston, cylinder head, valves, cylinder liner, and exhaust parts. These are leading to engines with higher energy efficiency.

Preignition Processes

Combustion in diesel engines is initiated by autoignition nuclei at numerous locations in the combustion chamber. There is, thus, a distinct difference between the spark ignition process in gasoline engines and the autoignition process in diesel engines. Henein (1976) has separated the preignition processes in diesel engines into physical and chemical processes. The physical processes are:

1. Disintegration of liquid jet and droplet formation
2. Heat transfer to the spray and droplet evaporation
3. Diffusion of the fuel vapor into the air to form a combustible mixture

The chemical processes are:

1. Decomposition of heavy hydrocarbons into lighter components
2. Preignition chemical reactions between the decomposed components and oxygen.

At the very early stages of injection, there is insufficient mass of fuel vapor to cause any detectable combustion. The early stages of the preignition process are dominated by physical processes leading to the formation of pockets of combustible mixture. The later stages of the injection process become dominated by the chemical changes, which lead to autoignition.

Ignition Delay

In diesel engines, the ignition delay is defined as the time between the start of injection and the formation of a detectable number of ignition nuclei. Ignition delay plays an important role in the combustion process and, consequently, has a great effect on mechanical stresses, engine noise, and exhaust emissions. Two methods are used in practice for determining ignition delay, one based on pressure measurements and the second on illumination. Measurements of cylinder pressure as a function of time shows a distinct point on the record where the pressure suddenly increases, due to the onset of combustion. At this point, sufficient exothermic reactions have occurred to cause a detectable increase in the gas pressure. Ignition delay can also be determined by measuring the time between the start of injection and the emission of visible radiation, due to chemiluminescence or thermal radiation. Chemiluminous radiation results from the excitation of formaldehyde molecules, and thermal radiation is due to the high-temperature carbon atoms in the flame. In high-speed direct injection diesel engines, the pressure increase due to combustion occurs after a delay of 1 ms, while the illumination delay is detected at 1.2 ms.

A record of cylinder pressure as a function of time can be obtained with relative ease, and ignition delays determined from pressure records are generally reproducible. Since the sudden pressure increase is directly related to stress and noise, the pressure records are the main source of information for determination of ignition delay. For diesel fuel, ignition delay is decreased from 0.75 ms at an air intake temperature of 313 K to 0.46 ms at an air intake temperature of 671 K. Henein (1976) has also shown that fuel with high volatility has a longer ignition delay. At an intake temperature of 313 K, gasoline has an ignition delay of 2.14 ms, while diesel number 2 fuel has a 0.75-ms delay. Since gasoline is more volatile, the rate of evaporation and diffusion is higher than that of the diesel fuel. If physical processes control the ignition delay, the ignition delay for gasoline should be shorter than that for diesel fuel. Henein (1976) concluded from his results that the rate-controlling process must be chemical rather than physical during the delay period in high-speed diesel engines. Measurements of the autoignition of fuel sprays in a constant volume bomb showed that the ignition delay was primarily affected by the air temperature, pressure, and oxygen concentration. All these factors affect the rate of chemical reactions. Meanwhile, the effect of the amount of fuel injected, the rate of fuel injection, injector-needle opening pressure, and nozzle size (all of which affect the physical processes) were found to be slight. Other factors which affect the physical process, such as air velocity, air-fuel rates, and intensity of turbulence, were found to have a negligible effect on the ignition delay. Physical processes become important as the droplet size increases. This was confirmed by Pedersen and Qvale (1974), who showed that the ignition delay period could be changed when variations were made using low injection pressures and large nozzle areas. Turbocharging has reduced the influence of ignition delay and, hence, diesel engine knock.

Preignition Chemical Reactions

Two types of chemical reactions have been found to take place prior to ignition. Garner et al. (1961) found that, for four different liquid hydrocarbons, peroxides and aldehydes were formed during the ignition delay and reached their peak concentration just before the start of the pressure increase due to combustion. For high cetane fuel numbers, the peak of concentration of these intermediate compounds occurred earlier than with lower cetane number fuels. Preignition reactions take place in two stages. First very slow reactions occur and form intermediate compounds, such as peroxides and aldehydes. Second, once a critical concentration of these intermediate compounds has been reached, very fast chain reactions occur and lead to autoignition. If the critical concentration is not reached in parts of the spray where mixtures are very lean or very rich, partial oxidation products, such as aldehydes and carbon monoxide, may become frozen without further reaction. The ignition delay may be considered to have ended when the critical concentrations for the intermediate compounds are reached.

The critical concentration of the intermediate compounds, which is enough to start chemical reaction and cause a detectable pressure increase, depends upon the total mass and heat capacity of the charge in the combustion chamber. The ignition delay, ID, in a diesel engine can be determined from the following equation

$$ID = \frac{A \exp (E_a/R_0 T)}{p^n} \tag{9.10}$$

where A is a constant dependent upon combustion chamber geometry, E_a is a global activation energy, R_0 is the universal gas constant, T is the absolute temperature, and p is the absolute pressure, with n an exponent. Calculations of ignition delay as a function of temperature show that gasoline fuel has a longer delay than diesel fuel. In the comparison of ignition delays between gasoline and diesel fuels, it is seen that the activation energy and, hence, the chemical reactions play a much more significant role than volatility. Global activation energies increase with the decrease in cetane numbers of fuels, thus, low cetane number fuels are more stable and resist autoignition more than high cetane number fuels.

Ignition delays can be calculated for various fuels by use of the equation

$$ID = Ap^n \phi^C e^{D/T} \tag{9.11}$$

where $\phi = p_o/(0.21 \times p_{total})$, ϕ has the value 1 for atmospheric air, ID is in milliseconds, p in bars, and T in kelvins. The values for constants A, C, D, and exponent n are given in Table 9.6.

Table 9.6 Empirical constants for calculation of ignition delay [Eq. (9.11)] (Henein, 1976)

	A	n	C	D
Kerosene	2.76×10^{-2}	-1.23	-1.60	7280
n-heptane	7.48×10^{-1}	-1.44	-1.39	5270
n-dodecane	8.45×10^{-1}	-1.31	-2.02	4350
n-$C_{16}H_{34}$	8.72×10^{-1}	-1.24	2.10	4050

The introduction of synthetic fuels derived from coal which have high aromatic content and low cetane numbers will result in increased operational problems associated with ignition and preignition.

Flame Propagation in Diesel Engines

In diesel engines which are not supercharged, compression ratios are between 17 and 22. Prior to combustion, toward the end of the compression stroke, pressures are between 40 and 70 bar, with gas temperatures between 800 and 1000 K. For diesel fuels, autoignition temperatures are around 650 K, so that gas temperatures are well above the fuel autoignition temperature. Injection takes place just before top dead center (TDC), so that, after ignition delay, combustion commences at TDC. During the ignition delay, the smallest droplets fully vaporize, the vapor is heated above the autoignition temperature, and combustion takes place spontaneously at each location where vapor temperature exceeds the autoignition temperature. The process of mixing determines the subsequent rate of combustion. For very large, slow-running diesel engines, there is ample time (~ 200 ms) for combustion to be completed. In this case, mixing between fuel and air is achieved by the momentum of the spray. In smaller engines (200-mm bore), there is sufficient time but not sufficient distance to complete mixing between the fuel jet and air. Swirl is introduced in order to increase the fuel jet distance by deflecting the jet away from the wall.

Since the gas temperature is higher than the self-ignition temperature, fuel-air mixing is the controlling parameter in chemical energy release. Fuel-air mixing on the macroscale is thought to be the energy release rate determining process, rather than turbulent transport phenomena on the microscale. Increasing attention is being directed toward understanding the influence of large- and small-scale turbulence on the mixing process. This requires the determination of instantaneous spatial and temporal variation of temperature, velocity, and species concentration.

Unburned Hydrocarbons

Unburned hydrocarbons in diesel exhaust consist of either original or decomposed fuel molecules or recombined intermediate compounds. A small portion of these hydrocarbons originate from lubricating oil. In diesel engines, the mass of air per cycle is almost constant. Change in load is accomplished by varying the amount of fuel injected. This, in turn, produces variations in atomization quality, injection duration and cylinder gas pressure and temperature, and the amount of fuel deposited on the walls. As the quantity of fuel is increased, more fuel becomes concentrated in the core and impinges on the walls. Increase in fuel flow results in longer periods of injection and, if injection timing and rate are kept constant, more fuel is injected later in the cycle. The time available for reaction for fuel injected near the end of the cycle is reduced while, at the same time, fuel concentrations are increased. These two factors tend to decrease the rate of chemical reactions. On the other hand, final gas temperatures are higher because more fuel is burned and also because there is a drop in the percentage heat losses to the coolant. These higher temperatures lead to a higher reaction rate.

At very light loads and idling conditions, fuel does not reach the walls and its concentration in the core is small. Under these conditions, the unburned hydrocarbon emissions originate mainly from the downstream, lean edge of the spray. Reaction rates are low, due to low temperatures and lean mixture ratios. The ratio of unburned hydrocarbons formed in this region to total fuel injection is the highest at idling.

At part loads, fuel/air ratio is increased, leading to higher concentrations in the core and more fuel deposition on the walls. Unburned hydrocarbons are formed under these conditions but, subsequently, there is sufficient oxygen in the mixture, together with the increased temperature, to promote oxidation reactions with the net result that hydrocarbon emissions are reduced.

At full-load and overload conditions, the increase in fuel/air ratio results in the formation of more unburned hydrocarbon molecules in the core and near the walls. This can lead to an increase in the emissions of unburned hydrocarbons. The effect of fuel/air ratio on the unburned hydrocarbon emissions in both direct injection and indirect injection engines is shown in Fig. 9.27.

With turbocharging, the average gas temperature is increased over the whole

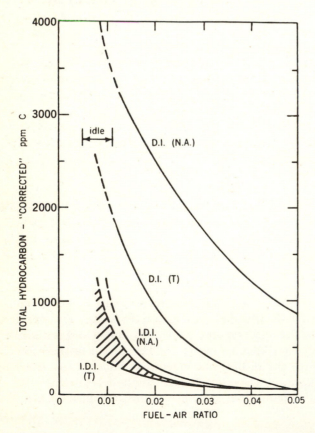

Figure 9.27 Effect of fuel/air ratio on the unburned hydrocarbon emissions in diesel engines. DI, direct injection; IDI, indirect injection; NA, naturally aspirated; T turbocharged. (*Perez and Landen, 1968.*)

cycle. This leads directly to an increase in the rate of oxidation reactions and reduction in hydrocarbon emissions. Further oxidation reactions occur in the exhaust manifold and the turbocharger. With turbocharging, higher exhaust temperatures are reached, the reaction time is increased, and mixing is improved. Advance in injection timing increases the ignition delay and can lead to an increase in unburned hydrocarbon emissions.

The degree of swirl in the DI engine should be carefully controlled. Initially, increase in the degree of swirl improves the mixing process and increases the hydrocarbon oxidation reactions. Excessive swirl can widen the spray or lead to an overlap of sprays, resulting in an increase in unburned hydrocarbon emissions. The swirl in DI engines can be most effectively varied by changing the ratio of the bowl diameter to its depth. Deep-bowl pistons tend to have a higher degree of swirl than shallow-bowl pistons, due to the conservation of moment of momentum.

Carbon Monoxide

During the early stages of spray combustion in DI engines, CO is believed to be formed in the lean-flame region. Since the local temperature is not sufficiently high, very little oxidation takes place. In the core of the spray and near the walls, CO is formed at a high rate. The rate of its elimination depends mainly upon the local oxygen concentration, mixing, local gas temperature, and the available time for oxidation. The effects of fuel/air ratio on carbon monoxide emissions are shown in Fig. 9.28. At light loads, CO formation is high because the gas tempera-

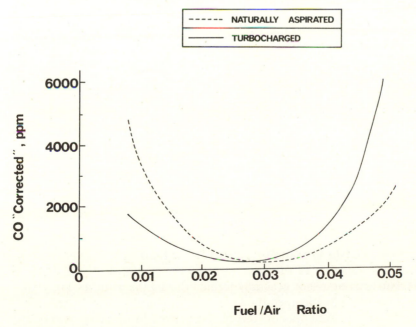

Figure 9.28 Carbon monoxide emissions in a diesel engine as a function of fuel/air ratio. (*Perez and Landen, 1968.*)

ture is low and very little oxidation takes place. As the load is increased, CO emissions decrease, due to the increase in gas temperatures. When the fuel/air ratio is increased beyond a certain limit, the elimination reactions may be reduced, despite the increase in temperature, due to the low oxidant concentration and short reaction time. The degree of swirl requires to be optimized, so as to provide the best economy, which is achieved when CO emissions are at a minimum.

Smoke and Other Particulates

Particulates emitted from diesel engines are of the following forms:

1. Liquid particulates appear as white/blue clouds of vapor emitted under cold starting, idling, and low loads. These consist mainly of fuel and a small portion of lubricating oil emitted without combustion; they may be accompanied by partial oxidation products. The white/blue clouds disappear as the load is increased and the cylinder walls become warmer.
2. Soot or black smoke is emitted as a product of the incomplete combustion process, particularly at maximum loads.
3. Other particulates include lubricating oil and fuel additives.

Black smoke emission consists of irregularly shaped agglomerated fine carbon particles. In the presence of oxygen, pyrolysis of the fuel molecules may take place to form carbon. These carbon particles may subsequently burn if there is sufficient oxygen and the temperature is sufficiently high.

In the spray core, especially under heavy loads, the oxygen concentration is low and the concentration of the high boiling point components is high. Pyrolysis of the molecules may take place and lead to the formation of acetylene and hydrogen. The simultaneous condensation and dehydrogenation of acetylene results in solid carbon. Due to the presence of oxygen, partial oxidation reactions may take place and result in a high concentration of CO. Carbon may be formed if the ratio of carbon monoxide to carbon dioxide exceeds an equilibrium constant. Under heavy loads, the spray impingement onto the cylinder walls leads to the formation of carbon. These carbon particles can, subsequently, be swept away from the walls and are either burned or emitted through the exhaust.

Nitric Oxide

In diesel combustion, NO formation is related to the local oxygen atom concentration, which is a function of the concentration of oxygen molecules and the temperature. Nitric oxide is not generally formed during the compression stroke because of the relatively low temperatures. As in other engines, effective reduction of NO is related to reduction of temperature, particularly for mixture ratios close to stoichiometric.

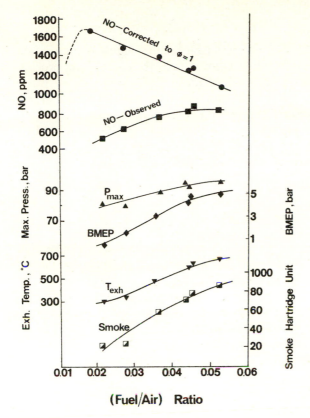

Figure 9.29 The effect of fuel/air ratio on nitric oxide emission and other performance parameters in a DI engine. (*Henein, 1976.*)

The effects of the overall fuel/air ratio on NO emission and related engine parameters is shown in Fig. 9.29. As the fuel/air ratio is increased from 0.02 to 0.06, the observed NO, smoke, and exhaust temperatures all increase. Late injection has been found to be an effective way to reduce NO emissions, but this results in loss in power and fuel economy.

Under the conditions of high pressure, low temperature (compared to premixed combustion), and slow mixing rates, superequilibrium concentrations of radicals do not reach significant levels. The mixing history in a diesel engine indicates that combustion mainly takes place at near stoichiometric conditions and, hence, formation of prompt NO is negligible. The Zeldovich mechanism for oxygen and nitrogen atom reactions is sufficient to describe the NO kinetics in a diesel engine. Figure 9.21 shows that NO is formed in a narrow band of fuel/air ratios around stoichiometric conditions. Use of very high fuel injection pressures leads to higher local temperatures earlier in the expansion process, resulting in a far more rapid NO-formation rate. Attempts to redistribute enthalpy over the working gases, and hence reduce temperatures by exhaust gas recirculation, has led to some reduction in NO emissions but has had the far greater consequence of significantly increasing the emission of soot.

Noise

In diesel engines, the major exciting forces, which set the engine structure into vibration and result in noise, are mainly produced from the combustion process. Reductions in combustion-generated noise in the critical midfrequency range (400–3000 Hz) have been achieved by shortening the ignition delay period, which results in a lower rate of pressure increase and maximum pressure, and in a smoother pressure trace. Methods which have been found to shorten delay periods and reduce engine noise are injection retard, injection rate modifications, turbocharge, increased compression ratio, the use of pilot injection, and the use of higher cetane number fuels.

Odor

Diesel exhaust odor is divided into two characteristics types, oil-kerosine and smoky-burnt. Mono- and polyoxygenated partial oxidation products and certain fuel fractions have been identified as odor-producing compounds. The principal components responsible for the characteristic oil-kerosine portion of diesel exhaust odor are alkyl benzenes, indans, tetralins, and indenes. Unburned fuel in the exhaust mainly contributes to the oil-kerosine odor. Oxygenated aromatic structures have been related to the smoky-burnt odor quality. Quenching of combustion reactions can lead to the emission of gases with odor.

Emission Controls

As in other combustion chambers, it has been found in diesel engines that many controls which reduce the concentration of incomplete combustion products cause an increase in NO emission. Many of the controls which have been introduced for reduction of NO have reduced the fuel economy, which is one of the major attractions for the diesel engine. Catalysts have not been found to be effective in diesel engines because of the presence of excess oxygen in the exhaust. Effecting reduction of NO emissions has mainly been achieved by reducing the maximum temperatures or oxygen concentration during the combustion process. This has been achieved by retarding injection timing, water addition, and exhaust gas recirculation.

Adding water changes the lean ignition limit and reduces maximum temperatures, resulting in increases in hydrocarbons and CO emissions, but reductions have been obtained in NO emissions. Exhaust gas recirculation leads to a reduction in NO emissions due to the increase in heat capacity of the charge. Exhaust gas recirculation affects combustion and emissions in a manner similar to water injection. Reductions have been achieved in both hydrocarbons and NO by use of exhaust gas recirculation. In order to achieve low NO emssions without increasing the other pollutants, the amount of exhaust gas recirculation needs to be optimized and decreased as the engine load is increased.

In the past, diesel engines were mainly used for heavy-duty vehicles but the superior fuel economy of the diesel engine is now making it an attractive competitor to the spark ignition engine for use in automobiles. Automobiles powered by diesel engines can meet the U.S. Federal Standards for CO and hydrocarbons but they have, as yet, not been able to meet the 0.4 g/mile standard for NO_x.

Control of Smoke Intensity in Diesel Exhaust

Reduction in the intensity of smoke emitted by diesel engines can be achieved by controlling or changing the following:

1. *Fuel type.* Reduction of the cetane number of the fuel results in lower rates of carbon formation and also lower rates of smoke emission.
2. *Injection timing.* Advancing the start of injection results in longer delay periods, more fuel injected before ignition, higher temperatures in the cycle, and a more rapid combustion process. The net result has been found to reduce the smoke intensity in the exhaust. Very late ignition can also lead to a reduction in smoke emission. This is related to the reduced rate of formation of carbon when the temperature of the flame is lowered.
3. *Rate of injection.* Increasing the rate of injection leads to a shorter time duration for the injection process and this has been found to be effective in reducing exhaust smoke.
4. *Fuel atomization.* Improved fuel atomization results in reduction of initial droplet size and, subsequently, reduced smoke emission.
5. *Inlet air temperature.* Increase in inlet air temperature results in higher gas temperatures during the entire cycle. The higher levels of temperature lead to faster evaporation, as well as to increased rates of chemical reaction. The net effect of higher gas temperatures on smoke emission is dependent upon the volatility of the fuel. For the more volatile fuels, increase in the gas temperature reduces the penetration and the spray cone angle, leading to high local concentration of droplets and over-rich mixtures near the nozzle. For the more volatile fuels, it has, therefore, been found that increasing inlet air temperature increases smoke intensity. On the other hand, for less volatile fuels, the increase in intake air temperature leads to acceleration of the oxidation reaction at a higher rate than the decomposition reactions, leading to a reduction in smoke intensity.
6. *Afterignition.* Afterignition, secondary injection, or dribbling due to fuel leakage all have very bad effects on smoke intensity.

TEN

LASER DIAGNOSTIC TECHNIQUES

10.1 NONINVASIVE PROBING OF FLAMES

Measurements of flame properties can be made by the insertion of probes into the flame, optical methods based upon emission from the flames, or by laser probing. Fristrom and Westenberg (1965) reviewed the techniques for probing in flames, including temperature measurement by thermocouples, velocity measurement by pressure impact probes and particle-track photography, and species concentration measurements using suction probes followed by gas analysis. Weinberg (1963) has reviewed optical methods for making measurements in flames, including interferometry, schlieren, and photographic methods. Gaydon (1974) has reviewed the spectroscopy of flames and has shown how flame characteristics can be determined from spectroscopic analysis of radiation emitted from flames. All of the probing methods have involved some form of disturbance to the fluid flow, heat transfer, and species concentration as a direct consequence of the insertion of a physical probe into the flame. In order to minimize these disturbance effects, fine-wire thermocouples and microsuction probes were used, made of materials which could be heated within the flame to temperatures approaching those at the measuring point or, alternatively, attempting to achieve a rapid quenching of reaction within the probe. Many of these probe instruments only allow time-average measurements and, in the case of gas sampling, averaging may require to be made over a period of several seconds or even minutes. Many of the optical methods provide information which is integrated over the optical path length of the instrument and the spatial resolution is poor.

The advent of the laser has radically altered the scope of making measurements in flames and, in the 1970s, there has been rapid development and extensive research into methods of using lasers for noninvasive probing of flames (Zinn, 1977).

The special advantages of using lasers in combustion systems are as follows:

1. Visible laser beams experience negligible attenuation in combustion gases unless there is a high concentration of particulates.
2. The laser beams cause no perturbation or disturbance to the system.
3. The coherence of laser light allows focusing of the beams on small measurement volumes.
4. Measurements can be made at large distances from the laser.
5. Radiation from the flame can be excluded by use of spatial and narrow-band spectral filtering.

Lasers provide high intensity, monochromatic light sources which allow measurements to be made with spatial resolutions of less than 1 mm^3. The special interest in laser probing has been due to the high-frequency response of an instrument such as a laser anemometer, which allows turbulence characteristics to be measured. The laser anemometer has become the most accurate means of measuring velocity of fluids and particles and has replaced the pitot tube and hot-wire anemometer in many aerodynamic and fluid-flow applications. In the field of combustion, it has proved to be the most accurate and reliable method of velocity measurement.

The measurement of temperature and species concentration in flames can be carried out by laser spectroscopy. By the use of pulsed tunable high-power lasers and high-resolution spectrometers, measurements can be made simultaneously of temperature and species concentration in flames. The frequency response of laser spectrometers is governed by the pulse rate of the laser, and further developments are required in laser technology in order to allow high-frequency response spectrometry measurements in turbulent flames. In this chapter, the basic principles of laser anemometry will be presented, followed by discussion of various optical arrangements, allowing measurements to be made in forward- and backscatter modes. The simultaneous measurement of three velocity components in systems with flow reversal and high turbulence intensity is made by the use of frequency shifting.

Electronic signals are processed and diagnosed by spectrum analyzers, frequency trackers, single-particle counters, photon correlators, and filter banks. The relative advantages of these processors are discussed, together with the sources of error in measurement, as well as in interpretation of time-averaged quantities. With systems which are online with a computer, an overall frequency response of 1 MHz can be achieved so that measurements using conditional sampling, intermittency factors, and examination of coherent structures in turbulent gas flows have become possible.

The special problems associated with the use of the laser anemometer in flames with high-temperature and concentration gradients are discussed and re-sults are presented of the measurements made under both laboratory and indus-trial conditions in gaseous and particle-laden flames. The measurement of velocity components and turbulence characteristics in gaseous flames is now a well estab-lished experimental technique, but difficulties are still experienced in making measurements in particle-laden flames. It is shown that the laser anemometer can be used for simultaneous measurement of velocity and particle size in two-phase flow systems.

Schwar and Weinberg (1969) discuss the general use of laser techniques in combustion research and Penner and Jerskey (1973) show how lasers can be used for local measurements of velocity components, species densities, and tempera-tures. The fundamental principles and practice of laser Doppler anemometry are comprehensively covered by Durst et al. (1976) and reviews of developments in laser velocimetry for flow measurements have been made by Durst et al. (1972), Trolinger (1974), Stevenson and Thompson (1972), and Thompson and Stevenson (1974). The specific applications of laser anemometry for combustion research have been reviewed by Self and Whitelaw (1976) and Chigier (1976a and 1976b). General reviews on laser spectroscopy have been made by Lapp and Penney (1974), Hartley (1975), Lederman (1977), and Eckbreth (1979a).

10.2 PRINCIPLES OF LASER ANEMOMETRY

The basic concept of applying lasers for velocity measurements was first put forward by Yeh and Cummins (1964). Initial experiments were carried out on laminar liquid systems but, with the development in optical configurations and electronic processing systems, measurements of velocity can now be made in the range between fractions of a millimetre per second and supersonic velocities in laminar and turbulent liquid and gas flow systems with the only proviso that there is an optically clear path to allow penetration of the laser beams. A laser anemometer is made up of three components: a laser light source, an optical arrangement, and an electronic signal processing system.

When a light wave interacts with a moving particle the frequency of light scattered is altered, and laser anemometry requires the measurement of this frequency which is directly proportional to the instantaneous velocity of the particle.

There are two possible means of explaining the basic physical principles of laser anemometry. When two laser beams intersect to form a measuring volume and a particle passes through this volume, the laser light is scattered and part of the scattered light is collected and transmitted to a photodetector. When the differential frequency of the scattered light is measured, it can be considered as being due to the Doppler shift of light scattered by the particle traversing the volume or, alternatively, the two beams can be considered to form a set of interfer-ence fringes and that light is scattered by the particles as they cross each interfer-ence fringe. Since, in both cases the differential frequency is measured, either of the two basic principles may be invoked.

The Doppler Effect

The Doppler effect is explained in terms of the wave theory of light. When monochromatic light passes through a fixed optical system, the frequency is conserved. The Doppler effect is the observed change in frequency of a train of waves caused by relative motion between the source and detector. If a source emits waves while in motion, the waves ahead of it will be more crowded and those behind it more widely spaced than if the source were stationary. A stationary detector measures waves of higher frequency when the source approaches and a lower frequency when the source recedes from the detector.

If a stationary source emits v waves in unit time in all directions and the waves travel with velocity c, a stationary detector monitors with wavelength $\lambda = c/v$. When the source moves toward the detector with velocity v, the v waves it emits in unit time will be crowded into a distance $(c - v)$. These waves pass the detector at a velocity unaffected by the motion, but will appear to have a wavelength $(c - v)/v$. The apparent frequency is thus $vc/(c - v)$.

A laser source provides a coherent beam of light with an almost uniform frequency, f_0. When a particle crosses this light beam, light is scattered. The frequency of the scattered light observed by a stationary object is $f_0 + f_D$, where f_D is the Doppler frequency shift of the light. The frequency shift occurs because the particle has a velocity component in the $\bar{k}_0$ direction, so that wavefronts pass the particle with frequencies higher or lower than f_0. The particle has a velocity component with respect to the detector which causes either a compression or spreading of the wavefronts as the wave is scattered in the direction of the detector, $\bar{k}_s$.

The observed phase scattered by a stationary particle changes at a rate w_0 (or $2\pi f_0$) radians per second. The phase at the moving particle changes at a rate $w_0 - \bar{v} \cdot \bar{k}_0$, while the rate of change of phase seen by the observer is $w_0 - \bar{v} \cdot \bar{k}_0 + \bar{v} \cdot \bar{k}_s$. The Doppler frequency is, therefore,

$$f_D = f_s - f_0 = \frac{1}{2\pi}(w_s - w_0) = \frac{\bar{v}}{2\pi} \cdot (\bar{k}_s - \bar{k}_0) \tag{10.1}$$

The Doppler frequency is directly proportional to the velocity component along the vector $\bar{k}_s - \bar{k}_0$. The direction of this vector is chosen so as to provide a Doppler frequency in a range that can be conveniently measured. Optical frequencies are usually in the range 10^{14}–10^{15} Hz and cannot be measured directly with electronic devices.

Optical Heterodyning

When two waves of slightly different frequency are mixed, heterodyning occurs and a "beat" frequency is set up. The resulting intensity (averaged over times large with respect to the optical wave period) is

$$I_s = A_1^2 + A_2^2 + 2A_1 A_2 \cos 2\pi(f_1 - f_2)t \tag{10.2}$$

where A_1, A_2 are the amplitudes and f_1, f_2 are the frequencies of the two waves.

This provides a time-varying intensity with frequency $f_1 - f_2$, which can be sensed by photodetectors and electronically analyzed. Electronic processors are currently limited to signals below 100 MHz and this provides an upper limit to velocity measurements.

At least one of the heterodyned light waves is scattered and Doppler shifted by the particle. If this is compared with a light wave of fixed frequency, this is referred to as the reference-scatter mode. When two light waves are Doppler shifted and the difference $f_1 - f_2$ is the difference between two Doppler shifts, this is referred to as the dual-scatter mode.

Scattering may be viewed in the forward or backward direction. For particles of the order of 1 μm, the intensity of scattered light in forward scatter can be considerably higher than that in back scatter. The lower intensities found in back scatter can be compensated for by increasing the intensity of the laser light at source or by slight increase in the size of particles. In systems which are unconfined, and where the viewing object is sufficiently small that it can easily be traversed, forward scatter is generally selected. When the viewing object is large and cannot be moved, as well as in cases where optical access is only possible from one side, the back-scatter mode must be adopted. Back-scatter radiation is usually received over smaller angles and difficulties may arise when high spatial resolution is required. For forward scatter, optical systems on either side of the viewing object must be carefully aligned.

The light scattered from the two-beam heterodynes (beats) and the beat frequency is given by

$$\delta v = \frac{2u}{\lambda} \sin (\theta/2) \tag{10.3}$$

where δv is the beat frequency, u is the velocity component in the plane of the laser beams and normal to the optical axis, and θ is the angle of intersection between the two incident beams. The beat frequency is independent of the viewing direction and this allows the aperture of the collection optics to be enlarged without affecting the velocity measurement.

The fringe spacing is given by

$$d_F = \lambda/2 \sin (\theta/2) \tag{10.4}$$

The intensity of light across a laser beam is not uniform but has a distribution which is approximately gaussian. It also diverges with a half-angle proportional to λ/a, where a is the beam radius. When the beam is focused in a homogeneous medium the intensity distribution is

$$I(r, z) = \frac{2P}{\pi a^2(z)} \exp -2\{r/a(z)\}^2 \tag{10.5}$$

where

$$a(z) = a_0\{1 + (\lambda z/\pi a_0)^2\}^{1/2} \tag{10.6}$$

where I is the intensity, P is the laser power, r and z are coordinates, with the center of the coordinate system coinciding with the geometrical focus, and $a(z)$ is the radius where the intensity is $(1/e^2)$ of its value on the axis. The minimum diameter (at the waist) is given by

$$2a_0 = (4/\pi)F\lambda \qquad (10.7)$$

where F is the F-number of the converging/diverging beam, given by $f/2a_L$, where f is the focal length, and $2a_L$ is the diameter of the beam at the lens. The half-angle of the converging beam is given by

$$\alpha/2 = \tan^{-1}(1/2F) \qquad (10.8)$$

The length between points of half intensity is given by

$$2l_0 = (8/\pi)F^2\lambda = 2F(2a_0) \qquad (10.9)$$

Laser beam diameters will be typically of the order of 1 mm with focal lengths in the range of 0.1–1 m. Beam waist diameters are in the range 0.1–1 mm and this governs the size of the control volume. Laser beams may be expanded in a telescope to a larger diameter before focusing in order to achieve higher spatial resolution. As the F-number is reduced, the quality of optical lenses must be increased in order to achieve small waist diameters.

In systems where the density is not uniform, the spherical wave fronts in the converging beam become distorted and lead to enlargement of the waist diameter. When there are temperature fluctuations and corresponding fluctuations in the refractive index, this can lead to jitter of the focal point.

The measuring volume is defined by the region of intersection of the two beams at their common focus. The geometrical detection volume, defined by the surface where the contrast of the fringes falls to $(1/e^2)$ of the center value, is ellipsoidal. In the focal plane, the elliptic cross section has major and minor diameters

$$d_1 = 2a_0 \qquad (10.10)$$

$$d_2 = 2a_0 \sec(\theta/2) \qquad (10.11)$$

The maximum length of the common volume is

$$2l = 2a_0 \operatorname{cosec}(\theta/2) \qquad (10.12)$$

The maximum number of (bright or dark) fringes, in the focal plane, is

$$N_{max} = d_2/\lambda_F = (8/\pi)F \tan(\theta/2) \qquad (10.13)$$

The width of the common volume in the plane of the incident beams, and hence the number of fringes, decreases from N_{max} in the focal plane to zero at the ends of the common volume.

The F-number determines the spatial resolution and the maximum number of fringes. The control volume dimensions affect the peak signal power, signal duration, the energy scattered per particle transit, and the data rate. Focusing the beams to a small waist diameter increases the peak signal power and allows detection of smaller diameter particles but, at the same time, reduces the signal rate due to reduction of the capture area of the control volume projected normal to the average flow direction. The data rate of utilized signals is dependent on the distribution and flux of particles as well as the threshold for detection.

Particle Seeding

Laser anemometry requires the presence of particles in the fluid which are sufficiently distinct from the fluid and sufficiently large to act as a scattering center. The laser anemometer measures the velocity of these particles and, if the particles have a high drag-to-momentum ratio, the particle velocity will be the same as that of the local fluid velocity. Particles in the size range 0.1–1 μm are generally recommended but, for measurements in low-speed flows, larger particle sizes are acceptable and, for measurements of high-frequency components in high-speed turbulent flows, it is most important that particle sizes are kept to the order of 0.1 μm. The concentration and size of dust particles in the environment can vary as a result of particulate pollution and dust in the atmosphere.

Highly sensitive anemometers have been developed to make measurements in unseeded gaseous flow where scattering of light is dependent on particles naturally present in the atmosphere. The concentration of particles in the atmosphere is generally very low and, in most practical gaseous flow systems, the flow field requires to be seeded with small particles. The particles should be as small as possible but care needs to be taken that seeding does not lead to corrosion, abrasion, chemical reaction, or agglomeration to sizes which cause the particles to deviate from the flow.

The extent to which particles that are suspended in a gas flow will follow turbulent fluctuations in the flow is dependent upon the balance between acceleration forces on the particle and the displaced fluid. This is given by

$$\frac{\pi d_p^3}{6} \rho_p \frac{dU_p}{dt} = \frac{\pi d_p^3}{6} \rho_f \frac{dU_f}{dt} - \frac{1}{2} \frac{\pi d_p^3}{6} \rho_f \frac{dV}{dt}$$

$$- 3\pi v \rho_f \, d_p V - \tfrac{3}{2} d_p^2 \rho_f \sqrt{\pi v} \int_{t_0}^{t} \frac{dV}{d\xi} \frac{d\xi}{\sqrt{t - \xi}} \qquad (10.14)$$

where d_p is the particle diameter, ρ_p and ρ_f are the densities of the particle and fluid respectively, v is the kinematic viscosity of the fluid, and V is the differential velocity between the particle and fluid, that is, $U_p - U_f$, and t and ξ are time, with ξ as dummy variable.

The four terms on the right-hand side of Eq. (10.14) represent, respectively, the acceleration of the fluid, the resistance of an inviscid fluid to the accelerating sphere, the Stokes drag force, and a drag force which takes account of the unsteady motion of the sphere. This equation may be applied to a turbulent flow on the assumption that (1) the turbulence is homogeneous and statistically steady, (2) the particles are much smaller than the turbulence microscale, (3) Stokes' drag law applies to the relative motion of the particle and fluid, and (4) there is no interaction between particles.

Table 10.1 Maximum particle diameter for given frequency response

Particle	Fluid	Density ratio	Viscosity (kg/m s)	Diameter (μm) $f = 1$ kHz	$f = 10$ kHz
Silicone oil	Atmospheric air	900	1.8×10^{-5}	2.6	0.8
TiO_2	Atmospheric air	3.5×10^3	1.8×10^{-5}	1.3	0.4
MgO	Methane-air flame (1800 K)	1.8×10^4	5.9×10^{-5}	2.6	0.8
TiO_2	Oxygen plasma (2800 K)	3.0×10^4	1.1×10^{-4}	3.2	0.8

For particles suspended in gases, the density ratio ρ_p/ρ_f is normally of the order of 10^3 and, provided the Stokes number $N_S \equiv \sqrt{v/\omega d_p^2}$ (where ω is the angular frequency of turbulence fluctuations) is greater than 8, Eq. (10.14) reduces to

$$\frac{dU_p}{dt} + \frac{18v}{d_p^2}\frac{V}{(\rho_p/\rho_f)} = 0 \tag{10.15}$$

Typical results obtained by solving Eq. (10.15) and requiring that the particles follow the fluid motion within one percent are given by Melling and Whitelaw (1973) in Table 10.1.

Methods for generation of aerosols and particle size measurement are discussed by Melling and Whitelaw. They conclude that the optimum form of seeding in terms of size, material, and concentration of aerosol particles will depend on factors such as the fluid temperature, fluid-flow rate and signal-processing system. Equation (10.15) provides a convenient means of estimating the maximum permissible particle diameter. In view of the polydisperse nature of aerosols and the small proportion of turbulence energy above 1 kHz, particles with a mean diameter of approximately 1 μm will be suitable for most purposes.

For nonburning systems, liquid aerosols can be generated by using airblast atomization. In combustion systems where large flow rates of solid particles are required, a fluidized-bed arrangement is recommended. Particle coagulation is unlikely to be important in seeded gas flows. Electrostatic-charge effects are also not considered to have any significant influence on laser anemometry measurements.

10.3 OPTICAL ARRANGEMENTS

The optical system is made up of two parts, the transmission and collection optics. In the transmission optics, the laser beam is split by a beam-splitter to produce two equal intensity parallel beams, which are passed through a lens and made to intersect at their common focus. Beam splitting may be accomplished in several ways and should provide equal path lengths, capable of adjustment to give a wide range of beam separations. Cemented achromatic doublets are quite adequate as lenses for laser anemometry and even simple lenses can be used with large F-numbers. The beam-splitter and focusing lens must be carefully prealigned and strongly secured so as to prevent relative motion induced by vibration. The focusing within the control volume can most conveniently be achieved by arranging for two incident laser beams to cross so that the control volume is formed at the intersection between the beams.

The collection optics is made up of the following items:

1. A collection lens
2. Aperture stops to block the direct unscattered beams in forward scatter
3. A spatial filter or pinhole placed at the image of the control volume to restrict the field of view

The transmission and collection optics must be fixed rigidly, preferably to a common mount, and either this is moved relative to the flow system, or the flow system is moved relative to the optics. The size and weight of the whole optical system and laser are, in many instances, sufficiently great that it becomes more convenient to move the flow system, and this is the practice which has been generally adopted for measurements made in many laboratories.

Signal amplitude is a function of the laser power, beam intersection angle, particle size, and scattering angle. In order to prevent saturation of the photomultiplier, the scattering angle is limited by reducing the aperture to a minimum value. In combustors containing a high concentration of particles, or when measurements need to be made close to surfaces, light can be scattered from particles or surfaces outside the control volume. This scattered light can be restricted by increasing the spatial filtering. The quality and sensitivity of the photomultiplier must be carefully selected when making measurements in combustion systems. The photomultiplier requires maximum quantum efficiency, high-frequency response, and a high gain to bring the signal to a convenient level without additional amplification.

Laser Light Source

Even though the principles of anemometry, using a light source, have been known for some time, no practical anemometer was developed until laser light sources became available. A laser light source provides a highly intense beam of coherent light of small enough diameter to allow focusing in a small control volume. The intensity of the incident light needs to be sufficiently large, so that the relatively small fractions of this incident light that are scattered can be collected and analyzed. The intensity of light required varies as the velocity of the particle. Durst and Whitelaw (1973) have calculated an approximate figure of 0.50 mW of laser power per meter per second of velocity but, in practical systems, 5 mW of power is required for low-velocity steady-flow systems and, in combustion systems with high turbulence intensities, argon ion lasers with powers of the order of 1 W are required.

Laser light is coherent and this enables the measurement of a signal which includes only one frequency. Turbulent fluctuations in the flow and noise in the photomultiplier and electronic system lead to a broadening of the signal, and the signal has to be specially treated in order to separate the true signal from the noise. The divergence of laser beams is generally very small, of the order of $2\,\mu$rad, so that, for optical distances of 1 m, the increase in diameter is considered to be negligible. For systems where optical paths are several meters long, account needs to be taken of the divergence of laser beams which can be of the order of 2 mrad.

Lasers are required to be continuous and two visible lasers have been generally used in anemometry. For low powers, the He–Ne 5 or 15-mW laser, which emits in the red (633 nm) has been found to be adequate in many laboratory experiments where the size of the system is small, the flow is reasonably steady, and particles are large enough to provide adequate scattering. For combustion systems, as well as for most wind tunnel and large-scale industrial usage, the more powerful argon ion lasers are used. The power output of a 4-W laser can be easily controlled down to 100 mW and is currently the most commonly used. Lasers with powers up to 20 W have also been constructed. The argon ion lasers emit in the green (515 nm) and blue (488 nm) and there is usually sufficient power to enable the green and blue lines to be used separately in order to simultaneously measure two velocity components. The output beams are plane polarized.

In the selection of the power of the continuous-wave laser that is used in a dual-laser anemometer, there is a conflict between the cost and health hazard associated with high-power lasers and the difficulty in obtaining good signal-to-noise ratios with low-power lasers. When measurements require to be made in combustion systems, there are special problems affecting the quality of signals. In luminous flames, the scattered laser light must be distinguished from the background flame radiation so that the laser powers are selected to be at least an order of magnitude greater than the background flame radiation at the laser wavelength. The efficiency of transmission and scattering of light is very poor when the system dictates the use of back scattering as a result of limited optical access to the combustion chamber. In many combustion systems, reverse flow conditions arise in recirculation zones as well as high turbulence intensities requiring the introduction of a frequency-shifting device, involving further laser power losses. In most systems, attempts are made to minimize the quantity and size of artificial seeding, and this also affects the intensity of scattered light. Practical experience has shown that it has been possible to make measurements where laser powers were moderately high—of the order of 1 W—whereas measurements have either been difficult or not possible when lasers of low power—of the order of 5 mW—have been used.

In examining laser power requirements, it needs to be recognized that only part of the total laser power is emitted at a particular wavelength and only part of this power reaches the control volume. Power losses through each optical unit are of the order of 10 to 20 percent and, for systems using diffraction gratings, overall power losses up to 50 percent occur between the laser exit and the control volume.

Under conditions of back scatter, the efficiencies of light scattering are greatly reduced so that laser powers require to be increased by a factor of 5 or 10, compared with those used under forward-scatter conditions. Despite the fact that it can be shown, in principle, that laser powers as low as 0.2 mW at the control volume would be sufficient to make measurements, it has been found that, in forward scatter, use of total laser powers less than 100 mW has led to significant difficulties in making measurements in combustion systems. It is recommended that laser powers in the region of 0.5 to 1 W be used in forward scatter, in order to obtain good signal-to-noise ratios using dust or seeding particles in the submicron range.

Dual-Beam Mode—Forward Scatter

The dual-beam mode has proved to be the most versatile of the various optical arrangements which have been tried for laser anemometry. The dual-beam mode in forward scatter is shown in Fig. 10.1. The laser beam passes through a beam-splitter and a series of mirrors, providing two beams which pass through the focusing lens and cross to form the measurement control volume. θ is the angle between the two incident beams. In forward scatter the collection optics and photodetector are placed on the opposite side to the laser focusing optics. This system can be used when the intensity of the scattered light is low. The scattered light can be collected over a wide angle since the measured frequency is independent of the direction of detection. In selecting an optical system, the dual-beam forward-scatter mode is generally preferred since measurements can be made with relatively low laser powers and low levels of seeding. It is, however, essential that the focusing and collection optics are carefully aligned and that optical access is provided on two sides of the system. The forward-scatter mode has been used in a wide range of laboratory-scale configurations but, generally, the flow system is moved while the optical system remains stationary.

Dual-Beam Mode—Back Scatter

The dual-beam back-scatter mode is shown in Fig. 10.2. The focusing optics remain essentially the same as that for the forward scatter mode. The collection optics are now placed on the same side as the focusing optics and a back-scatter

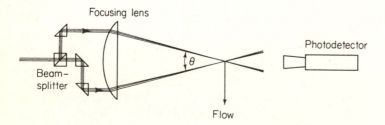

Figure 10.1 Laser anemometer dual-beam mode—forward scatter.

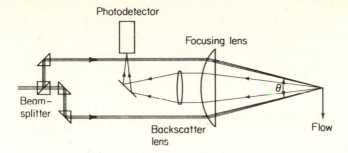

Figure 10.2 Laser anemometer dual-beam mode—back scatter.

lens collects the back-scattered light and transmits it via a mirror to the photodetector.

For most large-scale industrial systems, the back-scatter geometry is preferred because of the requirement to have access only on one side and also the fact that a compact optical unit can be prepared and readily moved. The scattering efficiency for back scatter is an order of magnitude less than that in forward scatter, and extra care also needs to be taken to avoid unwanted reflected or scattered light reaching the detector. Despite the sensitivity of the optical system, some exceedingly robust back-scatter laser anemometers have been constructed by rigidly screwing optical components to solid platforms.

Since the intensity of back-scattered light is usually much lower than that of forward-scattered light, either a higher concentration of scattering particles or higher laser power may be required.

Reference-Beam Mode

In the reference-beam mode, mixing takes place between an unscattered laser beam and a scattered Doppler shifted beam. The photodetector measures the sum of the two waves. In the reference-beam mode (Fig. 10.3), the photodetector is mounted coaxially with the reference laser beam. An adjustable neutral density filter is used to reduce the intensity of the reference beam until optimum Doppler signal quality is achieved.

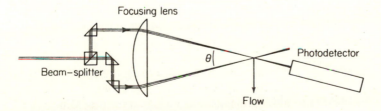

Figure 10.3 Laser anemometer reference-beam mode.

The reference-beam mode was the optical arrangement initially used in laser anemometry. It has the disadvantage that a minimum number of approximately ten particles is required in the control volume so that particle seeding densities require to be high. For systems which have heavy seeding, it may become necessary to use the reference-beam mode because, in this configuration, many particles can be in the measuring control volume at any one time, and this leads to a better signal-to-noise ratio. Integrated optical units have been developed by Durst and Whitelaw (1971), which are very convenient and versatile. These allow measurements to be made in forward- and back-scatter configurations and, by rotating the beam-splitter, velocity components can be measured perpendicular to the optical axis. It is also possible to arrange for the incident light to come from a single beam and view the scattered light from two angles. Such a system cannot produce any real fringes but, instead, virtual fringes exist and the analysis of the signals is essentially similar to that used in the dual-beam (fringe) mode.

Frequency Shifting

Optical anemometers have a 180° directional ambiguity so that, if no special devices are used, the sense (negative or positive) of the flow direction cannot be distinguished. For measurements in reverse flows, as occur in recirculation zones, it is necessary to distinguish the velocity sense but, also when local relative turbulence intensities exceed 30 percent, errors arise when negative velocities are not separated from their equivalent positive values. Errors result both in the measurement of mean and rms values, and these errors rise sharply for turbulence intensities above 30 percent. Elimination of directional ambiguity from laser anemometers is generally achieved by introducing a constant frequency shift on one or both of the laser beams to bias the beat frequency. With such an arrangement, measurements are made relative to a moving coordinate system (or moving set of fringes) and it becomes possible to measure zero velocity of a particle in the control volume, which becomes represented by the bias frequency. Positive and negative velocities are represented by frequencies above or below the frequency offset.

The real fringes in a differential-scatter system move perpendicular to their plane with velocity $u_F = \Delta v / \lambda_F$, where Δv is the induced frequency shift. By adjusting the magnitude and direction of the induced fringe velocity, motion of particles relative to the fringes can be made unidirectional, and Doppler frequency shifts δv (which may have either sign in a highly turbulent or pulsatile flow) are biased and observed as frequencies $(\Delta v + \delta v)$. Thus, the velocity probability density $P(u)$, which for a reversing flow is nonzero for u both more than and less than zero, is observed in a transformed velocity frame $u' = u + u_F$.

The principal means of obtaining frequency shifting are by the use of rotating gratings, acoustooptic devices, and electrooptic crystals. The cheapest and most convenient method of frequency shifting of a laser beam uses a rotating diffraction grating, mounted perpendicular to the direction of the incident light. The laser beam passing through the grating is split into a number (up to 32) of diffraction orders from which the $+1$ and -1 orders are selected. The light frequency in the $+1$- or -1-order diffracted beam is upshifted or downshifted, while the frequency of the zero-order beam remains unchanged. Disk-type rotating gratings have been used by Stevenson (1970), Baker et al. (1975), and Chigier and Dvorak (1975). The frequency shift is directly proportional to the rotational velocity of the disk and the total number of lines on the grating. By the use of synchronous motors and etched gratings, frequency shifts of up to 3 MHz are readily obtained and frequency can be readily followed by adding or subtracting the frequency shift. Particular care must be taken to reduce, as much as possible, the mechanical vibrations of the rotating grating, fluctuations in the rotational speed, and imperfections in the optical grating, which can lead to a broadening of the frequency shift. The efficiency of transmission of light is poor but, by the use of bleached gratings, efficiencies of 50 percent can now be achieved. Durst and Zaré (1974) provide a very useful survey of the various alternative methods of eliminating directional ambiguity.

Acoustooptic frequency shifting (Fig. 10.4) is based on the principle that ultrasonic energy, propagating through a liquid medium, forms a moving time-dependent three-dimensional diffraction grating as a result of the variations in the index of refraction of the liquid. In practice, acoustooptic Bragg cells are excited with ultrasonic waves of frequencies higher than 30 MHz. The use of a double Bragg cell with 15 and 25 MHz driving frequencies applied at perpendicular directions has been shown to be effective in laser velocimetry.

Shifting of light frequency by electrooptic means is achieved by generating a rotating electric field in one or several electrooptic cells. The incident circularly polarized light beam is either accelerated or decelerated, depending on the relative direction of rotation of the two fields. An upshift or downshift in the frequency of the outgoing light is thereby obtained. The electrooptic cell method has the advantages of no moving parts, accurate control of the frequency, and the possibility of obtaining rapid changes of shifted frequency. However, in order to obtain low megahertz frequency shifting with high conversion efficiencies, high voltage power supplies of several kilovolts are required. The other possible methods for obtaining frequency shift involve the employment of polarized light beams and the

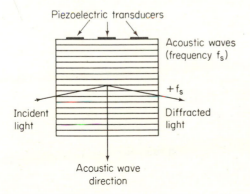

Piezoelectric transducers

Acoustic waves (frequency f_s)

$+f_s$

Incident light

Diffracted light

Acoustic wave direction

Figure 10.4 Frequency shifting by an acoustooptic modulator (Bragg cell).

use of multimode laser beams. Durst and Zaré (1974) have drawn up a table comparing the different methods of direction sensing in laser anemometry, and this can be used as a basis for selecting the most suitable method for any given application.

Measurement Control Volume (MCV)

The calculation of the size of the measurement control volume is dependent upon the assumption of the effective diameter of the laser beams. It is usual to assume that the light intensity across a laser beam has a gaussian distribution and that beyond the $1/e^2$ point the light intensity is too weak for measurements in laser anemometry. The waist diameter of the laser beam, length and height of control volume, fringe spacing, and number of fringes in the volume can be calculated from the following formulas:

Waist diameter

$$d_0 = \frac{4\lambda f}{\pi D_{1/e^2}} \tag{10.16}$$

Length of control volume

$$l_m = \frac{4\lambda f}{\pi D_{1/e^2} \sin \theta/2} = \frac{d_0}{\sin \theta/2} \tag{10.17}$$

Height of control volume

$$d = \frac{4\lambda f}{\pi D_{1/e^2} \cos \theta/2} = \frac{d_0}{\cos \theta/2} \tag{10.18}$$

Fringe spacing

$$\lambda_F = \frac{\lambda}{2 \sin \theta/2} \tag{10.19}$$

Number of fringes in volume

$$N_{fr} = \frac{8f \tan \theta/2}{\pi D_{1/e^2}} = \frac{2d_0 \tan \theta/2}{\lambda} \tag{10.20}$$

Actual height seen

$$D_{ph} = \frac{d}{M} \tag{10.21}$$

where λ = wavelength of light
λ_F = fringe spacing
f = focal length, focusing lens
D_{1/e^2} = diameter of laser beam at the $1/e^2$ point of its light intensity distribution
D_{ph} = aperture in front of photomultiplier
M = magnification of receiving optics

Table 10.2 Control volume and fringe spacing

θ	l_m	d	Δx	N_{fr}	X	f	d_0
Intersection angle (degrees)	Length of control volume (mm)	Height of control volume (mm)	Fringe spacing (μm)	No. of fringes	Beam separation (mm)	Focal length (mm)	Waist diameter (mm)
6	3.2	0.17	4.6	40	31.4	300	0.17
10	1.95	0.17	2.8	61	52.5	300	0.17
20	1.00	0.17	1.4	122	105	300	0.17
45	0.45	0.18	0.64	285	248	300	0.17
6	5.4	0.28	4.6	61	52.5	500	0.28
10	1.95	0.17	2.8	61	52.5	300	0.17
20	0.5	0.085	1.4	61	52.5	150	0.085
45	0.1	0.04	0.65	61	52.5	63	0.036

Variation of the angle of intersection of the laser beams is obtained by varying the beam separation or by varying the focal length of the focusing lens. When the focal length is maintained constant and the beam separation is varied, no significant change occurs in the height of the volume but the number of fringes can be very significantly increased by increasing the angle of intersection. When, on the other hand, the beam separation is maintained constant and the focal length of the lens is varied, then the number of fringes remains constant but the height of the control volume is very significantly reduced as the intersection angle is increased. Table 10.2 shows control volume dimensions and fringe spacing calculated for an argon laser with $\lambda = 0.488$ μm and a laser beam diameter, $D_{1/e^2} = 1.1$ mm. The intersection angle has been varied from 6 to 45° in the first instance by maintaining a constant focal length and varying the beam separation, while the second set of data is for the case where the beam separation is maintained constant and the focal length of the lens is varied.

Use of large beam intersection angles leads to increased spatial resolution, whereas small angles result in an increase in the length of the control volume. In practice it is found that a total angle between the laser beams of between 7 and 15° provides the best compromise between spatial resolution and accuracy. There still appears to be some doubt as to the effective size of a particular control volume. This can be determined by a controlled experiment in which the limits of the volume are measured in terms of a boundary within which the signal-to-noise ratio from a particle passing through the volume is sufficiently good for measurements to be made. This is equivalent to the distinction which is made between "in and out of focus" of particles in shadow photography.

In laser anemometry, filters are used so that measurements are made within a specific frequency bandwidth. In practice, it is desirable to make the minimum number of changes of filters and attempts are made to cover the span of frequencies associated with the span of velocities that require to be measured. Since the fringe spacing is effectively the proportionality constant between velocity and frequency, the selection of angle, and hence fringe spacing, needs to be taken into account in conjunction with the selection of the characteristics of the filter.

Simultaneous Measurements of Three Velocity Components

In three-dimensional flow systems where velocity components can be of the same order of magnitude, the complete determination of the flow field requires measurement of each of the three velocity components. Compared with other velocity measuring devices, the laser anemometer has the special advantage of only measuring the velocity component in the plane of the two intersecting laser beams, irrespective of the flow direction. The three velocity components, at a fixed "point" in space, can be measured simultaneously by focusing three sets of dual-laser beams which are in mutual orthogonal planes focused in the same point. In order to separate the components, the frequency of incident light in each of the three sets of beams should be different and a separate signal diagnostic system is required for each velocity component. In many flow systems there are two dominant velocity components and developments in laser anemometry have concentrated, initially, on measuring two transverse velocity components simultaneously.

Two pairs of incident beams of either different colors or orthogonal polarizations derived from the same laser are brought to a common focus so as to create

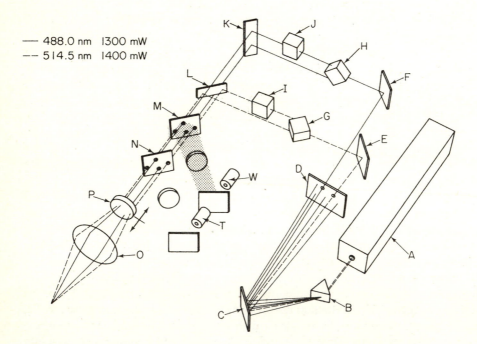

Figure 10.5 Two-color laser anemometer for simultaneous three-velocity component measurement. Back-scatter mode with provision for rapid scanning. (*Grant and Orloff, 1973.*)

two orthogonal real-fringe patterns in the same measurement volume. A single-collection optical system may be used in forward or back scatter and the scattered signals from the orthogonal fringe patterns divided into separate detection and signal-processing channels by using spectral filters or polarizers. Grant and Orloff (1973) developed a two-color system using the blue and green light emitted from an argon laser and this has been very effectively used for measurements in large wind tunnels and in the complex flow fields generated by helicopter blades. The back-scatter system is exceedingly robust and also allows continuous traversing of a moving lens. The system is shown in Fig. 10.5. Orloff and Logan (1973) have developed a scheme for simultaneously determining the third velocity component from the system which they had used for two-velocity-component measurements. They used the blue and green light from an argon laser for measuring velocity components normal to the optical axis. For the third component along the optical axis they used a single-scatter, reference-beam heterodyne arrangement. In almost all the systems which have been used when simultaneous measurement of velocity is required, it has been necessary to either duplicate or triplicate the signal-processing system. Because of the very high cost associated with this, there are still comparatively few such systems in existence. When measurements are restricted to time-average quantities, a single-signal processing system can be used by measuring each velocity component in sequence.

A three-dimensional laser anemometer can measure the following velocity components:

Instantaneous velocity	$\hat{U}_i$	$i = 1$ to 3
Mean velocity	$\overline{U}_i$	
Normal correlations	$\overline{u_i^2}, \overline{u_i^3}, \overline{u_i^4}, \ldots,$	
Stress correlations	$\overline{u_i u_j}, \overline{u_i u_j^2}, \overline{u_i u_j^3}, \ldots,$	
Space correlations	$\overline{u_{i_A} u_{j_B}}, \overline{u_{i_A} u_{j_B}^2}, \overline{u_{i_A} u_{j_B}^3}, \ldots,$	
Probability density distributions	$P(U_i)$	
Energy spectra	$E(U_i)$	

The measurement of instantaneous velocity is, by itself, of little value because of the variations with time. Some form of time averaging must be resorted to, but conditional sampling allows separate averaging of events. Ideally, this requires a continuous measurement of velocity with respect to time. The laser anemometer can only measure velocities of discrete particles so that, in principle, only a discontinuous variation of velocity with time can be measured. The high frequency response of the laser anemometer does, however, allow the time intervals between measurements to be made in the order of microseconds.

Time-averaged velocities have been the most common form of velocity measurement that has been made. If the whole flow system is basically steady, with only small perturbations superimposed on this steady flow, the time-average velocity has a significant and useful meaning. In flows with periodic variations, and flows in which there is a low-frequency large-scale movement superimposed on the main flow, it is more useful to separate the periodic flow from the main flow.

The normal stresses $\overline{u_i^2}$ are determined from a direct measurement of the rms of fluctuating components of velocity in each direction. They thus provide a measure of the extent to which there is deviation from the steady flow as given by the time-average velocity $\overline{U_i}$. The stress terms $\overline{u_i u_j}$ measure the shear in the flow and also provide indications of the degree of isotropy of the turbulence, which is governed by the difference in fluctuating components u_i and u_j.

The space correlations $\overline{u_{i_A} u_{j_B}}$ show the extent to which velocity fluctuations are correlated in space at two points, A and B. By increasing the separation between these two points, information is obtained concerning the scale and physical size of eddies within the flow. The higher-order correlations are less commonly used but there are special cases in which some physical interpretation can be found. Correlations of order greater than four add such a degree of complexity to the equations that they are seldom analyzed. The major emphasis in measurement has been, and probably will continue to be, on measurement of time-average values $\overline{u_i}$, $\overline{u_i^2}$, and $\overline{u_i u_j}$.

10.4 SIGNAL PROCESSING AND DIAGNOSTICS

Light scattered by particles crossing the control volume is collected by a photodetector, which produces a current. The signal from the photodetector requires to be processed and diagnosed in order to determine the velocity of the particles. The current produced at the photodetector is modulated at a frequency which is proportional to a component of the particles' instantaneous velocity. The signal quality depends upon laser power, ratio of fringe spacing to particle diameter, the optical geometry, and scattering properties of the particle. The choice of signal-processing system is governed by the signal quality and the intermittency of the signal. The sensitivity of the signal-processing system needs to be increased as the signal quality becomes poorer. When particle concentrations are sufficiently high, near-continuous signals are produced, which can be transformed into almost continuous records of instantaneous velocity. When particle concentrations are low, and the resulting signal becomes intermittent, special signal-processing techniques are required.

As photons from the scattered light enter the photomultiplier, a continuous signal is recorded of voltage as a function of time. If this is examined on an oscilloscope (Fig. 10.6), the continuous output will be seen to be made up of individual bursts which are associated with the passage of individual particles through the control volume. Except for conditions of very heavy seeding, signals are present for less than one percent of the time. When the signal is taken through a high-pass filter, the signals associated with individual bursts are found to contain both low- and high-frequency components. The high-frequency components

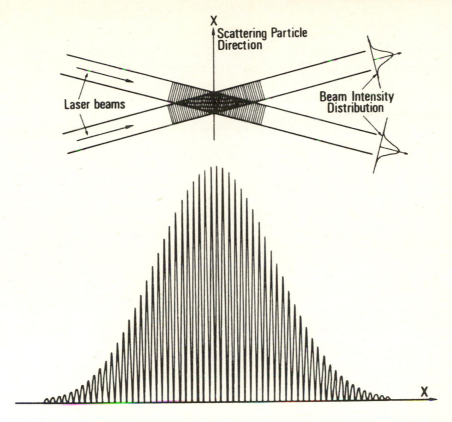

Figure 10.6 An oscilloscope signal for single-particle crossing interference or fringe pattern of two intersecting laser beams.

correspond to Doppler signals and are referred to as ac signals. The low-frequency component is referred to as the dc signal; it contains no information about the instantaneous velocity. The frequency of the Doppler burst is directly proportional to the instantaneous velocity component of the particle in the plane of the laser beams. The amplitude of the signal is a function of the intensity of the incident light, the beam intersection angle, the scattering angle, and the size, location, and scattering properties of the particle within the control volume.

Signals contain both photomultiplier and electronic noise and each individual system is designed so as to improve the signal-to-noise ratio. Increasing the laser intensity and increasing the sensitivity of the photomultiplier are ways of improving signal-to-noise ratio in addition to carefully aligning the optical components.

The intensities of the two transmitted light beams should be equal. Beam intersection angles must be accurately determined and precise knowledge is required of the measuring control volume and its magnification upon the photodetector. Experience has shown that the optimum ratio between fringe spacing and particle diameter is approximately equal to two.

Frequency Analysis

Frequency analysis requires the use of a filter window, with variable center frequency, which is swept across the signal at different frequencies to produce a probability function. The bandwidth of the filter can be varied and signals which do not occur at the filter setting are rejected. A sufficient amount of time is required in order to build up a probability function and the number of rejected signals may be very large. The average Doppler frequency is given by

$$\bar{v}_D = \frac{\int v_D P \, dv_D}{\int P \, dv_D} \tag{10.22}$$

where P is the probability function of frequency. Root mean square values are obtained from

$$\overline{(v_D - \bar{v})^2} = \frac{\int (v_D - \bar{v})^2 P \, dv_D}{\int P \, dv_D} \tag{10.23}$$

Mean velocities and one-point correlation functions, $\overline{u_i^2}$, $\overline{u_i^3}$, and $\overline{u_i^4}$, can readily be obtained from a probability distribution. The probability function can be digitized and processed on a computer.

This system of frequency analysis is inefficient in the use of available signals and comparatively long periods of time are required in order to obtain time averages. The technique is particularly imprecise when the mean frequency tends toward zero, while the range of frequencies is finite.

The principal methods of signal processing and diagnostics which have been developed and used in laser anemometry are based upon the use of: (1) spectrum analyzer, (2) frequency tracker, (3) particle counter, (4) filter bank, and (5) photon correlator. All of these instruments are commercially available.

Frequency-Tracking Demodulator

The principle of operation of a frequency-tracking demodulator is shown in Fig. 10.7. The input signal is mixed with a voltage-controlled oscillator signal and

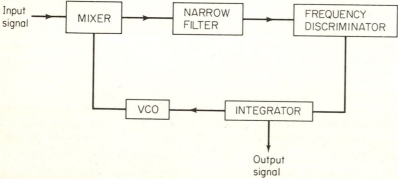

Figure 10.7 Frequency tracking demodulator.

fed through a narrow-band filter to a frequency discriminator. The discriminator converts frequency to voltage and, after integration, the output voltage is used as a measure of the velocity of the particle.

To take into account the periodicity of particles passing through the control volume resulting in a noncontinuous signal, a "dropout" control is used which holds a measured frequency until a new signal appears. When dropout times are greater than 20 percent this system can lead to significant errors. Frequency trackers are currently limited to an upper frequency of 50 MHz and there is also a limit on the rate of change of frequency.

Frequency trackers have been used effectively under conditions of high particle concentration. In many practical applications it has been found that particle concentrations are not sufficiently high for the requirements of a frequency tracker and, in addition, significant electronic noise occurs at the lower end of each frequency range. Many commercial frequency trackers have a number of frequency ranges and, where there are wide variations in velocity, these variations may not be covered within the single available frequency range. Frequency trackers are gradually being superseded by particle counters.

Single-Particle Counters

In the single-particle counter the time required for a particle to cross a given number of fringes is measured. This is equivalent to measuring the time required for the Doppler signal to complete a fixed number of cycles. The counter processor is composed of an input module, which acts as a discriminator and translator, and the output module, which calculates time and velocity from the signal. A Doppler burst from the receiving photodetector is fed directly into the processor. The low-frequency pedestal wave form is removed by high-pass filtering, followed by squaring and clipping the Doppler burst. The logic and output display module measures the time for the completion of a preset number of cycles, corresponding to the particle passage time through the interference fringes. The crystal-control clock, with frequencies as high as 100 MHz, is used for measuring the time required to complete a certain number of cycles. Provision is made to prevent overloading and reduce the probability of including measurements generated by multiple particles in the scattering volume. Fast logic circuits are used to validate or reject individual signals. In one type of processor, the times for five and eight cycles are compared and, unless the ratio is 5/8 to within some preset tolerance, the signal is rejected. Error logic tests such as these can be used to reject signals generated at the edges of the control volume with only a small number of Doppler cycles.

The output from the counter, in the form of digital signals, is fed to a fast-storage register and can then either be displayed visually or transmitted in digital form for interfacing with a computer for online data reduction. Data can also be recorded on magnetic tape for offline data reduction. The rate at which data can be handled can reach 10^6/s, provided that there is sufficient seeding to allow an almost continuous stream of particles to pass through the control volume. Maximum Doppler frequencies that can be reliably measured are currently 100 MHz and these can easily be extended to 200 MHz. Instantaneous accuracy for a single-particle measurement is 12 percent at 100 MHz but average readings can be accurate to better than one percent.

When particle concentrations are low and there are rapid changes of velocity, the counter system is considerably superior to the frequency tracker and normally does not require changes to be made for variations over the velocity range. With a counter it is easier to separate high and low velocities and, irrespective of the degree of seeding, measurements can be made of each single particle that crosses the control volume.

Photon Correlator and Filter Bank Processor

When particle concentrations are low, the resulting signal becomes intermittent and the more conventional signal-processing systems cannot be used. The photon correlator and filter bank techniques are then used for signal diagnostics.

A photon correlator correlates the emission of individual photoelectrons, which are multiplied in the dynode chain of the photomultiplier. Clipping techniques are used to replace the signal by "ones" and "zeros," according to the signal level before autocorrelation. The correlator digitally forms the auto-correlogram $R(\tau)$ of the signal $E(t)$ according to the formula

$$R(\tau) = \frac{1}{T} \int_0^T E(t)E(t + \tau) \, dt \qquad (10.24)$$

At a discrete number of delays, $\tau = n\tau_0$, (where $n = 1, 2, \ldots, n_{max}$ and n_{max} is a relatively large number) the minimum delay or resolving time, and τ_0 is chosen to be a suitably small fraction of the Doppler period. In order to obtain time-average turbulence properties, a large number of signals requires to be autocorrelated. If the velocity probability density is not gaussian, difficulties arise in the interpretation of the results. Photon correlators have been found to have special advantages when the intensity of scattered light is low and when the signal-to-noise ratio is low. Photon correlators have provided results in high-velocity flows and in highly turbulent unseeded flows, where difficulties have been found in using other signal processors.

In the filter bank processor, a number of filters are placed in series and individual signals are collected in the banks. This system avoids the necessity of tracking a frequency, which cannot be done when there are infrequent signals. Banks of up to 50 filters have been constructed in the frequency range 0.6–6 MHz, corresponding to a velocity range, after frequency shifting, from -2 m/s to 23 m/s. Velocity distributions are displayed on an oscilloscope and the stored integrated amplitudes corresponding to each filter are read from a digital voltmeter. Evaluations are made by computer of the true velocity-probability distribution, from which the mean frequency rms and skewness and flatness can be determined.

Sources of Error

The basic principle involved in laser anemometry is that of frequency measurement, and this is one of the most accurate forms of physical measurement that can be made. Wavelengths of lasers can be accurately determined and, thus, the degree of accuracy in measurement of velocity of single particles is dependent upon the accuracy of measurement of the angle of beam intersection. This angle can be measured with relative ease to better than 0.1 percent. Errors can arise, however, during the averaging process and these are affected by the shape of the observed velocity probability distribution.

The velocity determined from the signal in a laser anemometer is assumed to correspond with the velocity at the geometric center of the scattering volume, defined as the intersection point of the optical axes of the two incident laser beams. It is possible to measure such a velocity when a digital pulse counter is used and the particle passes through the intersection point. When averaging procedures are used for many particles, the mean velocity represents an average of the velocities of all the particles which pass through the whole of the scattering volume and contribute to the signal. In nonuniform flow systems, the velocity varies with position and, if this variation is significant within the scattering volume, or if the scattering volume is spatially asymmetrical or nonuniformly illuminated, the average velocity will reflect these variations. In practice, errors can arise in laser anemometer measurements due to the following:

1. Spatially nonuniform velocity distributions
2. Turbulent fluctuations
3. Truncation of the scattering volume by a wall or surface

Kreid (1974) analyzed these sources of error and presented a mathematical model for approximating the averaging process from which estimates of errors in measurement can be obtained, due to a nonuniform velocity distribution within the scattering volume. Kreid (1974) concluded that errors arising due to nonuniform velocity profiles are negligible, except from regions near to a bounding surface or any other regions in which the flow has extreme curvature. Errors can be minimized by reducing the size of the scattering volume and it is estimated that correction of the order of 4 percent is required for flows with turbulence intensity of the order of 20 percent. Care needs to be taken in using an anemometer in the proximity of surfaces as very significant errors can arise due to scattering volume truncation at a wall.

Errors in the spatial and temporal resolution of fluid velocity may occur due to the finite dimensions of the measurement control volume. When there is a velocity gradient across the control volume, this results in velocity-gradient broadening. This effect can be minimized by reducing the control volume dimensions or making corrections, if the velocity gradient is known. The value of the velocity component measured by the anemometer is a function of the effective probe volume and a random weight function to allow for the presence or absence of scattering particles. This is valid when a spectrum analyzer is used to process the signals. When tracking filters and frequency counters are used, additional biasing of the measured velocity component can arise, owing to the fact that more particles enter the probe volume when the flow is faster than the mean, and fewer when the flow is slower than the mean.

Particle distribution can influence the measured value of mean velocity, depending on the mode of averaging employed. If the tracking filter produces an analog voltage proportional to frequency, which is integrated over all time (even when there is no particle in the probe volume), the bias in mean velocity will, in general, be larger than if the voltage is integrated only over those periods when the particle is present in the probe volume. When a frequency counter is used to obtain an average from a number of "instantaneous" values (each of which corresponds to a small number of zero crossings), the bias in mean velocity will, in general, be larger if the average is obtained by dividing the total of a large number of zero crossings by the total crossing time.

In single-particle counters, when measurements are made with low particle concentrations, errors can occur in measuring the instantaneous flow velocity due to the following causes:

1. The tracer particle does not follow the flow exactly.
2. The distance between two fringes is not always constant.
3. Photomultiplier noise.
4. Electronic noise.
5. Inadequate high-pass filtering of the signal.
6. The counter has a ± 1 uncertainty.

If sufficient care is taken in designing and conducting experiments, the errors due to 1, 2, 5, and 6 can be minimized. Errors arising in single-particle counters are mainly caused by photomultiplier and electronic noise.

If the particles are not sufficiently small, particle velocities will not be the same as fluid velocities. This becomes a more serious problem under conditions of acceleration and deceleration. This effect is minimized by reducing particle size. In a turbulent flow with uniform concentration of particles, the flux of particles through the control volume is proportional to the instantaneous velocity. This can give rise to a statistical bias toward recording high velocities. Faster particles scatter, on average, fewer photons per pulse and this can lead to a lower probability of detection. If there is a correlation between the local instantaneous velocity and particle concentration, this can also lead to a bias.

Errors resulting in particle concentration, due to changes in the volume of fluid caused by temperature gradients or chemical reactions, have been shown to be negligible, except in the immediate vicinity of a reaction zone. When part of the flow is unseeded, the observed velocity probability is not the same as the true time-averaged distribution. This can lead to considerable errors near the edge of mixing layers. Temperature and density fluctuations result in fluctuations of turbulent refractive index. These can cause fluctuations in dimensions and position of the measuring volume and, in some cases, can result in the beams not crossing. Self and Whitelaw (1976) have concluded that such problems are only severe in large-scale combustors and that back-scatter geometry is less susceptible to error than forward-scatter.

10.5 LASER ANEMOMETER MEASUREMENTS IN FLAMES

The laser anemometer is the most reliable and accurate means for making velocity measurements in flames. It is the only means available for making comprehensive turbulence measurements in flames. The earliest reported measurements were those of Daschuk (1969, 1971). These were made at a time before the major developments had taken place in laser anemometry. Durst et al. (1972a), Baker et al. (1973), and coworkers have undertaken measurements under both laboratory and industrial conditions. Measurements have been made in both laminar and turbulent diffusion flames and velocity components have been measured in three-dimensional flows, including swirl and reverse flow measurements in recirculation zones, both with and without artificial seeding of the flow. Durst and Kleine (1973) measured turbulent-flame speed, Chigier and Dvorak (1975) have made measurements in the recirculation zone of flames with swirl, and Chigier and Styles (1975) have made measurements in liquid spray flames. The principal modifications which need to be made to a laser anemometer system when making measurements in flames is the provision of filtering in the collection optics so as to exclude radiation from the flame from reaching the photodetector. Under conditions of intense heat, special protection needs to be provided for the optical system. When seeding is required, metallic oxide particles, which can withstand the maximum temperatures in the flame, require to be used, and magnesium oxide powder has been found to be one of the more satisfactory materials. In all confined systems, optically clear windows must be provided to allow access for the laser beams. Quartz and sapphire windows can withstand high-temperature conditions and have been used in some combustion chambers. Good quality plate glass, provided with a screen of cold or warm gas for cooling and prevention of condensation on the windows, can also be used. It has been found useful to introduce shutters which are only opened at the time that measurements are being made. The laser anemometer provides information on flame noise, the interaction between turbulence and combustion, and is capable of providing information on correlations between velocity, temperature, and species concentration. Comparisons between axial, radial, and tangential fluctuating velocity components in a flame and a cold jet, as measured by Chigier and Dvorak (1975), are shown in Fig. 10.8. These results show that, as a direct consequence of combustion, accelerations occur to a greater extent with the axial velocity components than with the radial and tangential components. This study also showed increases in time-averaged velocity and change in the relative isotropy of the turbulence as a result of chemical reaction.

Many turbulence characteristics have been measured in flames, such as time

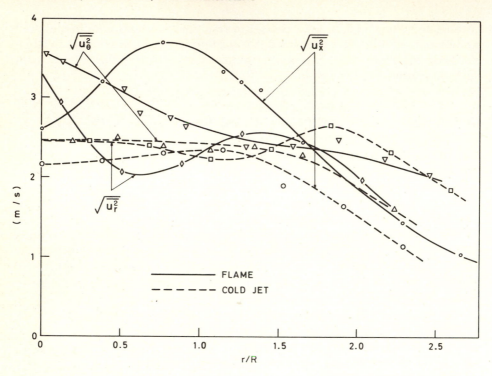

Figure 10.8 Axial, radial, and tangential fluctuating velocity components in a swirling flame and cold jet, measured by laser anemometer.

average, variance, probability density functions, kinetic energy, skewness, and flatness characteristics. The measurement of normal stress and shear stress from velocity component fluctuating correlations can be made with no significantly greater difficulty than under noncombustion conditions. A comprehensive review of laser anemometry with special application to various combustion flows is given by Self and Whitelaw (1976). A special examination (Chigier, 1975) of the suitability of laser anemometry for measurements in prototype gas turbine combustion chambers showed that the laser anemometer was sufficiently developed for such measurements to be undertaken. In internal combustion engines, transparent pistons have been constructed which allow, by mirror arrangements, laser beams to enter into the combustion chamber via the piston. Measurements have been made under both motoring and firing conditions.

Asalor and Whitelaw (1975) have shown that, in combustion systems, differences between true and measured probability density distributions can arise due to:

1. Temperature gradients which cause gradients of density and volumetric particle concentration.

2. Chemical reaction which may modify the volumetric particle concentration, as can arise when a volume of reactants containing a number of seeded particles reacts to form a different volume of products at the same pressure and temperature and with the same number of particles.
3. Fluctuations of temperature and velocity and, therefore, of particle concentration and velocity may be correlated in such a way that the number of particles crossing the control volume, with instantaneous velocities lower and higher than the true mean value, are not equal.

10.6 MEASUREMENT IN TWO-PHASE FLOWS

Two-phase flow systems arise in combustion, mainly due to the presence of solid or liquid particles in gas flows. Particle sizes can vary over a wide spectrum from minute dust particles to droplets of heavy liquid fuel or solid particles of pulverized coal of the order of 1 mm. In such two-phase flow systems, the particle velocities will differ from the local gas velocity according to the ratio of momentum to drag forces. For large particles in low-velocity gas streams, particles will be accelerating or decelerating. The laser anemometer can be used for measurement of velocity in two-phase flow systems, but it is necessary to measure the size of the particles as well as their velocity, and time average the velocity data according to the size of the particles. For combustion studies, there is interest in determining changes in size of particles as a function of distance and time since this provides information on the rate of evaporation and burning of these particles. Special instruments have thus been developed for the measurement of particle velocity, size distribution, and particle concentration using the laser Doppler system.

The application of laser anemometry for measurement of velocity in liquid spray flames has been shown by Styles and Chigier (1977) and by Owen (1977). Measurements were made of velocities of liquid droplets and the gas velocity, for particle sizes as large as 500 μm. Examination of individual probability density functions indicated a range of velocities which could be ascribed separately to liquid particles and the gas. Significant differences could be detected between the velocities of each phase. Spray boundaries could be identified and measured angles of the spray within the combustion chamber were found to be significantly different from the nominal spray cone angles. An example of velocity measurements made by laser anemometer in a kerosine spray flame is shown in Fig. 7.23.

Particle-Size Measurement

When a particle crosses the measurement control volume of a laser anemometer, the intensity of scattered light is a function of particle size. The intensity of scattered light is, however, also a function of incident laser power, optical geometry, location of the particle within the measurement control volume, and scattering properties of the particle and medium through which the scattered light passes. Measurement of particle size by laser anemometer is dependent upon maintaining these other factors constant, or monitoring their fluctuations so as to separate the effect of variations in particle size on scattered light intensity. Signals with good signal-to-noise ratio can be obtained when particles as large as 800 μm are passed through a control volume with fringe spacing of 5 μm.

When the oscilloscope trace of the laser Doppler signal is examined it is found to contain two components in amplitude: (1) a modulated scattered intensity of ac component, and (2) an average scattered intensity of dc component. The signal visibility is given by

$$\text{VIS} = \text{ac/dc} \qquad (10.25)$$

Farmer (1972) has examined the relationship between visibility and particle size and has shown that, for spherical particles of diameter D, the visibility is given by

$$\text{VIS} = 2J_1[\pi D/\lambda_F]/(\pi D/\lambda_F) \qquad (10.26)$$

where $J_1[\ ]$ is a first-order Bessel function of first kind and λ_F is the fringe spacing. For cylindrical particles of width W

$$\text{VIS} = \sin\,[\pi W/\lambda_F]/(\pi W/\lambda_F) \qquad (10.27)$$

The visibility function is resolution limited for small particles so that the visibility for a particle with a diameter of one-tenth fringe spacing equals 1.98. The visibility function is also limited for large particles, due to ambiguities for particles greater than the fringe spacing. Meyers and Walsh (1974) have carried out a computer simulation to examine the effect of particle size and composition on light-scattering characteristics. They show that the amplitude and visibility of the Doppler burst are determined by the size of the particle, which, when combined with the trajectory effect, determines the signal characteristics. Farmer (1972) has developed an instrument for the simultaneous measurement of particle size and velocity. Particle sizes in the range 0.2–500 μm in diameter can be measured at a rate of 20,000 particles per second.

Durst and Umhauer (1975) have used two basically independent optical systems—a dual-beam laser Doppler system for measurement of velocity and a white light source for particle sizing by light scattering. A photodetector measures the amplitude of light scattered from the white beam and this is related to particle size by the use of Mie's theory. Particles were in the order of 1 μm in diameter and concentrations of the order of $10^3/m^3$. Chou and Waterston (1975) have shown theoretically that the forward/back-scatter ratio is a monotonic function of particle size over the range 0.015–0.5 μm. Forward- and back-scattered light was collected by small mirrors and directed through a monochromator to a photo-multiplier. The ratio of measured forward- and back-scattered light was used to predict particle size which corresponded with measurements of particle size by other means. An upper limit of measurable particle size by this method is assumed to be equivalent to a width of a single fringe.

When spherical particles in the size range 100–500 μm are passed through the center of the control volume, the amplitude of the signal is found to increase almost linearly with particle diameter, provided that no change is made to the optical system and laser power is maintained constant. A family of calibration curves of particle diameter versus signal amplitude can be obtained for each particular anemometer system as a function of the location of the traverse through the control volume. Measurement of particle size, using a laser anemometer, can be made from the measurement of amplitude, visibility, signal duration, and shape of the signal, at the same time that velocity is measured as particles traverse the measurement control volume.

Simultaneous Measurement of Droplet Size and Velocity

The feasibility of using the laser anemometer as a means of simultaneously measuring velocity and particle size has been demonstrated by using a forward-scattering laser anemometer for the measurement of particle diameters larger than the anemometer fringe spacing (Yule et al., 1977). The optical arrangement for the calibration experiments is shown in Fig. 10.9. Particles were attached to a transparent section of a rotating disk so that individual particles of known size could be traversed through various sections of the measurement volume. The signals produced by a 70-μm transparent sphere are shown in Fig. 10.10. The traverse through $z = 0$ is through the center of the measurement volume and shows a gaussian distribution for both the pedestal and envelope of the signal. The mean of the signal is determined electronically and the measured peak of the mean E_{P_0} can be related to the particle size for a fixed optical arrangement. The signal shown for $z = 2$ mm is for a traverse made near the outer edge of the measurement volume.

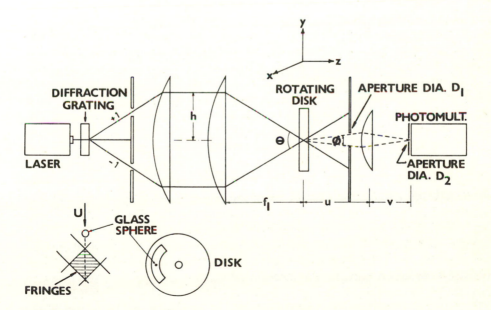

Figure 10.9 Optical system for particle-size measurement by laser anemometer. (*Yule et al., 1977.*)

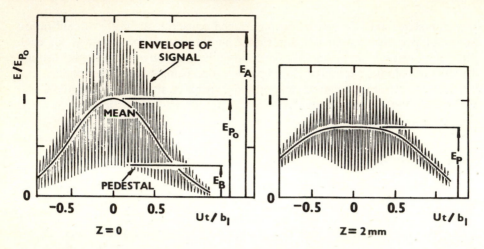

Figure 10.10 Laser anemometer signals for 70 μm transparent-particle traversing through (a) center of measurement volume, (b) 2 mm off the center of the measurement volume. (*Yule et al., 1977.*)

As the particle trajectory moves away from the center of the measurement volume, the amplitude of the signal decreases but the location of the traverse can be detected from the shape of the pedestal, which changes from a gaussian distribution to a double-peak signal with a central trough.

Figure 10.11 shows the measured peak amplitude as a function of particle diameter for a range of collection angles for both transparent and opaque spheres. These calibrations were determined by traversing individual particles within the size range from 30 to 300 μm through the center of the measurement volume. These results show that, if the optical geometry is selected and fixed and it is known whether the particles are opaque or transparent, it is possible to measure particle size when particles pass through or near the center of the measurement volume. The effective size of the measurement volume can be reduced by narrowing the field of view. For example, by the use of an additional collection system and photomultiplier at right angles to the axis, a discrimination system can be fitted to the logic of the signal diagnostics so that only the particles passing within 100 μm of the center of the measurement volume are measured.

The optical system shown in Fig. 10.12 is an off-axis forward-scatter mode with a narrow vertical slit photomultiplier aperture (Chigier et al., 1979). Off-axis collection reduces the dimensions of the region in which particles produce scattered light at the collecting photomultiplier, so that the effective length of the MCV is reduced. A collection angle of 6.8° was selected (Chigier et al., 1979) to preserve signal-to-noise ratios sufficiently high to obtain accurate velocity measurements and yet large enough to produce a significant reduction in the length of the MCV in which particles are detected. A vertical slit aperture, 85 μm wide, resulted in signals being accepted in a diagonal 230-μm wide slice of the MCV. The reduction in dimensions of the MCV is not as great as can be achieved by using the gate photomultiplier at 90°, but the off-axis geometry has the advantage

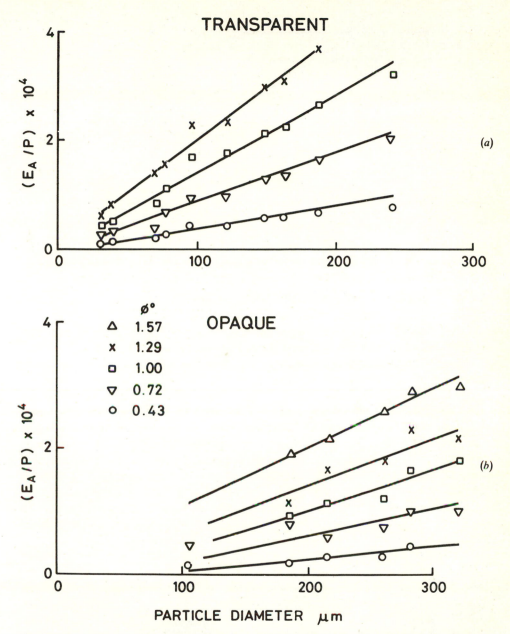

Figure 10.11 Calibration of laser anemometer for particle sizing. Peak signal amplitudes as function of particle diameter for a range of collection angles for (*a*) transparent, and (*b*) opaque spheres, traversing through the center of the measurement volume. (*Yule et al., 1977.*)

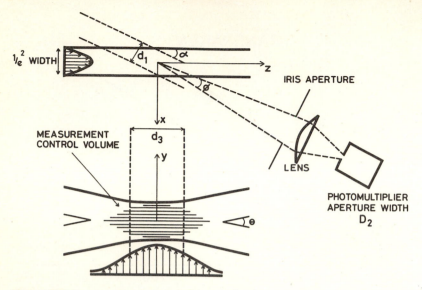

Figure 10.12 Laser anemometer optical system showing measurement control volume with "off-axis" collection for particle sizing. (*Chigier et al., 1979.*)

of eliminating the requirement of a second photomultiplier. This off-axis geometry enabled high particle number densities of at least $10^{10}/m^3$ to be measured without the statistical probability of there being more than one particle being measured at the same time. This density limit is sufficiently high for most low- to medium-throughput fuel sprays, except very close to the atomizer.

Figure 10.13 shows the variation of mean peak amplitude as a function of particle diameter for a beam angle of 6.71° and a collection angle of 1.21°. This

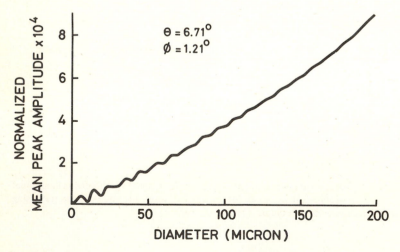

Figure 10.13 Mean peak amplitude E_{P_0} of Doppler signal as function of particle diameter. (*Chigier et al., 1979.*)

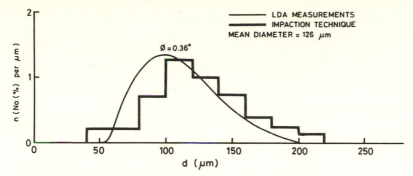

Figure 10.14 Comparison of size distribution measured by laser anemometer LDA and impaction technique. (*Ungut et al., 1978.*)

shows that, for particles larger than the fringe spacing in the range 30–200 μm, a one-to-one relationship between peak signal and size is obtained. This relation is different for transparent and opaque spheres, so that calculations and calibrations require to be made, based upon the optical properties of the particles.

In addition to the calibrations made by traversing individual particles through the measurement volume, further calibrations have been made by the use of mono- and polydisperse sprays. A direct comparison was made of size distributions measured by the laser anemometer with those measured by a standard impaction technique (Fig. 10.14). The indentations made by particles impinging on films on glass slides were examined by microscope and measured with a Quantimet image-analysis computer. This comparison showed reasonable agreement between the two particle-size methods.

Figure 10.15 shows comparisons of particle-size distributions measured in

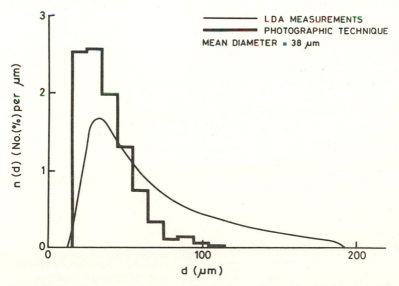

Figure 10.15 Comparison between size distribution measured in kerosine spray by laser anemometer and spark photography. (*Ungut et al., 1978.*)

kerosine sprays by the laser anemometer and spark photographic techniques. In the impaction techniques, small particles are deflected around the impactor and inaccuracies arise in analysis of dimensions of individual impactions. In the photographic technique, inaccuracies arise in determination of whether the particles are in focus. The statistical averaging procedures used in the impaction, photographic, and laser anemometer measurements are not the same, so that complete agreement is not found in particle-size distributions obtained with these various techniques. The laser anemometer has the advantage of being a non-invasive, high-frequency response system, which allows simultaneous measurement of particle size and velocity, which can be processed and analyzed by digital computer.

Figure 10.16 shows the temporal size distributions for measurements made at the same position in cold and burning kerosine sprays by the laser anemometer (Chigier et al., 1979). The cold spray was found to have a wider size distribution over the size range measured between 15 and 50 μm. These changes can be explained due to the preferential evaporation of small droplets, leading to total evaporation of the smallest droplets with a residue of larger droplets. The calculated local volume flux of droplets after ignition was found to be greatly reduced, due to combustion and evaporation. The mean droplet velocity and the

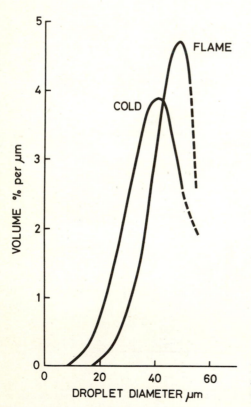

Figure 10.16 Particle-size measurement by laser anemometer in cold and burning kerosine sprays. (*Chigier et al., 1979.*)

variance of droplet velocity were measured as functions of droplet diameter. These results show that, for the cold-spray experiments, the mean velocities of larger droplets are higher than those for the small droplets. This demonstrates that the relatively small droplets lose most of their momentum soon after leaving the nozzle exit, due to their higher drag/inertia ratios, but the larger droplets are less affected because of their lower drag/inertia ratios. These measurements demonstrate the capability of the laser anemometer to measure local particle-gas velocity differentials, from which local Reynolds numbers and drag coefficients can be determined under conditions of vaporization and burning.

Mean Particle-Size Measurement by Laser Diffraction

A variety of techniques have been developed for the measurement of average size distributions of clouds of particles by measuring the properties of the scattered light. Very few of these techniques have been tested in heterogeneous combustion systems but one of the more promising techniques is the laser diffraction particle-size analyzer developed by Swithenbank et al. (1976), shown in Fig. 10.17. The technique is based on the Fraunhofer diffraction of a parallel beam of monochromatic laser light by moving particles. When a parallel beam of monochromatic light interacts with a particle, a diffraction pattern is formed whereby some of the light is deflected by an amount depending on the size of the particle. A Fourier transform lens is used to focus a stationary light pattern onto a multielement photodetector to measure the diffracted light energy distribution. When particles of different diameters are present in the light beam, a series of focused rings are generated at various radii, each focused light ring being a function of the particular particle size. These focused rings are detected by a special multielement detector, the output of which is multiplexed through an analog-to-digital converter.

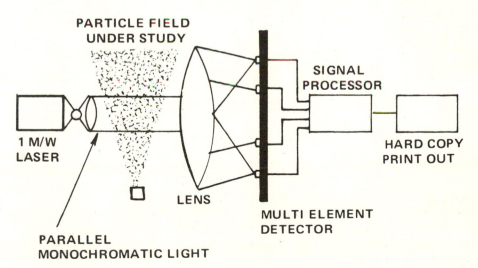

Figure 10.17 Laser diffraction particle-size distribution analyzer. (*Swithenbank et al., 1976.*)

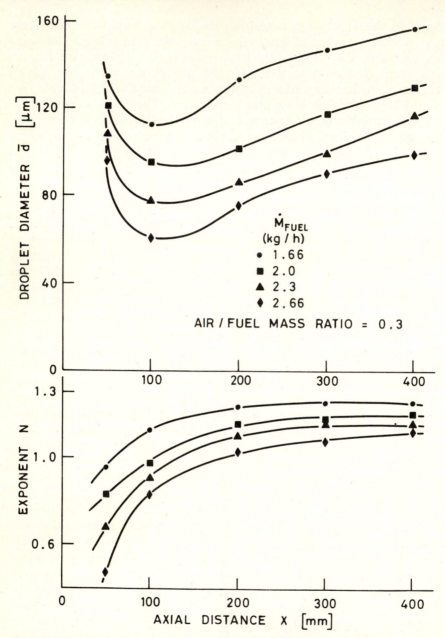

Figure 10.18 Variation of mean droplet diameter and exponent N in Rosin-Rammler size distribution equation with axial distance downstream in nonburning spray with variations in fuel flow rate, as measured by laser diffraction technique. (*Styles and Chigier, 1977.*)

The analysis of the measured light energy distribution into particle-size distribution is carried out by the computer, providing an immediate display of the measured size distribution. One particular advantage of this technique of measurement is that the diffraction pattern generated by particles is independent of the position of the particle in the beam. Hence, measurements can be made with the particles moving at any speed. Also for a group of particles, the combined diffraction pattern is directly related to their size distribution. Figure 10.18 shows measurements of size distributions made in air-assist fuel sprays using the laser diffraction analyzer.

For measurements in flames with high ambient lighting, parasitic effects become significant and the photodetector needs to be protected by an appropriate filter for the particular laser light. The use of high-power lasers, followed by attenuators before the detector or the application of a chopper to modulate the laser beam before it is passed through the flame, used in conjunction with a synchronous detector, have been proposed.

10.7 LASER SPECTROSCOPY

Laser spectroscopy for measurement of temperature and species concentration in combustion systems uses one of the following techniques: spontaneous Raman scattering, Rayleigh scattering, coherent anti-Stokes Raman spectroscopy (CARS), stimulated Raman spectroscopy, and laser excited fluorescence. Line-of-sight measurements and spatially resolved point measurements have been made and the feasibility of using tomography to provide spatially resolved field mapping has been demonstrated. Lasers require to be high-powered, coherent with high spectral purity. For measurements in fluctuating and turbulent flows, very high energy, very short pulse time, pulsed lasers are required. Spontaneous Raman scattering is the most highly developed and applied of laser spectroscopic techniques for general use. It is limited to flows without particles. Rayleigh scattering can only be used for total number density measurements. CARS has proved itself to be the most effective diagnostic technique for measurements in practical combustion systems normally laden with soot and other particulate matter. CARS is used primarily for major species measurements, while minor species measurements are made using laser excited fluorescence.

All optical phenomena are governed by Maxwell's equations, which yield the wave equation relating the electric field E of the incident electromagnetic light wave to the speed of light in vacuum and the generalized electric polarization P. When E and P are expanded into Fourier components, the induced polarization can be expressed as a power source of E. $X^{(1)}$ is the linear susceptibility of the medium, and $X^{(n)}$ are the nth order nonlinear susceptibilities, which express the extent to which the medium is susceptible to being polarized. Effects of nonlinearities in polarization become manifest only at very high laser intensities. Induced linear polarization modifies the propagation of the light wave through the medium. Raman and Rayleigh scattering arise from the oscillating polarization induced through the linear susceptibility. Polarizability is directly related to the susceptibility through the permittivity of free space.

Rayleigh scattering arises from the induced polarization oscillating at the same frequency as the incident radiation. Polarizability and induced polarization are dependent upon the nuclear positions of the molecule and modulated by rotation and vibration of molecules. In Raman scattering, the incident light frequency is shifted by the rotational-vibrational frequency, which may be interpreted as the beat frequency between the incident radiation and nuclear motions. Higher-order polarizations are considerably weaker and are related to the intra-atomic electric field, of order $3(10^8)$ V/cm. At high laser intensities of 10^9 W/cm^2, the ratio of succeeding polarizations is small, about 10^{-3}.

Raman Scattering

Laser Raman spectroscopy has a number of advantages for combustion diagnostics. All the species of interest can be monitored with the use of only a single laser. The laser can operate at any wavelength without the necessity of being tuned to the resonances of specific molecules. Visible wavelengths are favored; many species can be monitored simultaneously; Raman scattered intensities are unaffected by collision or quenching. Absolute calibration is readily achieved by comparing the scattered signal from a particular species with that of nitrogen. Spontaneous Raman scattering is very weak, with vibrational Raman cross sections of the order of 10^{-30} cm^2/sr.

Photons of light colliding with molecules may scatter elastically or inelastically from the target molecules. Elastic scattering is termed *Rayleigh scattering* and is unshifted in frequency from the incident light frequency.

When monochromatic radiation is incident on dust-free transparent gases, most of it is transmitted without change but, in addition, some scattering of the radiation occurs. If the wave number of the incident radiation is $\tilde{v}_0$, the scattered radiation shows pairs of new wave numbers, $\tilde{v}' = \tilde{v}_0 \pm \tilde{v}_M$ in addition to $\tilde{v}_0$. The wave numbers $\tilde{v}_M$ lie principally in the range associated with transitions between rotational, vibrational, and electronic levels. The scattered radiation usually has polarization characteristics different from those of the incident radiation. Both the intensity and the polarization of the scattered radiation depend upon the direction of observation. Such scattering of radiation with change of frequency is called *Raman scattering*. The phenomenon was first observed by Raman and Krishnan in 1928.

Raman scattering is an inelastic process with the incident photons either losing or gaining energy from the target molecules. When molecules gain energy and become excited, the process is termed *Stokes scattering* and, when molecules lose energy by changing from an excited state prior to the interaction, to a deexcited state, this process is termed *anti-Stokes scattering*. The fraction of molecules involved in the scattering process is so small that the perturbation to the molecular energy distribution is negligible. Energy exchanges are quantized in accordance with energy levels allowed by the laws of quantum mechanics. Since each molecular species possesses a characteristic set of energy states, the spectral distribution of Raman scattering is uniquely determined by the incident laser wavelength and the species from which scattering occurs. If no change in vibrational quantum number occurs, the scattering is termed *rotational Raman*, otherwise it is termed *vibrational-rotational*, or *vibrational*. The allowed changes in quantum number are governed by certain selection rules for the process. In diatomic molecules, vibrational quantum number changes are restricted to 0, ± 1, while rotational number changes of 0, ± 2 are permitted. If no change in rotational quantum number occurs, the Raman scattering is termed *Q-branch*, while for ± 2, it is termed *S and O*.

In purely rotational Raman scattering, only small energy changes occur and the spectra reside in close proximity to the laser frequency. In multicomponent gas mixtures, Raman spectra of the various gases overlap, making it difficult to detect any one species. Vibrational Raman bands are well displaced spectrally from the exciting frequency, and interferences between vibrational Raman spectra are rare.

The new wave numbers are termed *Raman lines*, or *bands*, and collectively constitute a Raman spectrum. Raman bands at wave numbers less than $\tilde{v}_0$ are referred to as *Stokes bands*, and those at wave numbers greater than $\tilde{v}_0$ as *anti-Stokes bands*.

Scattering of radiation without change of frequency from scattering centers like molecules, which are very much smaller than the wavelength of the incident radiation, is called *Rayleigh scattering*. Rayleigh scattering always accompanies Raman scattering. Within the spectrum of scattered radiation, the Rayleigh band, at wave number $\tilde{v}_0$, serves as a reference band for the determination of $\tilde{v}_M$ from $\tilde{v}'$. Presence of larger scattering centers, like dust particles, also leads to scattering of radiation without change of frequency—this is known as *Mie scattering*.

The relative intensities of Rayleigh and Raman scattering depend on many factors, including the physical state, the chemical composition, and the direction of observation relative to the direction of illumination. The intensity of Rayleigh scattering is generally about 10^{-3} of the intensity of the incident exciting radiation. The intensity of strong Raman bands is generally about 10^{-6} of the intensity of the incident exciting radiation. In flames, the intensity of Rayleigh scattering is typically 10^{-10} and the intensity of Raman scattering is 10^{-13} to 10^{-14} of the intensity of the incident-exciting radiation. Lasers have proved themselves to be excellent sources of monochromatic radiation and Raman spectra can be readily determined when scattered radiation is detected photoelectrically.

Energy Transfer Model

The origin of the modified frequencies of Raman scattering is explained in terms of energy transfer between the scattering system and the incident radiation. Incident radiation may cause an upward transition from a lower energy level E_1 to an upper energy level E_2. The energy difference $\Delta E = E_2 - E_1$ is equal to $hc\tilde{v}_M$. This energy may be regarded as being provided by the annihilation of one photon of the incident radiation of energy $hc\tilde{v}_0$ and the simultaneous creation of a photon of lower energy $hc(\tilde{v}_0 - \tilde{v}_M)$. If the system is already at an excited level, the radiation may cause a downward transition from E_2 to E_1. Under these conditions, a photon of the incident radiation of energy $hc\tilde{v}_0$ is annihilated with a simultaneous creation of a photon of higher energy $hc(\tilde{v}_0 + \tilde{v}_M)$.

In Rayleigh scattering, there is no resultant change in the energy state of the system. One photon of incident radiation is annihilated and a photon of the same energy is created simultaneously with no change in wave number.

A Raman band is characterized by the magnitude of its wave-number shift $|\Delta\tilde{v}|$ from the incident wave number where $|\Delta\tilde{v}| = |\tilde{v}_0 - \tilde{v}'| = \tilde{v}_M$. These wave-number shifts are referred to as Raman shift numbers. $\Delta\tilde{v}$ is positive for Stokes scattering and negative for anti-Stokes scattering. The intensity of anti-Stokes relative to Stokes Raman scattering decreases rapidly with increase in the wave-number shift. This is because anti-Stokes Raman scattering involves transitions to a lower energy state from a populated higher energy state. The thermal population of such higher states decreases exponentially as their energy, $hc\tilde{v}_M$, above the lower state increases.

Raman Spectra

The form of an observed Raman spectrum depends on the energy levels in the scattering system, the transitions permitted between them, and the conditions of observation. The resolving power of the instrument used to analyze the scattered radiation will also affect the observed spectrum. If the resolving power is great enough, the Raman spectrum for an assembly of free molecules is found to consist of a series of finely spaced lines of relatively small wave-number shift (Fig. 10.19). These fine lines arise from transitions between the closely spaced rotational levels of the molecules in the ground vibrational state. A number of other lines of larger wave-number shift are also seen—these arise from transitions which involve either a change of vibrational state or a simultaneous change of vibrational and rotational states. The Stokes and anti-Stokes Raman lines arising from transitions between a given pair of rotational levels will have similar intensities since the separation of the rotational energy levels is small. In contrast, the anti-Stokes

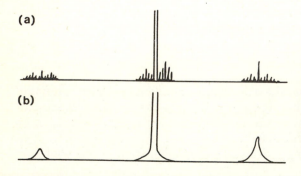

(a)

(b)

Figure 10.19 Diagrammatic representation of Raman spectra for (a) gas under high resolution showing resolved Stokes and anti-Stokes rotation and vibration-rotation lines; (b) gas under low resolution showing unresolved Stokes and anti-Stokes rotation and vibration-rotation bands.

Raman lines arising from transitions involving a change of vibrational quantum number will be very much weaker than the corresponding Stokes Raman lines, except for relatively small values of the Raman shift. The Stokes Raman spectrum originating from vibrational transitions will generally be more extensive and more intense than the corresponding anti-Stokes spectrum. In a flame, the anti-Stokes signal can exceed the Stokes signal due to wavelength scaling of the cross section and the strength of the vibrational matrix elements. For moderate resolving power, the Raman spectrum of a gas will usually show only Raman lines arising from vibrational transitions, as shown in Fig. 10.19b. For higher resolving powers, the more finely spaced lines associated with changes of rotational quantum number can be detected, as shown in Fig. 10.19a.

Raman spectroscopy has important practical advantages over infrared spectroscopy as a method for studying rotational and vibrational transitions. For all rotational, and nearly all vibrational levels, $\tilde{\nu}_M$ will lie in the range $0-35 \times 10^4$ m^{-1}. The complete Stokes Raman spectrum lies in the visible region of the spectrum for any exciting radiation in the range of wavelengths 400–600 nm. Consequently, in Raman spectroscopy, only one dispersing system and one detector are required to study the whole range of molecular rotational and vibrational wave numbers. Further, sample cells can be constructed of glass.

Raman spectroscopy has been dramatically transformed as a result of the availability of a range of gas, solid, and dye lasers. The output from a laser forms a self-collimated beam with a defined state of polarization; this greatly facilitates the study of intensities and polarizations. Lasers can provide a number of highly monochromatic excitation lines, ranging over the whole visible region; the frequency output of some lasers can be continuously varied over restricted ranges of the spectrum. Suitable excitation wave numbers, well removed from an absorption band, can be found for observation of ordinary Raman scattering. It is also possible to study resonance Raman scattering, which is produced when the excitation wave number is close to an absorption band.

Substantial developments have taken place, and continue to take place, in the dispersing and detection systems for Raman spectroscopy. The use of double and even triple monochromators provides such good discrimination against stray radiation and interferences that Raman spectra can be readily obtained from samples of poor quality. Detectors and associated electronics allow signals from very weak scattering to be effectively recorded.

Detection of Raman Spectra

Detection of Raman spectra requires a source of monochromatic radiation, a system for dispersion of the scattered radiation, and a detection device. The source for the excitation of the Raman spectra needs to be highly monochromatic (narrow line width) and be capable of providing a high irradiance at the sample. A gas laser meets these requirements and, in addition, provides radiation which is self-collimated and plane polarized. Most gas lasers provide a number of discrete wave numbers of varying power, and dye lasers provide excitation numbers which are continuously variable over a limited range. Laser beams may be focused to produce beams of much smaller diameter by using lenses. The region in which the beam is most concentrated is referred to as the focal cylinder, which, for diffraction-limited plane waves, has the following dimensions

$$D = \frac{4\lambda f}{\pi d} \qquad\qquad L = \frac{16\lambda f^2}{\pi d^2}$$

where λ is the wavelength of the laser radiation, d the diameter of the unfocused laser beam, and f the focal length of the focusing lens. Focused laser beams are normally used for the production of Raman spectra. The length resolution in Raman experiments is generally determined by the monochromator slit height.

The output of a laser is nominally described as having a given wave number; actually it consists of a number of modes of slightly different wave numbers which together form a band envelope whose width determines the observed width of that laser wave number. Very much narrower laser line widths can be achieved by using a mode-selecting etalon inside the laser cavity. An argon laser beam at 514 nm normally has a width of 0.15 cm^{-1}; by using an etalon, this can be reduced to 0.001 cm^{-1}. The power in the single-mode output is reduced to about 50 percent of the multimode output.

Filters and other optical devices may be inserted into the incident laser beam or the scattered radiation beam. Interference filters may be used to suppress unwanted laser frequencies, and a half-wave plate can be introduced to rotate the plane of polarization.

Laser light of power Q_i is directed through the medium; Raman scattering is collected at angle θ (typically 90°) to the incident beam over solid angle Ω over an extent l along the length of the laser beam (l is determined by an aperture stop such as a monochromator slit).

Raman scattering is not intensity dependent. The power of the scattered Raman radiation Q_r is a function of the initial laser power Q_i, the number density of the molecules in the appropriate initial quantum states for scattering to be observed, and n is the Raman scattering cross section—dependent on laser polarization and viewing direction, and the total optical collection efficiency. In atmospheric pressure flames, Q_r/Q_i is of the order 10^{-14}.

The characteristics of the dispersing system depend on whether it is to be used for the resolution of individual lines in rotation and vibration-rotation Raman bands, or for the study of vibrational bands under conditions of moderate resolution. The dispersing system is based on a diffraction grating in which front-surfaced concave mirrors are used to collimate the radiation before dispersion and to refocus it, after dispersion, onto the detector. The primary requirements are high resolving power and high reciprocal linear dispersion. A double monochromator grating dispersing system is used for the study of vibrational Raman spectra under medium resolution. The dispersed radiation is detected photoelectrically. The exit slit allows only a narrow wave-number band to reach the photomultiplier detector; rotation of the diffraction gratings allows successive bands to reach the detector.

The spectral purity of the monochromator system must be sufficient to enable radiation in the narrow waveband number $\tilde{\nu} \pm \delta\tilde{\nu}$, to which it is set, to be distinguished from radiation of other wave numbers. The intensities of spectral lines may be similar; the ability of the monochromator to distinguish between different wave numbers depends upon the resolving power, dispersion, and slit width of the monochromator.

Each monochromator system of a typical double monochromator for Raman spectroscopy would have a 10 cm × 10 cm plane reflectance grating, with 1200 grooves mm^{-1} used in the first order of diffraction, and collimating and focusing mirrors of 12-cm diameter and 75-cm focal length. In a two-component monochromator the dispersions are additive, giving an overall reciprocal linear dispersion of 20 cm^{-1} mm^{-1} and a resolution of 1 cm^{-1} in the region of 500 nm for a 50 μm slit width.

In scanning spectrometers the wave number is read directly from a scale which translates the grating orientation into wave numbers. Scanning speeds are usually in the range from 500 to 0.5 cm^{-1} per minute. More detailed information on the principles and applications of Raman spectroscopy can be found in the book by Long (1977).

Temperature and Species Concentration in Flames

Laser spectroscopy allows the measurement of temperature and species concentration at a "point" in combustion systems. When a high-powered laser light source is passed through a combustion gas and the scattered light is examined by a spectroscope, information can be obtained of temperature and species concentration levels. For an equilibrium system the intensity ratio of the Stokes to anti-Stokes lines is dependent upon the temperature of the molecule. For non-resonant scattering, the intensity of a Raman line is proportional to the frequency change, the number density of molecules, and a transition probability for the energy change. The number density is determined from a calibration experiment carried out under the same conditions of temperature and pressure as those in the test experiment. High laser intensities are required with the exciting radiation at high frequencies.

Laser selection depends primarily on the temporal resolution required of the measurement and the luminosity of the flame. For steady-state flames with low levels of luminosity, continuous-wave lasers are used and Raman scattered photons are counted. The laser beam is chopped and photons are detected by gating, which permits flame- or laser-induced background radiation to be sampled and subtracted. The Raman signal is enhanced by employing multipass cells or placing the experiment inside the laser cavity. In turbulent flows, time-dependent measurements need to be made. Averaged density measurements can be made by using broad detection bandwidths, but temperature measurements cannot be made readily in unsteady flows. Species probability distribution functions and power spectra can be obtained when the count rate becomes sufficiently high during the temporal resolution period. These techniques have not been proved in flames.

Pulsed lasers are employed for making instantaneous measurements and for probing in luminous environments. For luminous environments, high-peak laser powers are required while, for instantaneous measurements, high laser energies are required. In turbulent systems, single-pulse measurements allow determination of pdf's, from which true medium average properties and the magnitude of turbulent fluctuations can be ascertained. Averaging pulsed Raman data may not be representative of true medium averages, particularly with respect to temperature. Errors in averaging increase with the magnitude of the fluctuations. For fluctuations below five percent, and for small cycle-to-cycle fluctuations in repetitive processes, the error in average measurements is small.

Problems associated with background radiation are overcome by using pulsed lasers and time gating. By selection of an appropriate pulsed laser, peak Raman powers can be obtained in excess of the background radiation. Lasers include (1) mode-locked, cavity dumped argon ion lasers, (2) pulsed N_2 lasers, and (3) frequency-doubled neodynium:YAG ($2 \times$ Nd). In the presence of soot or dust particles, laser-induced or modulated particulate incandescence cause very serious broadband interference if the focal flux exceeds 10^6 W/cm^2, which is generally the case. These incandescent interferences are generally in phase temporally with the laser pulse and cannot easily be separated.

Single-pulse measurements have been obtained with large pulse energy lasers such as Q-switched ruby dye lasers and $2 \times$ Nd:YAG lasers. If the particle density is sufficiently high, low energy-pulse N_2 lasers and cw lasers can be employed to obtain time-resolved information. In flames, laser pulse energies of the order of 1 joule are required. Even with these energies, Raman scattering of only several thousand photons from the dominant constituent, generally N_2, are obtained.

The spectral distribution of Raman scattering is a display of state populations with appropriate weighting. Temperature is determined from the shape of the spectral distribution in one of the following ways: (1) from spectral band contours; (2) line intensity; or (3) band peak-height ratios. Contour fitting provides the highest measurement accuracy. The ratio of the anti-Stokes to Stokes vibrational scattering produced by a single pulse provides a reasonably accurate measure of the temperature. This ratio is independent of laser power and number density of particles.

The relatively much more intense scattered Rayleigh radiation at the exciting frequency is removed by filtering. Filtering can be accomplished with a high resolution Raman spectrometer using double monochromators. Filtering can also be accomplished with high-resolution interference filters or by polarizing the exciting frequency.

The basic equation for the number E_S of scattered photons received by a laser Raman sensor is

$$E_S = E_I \sigma_I (v_I/v_S) N L \, \Omega e \qquad (10.28)$$

where E_I is the number of photons from the laser source illuminating the volume of gas under observation

L is the length of the sample in the direction of the laser beam

N is the number density of the particles within the volume
σ is the Raman scattering cross-section at the laser frequency
Ω is the solid angle subtended by the receiving area of the sensor
e is the optical efficiency of the system
v_l is the laser frequency
v_S is the scattered light frequency

Following Goulard (1974), Eq. (10.28) can be regrouped so that terms on the left-hand side relate to the laser-detection system and those on the right-hand side to the geometry and nature of the experiment. In order to separate out the fourth-power, frequency dependence of Raman cross section on light frequency, let

$$\sigma = \sigma_0/q^4 \qquad (10.29)$$

where σ_0 is the cross section at a convenient reference frequency v_0 and

$$q = (v_0/v_l) \qquad (10.30)$$

Then, neglecting the ratio $(v_l/v_S) \simeq 1$, Eq. (10.28) can be written in the form

$$q^4 \frac{E_S}{E_0} = NL\sigma_0 \, \Omega e \qquad (10.31)$$

Lapp et al. (1972) and Lapp (1974) observed temperature-dependent effects in the spectral distribution of the Stokes Q-branch vibrational scattering. These effects arise predominantly from the vibration-rotation interaction and from significant population of excited vibrational levels. Upper-state bands originate from these excited levels and these are usually shifted toward the blue region of the spectrum. Observations were made using an argon ion laser, operated at 1.5 W at 488 nm. The scattered light was analyzed by a double monochromator with 500-nm blazed gratings. Scattering data for H_2O and O_2 were obtained from lean H_2-O_2 flames and data for N_2 were obtained from a lean H_2-air flame. For the diatomic molecules considered in these experiments the Stokes Q-branch fundamental series profiles are calculated. These calculated profiles are used to fit experimental profiles in order to determine the scattering gas temperature.

Figure 10.20 shows calculated Stokes Q-branch fundamental intensities for nitrogen over a range of temperatures from 300 to 3500 K. Vibrational temperatures are proportional to the integral of intensity for particular bands, while rotational temperatures are proportional to the profile on the short wavelength side of each band via the influence of the vibration-rotation interaction. For calculations made of nitrogen profiles, the spectral width can be seen to increase relatively with increase in temperature (Fig. 10.20). The N_2 temperature was determined by fitting theoretical profiles to experimental profiles obtained from the flame. An example of this fitting procedure is shown in Fig. 10.21. The Raman scattering signatures are direct measurements of the relative populations of the molecular internal modes and, for equilibrium situations, these relative populations correspond to the fundamental definition of temperature. This form of temperature diagnostics has the potential for becoming the most fundamentally

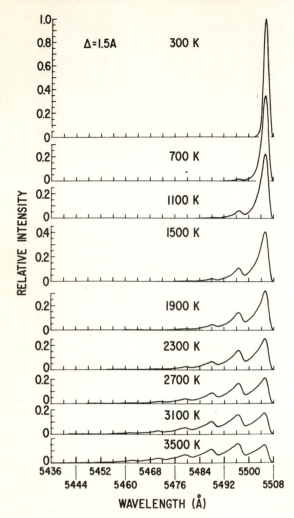

Figure 10.20 Calculated Stokes Q-branch fundamental intensities for nitrogen, used for the determination of temperature by laser Raman scattering. (*Lapp, 1974.*)

accurate scheme for nonperturbing measurements. Temperatures measured by Raman spectroscopy are shown to agree with independently measured temperatures utilizing a fine-wire thermocouple to within two percent.

Spontaneous Raman scattering has been successfully measured in a number of practical devices in which the particle concentrations are low. Johnston (1979) used cw laser Raman scattering to map out fuel-air distributions in an internal combustion engine prior to ignition. Time-averaged equivalence ratio measurements were made as a function of radius and crank angle for a motored, nonfiring, stratified charge engine with propane fuel. Under combustion conditions, strong laser-induced interferences precluded measurements under fuel-rich conditions, but some measurements were possible for lean engine operating conditions. Smith (1980) has used pulsed Raman spectroscopy to perform single-pulse temperature and density measurements in a homogeneous charge engine under combustion

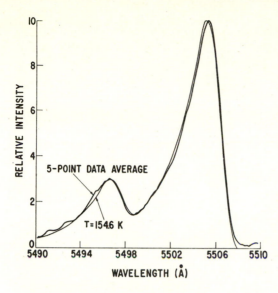

Figure 10.21 Determination of temperature using laser Raman scattering. Comparison of measured and theoretically calculated profile for nitrogen at 1546 K. (*Lapp, 1974.*)

conditions. Histograms were obtained at several locations over a variety of crank angles and equivalence ratios. Raman temperatures were deduced from the N_2 anti-Stokes to Stokes ratio, using a frequency-doubled neodynium laser. Attempts to make measurements by spontaneous Raman scattering in pulverized fuel utility boilers and in swirl turbulent propane diffusion flames have failed, due to the strong laser modulated particulate incandescent interferences.

The special advantages of using Raman scattering probe techniques are summarized by Lapp and Penney (1974):

1. Specific species measurement. Each line in the Raman spectrum can be identified with a particular type of molecule and its level of excitation.
2. The intensity of a Raman scattering line is directly proportional to the number density of corresponding molecules and independent of the density of other molecules.
3. Three-dimensional resolution. Measurements can be made within control volumes of dimensions smaller than 1 mm³.
4. Instantaneous time resolution. With the use of pulsed lasers information can be time-resolved to the order of nanoseconds.
5. Nonperturbing. The coupling interactions between optical radiation and non-absorbed gases is extremely weak. Consequently, over a wide range of experimental conditions, the state of the observed gas is not perturbed unless the exciting light is strongly absorbed.
6. Remote in situ capability. Measurements can be made in systems which allow optical access of moderate to good quality.
7. Accessibility of temperature information for gases in thermal equilibrium. The Raman spectrum depends on both density and temperature. However, the temperature dependence is independent of density and is sufficiently strong in appropriate spectral regions to allow sensitive temperature measurements.

8. Accessibility of density information. Density information from Raman scattering is effectively independent of temperature at temperatures below that at which appreciable vibrational excitation occurs in the observed species. At higher temperatures, the detection bandwidth requires to be chosen properly in order to avoid making temperature corrections.
9. Capability to probe systems not in equilibrium.
10. Simultaneous measurement of different components in a gas.
11. Relative lack of interference. The vibrational Raman lines of many molecules are well separated.
12. Wide sensitivity. Each molecule has at least one allowed Raman band, in contrast to infrared absorption, which is not sensitive to molecules without a permanent dipole moment.

The principal disadvantage of laser Raman spectroscopy is the weakness of the Raman effect. Raman scattering cross sections are three orders of magnitude smaller than Rayleigh scattering cross sections and ten to sixteen orders of magnitude smaller than cross sections for unquenched fluorescence. This disadvantage has been overcome by the use of high-powered lasers and highly sensitive spectrometers.

Coherent Anti-Stokes Raman Spectroscopy (CARS)

Coherent anti-Stokes Raman spectroscopy (CARS) is a nonlinear light wave mixing process, which provides information on temperature and species concentration with high spatial and temporal resolution. CARS has superseded spontaneous Raman scattering for measurements in practical combustion environments where the presence of small particles of soot, dust, or fuel cause laser-induced incandescence or fluorescence, which swamp the signals in spontaneous Raman scattering systems. The effect was originally discovered by Maker and Terhune (1965). Taran and his coworkers at ONERA in France (Régnier and Taran, 1974) pioneered the application of CARS for measurements in flames. Experiments demonstrated the feasibility of using cw CARS and pure rotational CARS. Roh et al. (1976) generated broadband CARS in a single pulse, thereby permitting the measurement of nearly instantaneous properties. Eckbreth (1979, 1980) has made major contributions to the development and proving of CARS in sooting and turbulent flames in practical combustion environments. He used crossed-beam phase matching to provide high spatial resolution.

Comparison Between CARS and Spontaneous Raman Scattering

1. CARS permits diagnostic probing in the high interference environments of practical combustion systems.
2. CARS signal levels are several orders of magnitude larger than those for Raman scattering.
3. CARS signals are coherent, i.e., beamlike.

4. All the CARS radiation can be collected.
5. Since CARS radiation can be collected in an extremely small solid angle, discrimination against background luminosity and laser-induced particulate interferences, such as incandescence and fluorescence, is greatly facilitated.
6. Discrimination against laser-induced interferences is improved because CARS signals are shifted in frequency from the laser frequency.
7. Signal to interference ratio (S/I) is improved by orders of magnitude over a broad range of operating conditions.
8. CARS is best suited for thermometry and major species concentration measurements.
9. CARS is complementary to laser-induced fluorescence, which is more appropriate to trace radicals and minority species.
10. Minority species measurements, using CARS, are possible if sacrifices are made in spatial and temporal resolution.

Theory and Application of CARS

The principle of CARS measurement is shown in Fig. 10.22.

Two incident laser beams, at pump frequency ω_1 and Stokes frequency ω_2, interact through the third-order nonlinear susceptibility of the medium to generate a polarization field, which produces coherent radiation at frequency $\omega_3 = 2\omega_1 - \omega_2$. CARS is thus referred to as "three-wave mixing." When the frequency difference $(\omega_1 - \omega_2)$ is close to the frequency of a Raman active resonance ω_v for third-order nonlinear susceptibility, the polarization exhibits a cubic dependence on the optical electric field strength. In isotropic media, such as gases, the third-order susceptibility is the lowest-order nonlinearity that is exhibited.

For efficient signal generation, the incident beams must be so aligned that the three-wave mixing process is properly phased. Since gases are virtually dispersionless, i.e., the refractive index is nearly invariant with frequency, the photon energy conservation condition $\omega_3 = 2\omega_1 - \omega_2$ indicates that phase matching occurs when the input laser beams are aligned parallel or collinear to each other. Collinear phase-matching leads to poor and ambiguous spatial resolution because the CARS radiation undergoes an integrative growth process. This difficulty is circumvented by employing crossed-beam phase matching, such as BOXCARS. In this approach, the pump beam is split into two components which, together with the Stokes beam, are crossed at a point to generate the CARS signal. CARS generation occurs only where all three beams intersect, resulting in high spatial precision.

Measurements of temperature and concentration are obtained from the shape of the spectral signature and/or intensity of the CARS radiation. Nonresonant contributions can be reduced or cancelled by proper orientation of the mixing wave field and CARS viewing polarizations. CARS computer codes to synthesize spectra have been developed and validated experimentally for a number of molecules.

● APPROACH

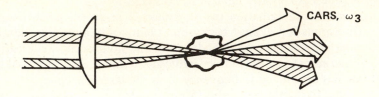

● ENERGY LEVEL DIAGRAM

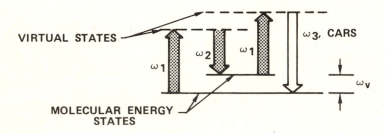

● SPECTRUM

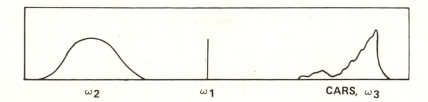

Figure 10.22 Coherent anti-Stokes Raman spectroscopy (CARS). (*Eckbreth, 1980.*)

Signal Detection

The CARS spectrum can be generated by employing a narrow band Stokes source, which is scanned to generate the CARS spectrum piecewise. This approach provides high spectral resolution, strong signals, and eliminates the need for a spectrometer. Resolution is limited only by the line widths of the lasers, which can be made very narrow. Very high resolution can be attained by using coherent spectroscopy. Continuous-wave CARS has been employed to fully resolve the Q-branch lines of H_2, O_2, and CH_4 with a spectral resolution of 30 MHz. This is nearly two orders of magnitude better resolution than that attainable with the finest Raman monochromators. Tunable Stokes wave CARS generation is unsuitable for turbulent combustion diagnostics, due to the nonlinear behavior of CARS on temperature and density. For time-resolved diagnostics, a broadband Stokes source is employed. This leads to weaker signals, due to the energy partitioning over a broad spectral region, but generates the entire CARS spectrum with each pulse, permitting instantaneous measurements of medium properties if broadband detection is employed. Repeating these measurements a statistically significant number of times permits determination of the pdf, from which true median averages and the magnitude of turbulent fluctuations can be ascertained.

High-intensity pulsed lasers are required to provide a statistically significant numbers of CARS photons in each spectral resolution interval and to generate CARS signals in excess of the various sources of interference. The frequency doubled neodynium YAG laser is preferred because of its high pulse repetition frequency and more favorable spectral range.

CARS measurements have been made of N_2, H_2, CO, O_2, and H_2O in flames. CARS measurements have been successfully made in highly sooting flames and for detailed temperature field mapping in laminar propane diffusion flames, as shown in Fig. 10.23. Each temperature point from broadband N_2 CARS spectra was averaged for 15 seconds, the spatial resolution was 0.3×1 mm. Single-shot (10 ns) spectra have also been obtained.

A unique feature of CARS is its potential for species concentration measurements from the shape of the spectrum (Fig. 10.24). It has been used for measurements of molecular species, with concentrations in the range from 0.1 to 20 percent.

When the nonresonant susceptibility is small relative to the resonant susceptibility of the species of interest, concentration measurements are performed from

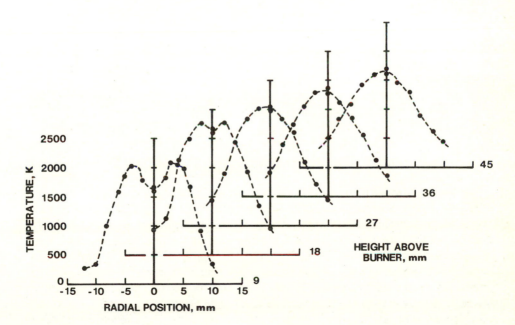

Figure 10.23 Temperature profiles in laminar propane diffusion flames measured by CARS from N_2 spectra. (*Eckbreth, 1980.*)

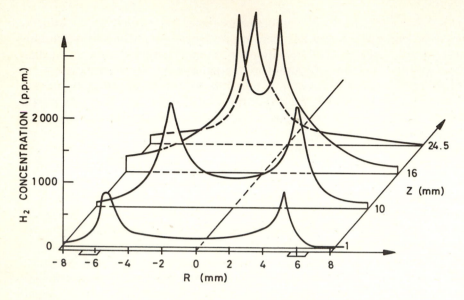

Figure 10.24 Concentration distributions of hydrogen in a natural gas flame measured by laser coherent Raman anti-Stokes scattering. (*Régnier and Taran, 1974.*)

the absolute intensity of a portion of, or all of, the spectrally integrated CARS signature. In the presence of a strong nonresonant background, it must be first cancelled, using polarization approaches or time delay.

Density measurements are generally made by comparing the CARS signal from the measurement volume with that generated simultaneously in a reference cell containing a gas at a known temperature and pressure. The reference cell can be placed either in series or in parallel with the sample volume. In atmospheric pressure flames, CARS concentration measurements can only be measured down to several tenths of a percent level. This limitation may be overcome by resonance enhancement.

CARS Measurements in Practical Combustion Chambers

CARS measurements have been made in internal combustion engines (Stenhouse et al., 1979), the exhaust of a kerosine-fired burner (Attal et al., 1980), and in liquid-fueled combustors housed in combustion tunnels (Switzer et al., 1979, and Eckbreth, 1980). In the internal combustion engine studies, both motored and gasoline-fired conditions were studied and CARS spectra of N_2 and propane were obtained. A narrow-band Stokes source was scanned synchronously with the engine cycle to generate the CARS spectrum piecewise and, subsequently, temporally resolved with respect to the engine cycle. Switzer et al. (1979) measured temperature and species concentration from N_2 and O_2 spectra in a bluff center-body combustor fueled with propane and various liquid fuels. Eckbreth (1980) made measurements in a swirl burner flame and in the exhaust of a gas turbine combustor can. Single-pulse thermometry was demonstrated both by Switzer and Eckbreth. Averaged CARS temperatures were found to be in good agreement with thermocouple measurements. Eckbreth (1980) introduced a 20-m long, 60-μm diameter fiber optic guide to pipe out the CARS signal from the burner test cell to an adjacent control room. Since this most recent experiment by Eckbreth has demonstrated that CARS can be used to make noninvasive measurements of temperature and species concentration in practical particle-laden turbulent combustion systems, the apparatus is described in detail.

The CARS instrument used by Eckbreth (1980) for measurements in a combustion tunnel is shown in Fig. 10.25. A frequency-doubled neodynium YAG laser emitted two beams at 532 nm by sequentially doubling the fundamental and residual laser beams. The primary beam energy had 200 mJ/pulse with a pulse duration of 10^{-8} s at a frequency of 10 Hz. The secondary beam energy was an order of magnitude lower than the primary and was used to optically pump the Stokes dye oscillator. The Stokes oscillator output was amplified in a dye cell; the amplifier was pumped by a fraction of the primary beam, split off at a 33 percent beam-splitter. The Stokes laser dye spectrum was centered at 606.7 nm, midway between the ground and first vibrational-state Raman shifts of nitrogen. The concentration of the Rhodamine 640 dye, dissolved in ethanol, was adjusted until the dye spectrum was centered in comparison with a neon lamp spectrum. N_2 is generally selected for CARS thermometry investigations in air-fired combustion systems because it is the dominant molecular constituent and is present everywhere in high concentration, independent of the extent of the chemical reaction.

The Stokes laser output was magnified by a factor of 2 or 3 and then passed through an optical flat which could be rotated to vary the BOXCARS phase-matching angle, and was then reflected off the dichroic D to the beam-crossing and focusing lens. The remaining primary beam was split into two components; these components were aligned parallel and sent to the crossing-focusing lens after transmission through the dichroic D. The pump beams were crossed at a half-angle of 0.77°; the measurement volume was approximately 0.5-mm diameter by 12 mm long, oriented transverse to the flow axis. A telescope was inserted in the primary beam to vary the wavefront curvature so that the beam waists coincided at the crossing point. All beams emerging from the combustor were recollimated by a lens, and the one pump component and adjacent Stokes beam were trapped.

The CARS beam was dispersed from the remaining overlapping pump component by the prisms shown in Fig. 10.25. This pump component was trapped and the CARS beam passed through cutoff and polarization filters. A microscope slide intercepted a fraction of the CARS signal which was sent to a photomultiplier tube fitted with a narrowband (1 nm) interference filter, which spectrally integrated the CARS signature. The output of the photomultiplier was sent to a BOXCARS averager. This integrated signal, after appropriate normalization, was employed for density determination. After passing the splitter slide, the CARS radiation passed through an electronic shutter to a lens which focused the CARS beam into an optical fiber, 60-μm diameter and 20 m long. The fiber transmitted the CARS beam to the spectrometer fitted with an optical multichannel analyzer

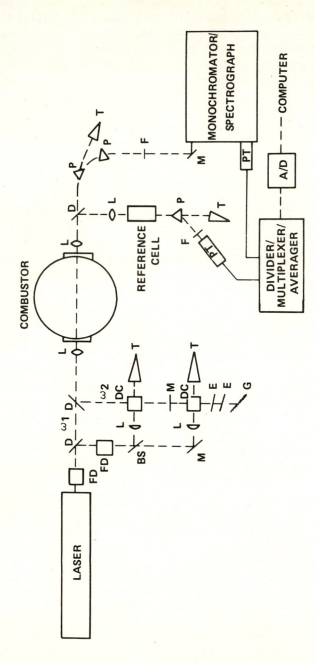

Figure 10.25 CARS diagnostic system. Code: FD, frequency doubler; M, mirror; D, dichroic; P, prism; T, trap; BS, beam-splitter; E, etalon; G, grating; DC, dye cell; PT, photomultiplier tube; L, lens; F, filter. (*Eckbreth, 1980.*)

located outside the test cell in the control room. The electronic shutter was used to sample individual CARS pulses from the 10 pulse/s train of signal pulses.

CARS spectra of N_2 were averaged on the optical multichannel analyzers (OMA) for 10–15 s, corresponding to 100–150 laser pulses. Since the CARS signal level does not vary linearly with temperature and density, it is not valid to average the CARS spectra to extract mean properties in turbulent-flow conditions. Averaged spectra show a ground state band due to Q-branch rotational-vibrational transitions between the ground and first vibrational states, which broadens with increasing temperature. A "hot" band from rotational-vibrational transitions between the first and second vibrational levels appears at temperatures above 900 K. Temperatures can be deduced from the full-width-at-half-height (FWHH) of the ground state band, and also from the ratio of the hot band to ground state band peak heights (PHR). Absolute values of temperature are determined by comparing spectral measurements with theoretical calculations. For the high-temperature data, Eckbreth (1980) showed that temperatures deduced from both the FWHH and PHR were in agreement within 50 K.

In the combustion tunnel experiments made by Eckbreth (1980) with liquid, Jet-A fuel, the flame was extremely luminous with high concentrations of soot. CARS measurements made through the liquid spray showed temperatures of the order of 900 K; downstream, in the highly turbulent and luminous region, temperatures of the order of 1500 K were measured. The spectrum was of very high quality, as fine as those previously produced in laboratory experiments. Provided that the windows are purged to prevent deposition, CARS measurements in liquid-fueled systems can be made with the same facility as in gaseous-fueled systems. In highly turbulent environments characterized by large fluctuations in density and temperature, the CARS spectrum must be captured and temperature information extracted with each laser pulse. Measurements are repeated a statistically significant number of times to obtain the temperature probability distribution function (pdf). The time-average and rms turbulent fluctuating temperature are determined at each location from the pdf. A comparison of a single CARS pulse and an averaged CARS spectrum, as measured by Eckbreth (1980), is shown in Fig. 10.26. The averaged CARS signals were obtained by inserting a neutral density filter to prevent the vidicon from the OMA from saturating.

From the experiments of Eckbreth and other investigators, CARS has been successfully demonstrated with both liquid and gaseous fuels in the primary zone, in the exhaust of continuous combustion chambers, and in internal combustion engines.

Laser-Induced Fluorescence Spectroscopy

Laser-induced fluorescence is used for measurement of minor species, such as flame radicals with concentrations at ppm or sub-ppm levels. Laser fluorescence measurements are made in a manner very similar to Raman scattering, with the exception that the laser must be tunable. Fluorescence is the spontaneous emis-

SINGLE PULSE (10^{-8} s)

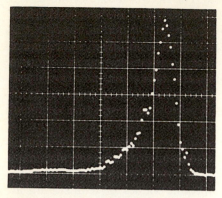

100 PULSE AVERAGE (10 s)

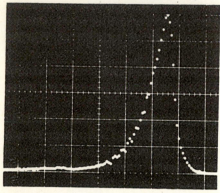

FREQUENCY ⟶
0.574 CM^{-1}/DOT

Figure 10.26 Single pulse and averaged N_2 CARS spectra in exhaust of gas turbine combustor burning Jet-A liquid fuel. (*Eckbreth, 1980.*)

sion of radiation from an upper electronic state. In laser-induced fluorescence spectroscopy, molecules are excited by absorption of laser radiation tuned to coincide with a particular molecule resonance. Fluorescence may occur at the same frequency as the exciting frequency, termed *resonance fluorescence*, or the frequency may be shifted when the molecule returns to a state other than that from which it originated. Resonance fluorescence suffers interference from elastic scattering processes, such as Mie or Rayleigh scattering, so that frequency shifted fluorescence is used, particularly when particles are present in the flame.

In order that fluorescence measurements can be made, the following information is required:

1. The emission spectrum of the molecular species must be known.
2. The molecular species must have absorption wavelengths accessible to a tunable laser source, i.e., in the range 0.2 to 1.5 μm.

3. The rate of radiative decay of the excited state must be known for quantitative measurements, since the fluorescence power is proportional to this rate.
4. The fluorescence efficiency needs to be evaluated.

Quenching, the process of deexcitation due to collisions, reduces the quantity of fluorescence. Chemical reaction can also change the fluorescence; the upper state may disappear. Corrections for quenching require to be made from data in flames where temperature and major species concentrations are known, together with the deactivation rates of the excited state for various collision partners, as a function of temperature. Measurements of hydroxyl concentration in premixed flames have shown that the fluorescence measurements are in excellent agreement with those determined by laser absorption, indicating that the corrections for quenching effects were adequate. In flames, the quenching rate can be very large, leading to very rapid decay. Stepowski (1979) succeeded in measuring OH concentrations by operating a flame at low pressure, 20 torr, to increase the quenching time. By using very short laser pulses, so that the laser pulse length is much less than the quenching time, fluorescence is independent of the quenching rate during the pulse.

Quenching corrections can be circumvented by using saturated fluorescence, in which the incident laser intensity is made sufficiently large, so that the absorption and stimulated emission rates are much greater than the collisional quenching rate. Under saturation, the fluorescence signal and, hence, the species detection sensitivity, is maximized. Measurement of minor species is of special interest for determination of emissions and atmospheric pollution. Measurements have been made of Na, Li, Zr, Tl, and In atomic species. The most widely investigated molecular species is NO_2, which has a broad absorption spectrum through the visible range and is, hence, easily excited by a variety of laser sources. Laser fluorescence is capable of detecting flame radicals such as OH, C_2, CH, CN, NH, NH_2, SH, and SO. Laser fluorescence has primarily been used in kinetic studies of flames, or flow reactors. It has the potential of being coupled to a CARS system for the measurement of minor species in practical combustion chambers.

BOOKS ON COMBUSTION,
FUEL SCIENCE,
AND TECHNOLOGY

Textbooks, monographs, bound volumes of review papers, reference books, and published proceedings of symposia dealing specifically with combustion, fuel science, and technology have been published for over four decades. The science and technology of combustion have advanced progressively and greatly expanded during the second half of the twentieth century. The Bibliography includes most, but not all, the books published, listed in alphabetical order. Many of the earlier books are out of print, almost all of the books currently in print were published in the 1970s. These cover most of the subject headings of previous books but do not necessarily supersede previously published books. The most comprehensive knowledge of coal science and technology can be found in books published before 1960—little progress was made in this subject during the hiatus of the 1960s and 1970s. Knowledge of combustion in engines was considered to be sufficient for design and operation of gas turbine and internal combustion systems on the basis of books published in the 1950s and 1960s. It was only with the advent of legislation on pollution and, subsequently, on fuel economy that the great lack of knowledge in this area was recognized. During the Second World War, Britain experienced a shortage of fuel and developed efficient usage of fuel by improved technology. The flood of low-cost oil and, to a lesser extent, natural gas since the Second World War led to a decline in interest and publications in the technology of combustion systems.

MONOGRAPHS

Most books listed in the Bibliography are monographs or advanced specialized books, dealing with combustion and fuel science. They have been written by research scientists working in specialized fields with basic backgrounds in chemistry, physics, or engineering. They suffer from lack of breadth and lack of pedagogical experience on the part of the authors. Lewis and von Elbe (1961) served for a long time as the standard reference for combustion.

TEXTBOOKS

Smith and Stinson (1952) wrote the first textbook for undergraduate engineering students concerning solid, liquid, and gaseous fuels and the problems associated with their combustion. Spalding wrote his first textbook on *Fundamentals of Combustion* in 1955. Subsequently, Williams (1965) wrote a textbook on *Combustion Theory*, while Strehlow (1968) covered a wider range of subjects on *Fundamentals of Combustion*, published by the International Textbook Company. Gaydon and Wolfhard published the first edition of their advanced book on *Flames—Their Structure, Radiation, and Temperature*, in 1953 and revised this each decade until the fourth edition in 1979. The monograph by Beér and Chigier in 1972 introduced the fluid and particle mechanics of flow fields in a book based on the authors' experience, mainly in the field of industrial furnaces. By the mid-1970s, the demand for textbooks on combustion began to increase. Courses on Combustion and Fuel Science and the number of students began to multiply and a general feeling was expressed by professors and teachers that there was no satisfactory textbook for general undergraduate and postgraduate course work. Most of the previously published books were out of print. Murty Kanury's book *Introduction to Combustion Phenomena* appeared in 1975, and Glassman's book on *Combustion* was published in 1977, while Gunther's book appeared in German in 1974, and Beér and Chigier's *Combustion Aerodynamics* was translated into Japanese in 1976. These books were immediately adopted as textbooks for course work. Spalding's book on *Combustion and Mass Transfer* (1979) reflected the growing interest in digital computational techniques.

By the end of the 1970s, a selection of textbooks had become available but the ever-increasing number of students and professors were seeking a comprehensive textbook with sufficient breadth and pedagogy to cater for the undergraduate and postgraduate need for textbooks.

REVIEW ARTICLES

During the 1970s, specialized journals and books containing reviews and state of the art articles began to be published in many fields of science and technology. This catered for the demand for publications intermediate between the highly specialized articles appearing in science and technology journals and textbooks which were in insufficient supply. *Progress in Energy and Combustion Science* was launched in 1975 as a quarterly journal, which is bound and sold as a book in separate volumes for each year of publication. In 1979, a special low-priced student edition, *Energy and Combustion Science*, was published in the Pergamon International Library. Articles were selected from the first three volumes of the journal on the basis of recommendations by professors and teachers of courses, prescribing review articles to students.

McGRAW-HILL SERIES OF TEXTBOOKS: ENERGY, COMBUSTION, AND THE ENVIRONMENT

A new series of textbooks, covering the wide spectrum of Energy, Combustion, and Environment, was initiated in 1980. The series is launched with the introductory textbook, *Energy, Combustion, and Environment*, by Chigier (1981), to be followed by a series of books written by the leading authors and teachers in the fields of Combustion Science and Fuel Technology.

BOOKS ON COMBUSTION, FUEL SCIENCE AND TECHNOLOGY

Beér, J. M., and N. A. Chigier: *Combustion Aerodynamics*, Applied Science, London, and Wiley, New York, 1972.

Benson, R. S., and N. D. Whitehouse: *Internal Combustion Engines*, vols. 1 and 2, Pergamon, 1979.

Bone, W. A., and D. T. A. Townend: *Flame and Combustion in Gases*, Longmans, Green and Co., London, 1927.

Bowman, C. T., and J. Birkeland (eds.): *Alternative Hydrocarbon Fuels: Combustion and Chemical Kinetics*, American Institute of Aeronautics and Astronautics, 1978.

Bradley, J. N.: *Flame and Combustion Phenomena*, Methuen, London, 1969.

Brame, J. S. S., and J. G. King: *Fuel*, Edward Arnold, London, 1955.

Chedaille, J., and Y. Braud: *Industrial Flames, 1, Measurement in Flames*, Arnolds, London, 1972.

Chigier N. A. (ed.): *Pollution Formation and Destruction in Flames, Progress in Energy and Combustion Science*, 1, Pergamon, 1976.

———— (ed.): *Energy from Fossil Fuels and Geothermal Energy, Progress in Energy and Combustion Science*, 2, Pergamon, 1977.

———— (ed.): *Energy and Combustion Science*, Selected Papers from *Progress in Energy and Combustion Science*, Student Edition, 1, Pergamon International Library, 1979.

De Soete, G., and A. Feugier: *Aspects Physiques et Chimiques de la Combustion*, Technip., Paris, 1976.

Edwards, J. B.: *Combustion*, Ann Arbor Science, Michigan, 1974.

Faraday, M.: *Chemical History of a Candle*, Viking Press, 1960.

Field, M. A., D. W. Gill, B. B. Morgan, and F. G. W. Hawksley: *Combustion of Pulverised Coal*, Institute of Fuel, London, 1967.

Francis, W.: *Fuels and Fuel Technology*, 1 and 2, Pergamon, London, 1965.

Fristrom, R. M., and A. A. Westenberg: *Flame Structure*, McGraw-Hill, New York, 1965.

Fuel and the Environment, The Institute of Fuel, London, 1973.

Gaydon, A. G.: *Spectroscopy and Combustion Theory*, Second Edition, Chapman and Hall, London, 1948.

——: *Spectroscopy of Flames*, 2d ed., Chapman and Hall, London, 1974.

—— and H. G. Wolfhard: *Flames—Their Structure, Radiation and Temperature*, 4th ed., Chapman and Hall, London, 1979.

Glassman, I., *Combustion*, Academic Press, New York, 1977.

Goodger, E. M., *Combustion Calculations*, Macmillan, London, 1977.

Goulard, R. (ed.): *Combustion Measurements—Modern Techniques and Instrumentation*, Hemisphere Publishing Corporation, 1976.

Gunther, R.: *Verbrennung und Feuerungen* (in German), Springer-Verlag, Berlin, 1974.

Harker, J. H., and D. A. Allen: *Fuel Science*, Oliver and Boyd, Edinburgh, 1972.

Himus, G. W.: *The Elements of Fuel Technology*, Leonard Hill (Books) Ltd., London, 1958.

Jost, W.: *Explosion and Combustion Processes in Gases*, McGraw-Hill, New York, 1946.

Kanury, A. Murty: *Introduction to Combustion Phenomena*, Gordon and Breach, New York, 1975.

Kennedy, L. A. (ed.): *Turbulent Combustion*. American Institute of Aeronautics and Astronautics, 1978.

Khitrin, L. N.: *The Physics of Combustion and Explosion*, Israel Program for Scientific Translations, Jerusalem, 1962.

—— (ed.): *Combustion in Turbulent Flow*, Transl. from Russian, Israel Program for Scientific Translations, Jerusalem, 1963.

Laffitte, P.: *La Propagation des Flammes dans les Mélanges Gazeux: les Déflagrations*, Hermann, Paris, 1939.

Lawton, J., and F. J. Weinberg: *Electrical Aspects of Combustion*, Oxford University Press, London and New York, 1969.

Lefebvre, A. H. (ed.): *Gas Turbine Combustor Design Problems*, Hemisphere Publishing Corporation, 1980.

Lewis, B., R. N. Pease, and H. S. Taylor: *Combustion Processes*, vol. 2 of *High Speed Aerodynamics and Jet Propulsion*, Oxford University Press, 1956.

—— and G. von Elbe: *Combustion, Flames and Explosions of Gases*, Academic Press, New York, 1961.

Lichty, L. C.: *Combustion Engine Processes*, McGraw-Hill, New York, 1967.

Minkoff, G. J., and C. F. H. Tipper: *Chemistry of Combustion Reactions*, Butterworths, London, 1962.

Ministry of Power: *The Efficient Use of Fuel*, Her Majesty's Stationary Office, London, 1958.

Oppenheim, A. K.: *Combustion and Propulsion*, Fourth AGARD Colloq., Pergamon, 1961.

Palmer, H. B., and J. M. Beér (eds.): *Combustion Technology*, Academic Press, New York and London, 1974.

Palmer, K. N.: *Dust Explosions and Fire*, Chapman-Hall, 1973.

Penner, S. S.: *Introduction to the Study of Chemical Reactions in Flow Systems*, Butterworths, 1955.

——: *Chemistry Problems in Jet Propulsion*, Pergamon, 1957.

——: *Chemical Rocket Propulsion and Combustion Research*, Gordon and Breach, 1962.

Predvoditelev, A. S. (ed.): *Gas Dynamics and Physics of Combustion*, Transl. from Russian, Israel Program for Scientific Translations, Jerusalem, 1962.

Shchelkin, K. I., and Y. K. Troshin: *Gasdynamics of Combustion*, Mono Book Corp., Baltimore, 1965.

Skinner, D. G.: *The Fluidized Combustion of Coal*, Institute of Fuel, London, 1970.

Smith, I. E. (ed.): *Combustion in Advanced Gas Turbine Systems*, Pergamon, 1968.

Smith, M. L., and K. W. Stinson: *Fuels and Combustion*, McGraw-Hill, New York, 1952.

Sokolik, A. S.: *Self-ignition, Flame and Detonation in Gases*, Israel Program for Scientific Translations, Jerusalem, 1963.

Spalding, D. B.: *Some Fundamentals of Combustion*, Academic Press, New York, 1955.

Spiers, H. M. (ed.): *Technical Data on Fuel*, 6th ed. British National Committee World Power Conference, London, 1962.

Stambuleanu, A.: *Flame Combustion Processes in Industry*, Abacus Press, England, 1976.

Strehlow, R. A.: *Fundamentals of Combustion*, International Textbook Co., Scranton, Pennsylvania, 1968.

Thring, M. W., J. Ducarme, and J. Fabri (eds.): *Selected Combustion Problems II*, Butterworths, London, 1956.

———: *The Science of Flames and Furnaces*, Wiley, New York, 1962.

Van Tiggelen, A.: *Oxydations et Combustions*, Technip., Paris, 1968.

von Karman, T., and S. S. Penner: *Selected Combustion Problems*, Butterworths, London, 1954.

Vulis, L. A.: *Thermal Regimes of Combustion*, McGraw-Hill, New York, 1961.

Weinberg, F. J.: *Optics of Flames*, Butterworths, London, 1963.

Williams, D. A., and G. Jones: *Liquid Fuels*, Pergamon, London, 1963.

Williams, F. A.: *Combustion Theory*, Addison-Wesley, Reading, Massachusetts, 1965.

Zeldovich, F. B., and A. S. Kompaneets: *Theory of Detonation*, Academic Press, 1960.

Zinn, B. T. (ed.): *Experimental Diagnostics in Gas Phase Combustion Systems*, American Institute of Aerodynamics and Astronautics, 1977.

SYMPOSIA

The Combustion Institute, Pittsburgh, Pennsylvania, has sponsored International Symposia on Combustion since 1928. Since 1952 they have been held regularly at two-yearly intervals, with a printed volume issued in the year following the Symposium. The *Proceedings of the Seventeenth Symposium* were printed in 1979.

JOURNALS

Papers on Combustion, Fuel Science, and Technology are found in a wide range of journals, covering the fields of Chemistry, Physics, and Engineering. The main journals, which are specifically devoted to Combustion and Fuel Technology are:

Combustion and Flame: the Journal of the Combustion Institute, published by Elsevier, New York. Six issues per year.

Combustion Science and Technology: published by Gordon and Breach, New York. Six issues per year.

Fuel: published by IPC Science and Technology Press Ltd., England. Monthly.

Fuel and Energy Abstracts: published by IPC Science and Technology Press Ltd., England. Bimonthly.

Journal of the Institute of Energy: published by the Institute of Energy, London. Quarterly.

Progress in Energy and Combustion Science: International Review Journal, published by Pergamon Press. Quarterly.

REFERENCES

Aldred, J. W., J. C. Patel, and A. Williams: *Combust. Flame*, **17**, 139–419 (1971).

Alperstein, M., and R. L. Bradow: *S.A.E. Trans.* **75**, Paper 660781 (1967).

Asalor, J. O., and J. H. Whitelaw: "The Influence of Combustion Induced Particle Concentration Variations on Laser Doppler Anemometry," LDA Symposium, Technical University of Denmark, Lyngby (1975).

Attal, B., M. Pealat, and J. P. Taran: "CARS Diagnostics of Combustion," AIAA Paper no. 80-0282 (1980).

Azelborn, N. A., W. R. Wade, J. R. Secord, and A. F. McLean: "Low Emissions Combustion for the Regenerative Gas Turbine, Part 2—Experimental Techniques; Results and Assessment," ASME Paper no. 73-GT-12 (1973).

Baker, R. J., P. J. Bourke, and J. H. Whitelaw: "Measurements of Instantaneous Velocity in Laminar and Turbulent Diffusion Flames Using an Optical Anemometer," *J. Inst. Fuel*, **46**, 388–395 (1973a).

———— ———— and ————: "Application of Laser Anemometry to the Measurement of Flow Properties in Industrial Burner Flames," *Fourteenth Symposium (International) on Combustion*, The Combustion Institute, pp. 699–706 (1973b).

———— P. Hutchinson, E. E. Khalil, and J. H. Whitelaw: "Measurements of Three Velocity Components in a Model Furnace with and without Combustion," *Fifteenth Symposium (International) on Combustion*, The Combustion Institute, pp. 553–559 (1975).

Ballal, D. R., and A. H. Lefebvre: *Acta Astronaut.*, **1**, 471–483 (1973).

———— and ————: "The Influence of Flow Parameters on Minimum Ignition Energy and Quenching Distance," *Fifteenth Symposium (International) on Combustion*, The Combustion Institute, pp. 1473–1481 (1975a).

———— and ————: "The Structure and Propagation of Turbulent Flames," *Proc. Roy. Soc., London*, **A334**, 217–234 (1975b).

Ballantyne, A., and K. N. C. Bray: "Investigations into the Structure of Jet Diffusion Flames Using Time/Resolved Optical Measuring Techniques," *Sixteenth Symposium (International) on Combustion*, The Combustion Institute, p. 777 (1977).

———— and J. B. Moss: *Combust. Sci. Technol.*, **17**, 63 (1977).

Ban, L. L., and W. M. Hess: *Carbon*, **7**, 723 (1969).

Barenblatt, G. I., J. B. Zeldovich, and A. G. Istratov: *PMTF*, **4**, 21 (1962).

Barker, R. F.: "The Availability and Quality of Future Fuels," Institution of Mechanical Engineers, London (1977).

Barr, J.: *Fuel*, **28**, 200 (1949).

Barrère, M., et al.: *Rocket Propulsion*, Elsevier, New York, 1960.

Batchelor, G. K.: *The Theory of Homogeneous Turbulence*, Cambridge University Press, 1953.

Beér, J. M., and N. A. Chigier: *Combustion Aerodynamics*, Applied Science, London; also Wiley, New York, 1972.

Bilger, R. W.: "Turbulent Jet Diffusion Flames," *Prog. Energy Combust. Sci.*, **1**, 87–109 (1976).

Boccio, J. L., G. Weilerstein, and R. B. Edelman: "A Mathematical Model for Jet Engine Combustor Pollutant Emissions, General Applied Science Lab., TR-781: also NASA CR-121, 208 (1973).

Böhm, G., and K. Clusius: *Naturforsch*, **4a**, 286 (1948).

Bowman, C. T.: "Kinetics of Pollutant Formation and Destruction in Combustion," *Prog. Energy Combust. Sci.*, **1**, 33–45 (1975).

Boyson, F., and J. Swithenbank: "Spray Evaporation in Recirculating Flow," *Seventeenth International Symposium on Combustion*, The Combustion Institute, pp. 109–111 (1978).

Bradshaw, P.: "The Effect of Initial Conditions on the Development of a Free Shear Layer," *J. Fluid Mech.*, **26**, 225–236 (1966).

———: *An Introduction to Turbulence and its Measurement*, Pergamon Press, 1971.

———: *Turbulence*, Topics in Applied Physics, 12, Springer-Verlag (1976).

Bray, K. N. C.: "Equations of Turbulent Combustion," University of Southampton, Report no. 330, 1973.

———: "Analytical and Numerical Methods for Investigation of Flow Fields with Chemical Reaction, Especially Related to Combustion," AGARD, NATO, Paris, p. II 2, 1975.

———: "The Interaction between Turbulence and Combustion," *Seventeenth International Symposium on Combustion*, The Combustion Institute, pp. 57–59 (1978).

——— and P. A. Libby: "Interaction Effects in Turbulent Premixed Flames," *The Physics of Fluids*, **19**, 1687–1701 (1976).

——— and J. B. Moss: University of Southampton, Dept. Aeronautics and Astronautics, AASU Report no. 335 (1974).

——— and ———: *Combustion and Flame*, **30**, 125 (1977).

Brown, G. L., and A. Roshko: "On Density Effects and Large Structure in Turbulent Mixing Layers," *J. Fluid Mech.*, **64**, 775–816 (1974).

Brzustowski, T. A.: *Can. J. Chem. Eng.*, **43**, 30–35 (1965).

Burgoyne, J. H., and L. Cohen: *Proc. Roy. Soc., London*, **A225**, 375 (1959).

Burke, S. P., and T. E. W. Schumann: *Industrial Engineering Chemistry*, **20**, 998–1004 (1928).

Bush, W. B., and F. E. Fendell: "Analytic Modeling of Turbulent Shear Flow with Chemical Reaction," in L. A. Kennedy (ed.), "Turbulent Combustion," *AIAA Prog. Astronaut. Aeronaut.*, **58**, 3–18 (1978).

Caretto, L. S.: "Mathematical Modelling of Pollutant Formation," *Prog. Energy Combust. Sci.*, **1**, 47–71 (1975).

Chigier, N. A.: "Laser Velocimetry," Workshop on Combustion Measurements in Jet Propulsion Systems, Project SQUID, Purdue University, 1975.

———: "The Atomization and Burning of Liquid Fuel Sprays," *Prog. Energy Combust. Sci.*, **2**, 97–114 (1976).

———: "Laser Velocimetry for Combustion Measurements in Jet Propulsion Systems," in R. Goulard (ed.), *Combustion Measurements, Modern Techniques and Administration*, Academic Press, pp. 67–90, 1976a.

———: "Combustion Diagnostics by Laser Velocimetry," AIAA Fourteenth Aerospace Sciences Meeting, Washington, D.C., Paper no. 76-32, 1976b.

———: "Instrumentation Techniques for Studying Heterogeneous Combustion," The Combustion Institute, Central States Section Technical Meeting on Fluid Mechanics of Combustion Processes, NASA Lewis Research Center, 1977.

——— and A. Chervinsky: "Aerodynamic Study of Turbulent Burning Free Jets with Swirl," *Eleventh Symposium (International) on Combustion*, The Combustion Institute, pp. 489–499, 1967.

———— and K. Dvorak: "Laser Anemometer Measurements in Flames with Swirl," *Fifteenth Symposium (International) on Combustion*, The Combustion Institute, pp. 573–585, 1975.

———— and C. G. McCreath: "Combustion of Droplets in Sprays," *Acta Astronaut.*, **1**, 687–710 (1974).

———— ———— and R. W. Makepeace: "Dynamics of Droplets in Burning and Isothermal Kerosene Sprays," *Combust. Flame*, **23**, 11–16 (1974).

———— and M. F. Roett: "Twin-Fluid Atomizer Spray Combustion," ASME Winter Annual Meeting, New York, Paper no. 72-WA/HT-25, 1972.

———— and A. C. Styles: "Laser Anemometer Measurements in Spray Flames," *Deuxieme Symposium European sur la Combustion*, The Combustion Institute, Orleans, France, pp. 563–568, 1975.

———— A. Ungut, and A. J. Yule: "Particle Size and Velocity Measurements in Flames by Laser Anemometer," *Seventeenth International Symposium on Combustion*, The Combustion Institute, 315–324, 1979.

———— and ————: "Particle Sizing in Flames with Laser Velocimeters, *Third International Workshop on Laser Velocimetry*, Hemisphere Press, 1979a.

Chiu, H. H., R. K. Ahluwalia, B. Koh, and E. J. Croke: "Spray Group Combustion," Sixteenth Aerospace Sciences Meeting, Huntsville, Alabama, Paper no. 78-75, 1978.

Chomiak, J.: *Combust. Flame*, **20**, 113 (1973).

————: "Basic Considerations in the Turbulent Flame Propagation in Premixed Gases," *Prog. Energy Combust. Sci.*, **5**, 207–221 (1979).

Chou, H. P., and R. M. Waterston: "The Measurement of Particle Size at the Intersection Points of a Crossed Beam Laser Doppler Anemometer," LDA Symposium, Technical University of Denmark, Lyngby, 1975.

Chu, B. T., and J. Y. Parlange: *J. Mecanique*, **1**, 293 (1962).

Clark, J. F.: *Proc. Roy. Soc., London*, **A296**, 519 (1967).

————: *Proc. Roy. Soc., London*, **A307**, 283 (1968).

————: *Proc. Roy. Soc., London*, **A312**, 62 (1969).

———— A. Melvin, and J. B. Moss: *Combust. Sci. and Tech.*, **4**, 17 (1974).

Cotton, D. H., N. J. Friswell, and D. R. Jenkins: *Combust. Flame*, **17**, 87 (1971).

Coward, H. F. and F. Brimsley: *J. Chem. Soc.*, **105**, 1895 (1914).

Damköhler, G.: *NACA*, **TM**, 1112 (1947).

Dashchuk, M.: "Measurement of Unburned Gas Velocities in Laminar and Moderately Turbulent Open Burner Flames Using a Laser Doppler Anemometer," *Amer. Phys. Soc. Bull.*, **14**, 1105 (1969).

————: "Preservation of Approach Flow Velocity and Turbulence in Bunsen Flames," *Thirteenth Symposium (International) on Combustion*, The Combustion Institute, pp. 659–665, 1971.

Dent, J. C., *SAE Trans.*, **80**, Paper 710571 (1971).

———— and J. A. Durham: *Proc. Inst. Mech. Eng.*, **188**, 21/74, 269, 280 (1974).

Dixon-Lewis, G., and A. Williams: "Some Observations on the Combustion of Methane in Premixed Flames," *Eleventh Symposium (International) on Combustion*, The Combustion Institute, pp. 951–958, 1967.

Donaldson, C. Du P.: "On the Modeling of the Scalar Correlations Necessary to Construct a Second-Order Closure Description of Turbulent Reacting Flows," in S. N. B. Murthy (ed.), *Turbulent Mixing in Nonreactive and Reactive Flows*, Plenum, New York, 131–153, 1975.

Dougherty, N. S., and R. A. Belz: *Holography of Reacting Liquid Sprays*, Arnold Engineering Development Center, Tennessee, 1971.

Dugger, G. L., R. C. Weast, and S. Heimel: "Effect of Preflame Reaction on Flame Velocity of Propane-Air Mixtures," *Fifth Symposium on Combustion*, Reinhold Pub. Co., New York, pp. 589–595, 1955.

Durst, F., and R. Kleine: "Velocity Measurements in Turbulent Premixed Flames by Laser Anemometry," *Gas Warme International*, **22**, no. 12 (1973).

———— and ————: *VDI Bereicht*, **211**, 79 (1974).

———— A. Melling, and J. H. Whitelaw: "The Application of Optical Anemometry to Measurement in Combustion Systems," *Combust. Flame*, **18**, 197–201 (1972).

—— —— and ——: "Laser Anemometry: A Report of Euromech 36," *J. Fluid Mech.*, **56**, 143–160 (1972*a*).

—— —— and ——: *Principles and Practice of Laser-Doppler Anemometry*, Academic Press, 1976.

—— and H. Umhauer: Local Measurements of Particle Velocities, Size Distribution and Concentration with a Combined Laser Doppler, Particle Sizing System," *LDA Symposium*, Technical University of Denmark, Lyngby, 1975.

—— and J. H. Whitelaw: "Integrated Optical Units for Laser Anemometry," *J. Phys. E.*, **4**, 804–808 (1971).

—— and ——: "Light Source and Geometric Requirements for the Optimization of Optical Anemometry Signals," *Opto-Electronics*, **5**, 137 (1973).

—— and M. Zaré: "Removal of Pedestals and Directional Ambiguity of Optical Anemometer Signals," *J. Appl. Opt.*, **13**, 2562 (1974).

Eckbreth, A. C.: "CARS Investigations in Flames," *Seventeenth International Symposium on Combustion*, The Combustion Institute, pp. 975–983, 1979.

——: "Combustion Diagnostics by Laser Raman and Fluorescence Techniques," *Prog. Energy Combust. Sci.*, **5**, 253–322 (1979*a*).

——: "CARS Thermometry in Practical Combustors," *Combust. Flame* (1980).

—— and R. J. Hall: "CARS Thermometry in a Sooting Flame," *Combust. Flame*, **36**, 87–98 (1979).

Eckhaus, W.: *J. Fluid Mech.*, **10**, 81 (1961).

Edelman, R. B., and O. Fortune: "A Quasi-Global Chemical Kinetic Model for the Finite Rate Combustion of Hydrocarbon Fuels," AIAA Paper 69–86, (1969).

—— and P. T. Harsha: "Some Observations on Turbulent Mixing with Chemical Reactions," in L. A. Kennedy (ed.), "Turbulent Combustion," *AIAA Prog. in Astronaut. Aeronaut.*, **58**, 55–202 (1978).

—— and ——: Laminar and Turbulent Gas Dynamics in Combustors—Current Status, *Prog. Energy Combust. Sci.*, **4**, 1–62 (1978*a*).

—— A. Turan, P. T. Harsha, E. Wong, and W. S. Blazowski: "Fundamental Characterization of Alternative Fuel Effects in Continuous Combustion Systems," *Combustor Modeling*, AGARD-CPP-275 (1979).

Einbinder, H.: *J. Chem. Phys.*, **21**, 480 (1951).

Eisenklam, P., S. A. Arunachalan, and J. A. Weston: "Evaporation Rates and Drag Resistance of Burning Droplets," *Eleventh (International) Symposium on Combustion*, The Combustion Institute, pp. 715–728 (1967).

Faeth, G. M.: "Current Status of Droplet and Liquid Combustion," *Prog. Energy Combust. Sci.*, **3**, 191–224 (1977).

Farmer, W. M.: "Measurement of Particle Size, Number Density and Velocity Using a Laser Interferometer," *J. Appl. Opt.*, **11**, 2603 (1972).

Favre, A. J., J. J. Gaviglio, and R. Dumas: "Space-Time Double Correlations and Spectra in a Turbulent Boundary Layer," *J. Fluid Mech.*, **2**, 313–342 (1957).

Fay, J. A.: *J. Aeronaut. Sci.*, **21**, 681 (1954).

Felton, P. C., J. Swithenbank, and A. Turan: "Progress in Modelling Combustors," University of Sheffield, Dept. Chemical Engineering and Fuel Technology, Report HIC 300, 1978.

Fenimore, C. P.: *Thirteenth Symposium (International) on Combustion*, The Combustion Institute, p. 373, 1971.

—— *Combust. Flame*, **19**, 289 (1972).

Flagan, R. C., S. Gallant, and J. P. Appleton: *Combust. Flame*, **22**, 299 (1974).

Fristrom, R. M.: "Radical Concentrations and Reactions in a Methane-Oxygen Flame," *Ninth Symposium (International) on Combustion*, Academic Press, pp. 560–575, 1963.

—— and A. A. Westenberg: *Flame Structure*, McGraw-Hill, New York, 1965.

Galyun, I. I., and Yu. A. Ivanov: *Phys. Com. Expl.*, *(USSR)*, **6**, 237 (1970).

Garner, F. H., F. Morton, and J. B. Saundy: *J. Inst. Petrol*, **47**, 175–193 (1961).

Gaydon, A. G.: *The Spectroscopy of Flames*, Chapman and Hall, London, 1974.

—— and H. G. Wolhard: *Flames—Their Structure, Radiation and Temperature*, Chapman and Hall, London, 1970.

Gelinas, R. J.: "Ignition Kinetics of C_1 and C_2 Hydrocarbons," Science Applications Inc., California, Report No. SAI/PL/C279, 1979.

Gollahalli, S. R., and T. A. Brzustowski: "Experimental Studies on the Flame Structure in the Wake of a Burning Droplet," *Fourteenth Symposium (International) on Combustion*, The Combustion Institute, pp. 1333–1344, 1973.

Goulard, R.: "Laser Raman Scattering Applications," in M. Lapp and M. Penney (eds.), *Laser Raman Gas Diagnostics*, Plenum, New York, pp. 3–14, 1974.

Grant, H. L.: "The Large Eddies of Turbulent Motion," *J. Fluid Mech.*, **4**, 149–190 (1958).

Grant, G. R., and K. L. Orloff: "Two-Colour Dual-Beam Backscatter Laser Doppler Velocimeter," *Appl. Opt.*, **12**, 2913 (1973).

Griffith, L.: "A Theory of Size Distribution of Particles in a Comminuted System," *Can. J. Res.*, **21**, sec. A, no. 6, 57–64 (1943).

Hammond, D. C., and A. M. Mellor: "Analytical Calculations for the Performance and Pollutant Emissions for Gas Turbine Combustors," *Comb. Sci. Tech.*, **4**, 101 (1971).

Harsha, P. T.: "A General Analysis of Free Turbulent Mixing," Arnold Engineering Development Center, TR-73-177, 1974.

Hartley, D. L., D. R. Hardesty, M. Lapp, J. Dooher, and F. Dryer: "The Role of Physics in Combustion, Efficient Use of Energy," *AIP Conf. Proc.*, no. 25 (1975).

Hawthorne, W. R., D. S. Weddell, and H. C. Hottel: "Mixing and Combustion in Turbulent Gas Jets," *Third Symposium on Combustion, Flame and Explosion Phenomena*, Williams and Wilkins, Baltimore, pp. 266–288, 1951.

Hayhurst, A., and I. Vince: "Nitric Oxide Formation from N_2 in Flames: The Importance of 'Prompt' NO," *Prog. Energy Combust. Sci.*, 1980.

Hedley, A. B., A. S. M. Nuruzzaman, and G. F. Martin: "Combustion of Single Droplets and Simplified Spray Systems," *J. Inst. Fuel*, **44**, 38–54 (1971).

Heimel, S., and R. C. Weast: "Effect of Initial Mixture Temperature on the Burning Velocity of Benzene-Air, *n*-Heptane-Air and Iso-Octane-Air Mixtures," *Sixth Symposium (International) on Combustion*, Reinhold Publishing Co., New York, pp. 296–302, 1957.

Henein, N. A.: "Analysis of Pollutant Formation and Control and Fuel Economy in Diesel Engines," *Prog. Energy Combust. Sci.*, **1**, 165–207 (1976).

Hestroni, G. and M. Sokolov: "Distribution of Mass, Velocity and Intensity of Turbulence in a Two-Phase Turbulent Jet," *Trans. ASME, J. Appl. Mech.*, **38**, 315–327 (1971).

Heywood, J. B.: "Pollutant Formation and Control in Spark-Ignition Engines," *Prog. Energy Combust. Sci.*, **1**, 135–164 (1976).

Hinze, J. O.: *Turbulence*, McGraw-Hill, 1959.

Hiroyasu, H., and T. Kadota: "Study on the Removal of the Diesel Engine Smokes; Part II, Measurement of the Fuel Droplet Distribution under High Pressure," *JARI*, TM2, 51–59 (1971).

Ho, C. M., K. Jakus, and K. H. Parker: "Temperature Fluctuations in a Turbulent Flame," *Combust. Flame*, **27**, 113–124 (1976).

Homann, K. H., and H. G. Wagner: *Eleventh Symposium (International) on Combustion*, The Combustion Institute, p. 371, 1967.

Hottel, H. C., and W. R. Hawthorne: "Diffusion in Laminar Flame Jets," *Third Symposium on Combustion, Flame and Explosion Phenomena*, Williams and Wilkins, Baltimore, pp. 254–300, 1949.

Ingebo, R. D., and H. H. Foster: "Drop Size Distribution for Cross Current Break-up of Liquid Jets in Airstreams," NACA Tech. Note 4087, 1957.

Istratov, A. G., and W. B. Librovich: *Prikl. Mat. Mech.*, **30**, 451 (1966).

Iverach, D., K. S. Basden, and N. Y. Kirov: *Fourteenth Symposium (International) on Combustion*, The Combustion Institute, p. 767, 1973.

Jessen, P. F., and A. G. Gaydon: *Twelfth Symposium (International) on Combustion*, The Combustion Institute, p. 481, 1969.

Johnston, S. C.: "Precombustion Fuel-Air Distribution in a Stratified Charge Engine using Laser Raman Spectroscopy," Paper 79-0433, SAE Automotive Engineering Congress and Exposition, 1979.

Kadota, T., and H. Hiroyasu: *Bulletin of the JSME*, **19**, 1515–1521 (1976).

Karlovitz, B., P. W. J. Denniston, and F. E. Wells: *J. Chem. Phys.*, **19**, 541 (1951).

Kennedy, I. M., and J. H. Kent: "Measurements of a Conserved Scalar in Turbulent Jet Diffusion Flames," Technical Note F-86, Dept. Mechanical Engineering, University of Sydney, 1978.

Kent, J. H., and R. W. Bilger: "Measurements of Turbulent Jet Diffusion Flames," Charles Kolling Research Lab., University of Sydney, TNF-41, 1972.

Khudyakov, J. N.: *Izv. Akad. Nauk. USSR. Otd. Tekhn. Nauk.*, **4**, 508–511 (1949).

Kolmogorov, A. N.: *Doklady, Akad. Nauk.*, *USSR.*, **30**, 299 (1941).

Kreid, D. K.: "Error Estimates for Laser-Doppler Velocimeter Measurements in Non-Uniform Flow: Error Estimates," Second International Workshop on Laser Velocimetry, Purdue University, 398–427, 1974.

Kuehl, D. K.: "Laminar Burning Velocities of Propane-Air Mixtures," *Eighth Symposium (International) on Combustion*, Williams and Wilkins, Baltimore, pp. 510–521, 1961.

Labowsky, M., and D. E. Rosner: "Conditions for Group Combustion of Droplets in Fuel Clouds," *Symposium on Evaporation-Combustion of Fuel Droplets*, Division of Petroleum Chemistry, American Chemical Soc., 1976.

Landau, L. D.: *Acta Phys. Chim. (USSR)*, **19**, 77 (1944).

—— and E. M. Lifschitz: *Mechanics of Continuous Media*, Moscow, 1953. Engl. transl., Addison-Wesley, 1959.

Lapp. M.: "Flame Temperatures from Vibrational Raman Scattering," in M. Lapp and C. M. Penney (eds.), *Laser Raman Gas Diagnostics*, Plenum, New York, pp. 107–146, 1974.

—— L. M. Goldman, and C. M. Penney: "Raman Scattering from Flames," *Science*, **175**, 1112–1115 (1972).

—— and C. M. Penney (eds.): *Laser Raman Gas Diagnostics*, Plenum, New York, 1974.

Launder, B. E., and D. B. Spalding: *Lectures in Mathematical Models of Turbulence*, Academic Press, New York, 1972.

Lavoie, G. A.: *Combust. Flame*, **15**, 97–108 (1970).

—— J. B. Heywood, and J. C. Keck: *Sci. Technol.*, **1**, 313–326 (1970).

Law, C. K.: "Motion of a Vaporizing Droplet in a Constant Cross Flow," *Int. J. Multiphase Flow*, **3**, 299–303 (1977).

Lederman, S.: "The Use of Raman Diagnostics in Flow Fields and Combustion," *Prog. Energy Combust. Sci.*, **3**, 1–34 (1977).

Lefebvre, A. H.: *Tenth Symposium (International) on Combustion*, The Combustion Institute, 1129–1137 (1955).

——: "Design Considerations in Advanced Gas Turbine Combustion Chambers," *Combustion in Advanced Gas Turbine Systems*, Pergamon, p. 3, 1968.

——: "Factors Controlling Gas Turbine Combustion Performance at High Pressure," *Combustion in Advanced Gas Turbine Systems*, Pergamon, pp. 211–226, 1968a.

——: "Pollution Control in Continuous Combustion Engines," *Fifteenth Symposium (International) on Combustion*, The Combustion Institute, pp. 1169–1180, 1975.

——: "The Performance of Prefilming Airblast Atomizers," U.S. Dept. of Energy, Workshop on Modeling of Combustion in Practical Systems, Los Angeles, 1978.

—— and R. S. Fletcher: "A Preliminary Study on the Influence of Fuel Staging on Nitric Oxide Emissions from Gas Turbine Combustors," Atmospheric Pollution by Aircraft Engines, AGARD CP no. 125, sec. 30, 1973.

—— and D. Miller: "The Development of an Airblast Atomizer for Gas Turbine Application," The College of Aeronautics, Cranfield, Report Aero no. 193, 1966.

—— and R. Reid: "The Influence of Turbulence on the Structure and Propagation of Enclosed Flames," *Combust. Flame*, **10**, 355–366 (1966).

Levich, V. G.: *Physicochemical Hydrodynamics*, Prentice-Hall, Englewood Cliffs, New Jersey, 1962.

Lewis, B., and G. von Elbe: *Combustion, Flames and Explosions of Gases*, Academic Press, New York, 1961.

Lewis, H. C., D. G. Edwards, M. J. Goglia, R. I. Rice, and L. W. Smith: "Atomization of Liquids in High Velocity Gas Streams," *Ind. Eng. Chem.*, **40**, 67–74 (1948).

Lieb, D. F., and L. H. S. Roblee Jr.: *Combust. Flame,* **16**, 385 (1970).

Long, D. A.: *Raman Spectroscopy,* McGraw-Hill, New York, 1977.

Longwell, J. P.: "Synthetic Fuels and Combustion," *Prog. Energy Combust. Sci.,* **3**, 127–138 (1977).

Lyn, W. T.: *Proc. Ninth Symposium of Combustion,* pp. 1069–1082, 1963.

McCreath, C. G., and N. A. Chigier: "Liquid-Spray Burning in the Wake of a Stabilizer Disc," *Fourteenth Symposium (International) on Combustion,* The Combustion Institute, pp. 1355–1363, 1973.

Magnussen, B. F., and B. H. Hertager: "On Mathematical Modeling of Turbulent Combustion with Special Emphasis on Soot Formation and Combustion," *Sixteenth Symposium (International) on Combustion,* The Combustion Institute, p. 719, 1977.

Maker, P. D., and R. W. Terhune: *Phys. Rev.,* **137**, A801–A818 (1965).

Manrique, J. A., and G. L. Borman: *Int. J. Heat Mass Transfer,* **12**, 1081–1095 (1969).

Margarvey, R. H., and C. S. MacLatchy: "The Disintegration of Vortex Rings," *Can. J. Phys.,* **42**, 684–689 (1964).

Markstein, G. H.: *J. Aero Sci.,* **18** (1951).

Masdin, E. G., and M. W. Thring: "Combustion of Single Droplets of Liquid Fuel," *J. Inst. Fuel,* 251–260 (1962).

Maxworthy, T.: *Phys. of Fluids,* **5**, 407 (1962).

Mayo, P. J., and F. J. Weinberg: *Proc. Roy. Soc., London,* **A319**, 351 (1970).

Melling, A., and J. H. Whitelaw: "Seeding of Gas Flows for Laser Anemometry," DISA Information, no. 15, pp. 5–14, 1973.

Mellor, A. M.: "Simplified Physical Model of Spray Combustion in a Gas Turbine Engine," *Combust. Sci. Tech.,* **8**, 101–109 (1973).

———: "Gas Turbine Engine Pollution," *Prog. Energy Combust. Sci.,* **1**, 111–133 (1976).

Mellor, R., N. A. Chigier, and J. M. Beér: "Hollow Cone Liquid Spray in Uniform Airstream," in E. R. Norster (ed.), *Combustion and Heat Transfer in Gas Turbine Systems,* Cranfield International Symposium Series, **11**, Pergamon, pp. 291–305, 1971.

Melton, R. B., Jr.: "Diesel Fuel Injection Viewed as a Jet Phenomenon," SAE Paper 710132, 1971.

Meyers, J. F., and M. J. Walsh: "Computer Simulation of a Fringe Type Laser Velocimeter," *Second International Workshop on Laser Velocimetry,* Purdue University, pp. 471–510, 1974.

Mickelson, W. R., and N. E. Ernstein: *Sixth Symposium (International) on Combustion,* The Combustion Institute, p. 325, 1957.

Milson, A., and N. A. Chigier: "Studies of Methane and Methane-Air Flames Impinging on a Cold Plate," *Combust. Flame,* **21**, 295–305 (1973).

Mitchell, R. E.: "Nitrogen Oxide Formation in Laminar Methane-Air Diffusion Flames," D.Sc. Thesis, Dept. Chemical Engineering, M.I.T., 1975.

Mizutani, Y., and A. Nakajima: "Combustion of Fuel Vapor-Drop-Air Systems," *Combust. Flame,* **21**, 343–357 (1973).

——— and M. Ogasawara: *J. Inst. Heat Mass Trans.,* **8**, 921 (1965).

Mulcahy, M. F. R., and I. W. Smith: "Kinetics of Combustion of Pulverized Fuel: A Review of Theory and Experiment," *Rev. Pure and Appl. Chem.,* **19**, 81 (1969).

Mullinger, P. J., and N. A. Chigier: "The Design and Performance of Internal Mixing Multijet Twin-Fluid Atomizers," *J. Inst. Fuel,* **47**, 251–261 (1974).

Natarajan, R.: "A Shadowgraphic Investigation of High Pressure Droplet Vaporization," *Combust. Flame,* **26**, 407 (1976).

——— and T. A. Brzustowski: "Some New Observations on the Combustion of Hydrocarbon Droplets at Elevated Pressures," *Combust. Sci. Tech.,* **2**, 259 (1970).

Newman, J. A., and T. A. Brzustowski: "Behavior of a Liquid Jet Near the Thermodynamic Critical Region," *AIAA J.,* **9**, 1595 (1971).

Nukiyama, S., and Y. Tanasawa: "Experiments on the Atomization of Liquids in an Airstream," *Trans. Soc. Mech. Eng. (Jpn),* **5**, 68–75 (1939).

——— and ———: "An Experiment on the Atomization of Liquid" (Fifth Report. Atomization Pattern of Liquid by Means of Airstream), *Trans. Soc. Mech. Engr. (Jpn),* **6**, S-7 (1940).

Odgers, J.: "Current Theories of Combustion within Gas Turbine Chambers," *Fifteenth Symposium (International) on Combustion,* The Combustion Institute, pp. 1321–1338, 1975.

Onuma, Y., and M. Ogasawara: "Studies on the Structure of a Spray Combustion Flame," *Fifteenth Symposium (International) on Combustion*, The Combustion Institute, pp. 453–465, 1975.

―――― ―――― and T. Inoue: *Sixteenth Symposium (International) on Combustion*, The Combustion Institute, 1977.

Oppenheim, A. K.: *Introduction to Gas Dynamics of Explosions*, Springer-Verlag, 1972.

Orloff, K. L., and S. E. Logan: "Confocal Backscatter Laser Velocimeter With On-Axis Sensitivity," *Appl. Opt.*, **12**, 2477 (1973).

Osgerby, I. T.: "Fuel Evaporation Rate in Intense Recirculation Zones," Arnold Engineering Development Center, Tennessee, 1974.

Owen, F. K.: "Laser Velocimeter Measurements of the Structure of Turbulent Spray Flames," AIAA Fifteenth Aerospace Sciences Meeting, Los Angeles, Paper no. 77–215, 1977.

Pagni, P. J., L. Hughes, and T. Novakov: "Smoke Suppressant Additive Effects on Particulate Emissions from Gas Turbine Combustors," AGARD Conf. Proc., no. 125 on Atmospheric Pollution by Aircraft Engines, pp. 28.1–28.11, 1973.

Pedersen, P. S., and B. Qvale: "A Model for the Physical Part of the Ignition Delay in a Diesel Engine," SAE Paper 740716, 1974.

Penner, S. S.: *Chemistry Problems in Jet Propulsion*, Macmillan, New York, pp. 276–296, 1957.

―――― and T. Jerskey: "Use of Lasers for Local Measurements of Velocity Components, Species Densities and Temperatures," *Ann. Rev. Fluid Mech.*, **5**, 9 (1973).

Perez, J. M., and E. W. Landen: "Exhaust Ignition Characteristics of Pre-Combustion Chamber Engine," SAE Paper no. 680421, 1968.

Di Piazza, J. T., M. Gerstein, and R. C. Weast: *Ind. Eng.-Chem.*, **43**, 2721 (1951).

Pischinger, F. F., and K. J. Klocker: "Single-Cylinder Study of Stratified Charge Process with Prechamber Injection," SAE Paper no. 741162, 1974.

Plee, S. L., and A. M. Mellor: "Characteristic Time Correlation for Lean Blowoff of Bluff-Body-Stabilized Flames," *Combust. Flame*, **35**, 61–80 (1979).

Polymeropoulas, C. E., and S. Das: *Combust. Flame*, **25**, 247 (1975).

Powell, N. H., and W. G. Browne: "Some Fluid Dynamic Aspects of Laminar Diffusion Flames," *Sixth Symposium (International) on Combustion*, The Combustion Institute, pp. 918–922, 1957.

Putnam, A. A., C. C. Miesse, and J. M. Piltcher: "Injection and Combustion of Liquid Fuels," WADC Tech. Rep. 56–344, Wright Air Development Center, Wright-Patterson Air Force Base, 1957.

Radcliffe, A.: "The Performance of a Type of Swirl Atomizer," *Inst. Mech. Engr., Proc.*, **169**, 93–106 (1955).

Ranz, W. E., and W. R. Marshall Jr.: *Chem. Eng. Proc.*, **48**, 141–146, 173–180 (1952).

Rao, K. V. L., and A. H. Lefebvre: "Minimum Ignition Engines in Flowing Kerosene-Air Mixtures," *Combust. Flame*, **27**, 1–20 (1976).

Régnier, P. R., and J.-P. E. Taran: "Gas Concentration Measurement by Coherent Raman Anti-Stokes Scattering," in M. Lapp and C. M. Penney (eds.), *Laser Raman Diagnostics*, Plenum, New York, pp. 87–103, 1974.

Rhodes, R. P., P. T. Harsha, and C. E. Peters: *Acta Astronaut.*, **1**, 443 (1974).

Richardson, L. F.: *Proc. Roy. Soc.*, **A110**, 709 (1926).

Rife, J., and J. B. Heywood: "Photographic and Performance Studies of Diesel Combustion with a Rapid Compression Machine," SAE Paper no. 740948, 1974.

Rizkalla, A. A., and A. H. Lefebvre: "Influence of Liquid Properties on Air Blast, Atomizer Spray Characteristics," *Trans. ASME, J. of Fluids Eng. for Power*, 173–177 (1975a).

―――― and ――――: "The Influence of Air and Liquid Properties on Air Blast Atomization," *Trans. ASME, J. of Fluids Eng.*, **97**, 316–320 (1975b).

Roh, W. B., P. W. Schreiber, and J. P. E. Taran: *Appl. Phys. Letts'*. **29**, 174 (1976).

Roshko, A.: "Structure of Turbulent Shear Flows: A New Look," AIAA Fourteenth Aerospace Sciences Meeting, Paper no. 76–78, 1976.

Rosin, P., and E. Rammler, "Laws Governing the Fineness of Powdered Coal," *J. Inst. Fuel*, **7**, 29–36 (1933).

Rosser, W. A., S. H. Inami, and H. Wise: "The Effect of Metal Salts on Premixed Hydrocarbon-Air Flames," *Combust. Flame*, **7**, 107–119 (1963).

Sanders, C. F., D. P. Teixeira, and N. B. de Volo: "The Effect of Droplet Combustion on Nitric Oxide Emissions by Oil Flames," Western States Section, The Combustion Institute, Paper no. 72–7, 1972.

Sass, R.: *Compressorless Diesel Engines*, Springer-Verlag, Berlin, 1929.

Savage, L. D.: *Combustion and Flame*, **6**, 77 (1962).

Sawyer, R. F., N. P. Cernansky, and A. K. Oppenheim: "Factors Controlling Pollutant Emissions from Gas Turbine Engines," AGARD Conference Proc. No. 125 on Atmospheric Pollution by Aircraft Engines, 1973.

Schwar, M., and F. J. Weinberg: "Laser Techniques in Combustion Research," *Combust. Flame*, **13**, 335 (1969).

Scorer, R. S.: *Environmental Aerodynamics*, Ellis Horwood, Chichester, England, and John Wiley & Sons, 1978.

Scott, W. M.: "Looking in on Diesel Combustion," SP-345, Society of Automotive Engineers, New York, 1968.

Scurlock, A. C., and J. H. Grover: *Fourth Symposium (International) on Combustion*, The Combustion Institute, 645, 1953.

Self, S. A., and J. H. Whitelaw: "Laser Anemometry for Combustion Research," *Combust. Sci. Tech.* **13**, 171 (1976).

Shahed, S. M., P. F. Flynn, and W. T. Lyn: "A Model for the Formation of Emissions in a Direct-Injection Diesel Engine," Combustion Modelling in Reciprocating Engines, International Symposium sponsored by General Motors Research Labs, 6–7 Nov., 1978.

Sjögren, A.: "Soot Formation by Combustion of an Atomized Liquid Fuel," *Fourteenth Symposium (International) on Combustion*, The Combustion Institute, pp. 919–927, 1973.

Smith, E. C. W.: *Proc. Roy. Soc.*, **174**, 110 (1940).

Smith, F. A., and S. F. Pickering: *Ind. Eng.-Chem.*, **20**, 1012 (1928).

Smith, J. R.: "Temperature and Density Measurements in an Engine by Pulsed Raman Spectroscopy," Paper 80–0137, SAE Automotive Engineering Congress and Exposition, 1980.

Smith, K. O., and F. C. Gouldin; "Experimental Investigation of Flow Turbulence Effects and Premixed Methane-Air Flames," in L. A. Kennedy (ed.), "Turbulent Combustion," *Prog. in Astronaut. Aeronaut.*, **58**, 37–54, AIAA (1978).

Smithells, A., and H. J. Ingle: *J. Chem. Soc.*, **61**, 204 (1892).

de Soete, G. G.: *Ind. Chim. Belg.*, **38**, 14 (1973).

————: *Fifteenth Symposium (International) on Combustion*, The Combustion Institute, 1975.

———— and A. Feugier: *Aspects Physiques et Chimiques de la Combustion*, Technip., Paris, 1976.

Spalding, D. B.: "The Combustion of Liquid Fuels," *Fourth Symposium (International) on Combustion*, Williams and Wilkins, Baltimore, pp. 847–864, 1953.

————: "A Theory of the Extinction of Diffusion Flames," *Fuel*, **33**, 255–273 (1954).

————: *Some Fundamentals of Combustion Processes*, Gas Turbine Series, vol. 2, Academic Press, New York, 1955.

————: "Combustion in Liquid Fuel Rocket Motors," *Aero Quart.*, **10**, 1–27 (1959).

————: "Mixing and Chemical Reaction in Steady Confined Turbulent Flames," *Thirteenth Symposium (International) on Combustion*, The Combustion Institute, pp. 649–657, 1971.

————: "The Theory of Turbulent Reacting Flows—A Review," Seventeenth Aerospace Sciences Meeting, New Orleans, AIAA Paper no. 79–0213, 1979.

Sparrow, E. M., and J. L. Gregg: *Trans. ASME*, **80**, 879–886 (1958).

Starkman, E. S., H. E. Stewart, and V. A. Zvunow: "An Investigation into the Formation and Modification of Emission Precursors," Paper 690020 presented at SAE Automotive Congress, Detroit, Mich., 1969.

Stenhouse, I. A., D. R. Williams, J. B. Cole, and M. D. Swords: *Appl. Opt.*, **18**, 3819 (1979).

Stepowski, D., and M. J. Cottereau: *Appl. Opt.*, **18**, 354 (1979).

Sternling, C. V., and J. O. L. Wendt: *AICh. J.*, **20**, 81 (1974).

Stevenson, W. H.: "Optical Frequency Shifting by Means of a Rotating Diffraction Grating," *Appl. Opt*, **9**, 649 (1970).

————and H. D. Thompson (eds.): "The Use of Laser Doppler Velocimeter for Flow Measurements," *Proc. of First International Workshop*, Purdue University, 1972.

Styles, A. C., and N. A. Chigier: "Combustion of Air Blast Atomised Spray Flames," *Sixteenth Symposium (International) on Combustion*, The Combustion Institute, pp. 619–630, 1977.

Subba Roa, H. N., and A. H. Lefebvre: "Ignition of Kerosene Fuel Sprays in a Flowing Air Stream," *Combust. Sci. Tech.*, **8**, 95–100 (1973).

Swithenbank, J., J. M. Beér, D. S. Taylor, D. Abbott, and C. G. McCreath: "A Laser Diagnostic for the Measurement of Droplet and Particle Size Distribution," *Fourteenth Aerospace Sciences Meeting*, AIAA Paper no. 76–69, 1976.

Switzer, G. L., W. M. Roquemore, R. P. Bradley, P. W. Schreiber, and W. B. Roh: *Appl. Opt.*, **18**, 2343 (1979).

Tabaczynski, R. J., J. B. Heywood, and J. C. Keck: *SAE Trans.*, **83**, Paper no. 72112 (1972).

Takeno, T., and Y. Kotani: "Transition and Structure of Turbulent Jet Diffusion Flames," in L. A. Kennedy (ed.), "Turbulent Combustion," *Prog. Astronaut. Aeronaut.*, **58**, 19–35, AIAA (1978).

Tennekes, H., and J. L. Lumley: *A First Course in Turbulence*, The M.I.T. Press, 1972.

Thompson, H. D., and W. H. Stevenson (eds.): *Proc. of Second International Workshop on Laser Velocimetry*, **I** and **II**, Purdue University, 1974.

Thompson, D., I. M. Vince, and N. A. Chigier: "Effect of Temperature Fluctuations on NO formation," *Second European Symposium on Combustion*, Orleans, France, pp. 303–308, 1975.

Tishkoff, J. M., and C. K. Law: "Application of a Class of Distribution Functions to Drop-Size Data by Logarithmic Least-Square Technique," *Trans. ASME, Eng. for Power*, **99**, 684–688 (1977).

Toor, H. L.: "Mass Transfer in Diluting Turbulent and Nonturbulent Systems with Rapid Irreversible Reactions and Equal Diffusivities," *Am. Inst. Chem. Eng. J.*, **8**, 70–78 (1962).

Townsend, A. A.: *The Structure of Turbulent Shear Flow*, Cambridge University Press, 1976.

Trolinger, J. D.: "Laser Instrumentation for Flow Diagnostics," AGARDograph no. 186, 1974.

Tucker, M.: NACA Report 1277, 1956.

Tuttle, J. H., M. B. Colket, R. W. Bilger, and A. M. Mellor: "Characteristic Times for Combustion and Pollutant Formation in Spray Combustion," *Sixteenth Symposium (International) on Combustion*, The Combustion Institute, pp. 209–219, 1977.

——— R. A. Shisler, R. W. Bilger, and A. M. Mellor: "Emissions from Aircraft Fuel Nozzle Flames," The Combustion Lab., Purdue University, Report no. PURDU-CL-75-04, 1975.

Ungut, A., A. J. Yule, D. S. Taylor, and N. A. Chigier: "Particle Measurement by Laser Anemometry," *AIAA J. Energy*, **2** (1978).

Van Krevelen, D. W.: *Coal Science*, Elsevier, 1961.

Vranos, A.: "Turbulent Mixing and NO$_x$ Formation in Gas Turbine Combustors," *Combust. Flame*, **22**, 253 (1974).

Wade, W. R., and W. Cornelius: "Emission Characteristics of Continuous Combustion Systems of Vehicular Power Plants—Gas Turbine, Steam, Sterling," in W. Cornelius and W. G. Agnew (eds.), *Emissions from Continuous Combustion Systems*, Plenum Press, New York, p. 375, 1972.

Watts, R., and W. M. Scott: "Air Motion and Fuel Distribution Requirements in High-Speed Direct-Injection Diesel Engines," *Diesel Engine Combustion Symposium*, Institute of Mechanical Engineers, London, pp. 167–177, 1970.

Weinberg, F. J.: "Location of the Schlieren Image in a Flame," *Fuel*, **34**, 84 (1955).

———: *Optics of Flames*, Butterworths, London, 1963.

———: *Proc. Roy. Soc.*, **307**, 195 (1968).

———: "The First Half Million Years of Combustion Research and Today's Burning Problems," *Prog. Energy Combust. ,Sci.*, **1**, 17–31 (1975).

Wentworth, J. T.: *SAE, Trans.*, **80**, Paper 710587 (1971).

Westenberg, A. E.: *Combust. Sci. Tech.*, **4**, 59 (1971).

Wigg, L. D.: "Drop Size Prediction for Twin-Fluid Atomizers," *J. Inst. Fuel*, **37**, 500–505 (1964).

Williams, A.: *Oxidation and Combustion Reviews*, **3**, 1–45 (1968).

———: "Combustion of Droplets of Liquid Fuels—A Review," *Combust. Flame*, **21**, 1–31 (1973).

———: "Fundamentals of Oil Combustion," *Prog. Energy Combust. Sci.*, **2**, 167–179 (1976).

Williams, F. A.: *Combustion Theory*, Addison-Wesley, Reading, Mass., 1965.

———: "Recent Advances in Theoretical Descriptions of Turbulent Diffusion Flames," in S. N. B. Murthy (ed.), *Turbulent Mixing in Nonreactive and Reactive Flows*, Plenum Press, New York, pp. 189–201, 1975.

Winant, C. D., and F. K. Browand: "Vortex Pairing: The Mechanism of Turbulent Mixing-Layer Growth at Moderate Reynolds Number," *J. Fluid Mech.*, **63**, 237–255 (1974).

Wise, H., J. Lorell, and B. J. Wood: *Fifth Symposium (International) on Combustion*, Reinhold, New York, pp. 132–141, 1955.

Wohl, K., C. Gazley, and N. Kapp: "Diffusion Flames," *Third Symposium (International) on Combustion*, The Combustion Institute, pp. 288–300, 1949.

Wright, F. J.: *Twelfth International Combustion Symposium*, p. 867, 1969.

—— and E. E. Zukoski: *Eighth Symposium (International) on Combustion*, The Combustion Institute, p. 933, 1962.

Wygnanski, I., and H. E. Fiedler: "Some Measurements in the Self-Preserving Jet," *J. Fluid Mech.*, **38**, 577–612 (1969).

Yagodkin, V. I.: *Izvestia, Ak. Nauk. Ser. Tech. (USSR)*, **7**, 101 (1955).

Yeh, Y., and H. Z. Cummins: "Localized Fluid Flow Measurements with an He–Ne Laser Spectrometer," *Appl. Phys. Lett.*, **4**, 176–178 (1964).

Yule, A. J.: "Observations of Late Transitional and Turbulent Flow in Round Jets," *Symposium on Turbulent Shear Flows*, Penn State University, 1977.

—— N. A. Chigier, and D. Thompson: "Coherent Structures in Combustion," *Symposium on Turbulent Shear Flows*, Penn State University, 1977.

—— —— S. Atakan, and A. Ungut: "Particle Size and Velocity Measurement by Laser Anemometry," AIAA Fifteenth Aerospace Sciences Meeting, Los Angeles, Paper no. 77–214, 1977.

Zinn, B. T. (ed.): "Experimental Diagnostics in Gas Phase Combustion Systems, *Am. Inst. Aeronaut. Astronaut.* (1977).

GLOSSARY

active species atoms or free radicals (molecular fragments) with at least one unpaired electron

adiabatic flame temperature flame temperature calculated on the basis of thermodynamics, assuming that the flame is an overall adiabatic process

apparent origin of jet the virtual point source of a jet, determined by extrapolation
back from the fully developed flow region

autoignition initiation of combustion by external heat but without a spark or flame

autoignition temperature the lowest temperature at which a combustible material
ignites spontaneously in air without an external spark or flame

barrel unit of measurement of liquids in the petroleum industry = 42 US standard
gallons = 35 imperial gallons = 0.159 m^3

blowoff the conditions at which a flame becomes unstable and is carried downstream by the flow

break point the point in a flame where transition between laminar and turbulent
flow becomes visible

bunker "C" fuel oil a heavy residual fuel oil used by ships, by industry, and for
large scale heating installations

burning a rapid, flaming, or incandescent oxidation process that produces heat

burning point the temperature at which a combustible material gives off sufficient
vapor to ignite and continue to burn

burning rate the velocity at which a solid or liquid is burned, measured in the
direction normal to the surface

burning velocity the velocity at which a flame propagates normally into unburned
gas.

carbonization heating of coal in the absence of air

cetane number the percentage by volume of cetane (cetane no. 100) in a blend with
alpha-methylnaphthalene (cetane no. 0) that matches the ignition properties of
a fuel under test at standard test conditions

char the carbonaceous material formed by incomplete combustion of an organic material

chemiluminescence the emission of light (not necessarily visible) by excited reaction products

cloud point the temperature at which paraffin wax begins to crystallize or separate from the solution, imparting a cloudy appearance to the oil as it is chilled under prescribed conditions

coherent structure a region of the flow which can be identified and which retains its identity during its motion downstream

combustion an exothermic chemical reaction with the characteristic ability to propagate through a combustible medium—usually a fuel and an oxidizer

combustion wave the reaction region of a propagating flame

compression ignition ignition in a diesel engine, in which the heat of compression ignites the fuel

conditional sampling selection of sections of a turbulent signal varying with time on the basis of signal amplitude or frequency satisfying preset conditions

cool flame a flame occurring in rich vapor-air mixture at temperatures below normal ignition temperatures for the particular fuel

deflagration a subsonic gaseous combustion process propagating through unreacted mixture

depolymerization a chemical reaction in which a polymer is broken down into monomers

detonation a self-propagating combustion reaction, or flame, that propagates at supersonic speed with the formation of a shock wave

diesel fuel a general term covering oils used as fuel in diesel and other compression ignition engines

diffusion flame a nonpremixed flame in which the fuel and oxidant are initially separated and the flame propagation is governed by the interdiffusion of the fuel and oxidizer

directional intermittency the portion of a velocity probability density distribution that is in a direction reversed to the main flow

double concentric jet jet emanating from a nozzle composed on a central orifice surrounded by an annular orifice separated by an interface

eddy a coherent region of fluid containing vorticity

equivalence ratio fuel/air mixture ratio normalized by the stoichiometric mixture ratio

explosion an exothermal combustion reaction associated with a rapid pressure rise (Explosions occur mainly in enclosed spaces but unconfined explosions can also occur when heat release is sufficient to induce large pressure gradients.)

fire uncontrolled combustion

fire point the lowest temperature at which a liquid emits sufficient flammable vapor to produce sustained combustion after removal of the ignition source

flame a rapid, gas-phase exothermic combustion process, characterized by self-propagation

flame front the mean reaction zone of a flame

flame ionization the imparting of electrical conductivity to hot combustion gases by electrically charged particles and electrons

flame propagation the motion or travel of a flame through a combustible mixture

flammability limit the maximum (rich) or minimum (lean) concentration of fuel in a fuel-oxidant mixture that will ignite

flashback propagation of a flame from an ignition source back to the fuel supply source

flash-point the minimum temperature at which vapors arising from a liquid fuel will ignite momentarily (i.e., flash) on application of a flame under specified conditions

free radical molecular fragments containing more than one atom, with each fragment having at least one unpaired electron

fuel oil a general term applied to an oil used for the production of power or heat (Residues or blends of distillates and residues used for burning in boilers and industrial furnaces.)

fully developed turbulent flow a turbulent fluid flow in which the distributions of all mean and all intensity components are independent of change in Reynolds number

gasification production of synthesis gas by reaction of coal with steam

gas oil a petroleum distillate having a viscosity and distillation range intermediate between those of kerosine and light lubricating oil

gasoline a refined petroleum distillate normally boiling within the limits of 30 to 200°C

glowing combustion oxidation of solid material without a visible gas phase flame

heat of combustion the quantity of heat evolved by the complete combustion of one mole of a substance

heterogeneous combustion combustion in which the reactants are not mixed uniformly and exist in different phases, usually applied to the burning of solid particles or liquid droplets in air

hydrogenation the increasing of hydrogen content in a gas (Increasing the calorific value of a gas by addition of hydrogen to low calorific value gas.)

hypergolic substances that ignite spontaneously when mixed with each other

ignition the initiation of combustion

ignition delay the time interval between the instant an oxidizer contacts a combustible substance and ignition takes place

ignition temperature the lowest temperature at which sustained combustion of a substance can be initiated

ignition time the time between the application of an ignition source to a mixture and the instant that self-sustained combustion begins

isotropic flow flow with local point characteristics that are the same in all directions

kerosine a refined petroleum distillate intermediate in volatility between gasoline and gas oil (Its distillation range generally falls within the limits of 150 to 300°C.)

knock the noise associated with self-ignition, ahead of the flame front, of a portion of the fuel-air mixture in the engine cylinder

laminar flow a fluid flow dominated by molecular viscosity in which transport of heat, mass, and momentum transverse to the flow is by molecular conductivity, density, and viscosity

liquefaction digestion of coal, using solvents (Production of solvent-refined coal in the form of petroleum substitute or Syncrude.)

mercaptan one of the organic compounds having the general formula, R–SH, i.e., the thiol group, –SH, is attached to a radical such as CH_3, C_2H_5, etc. (The simpler mercaptans have strong, repulsive, garlic-like odors which become less pronounced with increasing molecular weight and higher boiling points.)

middle distillate one of the distillates obtained between kerosine and lubricating oil fractions in the refining process (These include light fuel oils and diesel fuel.)

naphtha a distillate covering the end of the gasoline and beginning of the kerosine range and frequently used as a feed stock for reforming processes

octane number the octane number of a gasoline is a measure of its antiknock value (This quality is determined in a standard engine by matching for detonation the gasoline under test against a mixture of isooctane and normal heptane. The percentage by volume of isooctane in that mixture is the octane number.)

Octane numbers are measured under the following conditions:

motor   ASTM-CFR engine at 900 r/min
research ASTM-CFR engine at 600 r/min
road   determined by operating a car on the road or on a chassis dynamometer simulating highway conditions

oil shale a rock of sedimentary origin, with an ash content of more than 33 percent; the content organic matter yields oil when destructively distilled, but not appreciably when extracted with the ordinary solvents for petroleum

oxidant a substance that contains an atom or atomic group that gains electrons by chemical reaction (The most common oxidant is oxygen, but ozone, chlorine, hydrogen peroxide, nitric acid, metal oxides, the chlorates, and permanganates are all considered to be oxidants in combustion reactions.)

oxidation oxidation is the chemical reaction between oxidant and fuel (Slow oxidation involves reactions at a rate sufficiently slow not to merit the terms "combustion" or "flame." Rapid oxidation involves a sudden—almost discontinuous—increase in temperature reaction rate and pressure. Combustion and flames involve rapid oxidation, easily recognizable by chemiluminescence and chemi-ionization.)

oxidizer a substance that readily liberates oxygen without requiring replacement by another element (The term "oxidizer" is sometimes used more generally to define any substance that contains an atom or atomic group that gains electrons.)

pilot a small flame used to ignite gaseous fuel issuing from a large burner

pool fire a free-burning fire above a pool of combustible liquid or a large area of molten solid

potential core initial region of jet where conditions remain the same as at the nozzle exit

pour point the pour point of a petroleum oil is the lowest temperature at which the oil will pour or flow when it is chilled without disturbance under prescribed conditions

premixed flame a flame in which the fuel and oxidant are initially mixed prior to injection into burner or combustion chamber (Mixing may be complete when all fuel and oxidant are mixed at the molecular level. In incompletely mixed systems, mixing has been completed on the eddy level but not on the molecular level. In a partially premixed system, additional fuel or oxidant is required for stoichiometric combustion.)

pressure jet atomizer atomizer in which liquid breakup is achieved by pressurizing liquid, usually through a swirl nozzle, leading to the formation of a central air core

pyrolysis thermal cracking of hydrocarbon fuel when heated in the absence of oxidant

quenching distance the orifice diameter, wall separation, or mesh spacing just sufficient to prevent flame propagation

shale oil the distillate obtained when oil shale is heated in retorts

similarity the collapsing of profiles of temperature, velocity, and concentration, at different sections of a jet, into a single profile by normalizing with characteristic length and property scale

smoke a visible, nonluminous, airborne suspension of particles originating from combustion (Smoke consists largely of soot particles, ranging from 10 to 0.001 μm, produced by incomplete combustion of hydrocarbon fuels.)

soot carbonaceous particles resulting from incomplete combustion

spontaneous ignition initiation of combustion by internal chemical or biological reaction, producing sufficient heat for ignition

stoichiometric combustion combustion in which the quantities of fuel and oxidizer are chemically balanced, resulting in no excess fuel or oxidizer

streamlines lines joining points of constant stream function (Lines on stream tubes or annuli through which mass flow rate is constant.)

swirl bulk rotation of a fluid generated by flow over vanes, tangential injection of fluid into a duct, or mechanical rotation of a fixed wall or tube

swirl number axial flux of angular momentum divided by product of thrust and diameter of nozzle

thermal ignition initiation of combustion by supplying heat until the ignition point is reached

transition flow the flow regime between laminar and turbulent flow in which viscous and turbulent forces are of similar order of magnitude

transverse jets jets injected transverse to a mainstream flow

turbulence intensity ratio of rms of fluctuating component of velocity to time average velocity

turbulent burning velocity the velocity of flame propagation in a turbulent flow

turbulent flame a flame propagating through a turbulent stream

turbulent flow an irregular three-dimensional flow composed of eddies (Transport of heat, mass, and momentum is several orders of magnitude greater than by molecular conductivity, diffusivity, and viscosity.)

twin fluid atomizer use of high-pressure steam or air to effect breakup of liquid film in atomizer

unmixedness trapping of fuel and air pockets within combustion products, resulting in delay of mixing and, hence, of burning

vortex a rotational motion in a fluid (The flow induced by an orderly concentration of vorticity.)

vortex ring a toroidal vortex filament

vorticity degree of local rate of rotation within a fluid

Weaver flame speed factor an empirical parameter relating the maximum burning velocity of the gas and its stoichiometric air requirement (It is defined as being the burning velocity of a stoichiometric fuel-air mixture expressed as a percentage of the burning velocity of the same mixture of hydrogen-air.)

Wobbe number defined as the gross calorific value of the gas divided by the square root of its specific gravity with respect to air (It is proportional, with reasonable precision, to the heat input of a burner at a fixed pressure of fuel gas. It is also approximately inversely proportional to the amount of air premixed with the fuel prior to combustion.)